CONTEMPORARY MATHEMATICS

220

Homotopy Theory via Algebraic Geometry and Group Representations

Proceedings of a Conference on
Homotopy Theory
March 23–27, 1997
Northwestern University

Mark Mahowald
Stewart Priddy
Editors

AMERICAN MATHEMATICAL SOCIETY
FOUNDED 1888

This volume contains the proceedings of a conference on homotopy theory held at Northwestern University on March 23–27, 1997.

Support provided by the National Science Foundation and Northwestern University.

1991 *Mathematics Subject Classification.* Primary 55Pxx, 55Nxx, 55Qxx, 20Cxx; Secondary 55S10, 20J06.

Library of Congress Cataloging-in-Publication Data

Conference on Homotopy Theory (1997 : Northwestern University)

Homotopy theory via algebraic geometry and group representations : proceedings of a Conference on Homotopy Theory, March 23–27, 1997, Northwestern University / Mark Mahowald, Stewart Priddy, editors.

p. cm. — (Contemporary mathematics, ISSN 0271-4132 ; 220)

Includes bibliographical references.

ISBN 0-8218-0805-2 (alk. paper)

1. Homotopy theory—Congresses. 2. Geometry, Algebraic—Congresses. 3. Representations of groups—Congresses. I. Mahowald, M. E. (Mark E.), 1931– . II. Priddy, Stewart, 1940– . III. Title. IV. Series: Contemporary mathematics (American Mathematical Society) ; v. 220.

QA612.7.C647 1997

514′.24—dc21 98-27238

CIP

∞ The paper used in this book is acid-free and falls within the guidelines
established to ensure permanence and durability.

Visit the AMS home page at URL: `http://www.ams.org/`

10 9 8 7 6 5 4 3 2 1 03 02 01 00 99 98

This volume is dedicated to Mark Mahowald in gratitude for his many wonderful and continuing contributions to homotopy theory. Only the fact that he is an editor of this volume prevents me from saying more. For those who know Mark nothing more need be said.

S. Priddy

Contents

Preface

The academic year 1996–97 was designated as a special emphasis year in Algebraic Topology at Northwestern University. In addition to visitors and special courses, an international conference entitled "Current trends in algebraic topology with applications to algebraic geometry and physics" was held during the period March 23–27. A series of plenary lectures, listed on the following page, indicates the great breadth of the conference and the attempt made to promote interaction between various areas of mathematics. We gratefully acknowledge the generous support provided by the National Science Foundation and Northwestern University. Approximately 120 mathematicians participated in the conference.

Conference participants were invited to submit original research papers related to the theme of the conference. All submissions were refereed to usual journal standards. One special feature is the paper prepared by Charles Rezk which covers the topics of his emphasis year course. A set of problems presented at a special problem session is given at the end of the volume. We thank Ran Levi, who generously agreed to conduct this session and collect the problems.

Finally, we thank Ms. Melanie Rubin for her editorial support in preparing this volume and for her many hours of assistance in arranging the conference.

Mark Mahowald
Stewart Priddy

Plenary Talks

Lars Hasselholt, MIT, "Topological cyclic homology"

Ruth Charney, Ohio State University, "CAT(0) geometry and Artin groups"

Michael Hopkins, MIT, "Formal groups, spectra, and orientability"

Nicholas Kuhn, University of Virginia, "Cohomological stability and generic representation theory"

R. James Milgram, Stanford University, "The interaction of finite simple groups and stable homotopy theory"

William Dwyer, University of Notre Dame, "Duality in homotopy theory"

Douglas Ravenel, University of Rochester, "What we still don't know about loop spaces of spheres"

J. Peter May, University of Chicago, "Brave new worlds of stable homotopy theory"

Ulrike Tillmann, Oxford University, "Moduli space of Riemann surfaces—a K-theory approach"

Fred Cohen, University of Rochester, "Applications of group theory to classical homotopy theory"

Contemporary Mathematics
Volume **220**, 1998

Buildings, Group Extensions and the Cohomology of Congruence Subgroups

Alejandro Adem

Introduction

Let T_n denote the Tits Building associated to the discrete group $\mathrm{SL}_n(\mathbb{Z})$. In this note we will be interested in computing the equivariant cohomology $H^*_\Gamma(|T_n|, \mathbb{Z})$, where $|T_n|$ is the geometric realization of T_n, and Γ is a torsion-free discrete subgroup of finite index in $\mathrm{SL}_n(\mathbb{Z})$. In the special case $n = 3$, we will see that this equivariant cohomology in fact corresponds to the ordinary cohomology of a 4-dimensional Poincaré Duality group which has Γ as its quotient.

This is then applied to partially compute the cohomology of the level p congruence subgroups (p an odd prime) in $\mathrm{SL}_3(\mathbb{Z})$. In particular we obtain

THEOREM. *Let $\Gamma(p)$ denote the level p congruence subgroup of $SL_3(\mathbb{Z})$; then*

$$dim_{\mathbb{Q}}\ H^3(\Gamma(p), \mathbb{Q}) \geq \frac{1}{12}(p^3 - 1)(p^3 - 3p^2 - p + 15) + 1.$$

It turns out that this method can be used for other rank two groups such as $\mathrm{Sp}_4(\mathbb{Z})$ and $\mathrm{G}_2(\mathbb{Z})$. In §6 we apply this to the level p congruence subgroups in $\mathrm{Sp}_4(\mathbb{Z})$, obtaining the cohomology of the relevant parabolic subgroups and from this a lower bound for the fourth betti number:

THEOREM. *Let $\Gamma(p)$ denote the level p congruence subgroup in $Sp_4(\mathbb{Z})$; then*

$$dim_{\mathbb{Q}}\ H^4(\Gamma(p), \mathbb{Q}) \geq \frac{1}{24}(p^4 - 1)(2p^3 - 3p^2 - 2p + 27) + 1.$$

Our methods are mostly algebraic, involving techniques from group extensions and their cohomology. We offer a systematic approach which works with general coefficients and in addition we provide complete information on the cohomology of the parabolic subgroups. One of the key facts which we prove is that the congruence subgroups are homomorphic images of Poincaré duality groups with kernel a free, infinitely generated group. These groups are interesting in their own right, and probably deserve further attention. They can be realized as the fundamental groups of the Borel–Serre compactification; this is discussed in §7.

The application to the rational cohomology of congruence subgroups in $\mathrm{SL}_3(\mathbb{Z})$ is contained in the paper by Lee and Schwermer (see [LS]), although it is perhaps not as well known as it should be. Similarly there is overlap for $Sp_4(\mathbb{Z})$ (see [S]). In fact most of the work in this paper was done without knowledge of the existence of these papers. We recommend them to the reader.

In some sense this is an expository paper. One of its goals is to put computations for congruence subgroups within a simplified framework of group cohomology. Complete answers involve assembling a wide array of mathematical techniques beyond our scope, but at the very least one hopes this paper will motivate the reader to learn the more number–theoretic aspects of the subject (see [A], [AGG]).

The author is grateful to A. Ash and L. Solomon for helpful comments.

§0. Preliminaries

We recall that $\mathrm{SL}_n(\mathbb{Z})$ is a group of virtual cohomological dimension equal to $N = \frac{n(n-1)}{2}$. The module known as the Steinberg module St is a dualizing module for any torsion-free subgroup Γ of finite index in $\mathrm{SL}_n(\mathbb{Z})$; that is there are isomorphisms

$$H^i(\Gamma, M) \cong H_{N-i}(\Gamma, \mathrm{St} \otimes M)$$

where M is any $\mathbb{Z}\Gamma$ module, i any integer and these equivalences are given by taking cap product with a fixed element $z \in H_N(\Gamma, \mathrm{St})$; dually, we have isomorphisms

$$H_i(\Gamma, M) \cong H^{N-i}(\Gamma, \mathrm{Hom}(\mathrm{St}, M)).$$

Recall that $\mathrm{SL}_n(\mathbb{Q})$ acts naturally on the set of all proper non-trivial subspaces of $\mathbb{Q}^n$. This defines a partially ordered set T_n and hence its geometric realization $|T_n|$ which can be thought of as an $\mathrm{SL}_n(\mathbb{Z})$-CW complex of dimension $n-2$. Furthermore we have

$$H_*(|T_n|, \mathbb{Z}) \cong \begin{cases} \mathbb{Z} & *=0 \\ \mathrm{St} & *=n-2 \\ 0 & \text{otherwise.} \end{cases}$$

For details on these facts we refer to [Br1] and [Br2].

Next we recall the notion of equivariant cohomology: if G is a group acting on a space X, then

$$H_G^*(X) = H^*(X \times_G EG, \mathbb{Z})$$

where EG is the universal free contractible G space and $X \times_G EG = X \times EG/G$ (diagonal action).

We also recall that if X is a G-CW complex, then there exist two spectral sequences for computing $H_G^*(X)$, namely

(I) $E_2^{p,q} = H^p(G, H^q(X, \mathbb{Z})) \Rightarrow H_G^{p+q}(X)$

and

(II) $E_1^{p,q} = H^q(G, C^p(X)) \Rightarrow H_G^{p+q}(X)$

where $C^*(X)$ denotes the module of cellular p-cochains on X.

We remark here for later use that if

$$C_p(X) \cong \bigoplus_{i\in I} \mathbb{Z}[G/G_{\sigma_i}],$$

then

$$H^q(G, C^p(X)) \cong \bigoplus_{i\in I} H^q(G_{\sigma_i}, \mathbb{Z}).$$

§1. The Equivariant Cohomology of $|T_n|$.

We will prove the following

THEOREM 1.1. *Let $\Gamma \subseteq SL_n(\mathbb{Z})$ denote a torsion free subgroup of finite index, then there is a duality isomorphism*

$$H^i_\Gamma(|T_n|) \cong H^\Gamma_{\frac{n^2+n-4}{2}-i}(|T_n|).$$

Proof. To begin we observe that there is a fibration

$$\begin{array}{ccc} |T_n| & \longrightarrow & |T_n| \times_\Gamma E\Gamma \\ & & \downarrow \\ & & B\Gamma \end{array}$$

where $\dim|T_n| = n-2$, $\dim B\Gamma = N = \frac{n(n-1)}{2}$ and hence $\dim |T_n| \times_\Gamma E\Gamma = n-2+N = \frac{n^2+n-4}{2}$. Let us denote this number by M. Now in this case the spectral sequence of type (I) in homology degenerates to a long exact sequence

$$\cdots \to H_{i-n+2}(\Gamma, \mathrm{St}) \to H^\Gamma_i(|T_n|) \to H_i(\Gamma, \mathbb{Z}) \to H_{i-n+1}(\Gamma, \mathrm{St}) \to \cdots.$$

Note in particular that

$$H^\Gamma_M(|T_n|) \cong H_N(\Gamma, \mathrm{St}) \cong H^0(\Gamma, \mathbb{Z}) \cong \mathbb{Z}.$$

Hence there is a "fundamental class" $\mu \in H^\Gamma_M(|T_n|)$ which can be identified with the class $z \in H_N(\Gamma, \mathrm{St})$ which induces the duality isomorphisms $H^i(\Gamma, M) \cong H_{N-i}(\Gamma, \mathrm{St} \otimes M)$ via the cap product. We claim that

$$\cap\mu : H^i_\Gamma(|T_n|) \to H^\Gamma_{M-i}(|T_n|)$$

induces a duality isomorphism.

To prove this we look at the long exact sequence in cohomology:

$$\cdot \to H^i(\Gamma, \mathbb{Z}) \to H^i_\Gamma(|T_n|) \to H^{i-n+2}(\Gamma, \mathrm{St}^*) \to H^{i+1}(\Gamma, \mathbb{Z}) \to \cdots.$$

Cap product with μ induces maps from one long exact sequence to the other:

$$\begin{array}{ccccccc} H^i(\Gamma,\mathbb{Z}) & \to & H^i_\Gamma(|T_n|) & \to & H^{i-n+2}(\Gamma,\mathrm{St}^*) & \to & H^{i+1}(\Gamma,\mathbb{Z}) \\ \downarrow \cong & & \downarrow \cap\mu & & \downarrow \cong & & \downarrow \cong \\ H_{N-i}(\Gamma,\mathrm{St}) & \to & H^\Gamma_{M-i}(|T_n|) & \to & H_{M-i}(\Gamma,\mathbb{Z}) & \to & H_{N-i-1}(\Gamma,\mathrm{St}) \end{array}$$

Consider the special case when $n = 3$; then $|T_3|$ is an infinite graph, with free fundamental group F. In this case $|T_n| \times_\Gamma E\Gamma$ is aspherical and 4 dimensional; hence if $Q = \pi_1(|T_n| \times_\Gamma E\Gamma)$ we have an extension

$$1 \to F \to Q \to \Gamma \to 1$$

where Q is a 4-dimensional Poincaré Duality group with homology of *finite type*. In particular we have a long exact sequence:

$$0 \to H^1(\Gamma, \mathbb{Z}) \to H^1(Q, \mathbb{Z}) \to H^0(\Gamma, \mathrm{St}^*) \to H^2(\Gamma, \mathbb{Z}) \to H^2(Q, \mathbb{Z})$$

$$\to H^1(\Gamma, \mathrm{St}^*) \to H^3(\Gamma, \mathbb{Z}) \to H^3(Q, \mathbb{Z}) \to H^2(\Gamma, \mathrm{St}^*)$$

$$\to H^4(\Gamma, \mathbb{Z}) \to H^4(Q, \mathbb{Z}) \to H^3(\Gamma, \mathrm{St}^*) \to 0$$

which simplifies to yield
(1.2)

$$0 \to H^1(\Gamma, \mathbb{Z}) \to H^1(Q, \mathbb{Z}) \to H_3(\Gamma, \mathbb{Z}) \to H^2(\Gamma, Z) \to H^2(Q, \mathbb{Z}) \to$$

$$\to H_2(\Gamma, \mathbb{Z}) \to H^3(\Gamma, \mathbb{Z}) \to H^3(Q, \mathbb{Z}) \to H_1(\Gamma, \mathbb{Z}) \to 0$$

and $H^4(Q, \mathbb{Z}) \cong \mathbb{Z}$.

§2. Cohomology of Parabolic Subgroups

From now on we specialize to the case $n = 3$ and $\Gamma = \Gamma(p)$, a level p congruence subgroup, p an odd prime and $Q = Q(p)$. We will compute the integral cohomology of its intersections with the parabolic subgroups in $\mathrm{SL}_3(\mathbb{Z})$.

To begin let $B \subseteq \mathrm{SL}_n(\mathbb{Z})$ denote the subgroup of upper triangular matrices in $\mathrm{SL}_3(\mathbb{Z})$. Then

$$B \cap \Gamma(p) = \left\{ \begin{pmatrix} 1 & c_1 & c_2 \\ 0 & 1 & c_3 \\ 0 & 0 & 1 \end{pmatrix} \; p \text{ divides } c_1, c_2, \text{ and } c_3 \right\}.$$

The following elements generate this group

$$a = \begin{pmatrix} 1 & p & 0 \\ 0 & 1 & 0 \\ 0 & 0 & 1 \end{pmatrix}, \qquad b = \begin{pmatrix} 1 & 0 & 0 \\ 0 & 1 & p \\ 0 & 0 & 1 \end{pmatrix}, \qquad c = \begin{pmatrix} 1 & 0 & p \\ 0 & 1 & 0 \\ 0 & 0 & 1 \end{pmatrix}$$

and $aba^{-1}b^{-1} = c^p$.

In fact we may express $B \cap \Gamma(p)$ as a central extension

$$1 \to \mathbb{Z} \to B \cap \Gamma(p) \to \mathbb{Z} \oplus \mathbb{Z} \to 1$$

where $\langle c \rangle$ is the central subgroup and $\bar{a}$, $\bar{b}$ generate the quotient. We have an exact sequence in cohomology (over $\mathbb{Z}$):

$$0 \to \mathbb{Z} \oplus \mathbb{Z} \to H^1(B \cap \Gamma(p)) \to \mathbb{Z} \xrightarrow{d_2} \mathbb{Z} \to H^2(B \cap \Gamma(p)) \to \mathbb{Z} \oplus \mathbb{Z} \to 0$$

and $H^3(B \cap \Gamma(p)) \cong \mathbb{Z}$. From the extension data, coker $d_2 \cong \mathbb{Z}/p$ and hence we have proved

PROPOSITION 2.1.

$$H^i(B\cap\Gamma(p),\mathbb{Z})\cong\begin{cases}\mathbb{Z} & i\text{=0,3}\\ \mathbb{Z}\oplus\mathbb{Z} & i\text{=1}\\ \mathbb{Z}\oplus\mathbb{Z}\oplus\mathbb{Z}/p & i\text{=2}\\ 0 & \text{otherwise.}\end{cases}$$

□

This is a 3-dimensional Poincaré Duality group. Next we consider the parabolic subgroup $P_1\subseteq SL_3(\mathbb{Z})$ the subgroup with zeros in the second and third entries of its first column. We have

$$P_1\cap\Gamma(p)=\left\{\begin{pmatrix}1 & x & y\\ 0 & & \\ & & A\\ 0 & & \end{pmatrix}\middle|\ \begin{matrix}p\text{ divides }x\text{ and }y\\ A\in\Gamma_2(p)\end{matrix}\right\}.$$

Let

$$E=\left\{\begin{pmatrix}1 & x & y\\ 0 & 1 & 0\\ 0 & 0 & 1\end{pmatrix}\ p\text{ divides }x\text{ and }y\right\},$$

then $E\cong\mathbb{Z}\oplus\mathbb{Z}$ and if we embed $\Gamma_2(p)$ in $\Gamma(p)$ via $A\mapsto\begin{pmatrix}1 & \cdots\\ \vdots & \boxed{A}\end{pmatrix}$ we obtain a semidirect product decomposition

$$\boxed{P_1\cap\Gamma(p)=E\times_T\Gamma_2(p).}$$

If $A=\begin{pmatrix}a & b\\ c & d\end{pmatrix}\in\Gamma_2(p)$, the action on E is given by

$$\begin{pmatrix}1 & x & y\\ 0 & 1 & 0\\ 0 & 0 & 1\end{pmatrix}\overset{T_A}{\mapsto}\begin{pmatrix}1 & ax+cy & bx+dy\\ 0 & 1 & 0\\ 0 & 0 & 1\end{pmatrix}$$

or

$$x\mapsto ax+cy$$
$$y\mapsto bx+dy$$

and hence

$$T_A\begin{pmatrix}x\\ y\end{pmatrix}=\begin{pmatrix}a & c\\ b & d\end{pmatrix}\begin{pmatrix}x\\ y\end{pmatrix}=A^T\begin{pmatrix}x\\ y\end{pmatrix}.$$

We can now compute $H^*(P\cap\Gamma(p),\mathbb{Z})$; we use the Lyndon-Hochschild-Serre spectral

Note that as $\Gamma_2(p)$ is free (and hence 1-dimensional), there are no differentials and it collapses at E_2. We obtain

$$H^i(P_1 \cap \Gamma(p), \mathbb{Z}) \cong \begin{cases} \mathbb{Z} & i{=}0 \\ H^1(\Gamma_2(p), \mathbb{Z}) \oplus H^1(E, \mathbb{Z})^{\Gamma_2(p)} & i{=}1 \\ H^1(\Gamma_2(p), H^1(E, \mathbb{Z})) \oplus H^2(E, \mathbb{Z})^{\Gamma_2(p)} & i{=}2 \\ H^1(\Gamma_2(p), H^2(E), \mathbb{Z}) & i{=}3. \end{cases}$$

Now the $\Gamma_2(p)$ action on $H^1(E)$ is the natural action on $M = \mathbb{Z} \oplus \mathbb{Z}$, and hence that on $H^2(E)$ is via the determinant, from which $H^1(E, \mathbb{Z})^{\Gamma_2(p)} = 0$, $H^2(E, \mathbb{Z})^{\Gamma_2(p)} \cong \mathbb{Z}$ and we have proved

PROPOSITION 2.2.

$$H^i(P_1 \cap \Gamma(p), \mathbb{Z}) \cong \begin{cases} \mathbb{Z} & i{=}0 \\ H^1(\Gamma_2(p), \mathbb{Z}) & i{=}1 \\ H^1(\Gamma_2(p), M) \oplus \mathbb{Z} & i{=}2 \\ H^1(\Gamma_2(p), \mathbb{Z}) & i{=}3 \\ 0 & \textit{otherwise}. \end{cases}$$

□

Denote by M^* the dual 2-dimensional representation of $\Gamma_2(p)$. Then it is direct to verify that for the other parabolic subgroup P_2, we have

PROPOSITION 2.3.

$$H^i(P_2 \cap \Gamma(p), \mathbb{Z}) \cong \begin{cases} \mathbb{Z} & i{=}0 \\ H^1(\Gamma_2(p), \mathbb{Z}) & i{=}1 \\ H^1(\Gamma_2(p), M^*) \oplus \mathbb{Z} & i{=}2 \\ H^1(\Gamma_2(p), \mathbb{Z}) & i{=}3 \\ 0 & \textit{otherwise}. \end{cases}$$

□

Note that $\Gamma_2(p)$ is a free group of rank $1 + \frac{(p-1)p(p+1)}{12}$.

§3. Double Cosets and the Tits Building

We recall that $|T_3|$ is a 1-dimensional $\mathrm{SL}_3(\mathbb{Z})$–complex with 2 orbits of zero cells, and one orbit of 1-cells.* The respective isotropy subgroups are P_1, P_2 and B.

From the usual induction-restriction formula, we have

$$\mathbb{Z}[\mathrm{SL}_3(\mathbb{Z})/B]_{|\Gamma} \cong \bigoplus_{x_i \in I_0} \mathbb{Z}[\Gamma/\Gamma \cap x_i B x_i^{-1}]$$

where $I_0 = \Gamma \backslash \mathrm{SL}_3(\mathbb{Z})/B$ is the set of double cosets. Similarly we have

$$\mathbb{Z}[\mathrm{SL}_3(\mathbb{Z})/P_1]_{|\Gamma} \cong \bigoplus_{y_j \in I_1} \mathbb{Z}[\Gamma/\Gamma \cap y_j P_1 y_j^{-1}]$$

$$\mathbb{Z}[\mathrm{SL}_3(\mathbb{Z})/P_2]_{|\Gamma} \cong \bigoplus_{z_k \in I_2} \mathbb{Z}[\Gamma/\Gamma \cap z_k P_2 z_k^{-1}]$$

where $I_1 = \Gamma \backslash \mathrm{SL}_3(\mathbb{Z})/P_1$, $I_2 = \Gamma \backslash \mathrm{SL}_3(\mathbb{Z})/P_2$.

Note that as $\Gamma \lhd \mathrm{SL}_3(\mathbb{Z})$, we have $\Gamma \cap x_i B x_i^{-1} \cong \Gamma \cap B$, $\Gamma \cap y_j P_1 y_j^{-1} \cong \Gamma \cap P_1$, $\Gamma \cap z_k P_2 z_k^{-1} \cong \Gamma \cap P_2$, for all i, j and k.

Next note that the cardinality of the indexing can be determined as follows**:

$$\#I_0 = [\mathrm{SL}_3(\mathbb{Z})/\Gamma : B/\Gamma \cap B] = [\mathrm{SL}_3(p) : B(p)]$$

$$\#I_1 = [\mathrm{SL}_3(\mathbb{Z})/\Gamma : P_1/\Gamma \cap P_1] = [\mathrm{SL}_3(p) : P_1(p)]$$

$$\#I_2 = [\mathrm{SL}_3(\mathbb{Z})/\Gamma : P_2/\Gamma \cap P_2] = [\mathrm{SL}_3(p) : P_2(p)]$$

where $B(p)$, $P_1(p)$, $P_2(p)$ denote the corresponding subgroups in $\mathrm{SL}_3(p)$.

The orders of these subgroups can easily be determined to be

$$|B(p)| = 4p^3 \quad \text{and} \quad |P_1(p)| = |P_2(p)| = 2p^3(p^2 - 1).$$

From this we readily deduce that

$$\#I_0 = \frac{(p^3 - 1)(p^2 - 1)}{4} \quad \text{and} \quad \#I_1 = \#I_2 = \frac{p^3 - 1}{2}.$$

We now use this to describe the cellular chains on $|T_3|$ as Γ-modules. Given that over $\mathrm{SL}_3(\mathbb{Z})$ we have

$$C_0(|T_3|) \cong \mathbb{Z}[\mathrm{SL}_3(\mathbb{Z})/P_1] \oplus \mathbb{Z}[\mathrm{SL}_3(\mathbb{Z})/P_2]$$

$$C_1(|T_3|) \cong \mathbb{Z}[\mathrm{SL}_3(\mathbb{Z})/B]$$

---

*[illegible]

we obtain that as Γ-modules:

$$C_0(|T_3|) \cong \left(\bigoplus^{\frac{p^3-1}{2}} \mathbb{Z}[\Gamma/\Gamma \cap y_j P_1 y_j^{-1}]\right) \oplus \left(\bigoplus^{\frac{p^3-1}{2}} \mathbb{Z}[\Gamma/\Gamma \cap z_k P_2 z_k^{-1}]\right)$$

$$C_1(|T_3|) \cong \bigoplus^{\frac{(p^3-1)(p^2-1)}{4}} \mathbb{Z}[\Gamma/\Gamma \cap x_i B x_i^{-1}].$$

We note here that dividing out by the Γ-action we can describe $|T_3|/\Gamma$ in terms of the mod p building associated to $\mathrm{SL}_3(p)$. This a $(p+1)$-valent graph which can be described explicitly.

§4. Cohomology Calculations

We now use the second spectral sequence described in §0, applied to the 1-dimensional Γ-complex $|T_3|$; we obtain a long exact sequence

$$\cdots \to H^i_\Gamma(|T_3|) \to H^i(\Gamma, C^0(|T_3|)) \to H^i(\Gamma, C^0(|T_3|)) \to H^{i+1}_\Gamma(|T_3|) \to \cdots$$

Combining the results in §2, §3, we have

$$H^*(\Gamma, C^0(|T_3|)) \cong \left(\bigoplus^{\frac{p^3-1}{2}} H^*(P_1 \cap \Gamma(p), \mathbb{Z})\right) \oplus \left(\bigoplus^{\frac{p^3-1}{2}} H^*(P_2 \cap \Gamma(p), \mathbb{Z})\right)$$

$$H^*(\Gamma, C^1(|T_3|)) \cong \bigoplus^{\frac{(p^3-1)(p^2-1)}{4}} H^*(B \cap \Gamma(p), \mathbb{Z}).$$

We are now ready to substitute explicit values:

$$0 \to \mathbb{Z} = H^0_\Gamma(|T_3|) \to \mathbb{Z}^{p^3-1} \to \mathbb{Z}^{\frac{(p^3-1)(p^2-1)}{4}} \to H^1_\Gamma(|T_3|) \to \mathbb{Z}^{\left[\frac{12+(p-1)p(p+1)}{6}\right]\left[\frac{p^3-1}{2}\right]}$$

(4.1)

$$\to \mathbb{Z}^{\frac{(p^3-1)(p^2-1)}{2}} \to H^2_\Gamma(|T_3|) \to \left[\bigoplus^{\frac{p^3-1}{2}} H^1(\Gamma_2(p), M \oplus M^*) \oplus \mathbb{Z} \oplus \mathbb{Z}\right] \to$$

$$\left[\bigoplus^{\frac{(p^3-1)(p^2-1)}{4}} \mathbb{Z} \oplus \mathbb{Z} \oplus \mathbb{Z}/p\right] \to H^3_\Gamma(|T_3|) \to \bigoplus^{p^3-1} \mathbb{Z}^{\left[\frac{12+(p-1)p(p+1)}{12}\right]} \to \mathbb{Z}^{\left[\frac{(p^3-1)(p^2-1)}{4}\right]}$$

$$\to H^4_\Gamma(|T_3|) \to 0.$$

We now derive some consequences from this sequence. Recall first that $\chi(B\Gamma(p))$ $= 0$ and hence from (1.2) we deduce $\chi(BQ(p)) = \chi(|T_3| \times_\Gamma E\Gamma) = 0$. Hence if $\mathbb{F}$ is a field of characteristic prime to p, we infer from (4.1) that

$$\dim_{\mathbb{F}} H^1(\Gamma_2(p), M_{\mathbb{F}} \oplus M_{\mathbb{F}}^*) = \frac{(p-1)p(p+1)}{3}$$

where $M_{\mathbb{F}}$ is the 2-dimensional natural representation of $\Gamma_2(p)$. In addition one can verify that if $(p,q) = 1$, then $H^1(\Gamma_2(p), M \oplus M^*)$ is q-torsion-free. This follows from looking at the long exact sequence associated to mod q reduction and the fact that if $\mathbb{F}$ is a field of characteristic prime to p, then $(M_{\mathbb{F}} \oplus M_{\mathbb{F}}^*)^{\Gamma_2(p)} = 0$. Reducing mod p, we obtain a long exact sequence of the form

$$0 \to (\mathbb{F}_p)^4 \to H^1(\Gamma_2(p), M \oplus M^*) \xrightarrow{\cdot p} H^1(\Gamma_2(p), M \oplus M^*) \to [\mathbb{F}_p]^{4(1+\frac{(p-1)p(p+1)}{12})} \to 0.$$

Observing that $[(M \oplus M^*) \otimes \mathbb{Z}/p^2]^{\Gamma_2(p)} \cong (\mathbb{Z}/p)^4$, we infer (now using the mod p^2 reduction sequence) that $H^1(\Gamma_2(p), M \oplus M^*)$ has no elements of order p^2 and so we deduce

Proposition 4.2.

$$H^1(\Gamma_2(p), M \oplus M^*) \cong [\mathbb{Z}]^{\frac{(p-1)p(p+1)}{3}} \oplus (\mathbb{Z}/p)^4. \qquad \square$$

Let β_3 denote the third betti number associated to the classifying space of a group. Looking at the final part of the sequence (4.1), we find that

$$\beta_3(Q(p)) \geq (p^3 - 1)\left[1 + \frac{(p-1)p(p+1)}{12}\right] - \frac{(p^3-1)(p^2-1)}{4} + 1$$

which, combined with (1.2) and the fact that $H^1(\Gamma, \mathbb{Z}) = 0$ (see [K] or [Se], pg. 122) yields

THEOREM 4.3.

$$\beta_3(\Gamma(p)) \geq \beta_3(Q(p)) \geq \tfrac{1}{12}(p^3-1)(p^3-3p^2-p+15) + 1. \qquad \square$$

§5. The Symplectic Group: Parabolic Subgroups

We will use the methods previously described to analyze the cohomology of the congruence subgroups in $\mathrm{Sp}_4(\mathbb{Z})$. The key fact here is that $\mathrm{Sp}_4(\mathbb{Z})$ is also a rank two group, hence its Tits Building will once again be a graph with an edge transitive action of the group. As before, the cohomology can be approached using the parabolic subgroups.

We proceed as before, now let $|W|$ denote the geometric realization of the Tits Building for $\mathrm{Sp}_4(\mathbb{Z})$; we denote $L(p) = \pi_1(|W| \times_{\Gamma(p)} E\Gamma(p))$, where now $\Gamma(p) \subset \mathrm{Sp}_4(\mathbb{Z})$ is a level p congruence subgroup, p an odd prime. In this instance, $\Gamma(p)$ has cohomological dimension equal to 4, and hence $L(p)$ has cohomological dimension equal to 5; as before it will be a Poincaré Duality group.

We need to describe the parabolic subgroups in some detail. If

$$\begin{pmatrix} 0 & 0 & 0 & -1 \end{pmatrix}$$

then $\mathrm{Sp}_4(\mathbb{Z})$ can be described as the matrices $A \in \mathrm{SL}_4(\mathbb{Z})$ such that $A^TJA = J$. It is convenient to write

$$A = \begin{pmatrix} A_1 & A_2 \\ A_3 & A_4 \end{pmatrix}$$

where A_1, A_2, A_3, A_4 are 2×2 matrices. Then, if we write

$$J = \begin{pmatrix} 0 & -Q \\ Q & 0 \end{pmatrix},$$

the conditions on A can be rewritten as

$$A_3^TQA_1 - A_1^TQA_3 = 0,\ A_3QA_2 - A_1^TQA_4 = -Q$$

$$A_4^TQA_1 - A_2^TQA_3 = Q,\ A_4^TQA_2 - A_2^TQA_4 = 0.$$

The parabolic subgroups in $\mathrm{Sp}_4(\mathbb{Z})$ can be obtained by intersecting with appropriate parabolics in $\mathrm{SL}_4(\mathbb{Z})$. To begin we consider the parabolic subgroup G_1 of symplectic matrices with $A_3 = 0$. Manipulating the defining equations one can easily see that if

$$S = \begin{pmatrix} I & 0 \\ 0 & Q \end{pmatrix},$$

then the conjugate group SG_0S^{-1} can be described as the set of matrices

$$\begin{pmatrix} D & B \\ 0 & (D^T)^{-1} \end{pmatrix}$$

where $D \in \mathrm{GL}_2(\mathbb{Z})$ and $BD^T = DB^T$.

Next we consider the symplectic matrices where the first column has zero entries except in the top corner. To simplify things we will assume that this non–zero entry is 1. After some manipulation we obtain a description of this group (denoted G_2) as the matrices of the form

$$\begin{pmatrix} 1 & (u_1\ u_2)\begin{pmatrix} 0 & 1 \\ -1 & 0 \end{pmatrix}C & & w \\ 0 & c_{11} & c_{12} & u_1 \\ 0 & c_{21} & c_{22} & u_2 \\ 0 & 0 & 0 & 1 \end{pmatrix}$$

where

$$C = \begin{pmatrix} c_{11} & c_{12} \\ c_{21} & c_{22} \end{pmatrix} \in \mathrm{SL}_2(\mathbb{Z}),\ u_1, u_2, w \in \mathbb{Z}.$$

Letting $C \in \mathrm{GL}_2(\mathbb{Z})$ we obtain an index 2 extension group which is evidently a maximal parabolic, denoted G_2.

The intersection $G_0 = G_1 \cap G_2$ can then be described as the determinant one matrices of the form

$$\begin{pmatrix} \pm 1 & x & y & z \\ 0 & \pm 1 & w & y - xw \\ 0 & 0 & \pm 1 & -x \\ 0 & 0 & 0 & \pm 1 \end{pmatrix}$$

where $x, y, z, w \in \mathbb{Z}$. As expected, this coincides with the intersection $U_T \cap \mathrm{Sp}_4(\mathbb{Z})$, where U_T is the subgroup of all 4×4 matrices which are upper triangular. We will not dwell on the parabolics themselves, rather we are interested in their intersections with the congruence subgroup $\Gamma(p)$.

5.1 The group $G_1 \cap \Gamma(p)$.

Using the fact that $\Gamma(p) \triangleleft \mathrm{Sp}_4(\mathbb{Z})$, this group can be described as the matrices of the form

$$\begin{pmatrix} D & B \\ 0 & (D^T)^{-1} \end{pmatrix}$$

where $D \in \Gamma_2(p) \subset \mathrm{SL}_2(\mathbb{Z})$, $B = pE$ and $ED^T = DE^T$. Let V denote the subgroup consisting of all matrices in $G_1 \cap \Gamma(p)$ of the form

$$\begin{pmatrix} I & B \\ 0 & I \end{pmatrix};$$

which one can easily check is a normal subgroup isomorphic to $\mathbb{Z}^3$. Using this it follows that

$$G_0 \cap \Gamma(p) \cong V \times_T \Gamma_2(p)$$

where $X \in \Gamma_2(p)$ acts via

$$\begin{pmatrix} I & B \\ 0 & I \end{pmatrix} \mapsto \begin{pmatrix} I & XBX^T \\ 0 & I \end{pmatrix}.$$

In fact this action can be expressed more generally as the restriction of a homomorphism $T : \mathrm{SL}_2(\mathbb{Z}) \longrightarrow \mathrm{SL}_3(\mathbb{Z})$ given by

$$\begin{pmatrix} a & b \\ c & d \end{pmatrix} \mapsto \begin{pmatrix} a^2 & 2ab & b^2 \\ ac & ad+bc & bd \\ c^2 & 2cd & d^2 \end{pmatrix}.$$

5.2 The group $G_2 \cap \Gamma(p)$.

In the description of G_2 given above, we simply require that $C \in \Gamma_2(p) \subset \mathrm{SL}_2(\mathbb{Z})$, $u_1, u_2, z \in p\mathbb{Z}$ and that the other 2 diagonal entries be equal to one. Let U denote the normal subgroup consisting of matrices in $G_2 \cap \Gamma(p)$ such that $C = I$. Then we can write the group as a semidirect product

$$U \times_T \Gamma_2(p)$$

where an element $C \in \Gamma_2(p)$ acts via

$$\begin{pmatrix} 1 & (u_1\ u_2)\begin{pmatrix} 0 & 1 \\ -1 & 0 \end{pmatrix} & z \\ 0 & I & \begin{pmatrix} u_1 \\ u_2 \end{pmatrix} \\ 0 & 0 & 1 \end{pmatrix} \mapsto \begin{pmatrix} 1 & (u_1\ u_2)\begin{pmatrix} 0 & 1 \\ -1 & 0 \end{pmatrix} C^{-1} & z \\ 0 & I & C\begin{pmatrix} u_1 \\ u_2 \end{pmatrix} \\ 0 & 0 & 1 \end{pmatrix}$$

5.3 The group $G_0 \cap \Gamma(p)$.

We conjugate to get this into suitable form:

$$\begin{pmatrix} I & 0 \\ 0 & Q \end{pmatrix} \begin{pmatrix} 1 & x & y & z \\ 0 & 1 & w & y-xw \\ 0 & 0 & 1 & -x \\ 0 & 0 & 0 & 1 \end{pmatrix} \begin{pmatrix} I & 0 \\ 0 & Q \end{pmatrix} = \begin{pmatrix} 1 & x & z & y \\ 0 & 1 & y-xw & w \\ 0 & 0 & 1 & 0 \\ 0 & 0 & -x & 1 \end{pmatrix}$$

where the group $\mathbb{Z}$ is generated by

$$t = \begin{pmatrix} 1 & p \\ 0 & 1 \end{pmatrix}, \quad T(t) = \begin{pmatrix} 1 & 2p & p^2 \\ 0 & 1 & p \\ 0 & 0 & 1 \end{pmatrix}.$$

§6. The Symplectic Group: Cohomology Calculations

We are now in a position to calculate the cohomology of the groups $G_i \cap \Gamma(p)$. Using the spectral sequence associated to the extension in 5.1, we have

THEOREM 6.1.

$$H^i(G_1 \cap \Gamma(p), \mathbb{Z}) \cong \begin{cases} \mathbb{Z} & i=0 \\ (\mathbb{Z})^{\frac{1+(p-1)p(p+1)}{12}} & i=1 \\ H^1(\Gamma_2(p), H^1(V, \mathbb{Z})) & i=2 \\ \mathbb{Z} \oplus H^1(\Gamma_2(p), H^2(V, \mathbb{Z})) & i=3 \\ (\mathbb{Z})^{1+\frac{(p-1)p(p+1)}{12}} & i=4. \\ 0 & \text{otherwise.} \end{cases}$$

□

To compute $H^*(G_2 \cap \Gamma(p), \mathbb{Z})$, we first observe that

$$H^i(U, \mathbb{Z}) \cong \begin{cases} \mathbb{Z} & i=0 \\ \mathbb{Z} \oplus \mathbb{Z} & i=1 \\ \mathbb{Z} \oplus \mathbb{Z} \oplus \mathbb{Z}/p & i=2 \\ \mathbb{Z} & i=3. \end{cases}$$

As a $\Gamma_2(p)$–module, the cohomology of $H^*(U, \mathbb{Z})$ is trivial in dimensions 0 and 3; the dual of the 2-dimensional natural representation M^* in H^1; and $M^* \oplus \mathbb{Z}/p$, where $\mathbb{Z}/p$ has a trivial action in H^2. From this information we can obtain

THEOREM 6.2.

$$H^i(G_2 \cap \Gamma(p), \mathbb{Z}) \cong \begin{cases} \mathbb{Z} & i=0 \\ H^1(\Gamma_2(p), \mathbb{Z}) & i=1 \\ H^1(\Gamma_2(p), M^*) \oplus \mathbb{Z}/p & i=2 \\ \mathbb{Z} \oplus H^1(\Gamma_2(p), M^*) \oplus H^1(\Gamma_2(p), \mathbb{Z}/p) & i=3 \\ (\mathbb{Z})^{1+\frac{(p-1)p(p+1)}{12}} & i=4 \\ 0 & \textit{otherwise.} \end{cases}$$

□

To compute $H^*(G_0 \cap \Gamma(p), \mathbb{Z})$ we use the semidirect product description provided in 5.3 to obtain

$$H^*(G_0 \cap \Gamma(p), \mathbb{Z}) \cong H^0(\mathbb{Z}, H^*(V, \mathbb{Z})) \oplus H^1(\mathbb{Z}, H^{*-1}(V, \mathbb{Z})).$$

These terms can be explicitly computed, yielding

THEOREM 6.3.

$$H^i(G_0 \cap \Gamma(p), \mathbb{Z}) \cong \begin{cases} \mathbb{Z} & i=0 \\ \mathbb{Z} \oplus \mathbb{Z} & i=1 \\ \mathbb{Z} \oplus \mathbb{Z} \oplus \mathbb{Z}/p \oplus \mathbb{Z}/2p & i=2 \\ \mathbb{Z} \oplus \mathbb{Z} \oplus \mathbb{Z}/p \oplus \mathbb{Z}/2p & i=3 \\ \mathbb{Z} & i=4. \\ 0 & \textit{otherwise.} \end{cases}$$

□

Just as before, the cohomology of these subgroups can be used to estimate the cohomology of $\Gamma(p)$. We will concentrate here on finding a lower bound for $\beta_4(\Gamma(p))$. For this we need to compute the indices $j_i = [\mathrm{Sp}_4(\mathbb{F}_p) : G_i(p)]$ where $G_i(p)$ denote the corresponding groups reduced mod p. The following can readily be established:

$$|\mathrm{Sp}_4(\mathbb{F}_p)| = p^4(p^4-1)(p^2-1), \ |G_0(p)| = 8p^4, \ |G_1(p)| = 2p^4(p^2-1) = |G_2(p)|.$$

From this we deduce

As in §4 we have a long exact sequence which can be used to estimate the cohomology of $\Gamma(p)$. In this case when we look at the end of the sequence we obtain a lower bound for $\beta_4(L(p))$ and hence we have

THEOREM 6.4.

$$\beta_4(\Gamma(p)) \geq \beta_4(L(p)) \geq \frac{1}{24}(p^4-1)(2p^3-3p^2-2p+27)+1.$$

□

Note that we have a complete *integral* calculation for the cohomology of the constituent subgroups. Hence a calculation over the field $\mathbb{F}_p$ is also possible. Also it is interesting to note the presence of 2–torsion in the cohomology of $G_0 \cap \Gamma(p)$.

§7. Final Remarks

Let $\mathcal{A} = P_1 *_B P_2$ denote the usual amalgamated product of parabolics in $\mathrm{SL}_3(\mathbb{Z})$. This group acts in the usual way on an infinite tree K with a single edge as a quotient. Now there is a natural surjection $\mathcal{A} \to \mathrm{SL}_3(\mathbb{Z})$ induced by the inclusions of the parabolic subgroups. If the kernel of this map is denoted by F, then it is the fundamental group of an infinite graph on which $\mathrm{SL}_3(\mathbb{Z})$ acts with quotient a single edge. In fact this graph can be identified with the Tits building, and from our previous constructions it is easy to see that the group $Q(p)$ is a finite index subgroup fitting into an extension

$$1 \to Q(p) \to \mathcal{A} \to \mathrm{SL}_3(\mathbb{F}_p) \to 1.$$

In other words, the amalgam $\mathcal{A}$ is a 4-dimensional virtual duality group, mapping onto $\mathrm{SL}_3(\mathbb{Z})$. In fact this is a purely algebraic construction of the Borel–Serre compactification for $\mathrm{SL}_3(\mathbb{Z})$. More precisely, it is easy to verify that if $\overline{X}$ is their construction, then $\pi_1(\partial(\overline{X}/\Gamma(p))) \cong Q(p)$. Observe that in this case $\partial(\overline{X}/\Gamma(p))$ is a closed 4-dimensional manifold, in fact homotopy equivalent to the Borel construction $|T_3| \times_{\Gamma(p)} E\Gamma(p) \cong BQ(p)$. This situation evidently generalizes to other rank 2 situations, such as $\mathrm{Sp}_4(\mathbb{Z})$ (discussed in §6) and $\mathrm{G}_2(\mathbb{Z})$. It is hoped that the purely cohomological approach which we have presented here will help yield some insight for understanding the cohomology of congruence subgroups, but it remains a difficult undertaking.

Detailed calculations for the cohomology of congruence subgroups with rational coefficients can also be found in the paper by Lee and Schwermer [LS], we refer the interested reader to this reference for further details. In their language, the image of the cohomology of $\Gamma(p)$ in the cohomology of $Q(p)$ corresponds to the 'cohomology at infinity' and the kernel can be identified with the 'cuspidal cohomology'.

References

[A] A. Ash, *Cohomology of Congruence Subgroups of $SL_n(\mathbb{Z})$*, Mathematische Annalen **249** (1980), 55–73.

[AGG] A. Ash, Grayson, D., and Green, P., *Computations of Cuspidal Cohomology of Congruence Subgroups of $SL_3(\mathbb{Z})$*, J. Number Theory **19** (1984), 412–436.

[BS] A. Borel and Serre, J.-P., *Corners and Arithmetic Groups*, Comm. Math. Helv. **48** (1973), 436–491.

[Br1] K. Brown, *Cohomology of Groups*, Graduate Texts in Mathematics, vol. 87, Springer-Verlag, 1982.

[Br2] K. Brown, *Buildings*, Springer–Verlag, 1989.

[K] D. A. Kajdan, *On the Connection of the Dual Space of a Group with the Structure of its Closed Subgroups*, Funct. Anal. and Appl. **1** (1967), 63–65.

[LS] R. Lee and Schwermer J., *Cohomology of Arithmetic Subgroups of $SL_3(\mathbb{Z})$ at Infinity*, J. Reine Angew. Math. **330** (1982), 100–131.

[S] J. Schwermer, *On Arithmetic Quotients of the Siegel Upper Half Space of Degree Two*, Compositio Math. **58** (1986), 233–258.

[Se] J.-P. Serre, *Arithmetic Groups*, Homological Group Theory, London Math. Soc. (C.T.C. Wall, ed.), vol. LNS 36, Cambridge University Press, 1979, pp. 105–136.

MATHEMATICS DEPARTMENT, UNIVERSITY OF WISCONSIN, MADISON, WI 53706
E-mail address: `adem@math.wisc.edu`

Contemporary Mathematics
Volume **220**, 1998

Hecke algebras acting on elliptic cohomology

ANDREW BAKER

Introduction.

In our earlier papers [**2,3,4,5,6**], we investigated stable operations and cooperations in elliptic cohomology and its variants, relating these to known operations on rings of modular forms. The purpose of this article is to give an introduction to these stable operation algebras, in particular explaining the connections with Hecke algebras and Morava stabilizer algebras; further details will appear in [**7**]. In [**9**] we make use of some of the Hecke operations described here to calculate parts of the Adams E_2-term based on elliptic homology. In work currently in progress we describe the operation algebra for the supersingular elliptic cohomology of §4 in terms of isogenies of supersingular elliptic curves over finite fields and use this to study the v_2-periodic part of elliptic cohomology Adams E_2-term which was determined in [**8**].

I would like to thank Jack Morava for providing the original stimulus for this paper and also to offer felicitations and thanks to Mark Mahowald, who always knows a good operation.

§1 Recollections on elliptic cohomology.

We refer to [**17,18,19**] for detailed expositions of the basic notions of elliptic cohomology; for discussions of reductions modulo prime ideals see [**3,4**]. For actions of Hecke algebras see [**2,3**].

In the present exposition we will concentrate on elliptic cohomology of level 1, having coefficient ring $E\ell\ell_* = \mathbb{Z}[1/6][E_4, E_6, \Delta^{-1}]$, the ring of modular forms for the full modular group $\mathrm{SL}_2(\mathbb{Z})$ which are meromorphic at infinity and have q-expansion coefficients in the ring $\mathbb{Z}[1/6]$. Here, as usual, the topological grading

1991 *Mathematics Subject Classification.* Primary 55N22, 14L05, 11F11, 11F25.

Key words and phrases. Elliptic cohomology, Hecke algebra.

The author would like to acknowledge the support of the EPSERC, EU and NSF whilst this work was pursued.

is twice the weight. By definition,

$$E\ell\ell^*(X) = E\ell\ell_* \underset{MU_*}{\otimes} MU^*(X)$$

for any finite CW-complex X. The functor $E\ell\ell^*(\ )$ is a multiplicative complex oriented cohomology theory and for our purposes is the 'universal example' of an elliptic cohomology theory (at least for those which are localised away from the primes 2 and 3).

In **[2]**, we extended the natural action of the Hecke operator T_N on the coefficient ring $E\ell\ell_*$ to a stable operation

$$\mathrm{T}_N\colon E\ell\ell^*(\ ) \longrightarrow E\ell\ell[1/N]^*(\ ),$$

where the target is elliptic cohomology with N inverted. These operations satisfy analogues of the usual relations amongst the Hecke operators and this characterises our extensions. These operations are $\mathbb{Z}[1/6N]$-linear but *not* multiplicative, although by construction they are symmetrisations of multiplicative stable operations into higher level versions of elliptic cohomology.

§2 Recollections on Hecke algebras and a twisted analogue.

For basic definitions of *Hecke algebras* and their structure we refer the reader to **[16]**. However, we will need a twisted version which so far as we can determine does not explicitly appear in the literature.

Let $\Bbbk$ be a commutative unital ring, L be a commutative $\Bbbk$-algebra, and G be a subgroup of the $\Bbbk$-algebra automorphism group of L. Let $H \leqslant G$ be a subgroup satisfying the *Hecke condition*, i.e., for each $g \in G$, the indices

$$[H, gHg^{-1} \cap H] \quad \text{and} \quad [gHg^{-1}, gHg^{-1} \cap H]$$

are finite. Then we can define the Hecke algebra $\Bbbk[H\backslash G/H]$ as in **[16]**.

Now let L^H denote the subalgebra of L fixed by H. If we interpret G as acting trivially on L^H, we can also form the Hecke algebra $L^H[H\backslash G/H]$. However, we wish to form a *twisted Hecke algebra* $L^H\{H\backslash G/H\}$ which in the case where $H = N$ is normal in G, agrees with the usual *twisted group ring* $L^N\{G/N\}$, well-known to group theorists.

We define the underlying $\Bbbk$-module of $L^H\{H\backslash G/H\}$ to be

$$\left\{ \sum_{g\colon G/H} \ell_g gH : \forall h \in H,\ \forall g,\ {}^h\ell_g = \ell_{hg} \right\},$$

where ℓ_g depends only on the coset gH and the sum has finitely nonvanishing terms. We also use the notation of **[16]** in which $g\colon G/H$ means that g ranges over a complete set of representatives of cosets of H in G; similarly $g\colon H\backslash G/H$ will mean that gH ranges over a complete set of (left) representatives of the double coset HgH, etc. $L^H\{H\backslash G/H\}$ can also be identified with the set of left invariant elements in the free left L-module $L[G/H]$ under the left $\Bbbk$-linear action of H given by

$$h \cdot \sum_{g : G/H} \ell_g gH = \sum_{g : G/H} {}^h\ell_g hgH.$$

Clearly $L^H\{H\backslash G/H\}$ is a left L^H-module. In order to define a product on $L^H\{H\backslash G/H\}$, it suffices to define the product of generators of the form

$$\lambda_g = \sum_{k :\ H/gHg^{-1}\cap H} \ell_{kg} kgH$$

and

$$\lambda_{g'} = \sum_{k' :\ H/g'Hg'^{-1}\cap H} \ell_{k'g'} k'g'H.$$

Therefore we set

$$\lambda_g \cdot \lambda_{g'} = \sum_{\substack{k :\ H/gHg^{-1}\cap H \\ k' :\ H/g'Hg'^{-1}\cap H}} \sum \ell_{kg} {}^{kg}\ell_{k'g'} kgk'g'H.$$

It can be verified that this product does indeed make $L^H\{H\backslash G/H\}$ into a $\Bbbk$-algebra, but *not* always an L^H-algebra. In fact, there is far more structure in this object. The most transparent way to understand this is to pass to a dual object,

$$\mathrm{Map}^{\mathrm{f}}_H(G/H, L) = \{f \colon G/H \longrightarrow L : \forall g \colon G/H, \forall h \in H,\ f(hgH) = {}^h f(gH)\},$$

where $\mathrm{Map}^{\mathrm{f}}$ denotes the set of finitely supported maps $G/H \longrightarrow L$.

It is then possible to define a Hopf algebroid structure upon $\mathrm{Map}^{\mathrm{f}}_H(G/H, L)$, as explained in [**7**]. In examples of interest to topologists, this gives twisted Hecke algebras closely related to cooperation algebras in (co)homology theories. For example, in the case of elliptic cohomology we have the twisted Hecke algebra

$$E\ell\ell_* \otimes \mathbb{Q}\{\mathrm{SL}_2(\mathbb{Z})\backslash \mathrm{GL}_2(\mathbb{Q})/\mathrm{SL}_2(\mathbb{Z})\} = (\widetilde{E\ell\ell_*} \otimes \mathbb{Q})^{\mathrm{SL}_2(\mathbb{Z})}\{\mathrm{SL}_2(\mathbb{Z})\backslash \mathrm{GL}_2(\mathbb{Q})/\mathrm{SL}_2(\mathbb{Z})\},$$

where $\widetilde{E\ell\ell_*}$ is the union of a tower of finite Galois extensions of $E\ell\ell_*$. This is closely related to the structure of the cooperation Hopf algebroid $E\ell\ell_* E\ell\ell$, the details appearing in [**6,7**].

§3 Ordinary reduction.

For any prime p greater than 3, there is a reduction of elliptic cohomology modulo p, whose coeeficient ring is $E\ell\ell_*/(p)$. This coefficient ring is a graded

PID, since the ring $E\ell\ell_*$ has dimension 2. We can map this ring into the (ungraded) ring $\mathbb{F}_p((q))$ by sending the residue class $F \pmod p \in E\ell\ell_{2n}/(p)$ to its reduced q-expansion,

$$\tilde{F}(q) \pmod p.$$

However, this is not a monomorphism, since for example we have

$$\tilde{E}_{p-1}(q) \equiv 1 \pmod p. \tag{3.1}$$

Indeed, for any pair of modular forms

$$F \in E\ell\ell_{2n}, \quad G \in E\ell\ell_{2n+p-1}$$

satisfying

$$\tilde{F}(q) \equiv \tilde{G}(q) \pmod p,$$

the element $FE_{p-1} - G$ is in the kernel of this q-expansion modulo p map. We use this to define the *ordinary reduction at p* of elliptic cohomology. Note that the element E_{p-1} will be invertible as a consequence of Equation (3.1). At the other extreme, in §4 we will consider the *supersingular reduction at p*, in which E_{p-1} will be contained in a prime ideal.

We follow closely the ideas described by Serre in [**25,24**]. However, it should be noted that we are working with the ring of modular forms which are meromorphic at infinity, whereas he works in the ring of holomorphic modular forms. This difference is insignificant because of the following easily verified fact.

LEMMA 3.1. *Let p be a prime greater than 3. Then in the ring of holomorphic modular forms with q-expansion coefficients in $\mathbb{Z}_{(p)}$, the element E_{p-1} is not divisible by Δ; this continues to hold after reduction modulo p.*

We can view the image of $E\ell\ell_*/(p)$ in $\mathbb{F}_p((q))$ as a graded field, with grading group some quotient of $\mathbb{Z}/(p-1)$, since E_{p-1} is of weight $p-1$. The following result is discussed in [**25**] and was proved originally by Swinnerton-Dyer and Serre.

THEOREM 3.2. *For each prime $p > 3$, the image of $E\ell\ell_*/(p)$ in $\mathbb{F}_p((q))$ is the $\mathbb{Z}/2(p-1)$-graded ring*

$$E\ell\ell_*/(p, E_{p-1} - 1).$$

Moreover, this is a graded integral domain; equivalently, $E_{p-1} - 1$ is a prime element in $E\ell\ell_*/(p)$ viewed as an ungraded ring.

As a consequence, we can form the ordinary reduction of elliptic cohomology at p as the functor

$$E\ell\ell[1]^*(\ ) = E\ell\ell[1]_* \underset{E\ell\ell_*/(p)}{\otimes} (E\ell\ell/(p))^*(\ ),$$

where we set

$$E\ell\ell[1]_* = E\ell\ell/J_1,$$

with the ideal J_1 being generated by p together with all the homogeneous elements in the kernel of the map into $\mathbb{F}_p((q))$.

THEOREM 3.3. *The functor* $E\ell\ell[1]^*(\ )$ *is a multiplicative complex oriented cohomology theory on finite CW-complexes.*

The proof of this uses the Landweber Exact Functor Theorem in its modulo p version, together with the fact that working modulo p we can identify E_{p-1} with the Hazewinkel generator v_1 (i.e., the $[p]$-series of the canonical formal group law over $E\ell\ell_*/(p)$ begins with the term $E_{p-1}X^p$ up to a unit in $\mathbb{F}_p^\times$). It seems more natural to replace this $\mathbb{Z}$-graded theory with the $\mathbb{Z}/2(p-1)$-graded theory

$$E\ell\ell[1]^\bullet(\ ) = E\ell\ell[1]_\bullet \underset{E\ell\ell_\bullet/(p)}{\otimes} (E\ell\ell/(p))^*(\ )$$

where

$$E\ell\ell[1]_\bullet = E\ell\ell_*/(p, E_{p-1} - 1).$$

The even part of this grading is best viewed as coming from the subgroup of $(p-1)$st roots of unity in $\mathbb{Z}_p^\times$, the units of p-adic integers $\mathbb{Z}_p$. Indeed there is an integral version of this cohomology theory whose coefficient ring can be identified with the ring of *meromorphic p-adic modular forms* obtained from the holomorphic version of Serre **[24]** (with integral q-expansion coefficients); this is naturally graded on $\mathbb{Z}_p^\times$.

The Hecke algebra

$$H_{(p)} = \bigotimes_{\substack{\ell \text{ prime} \\ \ell \neq p}} \mathbb{Z}_{(p)}[\mathrm{T}_\ell, \mathrm{T}_{\ell,\ell}, \mathrm{T}_{\ell,\ell}^{-1}]$$

acts upon the cohomology theory $E\ell\ell[1]^*(\ )$. However, there are also two other operations $\mathrm{U}_p, \mathrm{V}_p$ acting, as described in **[3]**. Moreover, these do not commute, so there is a non-commutative stable operation algebra acting. To understand this, we need to recall a result of Serre **[24]**.

THEOREM 3.4. *Let F be a modular form for the congruence subgroup $\Gamma_0(p)$, whose q-expansion at each cusp has coefficients in $\mathbb{Z}_{(p)}$. Then F is a p-adic modular form for $\mathrm{SL}_2(\mathbb{Z})$. In particular, for each natural number $k > 0$, there exists a modular form G for $\mathrm{SL}_2(\mathbb{Z})$ with q-expansion coefficients in $\mathbb{Z}_{(p)}$ and an integer $t \geqslant 0$ such that*

$$\tilde{F}(q) \equiv \tilde{G}(q) E_{p-1}^{tp^{k-1}} \pmod{p^k}.$$

Thus when working modulo p or p-adically, we are in effect working with modular forms for $\Gamma_0(p)$ in a suitable sense, since any modular form for $\mathrm{SL}_2(\mathbb{Z})$ is automatically one for $\Gamma_0(p)$.

Now modular forms for $\Gamma_0(p)$ can be viewed as functions on the space of pairs (L, P), where L is a lattice and $P \leqslant \mathbb{C}/L$ is a subgroup of order p. Such a function has two q-expansions, corresponding to the two cusps $i\infty$ and 0, and these are interchanged by the matrix $\begin{pmatrix} 0 & -1 \\ 1 & 0 \end{pmatrix}$. Following the account in **[16]**, for each $N \geqslant 1$, we introduce the set of matrices

$$M_2^0(N) = \left\{ \begin{pmatrix} a & b \\ c & d \end{pmatrix} \in M_2(\mathbb{Z}) : c \equiv 0 \pmod{p},\ p \nmid a,\ \det \begin{pmatrix} a & b \\ c & d \end{pmatrix} = N \right\}$$

for $0 < N \in \mathbb{Z}$. Then the Hecke algebra $\mathbb{Z}[\Gamma_0(p) \backslash \bigcup_{p \nmid N} M_2^0(N)/\Gamma_0(p)]$ acts upon $E\ell\ell[1]_*$ and the cohomology theory $E\ell\ell[1]^*(\ )$.

§4 Supersingular reduction.

Let $p > 3$ be a prime. We will consider *supersingular elliptic cohomology at p*. This is described in detail in **[4]** and here we confine ourselves to a brief account.

In general, the graded ideal $J_2 = (p, E_{p-1}) \lhd E\ell\ell_*$ is not prime. However, an argument due to Serre and related to the proof of Theorem 3.2 gives the following.

PROPOSITION 4.1. *In $E\ell\ell_*/(p)$, the prime factors $\pi_1, \pi_2, \dots, \pi_d$ of E_{p-1} all have multiplicity* 1. *Hence, the quotient ring is a product of graded fields*

$$E\ell\ell_*/(p, E_{p-1}) \cong \prod_{1 \leqslant k \leqslant d} E\ell\ell_*/(p, \pi_k).$$

In **[4]** we showed there is a ring homomorphism $h_2 \colon K(2)_* \longrightarrow E\ell\ell_*/(p, E_{p-1})$ classifying the natural formal group law over the target (or rather its canonical p-typification), and so we can form the functor

$$(E\ell\ell/(p, E_{p-1}))^*(\ ) = E\ell\ell_*/(p, E_{p-1}) \underset{K(2)_*}{\otimes} K(2)^*(\ ),$$

on finite CW-complexes. Similarly, we can form theories

$$(E\ell\ell/(p, \pi_k))^*(\ ) = E\ell\ell_*/(p, \pi_k) \underset{K(2)_*}{\otimes} K(2)^*(\ ) \quad (1 \leqslant k \leqslant d).$$

THEOREM 4.2. *The functor $(E\ell\ell/(p, E_{p-1}))^*(\ )$ is a multiplicative complex oriented cohomology theory on finite CW-complexes. Moreover, there is a natural splitting of algebra theories over $K(2)^*(\ )$,*

$$(E\ell\ell/(p, E_{p-1}))^*(\ ) \cong \prod_{1 \leqslant k \leqslant d} (E\ell\ell/(p, \pi_k))^*(\ ).$$

We refer to all of these theories *supersingular reductions of elliptic cohomology*.

The p-local Hecke algebra $H_{(p)}$ acts on the theory $(E\ell\ell/(p, E_{p-1}))^*(\ )$ since the ideal (p, E_{p-1}) is invariant under the actions of all the Hecke operators. This can be seen using the fact that the q-expansion of E_{p-1} is congruent to 1 $(\mathrm{mod}\ p)$; indeed the supersingular quotient ring is investigated in **[22]**. It also follows from our construction of the Hecke operators since in $E\ell\ell_*/(p)$, E_{p-1} is the image of v_1 under the map $MU_* \longrightarrow E\ell\ell_*$ classifying the canonical formal group law. However, in general the prime ideals (p, π_k) are not preserved under the Hecke operators, as was shown explicitly in the MSc thesis of R. Field [**11**]. Thus we cannot expect that the Hecke algebra will preserve the decomposition of Theorem 4.2.

Let us analyse in detail one of the quotient graded fields of Theorem 4.1. We set $\pi = \pi_k$, for $1 \leqslant k \leqslant d$. Using ideas of [**4**], we define the element $W = W_\pi \in E\ell\ell_{2w}/(p, \pi)$ by

$$W = \begin{cases} E_6 & \text{if } \pi = E_4, \\ E_4 & \text{if } \pi = E_6, \\ E_6 E_4^{-1} & \text{otherwise.} \end{cases}$$

Here we also set

$$w = \begin{cases} 6 & \text{if } \pi = E_4, \\ 4 & \text{if } \pi = E_6, \\ 2 & \text{otherwise.} \end{cases}$$

Using results from [**4**] we obtain the following.

THEOREM 4.3. *There is an isomorphism of graded fields*

$$E\ell\ell_*/(p, \pi) \cong \mathbb{F}_{p^e}[W, W^{-1}],$$

where $e = 1, 2$ *depending upon* p *and* π. *Moreover, if* $e = 2$, *there are exactly two such isomorphisms.*

Proof. Only the last sentence needs comment. Let $\mathbb{F}$ be the degree 0 part of $E\ell\ell_*/(p, \pi)$. Then there is an $\alpha \in \mathbb{F}$ such that

$$E_4 = \alpha W^2, \quad \text{and} \quad E_6 = \alpha W^3.$$

Now we can identify $\mathbb{F}$ with $\mathbb{F}_{p^2}$ by identifying α with one of the roots of the minimal polynomial of α in $\mathbb{F}_{p^2}$, but of course there are exactly two ways to do this.

The details of this proof suggest that we consider the *Frobenius operation*

$$\mathrm{Fr}\colon E\ell\ell_*/(p, \pi) \longrightarrow E\ell\ell_*/(p, \pi)$$

defined by

$$\mathrm{Fr}(F) = F^{(p)} = F^p W^{(p-1)\,\mathrm{wt}\,F/2},$$

which agrees with the usual Frobenius operation Fr on $\mathbb{F}$.

We also recall a formula that holds in connection with the homomorphism

$$h_2\colon K(2)_* = \mathbb{F}_p[v_2, v_2^{-1}] \longrightarrow E\ell\ell_*/(p, E_{p-1}),$$

and in which $\left(\dfrac{-1}{p}\right)$ is a Legendre symbol.

THEOREM 4.4. *In $E\ell\ell_*/(p, E_{p-1})$ we have the relation*

$$h_2(v_2) = \left(\frac{-1}{p}\right) \Delta^{(p^2-1)/12}.$$

Now we can tensor the field $E\ell\ell_*/(p,\pi)$ with the Galois field $\mathbb{F}_{p^2}$, and obtain a homomorphism

$$\mathbb{F}_{p^2} \otimes K(2)_* \xrightarrow{1\otimes h_2} \mathbb{F}_{p^2} \otimes E\ell\ell_*/(p,\pi).$$

Notice that the target is either

$$\mathbb{F}_{p^2}[W, W^{-1}] \quad \text{if } e = 1,$$

or

$$\mathbb{F}_{p^2}[W, W^{-1}] \times \mathbb{F}_{p^2}[W, W^{-1}] \quad \text{if } e = 2,$$

and the latter has an action of the Galois group

$$G_2 = \mathrm{Gal}(\mathbb{F}_{p^2}/\mathbb{F}_p) \cong \mathbb{Z}/2,$$

generated by the Frobenius map Fr which interchanges the factors. But by Theorem 4.4, the composite

$$h_2' \colon K(2)_* \xrightarrow{\text{incl}} \mathbb{F}_{p^2} \otimes K(2)_* \xrightarrow{1\otimes h_2} \mathbb{F}_{p^2} \otimes E\ell\ell_*/(p,\pi)$$

is invariant under this action of G_2.

We can consider the natural transformation of cohomology theories

$$(E\ell\ell/(p,\pi))^*(\ ) \longrightarrow \mathbb{F}_{p^2} \otimes (E\ell\ell/(p,\pi))^*(\ )$$

and observe that the natural orientation (coming form $K(2)^*(\ )$) is invariant under the action of the group G_2 acting via the coefficients. The (strict) automorphism group of the corresponding formal group law over the ring $(E\ell\ell/(p,\pi))_*$ now acts as a group of multiplicative stable operations which are Galois invariant in the sense that

$$\mathrm{Fr}(\overline{\theta}(t)) = (\overline{^{\mathrm{Fr}}\theta})(\mathrm{Fr}(t)).$$

Here we recall that the group of strict automorphisms of $F^{K(2)}$ is the group of units $\theta \in \mathbb{D}_2$ (a certain central division algebra over $\mathbb{Q}_p$ of dimension 4) satisfying

$$|\theta|_p \leqslant 1$$

and

$$|1 - \theta|_p < 1,$$

with $^{\mathrm{Fr}}\theta = S\theta S^{-1}$ where $S \in \mathbb{D}_2$ satisfies $S^2 = p$. This is all explained in Ravenel's book **[21]**. We denote this group of strict automorphisms by $\mathbb{S}_2$. Using the subgroup $\widetilde{\mathbb{S}}_2 \leqslant \mathbb{D}_2^\times$ generated by $\mathbb{S}_2$ together with the central element $S^2 = p$, we can form the double coset space $\langle p\rangle\backslash\widetilde{\mathbb{S}}_2/\langle p\rangle$. Then the associated Hecke algebra $\mathbb{Z}_p[\langle p\rangle\backslash\widetilde{\mathbb{S}}_2/\langle p\rangle]$ acts continuously upon

$$^{\mathrm{Fr}}(\mathbb{F}_{p^2}\otimes E\ell\ell/(p,\pi))^*(\ )\cong(E\ell\ell/(p,\pi))^*(\ ).$$

Hence wc can form the twisted Hecke algebra

$$(\mathbb{F}_{p^2}\otimes E\ell\ell/(p,\pi)_*)^{\langle p\rangle}\left\{\langle p\rangle\backslash\widetilde{\mathbb{S}}_2/\langle p\rangle\right\}$$

and this acts continuously upon $(E\ell\ell/(p,\pi))^*(\ )$.

References

[1] J. F. Adams, *Stable Homotopy and Generalised Homology*, University of Chicago Press, 1974.

[2] A. Baker, *Hecke operators as operations in elliptic cohomology*, J. Pure and Applied Alg. **63** (1990), 1-11.

[3] A. Baker, *Elliptic cohomology, p-adic modular forms and Atkin's operator* U_p, Contemp. Math. **96** (1989), 33–38.

[4] A. Baker, *On the homotopy type of the spectrum representing elliptic cohomology*, Proc. Amer. Math. Soc. **107** (1989), 537–48.

[5] A. Baker, *Elliptic genera of level N and elliptic cohomology*, Jour. Lond. Math. Soc. **49** (1994), 581–93.

[6] A. Baker, *Operations and cooperations in elliptic cohomology, Part I: Generalized modular forms and the cooperation algebra*, New York J. Math. **1** (1995), 39–74.

[7] A. Baker, *Operations and cooperations in elliptic cohomology, Part II: Twisted Hecke algebras and stable operations*, in preparation.

[8] A. Baker, *On the Adams* E_2*-term for elliptic cohomology*, Glasgow University Mathematics Department preprint 97/15.

[9] A. Baker, *Hecke operations and the Adams* E_2*-term based on elliptic cohomology*, Glasgow University Mathematics Department preprint no. 97/22.

[10] A. Baker & U. Würgler, *Liftings of formal group laws and the Artinian completion of* $v_n^{-1}BP$, Proc. Camb. Phil. Soc. **106** (1989), 511–30.

[11] R. Field, *Hecke Operators on Quotients of the Ring of Modular Forms of Level 1*, MSc thesis, Manchester University, 1991.

[12] J. Igusa, *On the transformation theory of elliptic functions*, Amer. J. Math. **81** (1959), 436–52.

[13] J. Igusa, *On the algebraic theory of elliptic modular functions*, J. Math. Soc. Japan **20** (1968), 96–106.

[14] N. Katz, *p-adic properties of modular schemes and modular forms*, Lecture Notes in Mathematics **350** (1973), 69-190.

[15] N. Koblitz, *Elliptic Curves and Modular Forms*, Springer-Verlag, 1984.

[16] A. Krieg, *Hecke algebras*, Mem. Amer. Math. Soc. **87** No. 435 (1990).

[17] P. S. Landweber, *Elliptic cohomology and modular forms*, Lecture Notes in Mathematics **1326** (1988), 55–68.

[18] P. S. Landweber, *Supersingular elliptic curves and congruences for Legendre polynomials*, Lecture Notes in Mathematics **1326** (1988), 69–93.

[19] P. S. Landweber, D. C. Ravenel & R. E. Stong, *Periodic cohomology theories defined by elliptic curves*, Contemp. Math. **181** (1995), 317–337.

[20] H. Matsumura, *Commutative Ring Theory*, Cambridge University Press, 1986.

[21] D. C. Ravenel, *Complex Cobordism and the Stable Homotopy Groups of spheres*, Academic Press, 1986.

[22] G. Robert, *Congruences entre séries d'Eisenstein, dans le cas supersingular*, Invent. Math. **61** (1980), 103–158.

[23] J-P. Serre, *Cours d'Arithmétique*, Presses Universitaires de France, 1977.
[24] J-P. Serre, *Formes modulaires et fonctions zeta p-adiques*, Lecture Notes in Mathematics **350** (1973), 191-268.
[25] J-P. Serre, *Congruences et formes modulaires,(Sém. Bourbaki, Vol.* 24e*, no. 416, 1971/2)*, Lecture Notes in Mathematics **317** (1973), 319-38.
[26] J. Silverman, *The Arithmetic of Elliptic Curves*, Springer-Verlag, 1986.
[27] N. Yagita, *The exact functor theorem for* BP_*/I_n, Proc. Jap. Acad. **52** (1976), 1-3.

UNIVERSITY OF GLASGOW, GLASGOW G12 8QW, SCOTLAND.
E-mail address: a.baker@maths.gla.ac.uk

Contemporary Mathematics
Volume **220**, 1998

Some root invariants and Steenrod operations in $\mathrm{Ext}_A(F_2, F_2)$

ROBERT R. BRUNER

ABSTRACT. We give the results of computations of root invariants in Ext over the Steenrod algebra through the 25-stem, with partial information through the 45-stem. This allows the computation of some new Steenrod operations as well.

These calculations were started to provide evidence for or against the algebraic Bredon-Löffler conjecture. This is the conjecture that

$$\eta_k^* : \mathrm{Ext}_A^{s,t}(F_2, F_2) \longrightarrow \mathrm{Ext}_A^{s,t}(\Sigma L_{-k}, F_2)$$

is a monomorphism for $0 < t - s < k/2$. Here L is the Laurent series ring $F_2[x, x^{-1}]$ with its usual Steenrod algebra action, L_{-k} is the submodule of L consisting of those elements whose degree is greater than or equal to $-k$, and $\eta_k : \Sigma L_{-k} \longrightarrow F_2$ is the nonzero homomorphism. The calculations of η_k^* showed [**5**] that the conjecture holds for $k \leq 55$. With those calculations in hand, we needed only calculate the homomorphisms

$$i_k^* : \mathrm{Ext}_A^{s,t}(\Sigma^{1-k} F_2, F_2) \longrightarrow \mathrm{Ext}_A^{s,t}(\Sigma L_{-k}, F_2)$$

induced by the projections onto the bottom dimensions, $i_k : \Sigma L_{-k} \longrightarrow \Sigma^{1-k} F_2$, to calculate all the root invariants through the 25 stem, and about half of those in the 26 through 45 stems. This extends the calculations of Mahowald and Shick [**7**], who found all the root invariants through the 16 stem. Our calculations were done by first using the programs described in [**1, 2**] to compute resolutions of the modules L_{-k} and the chain maps induced by η_k and i_k. We then used MAGMA to compute the induced maps of Ext and the root invariants. A calculation of $\mathrm{Ext}_A(F_2, F_2)$ for $s < 40$ and $t \leq 140$ with complete information on the product structure was calculated in the same manner: using MAGMA to process the

1991 *Mathematics Subject Classification.* Primary 55Q45, 55Q35, 55Q91; Secondary 55P42, 55P91, 55T15 .

output of the minimal resolution programs [**1, 2**]. A preliminary version of this can be found in [**3**], where the notation for elements of $\mathrm{Ext}_A(F_2, F_2)$ used here is defined. Charts showing the results of the Ext calculations (but not the induced homomorphisms) can be found at http://www.math.wayne.edu/~rrb/cohom.

By [**6**, Proposition 2.5], we may conclude that $Sq^0(x) \in R(x)$ when $|Sq^0(x)| = |R(x)|$, that is, if $k = n + s + 1$. This allows us to use the calculations here to determine a number of Steenrod operations.

THEOREM 1. *In the cohomology of the mod 2 Steenrod algebra*
(i) $Sq^0(Ph_1) = h_2 g$
(ii) $Sq^0(Pc_0) = h_1 y$
(iii) $Sq^0(Pd_0) = g d_1$
(iv) $Sq^0(i) = h_2 C$
(v) $Sq^0(k) = h_2 h_5 n + \epsilon h_3 D_2$, *where* $\epsilon \in \{0, 1\}$
(vi) $Sq^0(q) = h_2 Q_3$
(vii) $Sq^0(l) = h_0 x_{6,47}$
(viii) $Sq^0(y) = h_4 Q_3 + h_2 x_1$

These are the Sq^0's which are not 'tautologous'. In addition, we have 12 more which allow us to identify, in the resolution produced by the machine, the elements named. That is, we know by definition that $e_1 = Sq^0(e_0)$, for example. However, we have no way *a priori* to determine which of two elements in the 38 stem is e_1 since we have no general mechanical method of computing Steenrod operations. Thus, the root invariant is quite useful, in that it unambiguously identifies e_1 for us. The elements identified in this manner are n_1, p_1, e_1, e_2, f_1, and f_2. We also determine m_1 and t_1 up to F_2 indeterminacy.

Finally, note that the root invariant $R(h_2^2 g) = r_1$ shows that the Strong Algebraic Bredon-Löffler Conjecture in [**4**] must have its $(t - s)$-intercept decreased by at least 3. This is an inessential modification, as the key observation in these calculations supporting the conjecture is the slope -1/2.

CONJECTURE 2. *(Strong Algebraic Bredon-Löffler Conjecture) The map* η_k^* *is a monomorphism if*

$$s < (k - n - 3)/2.$$

KEY TO THE TABLES: Each row in the table lists the filtration s and stem n of an element $x \in \mathrm{Ext}_A(F_2, F_2)$, its root invariant $R(x)$, the dimension k of the cell on which it occurs, whether $Sq^0(x) \in R(x)$, and a basis for the indeterminacy in $R(x)$. The root invariant $R(x)$ is in the $n + k - 1$ stem. By [**6**, Proposition 2.5], we may test whether or not $Sq^0(x) \in R(x)$ by checking whether or not $|Sq^0(x)| = |R(x)|$, that is, $k = n + s + 1$.

k	s	n	x	$R(x)$	$Sq^0 \in R(x)$?	Indeterminacy
2	1	0	h_0	h_1	yes	
3	2	0	h_0^2	h_1^2	yes	
4	3	0	h_0^3	$h_0^2h_2$	yes	
8	4	0	h_0^4	$h_0^3h_3$	no	
10	5	0	h_0^5	Ph_1	no	
11	6	0	h_0^6	h_1Ph_1	no	
12	7	0	h_0^7	$h_0^2Ph_2$	no	
16	8	0	h_0^8	$h_0^7h_4$	no	
18	9	0	h_0^9	P^2h_1	no	
19	10	0	h_0^{10}	$h_1P^2h_1$	no	
20	11	0	h_0^{11}	$h_0^2P^2h_2$	no	
24	12	0	h_0^{12}	h_0^5i	no	
26	13	0	h_0^{13}	P^3h_1	no	
27	14	0	h_0^{14}	$h_1P^3h_1$	no	
28	15	0	h_0^{15}	$h_0^2P^3h_2$	no	
32	16	0	h_0^{16}	$h_0^{15}h_5$	no	
34	17	0	h_0^{17}	P^4h_1	no	
35	18	0	h_0^{18}	$h_1P^4h_1$	no	
36	19	0	h_0^{19}	$h_0^2P^4h_2$	no	
3	1	1	h_1	h_2	yes	
5	2	2	h_1^2	h_2^2	yes	
5	1	3	h_2	h_3	yes	
6	2	3	h_0h_2	h_1h_3	yes	
7	3	3	$h_0^2h_2$	$h_1^2h_3$	yes	
9	2	6	h_2^2	h_3^2	yes	
9	1	7	h_3	h_4	yes	
10	2	7	h_0h_3	h_1h_4	yes	
11	3	7	$h_0^2h_3$	$h_1^2h_4$	yes	
12	4	7	$h_0^3h_3$	$h_0^2h_2h_4$	yes	
11	2	8	h_1h_3	h_2h_4	yes	
12	3	8	c_0	c_1	yes	
13	3	9	$h_1^2h_3$	$h_2^2h_4$	yes	
14	4	9	h_1c_0	h_2c_1	yes	
15	5	9	Ph_1	h_2g	yes	
21	6	10	h_1Ph_1	r	no	
20	5	11	Ph_2	$h_0^3h_4^2$	no	
22	6	11	h_0Ph_2	q	no	
23	7	11	$h_0^2Ph_2$	h_1q	no	
17	2	14	h_3^2	h_4^2	yes	
18	3	14	$h_0h_3^2$	$h_1h_4^2$	yes	
19	4	14	d_0	d_1	yes	

k	s	n	x	$R(x)$	$Sq^0 \in R(x)$?	Indeterminacy
20	5	14	h_0d_0	h_0p	yes	
25	6	14	$h_0^2d_0$	y	no	h_1x
17	1	15	h_4	h_5	yes	
18	2	15	h_0h_4	h_1h_5	yes	
19	3	15	$h_0^2h_4$	$h_1^2h_5$	yes	
20	4	15	$h_0^3h_4$	$h_0^2h_2h_5$	yes	
21	5	15	h_1d_0	h_2d_1	yes	
24	5	15	$h_0^4h_4$	$h_0^3h_3h_5$	no	
26	6	15	$h_0^5h_4$	h_5Ph_1	no	$h_0^2f_1$
27	7	15	$h_0^6h_4$	$h_1h_5Ph_1$	no	
28	8	15	$h_0^7h_4$	$h_0^2h_5Ph_2$	no	
19	2	16	h_1h_4	h_2h_5	yes	
23	6	16	$h_1^2d_0$	h_1x	yes	
24	7	16	Pc_0	h_1y	yes	
21	3	17	$h_1^2h_4$	$h_2^2h_5$	yes	
22	4	17	e_0	e_1	yes	
23	5	17	h_0e_0	h_1e_1	yes	
24	6	17	$h_0^2e_0$	$h_0^2f_1$	yes	
31	7	17	$h_0^3e_0$	$h_1^2h_5d_0$	no	
30	8	17	h_1Pc_0	N	no	
32	9	17	P^2h_1	$h_1h_5Pc_0$	no	
21	2	18	h_2h_4	h_3h_5	yes	
22	3	18	$h_0h_2h_4$	$h_1h_3h_5$	yes	
23	4	18	f_0	f_1	yes	
23	4	18	$h_0^2h_2h_4$	$h_1^2h_3h_5$	yes	
24	5	18	h_0f_0	$h_0^2c_2$	yes	
37	10	18	$h_1P^2h_1$	R_1	no	$h_0^2h_5i$
23	3	19	c_1	c_2	yes	
35	9	19	P^2h_2	h_5Pd_0	no	
36	10	19	$h_0P^2h_2$	$h_0^2h_5i$	no	
39	11	19	$h_0^2P^2h_2$	h_1Q_1	no	
25	4	20	g	g_2	yes	
26	5	20	h_0g	h_1g_2	yes	
28	6	20	h_0^2g	$h_0h_2g_2$	no	
25	3	21	$h_2^2h_4$	$h_3^2h_5$	yes	
27	5	21	h_1g	h_2g_2	yes	
27	4	22	h_2c_1	h_3c_2	yes	
31	8	22	Pd_0	gd_1	yes	
36	9	22	h_0Pd_0	$h_0^2Q_2$	no	
41	10	22	$h_0^2Pd_0$	$x_{10,27}$	no	$x_{10,28}, h_1X_1$
28	4	23	h_4c_0	h_5c_1	yes	

k	s	n	x	$R(x)$	$Sq^0 \in R(x)$?	Indeterminacy
29	5	23	h_2g	h_3g_2	yes	
30	6	23	h_0h_2g	$h_1h_3g_2$	yes	
31	7	23	i	h_2C	yes	
36	8	23	h_0i	$h_0^2D_2$	no	
38	9	23	h_0^2i	$h_0^2B_3$	no	
40	9	23	$h_0^2i + h_1Pd_0$	$h_0^7h_5^2$	no	
42	10	23	h_0^3i	$x_{10,32}$	no	$h_0^2h_3Q_2$
43	11	23	h_0^4i	$h_1x_{10,32}$	no	h_0B_{23}
44	12	23	h_0^5i	$h_0^2PD_2$	no	
30	5	24	$h_1h_4c_0$	$h_2h_5c_1$	yes	
43	10	24	$h_1^2Pd_0$	$B_5 + PD_2$	no	
44	11	24	P^2c_0	$x_{11,35} + h_0^2x_{9,40}$	no	
38	8	25	Pe_0	$x_{8,32} + x_{8,33}$	no	
39	9	25	h_0Pe_0	$h_1x_{8,32}$	no	$h_0^2x_{7,33}, h_0^8h_6$
40	10	25	$h_0^2Pe_0$	$h_0^2h_3Q_2$	no	
45	11	25	$h_0^3Pe_0$	h_2B_5	no	h_0PA
46	12	25	$h_1P^2c_0$	$h_2x_{11,35}$	no	
47	13	25	P^3h_1	$x_{13,34}$	no	$x_{13,35}$
41	6	26	h_2^2g	r_1	no	
36	7	26	j	h_0A'	no	
39	8	26	h_0j	h_3Q_2	no	h_0h_2A, h_0^2A''
40	9	26	h_0^2j	$h_0^2h_3D_2$	no	
51	13	27	P^3h_2	e_0B_4	no	$h_0^6x_{7,57}$
41	8	28	d_0^2	G_{21}	no	h_0h_3A'
42	9	28	$h_0d_0^2$	h_1G_{21}	no	$h_0h_2x_{7,40}$
44	10	28	$h_0^2d_0^2$	$h_0h_2G_{21}$	no	
37	7	29	k	h_2h_5n	yes	h_3D_2
40	8	29	h_0k	h_0h_3A'	no	
43	9	29	h_0^2k	h_2G_{21}	no	
33	2	30	h_4^2	h_5^2	yes	
34	3	30	$h_0h_4^2$	$h_1h_5^2$	yes	
35	4	30	$h_0^2h_4^2$	$h_1^2h_5^2$	yes	
36	5	30	$h_0^3h_4^2$	$h_0^2h_2h_5^2$	yes	
37	6	30	r	r_1	yes	
39	7	30	h_0r	$h_2^2H_1 + h_3A'$	no	
41	8	30	h_0^2r	$h_3x_{7,33}$	no	
44	9	30	h_0^3r	$h_0^2h_4D_2$	no	
49	10	30	h_0^4r	$P^2h_5^2$	no	e_0A'
50	11	30	h_0^5r	$h_1P^2h_5^2$	no	
51	12	30	P^2d_0	$x_{12,44}$	no	
33	1	31	h_5	h_6	yes	

k	s	n	x	$R(x)$	$Sq^0 \in R(x)$?	Indeterminacy
34	2	31	h_0h_5	h_1h_6	yes	
35	3	31	$h_0^2h_5$	$h_1^2h_6$	yes	
35	3	31	$h_1h_4^2$	$h_2h_5^2$	yes	
36	4	31	$h_0^3h_5$	$h_0^2h_2h_6$	yes	
37	5	31	n	n_1	yes	
40	5	31	$h_0^4h_5$	$h_0^3h_3h_6$	no	
42	6	31	$h_0^5h_5$	h_6Ph_1	no	
43	7	31	$h_0^6h_5$	$h_1h_6Ph_1$	no	$h_2^2Q_3$
40	8	31	d_0e_0	$h_3x_{7,33}$	yes	
44	8	31	$h_0^7h_5$	$h_0^2h_6Ph_2$	no	
45	9	31	$h_0d_0e_0$	h_3G_{21}	no	$h_0^2x_{7,53}$
48	9	31	$h_0^8h_5$	$h_0^7h_4h_6$	no	
46	10	31	$h_0^2d_0e_0$	$h_0x_{9,51}$	no	
50	10	31	$h_0^9h_5$	$h_6P^2h_1$	no	
51	11	31	$h_0^{10}h_5$	$h_1h_6P^2h_1$	no	h_0gA'
35	2	32	h_1h_5	h_2h_6	yes	
37	4	32	d_1	d_2	yes	
39	6	32	q	h_2Q_3	yes	
40	7	32	l	$h_0x_{6,47}$	yes	
43	8	32	h_0l	$x_{8,51}$	no	$h_0^2h_6Ph_2$
44	9	32	h_0^2l	$h_0^2x_{7,53}$	no	
37	3	33	$h_1^2h_5$	$h_2^2h_6$	yes	
38	4	33	p	p_1	yes	
39	5	33	h_0p	h_1p_1	yes	
45	7	33	h_1q	$x_{7,57}$	no	$h_1x_{6,53}, h_0^2h_6d_0$
37	2	34	h_2h_5	h_3h_6	yes	
38	3	34	$h_0h_2h_5$	$h_1h_3h_6$	yes	
39	4	34	$h_0^2h_2h_5$	$h_1^2h_3h_6$	yes	
41	6	34	h_2n	h_3n_1	yes	h_6Ph_2
43	8	34	d_0g	$h_1x_{7,53}$	yes	$h_0^2x_{6,53}$
44	9	34	h_0d_0g	$h_0^2m_1$	yes	
48	11	34	Pj	h_0gA'	no	
41	5	35	h_2d_1	h_3d_2	yes	
43	7	35	m	m_1	yes	$h_0^2h_6d_0$
44	8	35	h_0m	$h_0^2t_1$	yes	
47	9	35	h_0^2m	$h_3x_{8,51}$	no	
43	6	36	t	t_1	yes	$h_0^4h_4h_6$
41	3	37	$h_2^2h_5$	$h_3^2h_6$	yes	
43	5	37	x	x_1	yes	
44	6	37	h_0x	$h_0^2e_2$	yes	
45	7	37	h_1t	h_2t_1	yes	

k	s	n	x	$R(x)$	$Sq^0 \in R(x)$?	Indeterminacy
46	7	37	$h_0^2 x$	$h_0 h_2 x_1$	no	
46	8	37	$e_0 g$	$e_1 g_2$	yes	
48	8	37	$h_0^3 x$	$h_0^4 f_2$	no	
41	2	38	$h_3 h_5$	$h_4 h_6$	yes	
42	3	38	$h_0 h_3 h_5$	$h_1 h_4 h_6$	yes	
43	4	38	$h_0^2 h_3 h_5$	$h_1^2 h_4 h_6$	yes	
43	4	38	e_1	e_2	yes	
44	5	38	$h_0^3 h_3 h_5$	$h_0^2 h_2 h_4 h_6$	yes	
45	6	38	$h_1 x$	$h_2 x_1$	yes	
45	6	38	y	$h_4 Q_3 + h_2 x_1$	yes	
46	7	38	$h_0 y$	$h_1 h_4 Q_3$	yes	$h_0^2 h_6 g$
48	8	38	$h_0^2 y$	$h_0^5 c_3$	no	
43	3	39	$h_1 h_3 h_5$	$h_2 h_4 h_6$	yes	
44	4	39	$h_5 c_0$	$h_6 c_1$	yes	
45	5	39	$h_1 e_1$	$h_2 e_2$	yes	
49	7	39	$h_1 y$	$x_{7,74}$	no	
45	4	40	$h_1^2 h_3 h_5$	$h_2^2 h_4 h_6$	yes	
45	4	40	f_1	f_2	yes	
46	5	40	$h_0 f_1$	$h_1 f_2$	yes	
46	5	40	$h_1 h_5 c_0$	$h_2 h_6 c_1$	yes	
47	6	40	$h_0^2 f_1$	$h_1^2 f_2$	yes	
47	6	40	$h_5 P h_1$	$h_2 h_6 g$	yes	
45	3	41	c_2	c_3	yes	
46	4	41	$h_0 c_2$	$h_1 c_3$	yes	
47	5	41	$h_0^2 c_2$	$h_1^2 c_3$	yes	
49	4	44	g_2	g_3	yes	
49	3	45	$h_3^2 h_5$	$h_4^2 h_6$	yes	

References

1. Robert R. Bruner, "Calculation of large Ext modules", *Computers in Geometry and Topology* (M. C. Tangora, ed.), Marcel Dekker, New York, 1989, 79-104.
2. R.R.Bruner, "Ext in the nineties", pp. 71-90 in *Algebraic Topology*, Oaxtepec 1991, Contemp. Math. **146**, Amer. Math. Soc., Providence, 1993.
3. R.R.Bruner, "The Cohomology of the Mod 2 Steenrod Algebra", preprint, April 1997.
4. Robert R. Bruner, "Some remarks on the root invariant", *Proceedings of the Fields Institute Workshops on Homotopy Theory, January 1996, Contemp. Math.*, to appear.
5. Robert Bruner and John Greenlees, "The Bredon-Löffler conjecture", *Exper. Math.* **4** (1995), 101-109.
6. M.E.Mahowald and D.C.Ravenel, "The root invariant in homotopy theory", *Topology* **32** (1993), 865-898.
7. M.E.Mahowald and Paul Shick, "Some root invariants in $\mathrm{Ext}_A(Z/2, Z/2)$ ", *Proceedings of the Northwestern Homotopy Theory Conference, Contemp. Math.* **19** (1983), 227-231.

Contemporary Mathematics
Volume **220**, 1998

TOWARD HOMOLOGY OPERATIONS FOR MAPPING CLASS GROUPS

F. R. Cohen* and Ulrike Tillmann

Abstract. Independent of any infinite loop space structure, Araki-Kudo-Dyer-Lashof operations Q_n for all n are constructed for the stable homology of the mapping class groups. They are induced by an action of the cyclic groups and are non-trivial. On the other hand, a natural action of the homology of the pure ribbon braid groups on the homology of the mapping class groups is shown to be zero stably.

Introduction.

Let $F_{g,n}$ denote an oriented surface of genus g with n boundary components. The associated mapping class group is the discrete group $\Gamma_{g,n} := \pi_0 \mathrm{Diff}^+(F_{g,n}; \partial)$ of connected components of its orientation preserving diffeomorphisms that fix the boundary pointwise. Its classifying space $B\Gamma_{g,n}$ is homotopy equivalent to the moduli space $\mathcal{M}_{g,n}$ of Riemann surfaces of topological type $F_{g,n}$ when $n > 0$ and rationally homotopy equivalent when $n = 0$ and $g \geq 2$.

In [M] Miller considers the monoid $\mathcal{M} = \coprod_{g \geq 0} B\Gamma_{g,1}$ with a product $\star$ given by a pairs of pants construction. Furthermore, based on this, he sketches an argument that its group completion $B\Gamma_\infty^+$ has a double loop space structure. Bödigheimer [Bö 1] gives explicit constructions and also defines the corresponding first two Araki-Kudo-Dyer-Lashof operations Q_0, Q_1. These are induced by group homomorphism

$$\beta_p \wr \Gamma_{g,1} \longrightarrow \Gamma_{pg,1}$$

from the wreath product of $\Gamma_{g,1}$ with the braid group on p-strings. (See Section 1 for its definition.) Already for $p = 2$ and $\beta_2 = \mathbb{Z}$ it is easy to see that this map does

1991 *Mathematics Subject Classification.* Primary 20J05, 55S12, 57M20.
*Partially supported by the NSF.

not factor through the wreath product with the cyclic group $C_2 \wr \Gamma_{g,1}$ so that it does not induce higher homology operations. Indeed, there are non-vanishing secondary homology operations in more than one variable, Browder operations, which do not vanish [FS]. This is one of the reasons why the double loop space structure on $B\Gamma_\infty^+$ was believed to be best possible. However, recently an infinite loop space structure on $B\Gamma_\infty^+$ has been exhibited [T1], and therefore Araki-Kudo-Dyer-Lashof operations of every dimension must exist. One of the goals of this short paper is to understand this situation.

Recall, Harer [H1] proved that in homology (with simple coefficients) the natural map $\sigma : \Gamma_{g,1} \to \Gamma_{g+1,1}$ induced by extending the diffeomorphisms by the identity over a glued on torus is an isomorphism in dimensions less than $g/3$. This was later improved by Ivanov [I] to $g/2$. Thus we may speak of the stable part of the homology of $\Gamma_{g,1}$. The main point is that the existence of an infinite loop space structure on $B\Gamma_\infty^+$ gives us homology operations on the stable(!) homology only, while an obstruction to extending Miller and Bödigheimer's Q_0 and Q_1 is an obstruction to extending homology operations for all(!) homology classes of $B\Gamma_{g,1}$ for all g.

In this article, we construct explicit homology operations Q_n, $n \geq 0$ on the stable homology, extending the existing Q_0 and Q_1. Apart from understanding the above mentioned apparent contradiction, the explicit geometric construction of these operations should also be of interest as operations coming from the infinite loop space structure are difficult to describe and understand in concrete terms. This is because the infinite loop space structure "lives" naturally on the deloop of $B\Gamma_\infty^+$ rather than on itself. Indeed, the construction of the Q_n in Section 2 based on an action of the cyclic groups was already used for explicit calculations in [CC]. This cyclic action is compatible with the operad structure coming from the ribbon braid groups. This is explained in Section 1. These constructions are general and apply in other situations as described below. Section 3 contains some easy consequences. In Section 4 we prove that on stable homlogy, the action of the pure braid group is trivial. This result is possibly the best answer we can offer as a solution to the above apparent contradiction.

Open questions and remarks.

1.) In [T1] the infinite loop space structure was shown to have the same first deloop as the double loop space structure of Miller and Bödigheimer. It is not clear, though likely, that also the second deloops coincide.

2.) Similarly, we do not know whether the operations constructed here are the same as the ones corresponding to the infinite loop space structure. Only the Pontrjagin products Q_0 are the same as the first deloops are the same. If these operations were induced by the infinite loop space structure, then the Adem relations and the Cartan formula would follow immediately. At this point it remains an open question whether the Q_i's (constructed here) satisfy the Adem relation and Cartan formula (see Section 4), and whether they commute with the stabilization map $\sigma : \Gamma_{g,1} \to \Gamma_{g+1,1}$. However, it is shown that they are compatible with the operations in the homology of $BSP(\mathbb{Z})$, and the space $\mathrm{Im}J$; namely the operations constructed here project to the natural operations in the homology of $BSP(\mathbb{Z})$, and

ImJ via the associated natural maps as given by Proposition 3.3.

3.) In [FS] Fiedorowicz and Song show that the Browder operation applied to a class in $H_0 B\Gamma_{1,1}$ is non-trivial. This implies that Miller's monoid $\mathcal{M}$ does not support a little 3-cube action. However, this obstruction class lies in the first homology of $\Gamma_{2,1}$ and hence is unstable ($H_1 B\Gamma_\infty = 0$). It is therefore not an obstruction to a positive answer to the question raised in 2.). This unusual situation arises because the natural map of Miller's monoid $\mathcal{M}$ to its group completion is not an injection on homology.

4.) For simplicity, we work here with the groups $\Gamma_{g,1}$. However, our constructions work also for the mapping class groups $\Gamma^k_{g,n+1}$ of surfaces with n additional boundary components and k points. For Harer's stability theorem holds also for these groups [H1]. In particular, there are structure maps (compare Section 2)

$$\theta : C_p \wr \Gamma^k_{g,n+1} \longrightarrow \Gamma^{pk}_{pg,pn}.$$

It should be pointed out though, that while the stable homology of $\Gamma^k_{g,n}$ does not depend on n already the second homology groups vary with k: $H_2\Gamma^k_{\infty,n+1} = \mathbb{Z}^{k+1}$. The analogue of Miller's monoid is $\coprod_{g,k,n\geq 0} B\Gamma^k_{g,n+1}$. It should not be difficult to prove that its group completion has an infinite loop space structure by arguments similar to those used in [T1]. But the relation of its homology to the homology of each $\Gamma^k_{g,n}$ is not clear.

5.) Similarly, our constructions give also rise to homology operations for the spin mapping class groups $G^s_{g,1}$ of surfaces with spin structure of Arf invariant $s = 0, 1$. Harer [H2] proved a homology stability theorem in this case. Again this is homological evidence that the monoid $\coprod_{g\geq 0, s=0,1} BG^s_{g,1}$ group completes to an infinite loop space the homology of which is the stable homology of the spin mapping class groups. Bauer [B] shows that this is indeed the case.

Throughout this paper we fix a prime p. All chain complexes and homology groups have $\mathbb{Z}/p\mathbb{Z}$ coefficients.

Added in proof. The second author (*Higher genus surface operad detects infinite loop spaces*, preprint) has constructed a natural infinite loop space structure on $\mathbb{Z}\times B\Gamma^+_\infty$ which extends Miller's double loop space structure. The Araki-Kudo-Dyer-Lashof operations for this infinite loop space structure are the ones constructed in this paper. In particular this implies that the Cartan formula and the Adem relations hold!

1. A commutative diagram of groups.

The goal in this section is to set up Diagram 1.2 below. From it we will derive the definition of the homology operations in the next section and some of their

C_p the cyclic group of order p.

β_p the braid group on p strings.

Γ_g the mapping class group of a closed surface of genus g.

$\Gamma_{g,n}$ the mapping class group of a surface of genus g with n parametrized boundary components which are fixed (pointwise).

$\tilde{\Gamma}_{0,p}$ the mapping class group of a sphere with p parametrized boundary components which may be interchanged in a parametrization preserving way. By considering the action on the p boundary components, one sees that this group fits into a short exact sequence $\Gamma_{0,p} \to \tilde{\Gamma}_{0,p} \to \Sigma_p$.

$\tilde{\Gamma}_{0,p,1}$ the mapping class group of a sphere with $p+1$ parametrized boundary components, the first p of which may be interchanged in a parametrization preserving way. This group is also known as the p-th ribbon braid group.

Gluing a disk onto the $(p+1)$-th boundary component and extending the underlying diffeomorphisms by the idenity defines a map of groups $\flat : \tilde{\Gamma}_{0,p,1} \to \tilde{\Gamma}_{0,p}$. Consider the infinite cyclic group $\mathbb{Z}$ in $\tilde{\Gamma}_{0,p,1}$ generated by the diffeomorphism α that rotates the first p boundary components into each other but fixes the last (see Figure 1). The image of α under $\flat$ is carried into a diffeomorphism that is homotopic to the diffeomorphism that rotates the boundary components and fixes north and south pole of the sphere. The latter diffeomorphism is clearly of order p. Hence, we have a commutative diagram of groups

$$\begin{array}{ccc} \mathbb{Z} & \xrightarrow{incl} & \tilde{\Gamma}_{0,p,1} \\ {\scriptstyle\flat}\downarrow & & {\scriptstyle\flat}\downarrow \\ C_p & \xrightarrow{incl} & \tilde{\Gamma}_{0,p} \end{array}$$

We can identify $\tilde{\Gamma}_{0,p,1}$ with the ribbon braid group on p ribbons. With this identification, note that the top horizontal arrow may indeed be factored through the braid group β_p.

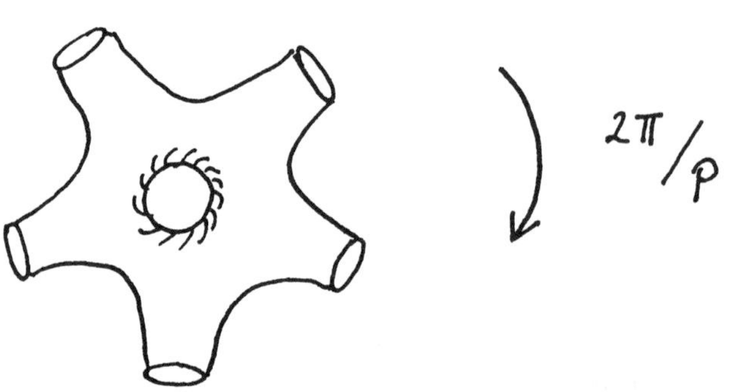

Figure 1: The generator α in $\mathbb{Z} \subset \tilde{\Gamma}_{0,p,1}$.

The groups C_p, $\mathbb{Z}$, β_p, $\tilde{\Gamma}_{0,p}$, and $\tilde{\Gamma}_{0,p,1}$ all have natural representations in Σ_p given by their action on the p boundary components as subgroups of the appropriate mapping class groups. Thus we can form their wreath products with the mapping class group $\Gamma_{g,1}$. Gluing surfaces and assembling diffeomorphisms (which can be done as the parametrizations of the boundary components are always preserved) defines structure maps $\eta : \tilde{\Gamma}_{0,p,1} \wr \Gamma_{g,1} \longrightarrow \Gamma_{pg,1}$ and $\tilde{\Gamma}_{0,p} \wr \Gamma_{g,1} \longrightarrow \Gamma_{pg}$. We can summarize the above observations in the following commutative diagram of groups of which the top right hand side map is the map mentioned in the introduction.

$$
\begin{array}{ccccc}
\mathbb{Z} \wr \Gamma_{g,1} & \longrightarrow & \beta_p \wr \Gamma_{g,1} & \longrightarrow & \Gamma_{pg,1} \\
\| & & \downarrow & & \| \\
\mathbb{Z} \wr \Gamma_{g,1} & \longrightarrow & \tilde{\Gamma}_{0,p,1} \wr \Gamma_{g,1} & \xrightarrow{\eta} & \Gamma_{pg,1} \\
\downarrow & & \flat\downarrow & & \flat\downarrow \\
C_p \wr \Gamma_{g,1} & \longrightarrow & \tilde{\Gamma}_{0,p} \wr \Gamma_{g,1} & \longrightarrow & \Gamma_{pg}
\end{array}
\tag{1. 2}
$$

2. Definition of homology operations.

We will now define the Araki-Kudo-Dyer-Lashof operations for the stable homology classes of the mapping class group. The construction is based on Diagram 1.2. But first we will recall the definition of Araki-Kudo-Dyer-Lashof operations for Γ-spaces and E_∞-spaces (compare [CLM, p. 7]).

Definition 2.1. Assume we have a map

$$\theta : E\Sigma_p \times_{\Sigma_p} X^p \longrightarrow X$$

for some space X where $E\Sigma_p$ denotes the universal Σ_p-bundle. Let W_* be the standard free minimal resolution of C_p over $\mathbb{Z}/p\mathbb{Z}$ with e_i the generator in dimension i, and let $W_* \to C_* E\Sigma_p$ be a morphism of C_p-complexes. This defines $\mathbb{Z}/p\mathbb{Z}$-homology operations

$$Q_i : H_n(X) \longrightarrow H_{pn+i}(X)$$

for $i = 0, 1, 2, 3, \ldots$ by setting $Q_i(x) := \theta_*(e_i \otimes x^{\otimes p})$.

Whenever X is a Γ-space in the sense of Segal [S] or equivalently an E_∞-space in the sense of May [Ma], then the group completion of X is an infinite loop space and θ is given by one of the structure maps. On the other hand, if X is only a E_n-space and its group completion is only an n-fold loop space, then the structure space $E\Sigma_p$ is replaced by a free Σ_p-space which is only $(n-2)$-connected, the Boardman-Vogt "little cubes" of appropriate dimension. Hence, only the operations Q_i for $i = 0, \ldots, n-1$ can be defined. When $n = 2$, $E\Sigma_p$ may be replaced by $E\beta_p$. This is why the double loop space structure of Miller and Bödigheimer gives rise to just the first two operations.

From the above definition, we see that we do not need a map θ but just a C_p-

Fact 1. Let G and K be discrete groups, $\rho : G \longrightarrow \Sigma_p$ be a representation, and $G \wr K$ be the corresponding wreath product. Then

$$B(G \wr K) \simeq B(\mathrm{Ker}\rho) \times_{\mathrm{Im}\rho} (BK)^p.$$

This is easy to see as both spaces are $\mathrm{Im}\rho$-bundles over $BG \simeq B(\mathrm{Ker}\rho)/\mathrm{Im}\rho$ with fiber $(BK)^p$ classified by the representation ρ. In particular, the shuffle map gives an equivalence of chain complexes

$$C_*EC_p \otimes_{C_p} (C_*B\Gamma_{g,1})^{\otimes p} \xrightarrow{\simeq} C_*B(C_p \wr \Gamma_{g,1}).$$

Fact 2. The index of C_p in Σ_p is $(p-1)!$ which is coprime to p. Hence, a simple transfer argument shows that for any group K the chain complex $C_*B(\Sigma_p \wr K)$ is a direct factor in the chain complex $C_*B(C_p \wr K)$ as we are working with $\mathbb{Z}/p\mathbb{Z}$ coefficients. (Here take $B\Sigma_p = E\Sigma_p/\Sigma_p$ and $BC_p = E\Sigma_p/C_p$.)

Fact 3 [N]. For any subgroup π of Σ_p, any space X and free π-complex V, there is an isomorphism of homology groups

$$H(V \otimes_\pi (H_*X)^{\otimes p}) = H(V \otimes_\pi (C_*X)^{\otimes p}).$$

Together with the map on the bottom row of Diagram 1.2 the first two facts give us a map of chain complexes

$$C_*E\Sigma_p \otimes_{\Sigma_p} (C_*B\Gamma_{g,1})^{\otimes p} \hookrightarrow C_*EC_p \otimes_{C_p} (C_*B\Gamma_{g,1})^{\otimes p} \longrightarrow C_*B\Gamma_{pg}.$$

Now, by Harer's stability theorem [H1] as improved by Ivanov [I], $\flat$ induces an isomorphism $H_*B\Gamma_{pg,1} \rightarrow H_*B\Gamma_{pg}$ for $2*+1 < pg$. Hence, in this stable range, we can define a map on homology groups using the third fact and the above chain map by setting

$$\theta_* : H(W_* \otimes_{C_p} (H_*B\Gamma_{g,1})^{\otimes p}) = H(W_* \otimes_{C_p} (C_*B\Gamma_{g,1})^{\otimes p}) \longrightarrow H_*B\Gamma_{pg} \simeq H_*B\Gamma_{pg,1}.$$

Finally, for every i, g and n with $2(pn+i)+1 < pg$ the formula $Q_i(x) := \theta_*(e_i \otimes x^{\otimes p})$ defines homology operations

$$Q_i : H_nB\Gamma_{g,1} \longrightarrow H_{pn+i}B\Gamma_{pg,1}.$$

2.2 Remark. For infinite loop spaces, most of the Q_i vanish uniformly. In the cases for which i is even, the only possible non-trivial ones occur when i is a multiple of $(p-1)$. We do not know this for the operations constructed here.

2.3 Remark. The constructions above are valid in much more generality. Indeed the surfaces to be glued to the genus zero surface could have extra boundary components and may all have numerous punctures as Harer's stability theorem applies in all of these cases [H1]. Similar statements also hold for the spin mapping class groups. In the latter case the range of stability has to be adjusted down [H2].

3. Immediate properties of Q_i.

Proposition 3.1. *The Q_i's are linear.*

Proof. Though the operations are not linear on the chain level, they are linear modulo boundaries. This is proved by standard arguments. ◇

Proposition 3.2. *When defined, Q_0 and Q_1 coinside with the Miller-Bödigheimer operations.*

Proof. Q_0 and Q_1 are defined via the top row of Diagram 1.2 in [Bö 1]. More precisely, Q_0 is in both cases just the Pontrjagin product induced by Miller's pair of pants construction. Q_1 is only non-zero when $p = 2$. In that case, it is induced via the cell that generates the first homology of $B\beta_2 \simeq S^1$. This cell corresponds to the cell $e_1 \in W_1$ under the map $\mathbb{Z} \to C_2$. Commutativity of Diagram 1.2 thus implies the proposition. ◇

As our homology operations are not derived from the structure maps of an E_∞-space, we can in general not assume that a map of E_∞-spaces induces a map in homology that commutes with the operations. We check this now by hand for the most important case. The action of the diffeomorphisms on the first homology of the underlying surface defines a representation $\tau : \Gamma_{g,1} \to \mathrm{SP}_{2g}\mathbb{Z}$ from the mapping class group into the symplectic group. Recall [T1] that the induced map $B\tau : B\Gamma_\infty^+ \to B\mathrm{SP}_\infty\mathbb{Z}^+$ is indeed a map of infinite loop spaces. Here $\Gamma_\infty = \lim_g \Gamma_{g,1}$ is the stable mapping class group with the limit taken via the stabilization map σ, and $+$ denotes Quillen's plus construction with respect to the commutator subgroups of the fundamental groups.

Proposition 3.3. *On stable homology, $B\tau$ commutes with the homology operations Q_i for $i = 0, 1, 2, \ldots$.*

Proof. The infinite loop space structure on $B\mathrm{SP}_\infty\mathbb{Z}^+$ is induced by matrix block addition, and its structure maps θ' are defined by the wreath product evaluation maps

$$\Sigma_p \wr \mathrm{SP}_{2g}\mathbb{Z} \longrightarrow \mathrm{SP}_{2pg}\mathbb{Z}.$$

The proposition follows if the following diagram of chain complexes commutes.

$$\begin{array}{ccccc}
\theta_* : C_*B(\Sigma_p \wr \Gamma_{g,1}) & \longrightarrow & C_*B(C_p \wr \Gamma_{g,1}) & \longrightarrow & C_*B\Gamma_{pg} \\
\Big\downarrow {\scriptstyle B(1\wr\tau)} & & \Big\downarrow {\scriptstyle B(incl\wr\tau)} & & \Big\downarrow {\scriptstyle B\tau'} \\
\theta'_* : C_*B(\Sigma_p \wr \mathrm{SP}_{2g}\mathbb{Z}) & = = & C_*B(\Sigma_p \wr \mathrm{SP}_{2g}\mathbb{Z}) & \longrightarrow & C_*(B\mathrm{SP}_{2pg}\mathbb{Z})
\end{array}$$

Here τ' denotes the representation of the mapping class group of the closed surface on the first homology. Note that $\tau = \tau' \circ \flat$. The top left map is split by the natural inclusion map, and so the left square commutes.

3.4 Remark. It was observed in [CC] that these operations are consistent with those for the general linear groups over finite fields. Indeed, the operations Q_i were shown to "fit" with the iterated wreath product of cyclic groups Z/pZ in $\mathrm{GL}(n, F_q)$ for appropriate values of q. This was the key in the proof of the main result in [CC] that Im J, the image of J, away from the prime two splits stably off $B\Gamma_\infty^+$. By construction, the operations here are compatible with this splitting and are therefore non-trivial. Furthermore, the authors proved that $B\Sigma_\infty^+$ localized away from two splits off $B\Gamma_\infty^+$ [C] [T2]. A close analysis of the proof shows that also this splitting is compatible with the operations Q_i. As the splitting is indeed a splitting of infinite loop spaces [T2], this is further evidence that the Q_i's constructed in Section 2 coincide with those supported by the infinite loop space structure on $B\Gamma_\infty^+$.

4. The action of the pure braid group.

The classical Araki-Kudo-Dyer-Lashof operations (for Γ-spaces or E_∞-spaces) are induced by an action of an operad involving the symmetric groups Σ_p. Because of this, they support additional symmetries which give rise to the Cartan formula and Adem relations [CLM]. The operations constructed in Section 2, however, are essentially induced by an action of an operad involving the ribbon braid groups $\tilde{\Gamma}_{0,p,1}$. Hence, one might expect error terms for the Cartan formula and Adem relation arising from the difference between the symmetric groups and the ribbon braid groups. The kernel of the natural representation $\eta : \tilde{\Gamma}_{0,p,1} \to \Sigma_p$ is the pure ribbon braid group $\Gamma_{0,p+1}$. The main result of this section, Corollary 4.4, states that the action of the pure ribbon braid group on stable classes is trivial through a range of dimensions.

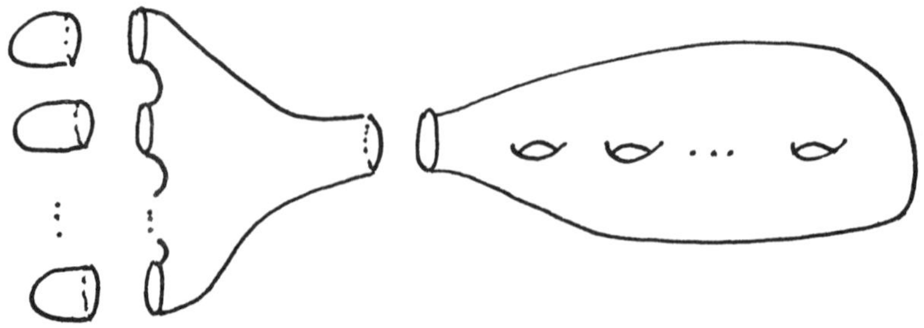

Figure 2

Let $\gamma : \Gamma_{g,n} \times \Gamma_{h,m} \to \Gamma_{g+h,n+m-2}$ denote the map induced by gluing two surfaces along one of their boundary components. Let $\flat : \Gamma_{g,n} \to \Gamma_{g,n-1}$ denote the map obtained when a disk is glued onto the surface and the diffeomorphisms are extended by the identity. Indeed, as the mapping class group of the disk is trivial, $\flat$ is just a special version of γ.

4.1 Key lemma. *The following diagram of groups commutes.*

$$\begin{array}{ccccc} \Gamma_{0,n+1} \times \Gamma_{g,1} & \xrightarrow{\gamma} & \Gamma_{g,n} & \xrightarrow{\flat^n} & \Gamma_g \\ {\scriptstyle 0\times id}\downarrow & & & & \Vert \\ \Gamma_{0,n+1} \times \Gamma_{g,1} & \xrightarrow{\gamma} & \Gamma_{g,n} & \xrightarrow{\flat^n} & \Gamma_g \end{array}$$

Proof. Gluing is associative. Hence, we may first apply $\flat^n$ to $\Gamma_{0,n+1}$ to get a map to the trivial group $\Gamma_{0,1}$, and then apply γ (now equivalent to $\flat$). The effect of this is the same as $\flat^n \circ \gamma$ and $\flat^n \circ \gamma \circ (0 \times id)$. See Figure 2. ◇

4.2. Corollary. $\gamma_* = \gamma_*(0 \times id)_*$ *on homology in dimensions* $2 * +1 < g$.

Proof. By Harer's stability theorem [H1] [I], $\flat_*$ is invertible on homology through a range of dimensions. This range is exactly given by $2 * +1 < g$. ◇

4.3 Digression on circle actions. Consider the case when $n = 1$. A geometric model of $B\Gamma_{g,1}$ is the moduli space $\mathcal{M}_{g,1}$ of Riemann surfaces of genus g with one marked point and a given unit tangent vector at this point. The circle action on $B\Gamma_{g,1}$ induced by the Dehn twist around the boundary component corresponds to rotating the tangent vector in $\mathcal{M}_{g,1}$. Bödigheimer [Bö 2] expects that this circle action is related to the cyclic structure that he finds on his model of $\mathcal{M}_{g,1}$ in term of slit domains in the complex plane.

Corollary 4.2 implies that the action of the circle $S^1 = B\Gamma_{0,1+1}$ is homologically trivial in the stable range. In particular, the generator of $H_1 S^1$ acts trivial on classes of dimensions $* < (g-3)/2$. This observation should simplify an analysis of the cyclic structure in the stable range.

4.4. Corollary. *The following diagram of groups commutes on homology in dimensions* $2 * +1 < g_1$.

$$\begin{array}{ccc} \Gamma_{0,n+1} \times (\Gamma_{g_1,1} \times \cdots \times \Gamma_{g_n,1}) & \xrightarrow{\gamma^n} & \Gamma_{g_1+\cdots+g_n,1} \\ {\scriptstyle 0\times id\times(id)^{n-1}}\downarrow & & \uparrow{\scriptstyle \gamma^{n-1}} \\ \Gamma_{0,n+1} \times \Gamma_{g_1,1} \times (\Gamma_{g_2,1} \times \cdots \times \Gamma_{g_n,1}) & \xrightarrow{\gamma\times(id)^{n-1}} & \Gamma_{g_1,n} \times (\Gamma_{g_2,1} \times \cdots \times \Gamma_{g_n,1}) \end{array}$$

Proof. This follows immediately from associativity of gluing, the Künneth theorem, and Corollary 4.2:

$$\gamma^n = \gamma^{n-1}(\gamma \times id^{n-1}),$$

This result can be interpreted to say that the stable homology of the mapping class group is a trivial module over the pure ribbon braid groups. In particular, if $a \in H_i B\Gamma_{0,n+1}$ then the map of homology groups

$$H_* B\Gamma_{g,1} \longrightarrow H_{nk+i} B\Gamma_{ng,1}$$

given by $x \mapsto \gamma_*^n(a, x, \ldots, x)$ is the n-th power map $x \mapsto x^n$ for $i = 0$, and is zero whenever $i > 0$ and $2(nk + i) + 1 < g$.

This is evidence that the action of the ribbon braid group on stable classes may factor through an action of the symmetric group. Expecting that the Q_i's satisfy the Adem relations and Cartan formula, one might therefore ask first whether analogue relations and formula are satisfied by the ribbon braid group action. Indeed, the analogue of the Adem relations follows because the ribbon braid groups $\tilde{\Gamma}_{0,p,1}$ form an operad and as such act on $\bigsqcup_{g \geq 0} \Gamma_{g,1}$ via η. In particular,

4.5. Proposition. *There is a commutative diagram of groups*

$$\begin{array}{ccc} \tilde{\Gamma}_{0,p,1} \wr \tilde{\Gamma}_{0,p,1} \wr \Gamma_{g,1} & \xrightarrow{\eta \wr id} & \tilde{\Gamma}_{0,p^2,1} \wr \Gamma_{g,1} \\ {\scriptstyle id \wr \eta} \downarrow & & \downarrow {\scriptstyle \eta} \\ \tilde{\Gamma}_{0,p,1} \wr \Gamma_{pg,1} & \xrightarrow{\eta} & \Gamma_{p^2 g,1} \end{array}$$

$\diamondsuit$

Hence any homology operations induced by η satisfy relations coming from the above diagram. Operations like this and the corresponding relations have been studied by many authors including Getzler, Kapranov, and Kontsevich. In case of the analogue for the Cartan formula one would like the ribbon braid group action to commute with Miller's pairs of pants product. This is not the case. But the following weaker statement holds.

4.6. Proposition. *The following diagram of groups commutes on homology in dimensions* $2 * +1 < \max(g, h)$.

$$\begin{array}{ccc} \tilde{\Gamma}_{0,p,1} \wr (\Gamma_{g,1} \times \Gamma_{h,1}) & \xrightarrow{\eta \times \eta} & \Gamma_{pg,1} \times \Gamma_{ph,1} \\ {\scriptstyle id \wr \star} \downarrow & & \downarrow {\scriptstyle \star} \\ \tilde{\Gamma}_{0,p,1} \wr \Gamma_{g+h,1} & \xrightarrow{\eta} & \Gamma_{p(g+h),1} \end{array}$$

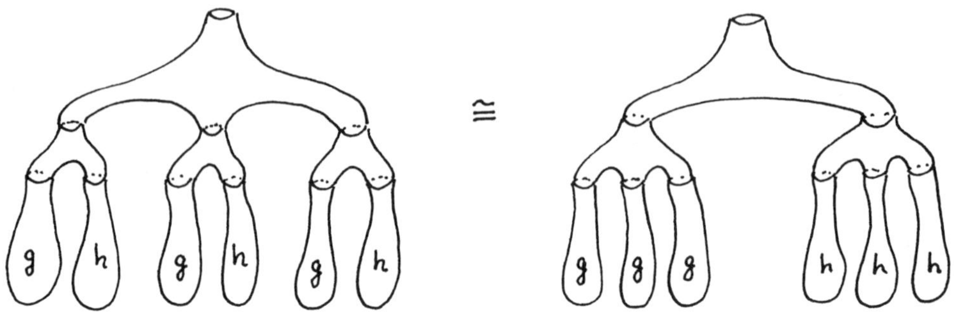

Figure 3

Proof. The two compositions $\star\circ(\eta\times\eta)$ and $\eta\circ(id\wr\star)$ are supported by two different but diffeomorphic surfaces. In both cases p copies of a surface $F_{g,1}$ and $F_{h,1}$ are glued to different genus zero surfaces with $p+1$ boundary components. See Figure 3 for an illustration. Consider the same diagram on the level of diffeomorphisms. After appropriate identification of the two genus $p(g+h)$ surfaces, an easy diagram chase proves that the diffeomorphisms one gets by going around the diagram in both ways differ by a diffeomorphism on the genus zero surface which fixes the boundary components. By Corollary 4.4 this is however irrelevant for the induced maps on homology in the specified dimensions. $\diamondsuit$

4.7 Remark. The above results hold also for the $\Gamma^k_{g,n+1}$ by Harer's stability theorem [H1], and the spin mapping class groups $G^s_{g,1}$ by Harer's stability theorem for these groups [H2] and its multi-boundary version [B].

References.

[B] T. Bauer, MSc-Thesis, Oxford University, 1997.

[Bö 1] C.-F. Bödigheimer, *On the topology of moduli spaces of Riemann surfaces. Part II: Homology operations*, preprint, Mathematica Gottingensis, Heft 23 (1990).

[Bö 2] C.-F. Bödigheimer, Cyclic homology and moduli spaces of Riemann surfaces, K-theory - Strasbourg 1992, Astérisque **226** (1994), 43–57.

[CC] R. Charney, F.R. Cohen, *A stable splitting for the mapping class group*, Michigan Math. J. **35** (1988), 269-284.

[C] F.R. Cohen, *On a splitting for the mapping class group*, preprint.

[CLM] F.R. Cohen, T.J. Lada, J.P. May, The Homology of Iterated Loop Spaces, Springer Verlag LNM **533**, 1976.

[FS] Z. Fiedorowicz, Y. Song, *The braid structure of the mapping class groups*, Sci. Bull. Josai Uni. **2** (1997), 21–30.

[H1] J.L. Harer, *Stability of the homology of the mapping class groups of orientable surfaces*, Annals Math. **121** (1985), 215–249.

[H2] J.L. Harer, *Stability of the homology of moduli spaces of Riemann surfaces with spin structure*, Math. Ann. **287**, 323–334.

[I] N.V. Ivanov, *On the homology stability for Teichmüller mudular groups: closed surfaces and twisted coefficients*, in Mapping Class Groups and Moduli Spaces of Riemann Surfaces, Contemp. Math. **150** (1993), 149–194.

[Ma] J.P. May, *E_∞ spaces, group completion, and permutative categories* in New Developments in Topology, LMS Lecture Note Series **11** (1974), 61–94.

[S] G. Segal, *Categories and cohomology theories*, Topology **13** (1974), 293–312.

[T1] U. Tillmann, *On the homotopy of the stable mapping class group*, to appear in Invent. Math.

[T2] U. Tillmann, *A splitting for the stable mapping class group*, preprint.

Department of Mathematics University of Rochester
Rochester, NY 14627
U.S.A
cohf@math.rochester.edu

Mathematical Institute
24-29 St Giles Street
Oxford OX1 3LB
UK
tillmann@maths.ox.ac.uk

Contemporary Mathematics
Volume **220**, 1998

Spanier-Whitehead Duality and Critical Points

Octavian Cornea

ABSTRACT. We study the relative attaching map associated to two consecutive critical points in a context considerably more general than the Morse one. We prove that this map behaves well when suspended and that the relative attaching maps with respect to the direct and reverse flows are Spanier-Whitehead duals modulo some twisting coming from the normal structure.

1. Introduction

Passage through a non-degenerate critical point of a smooth function f corresponds to the attachment of a cell. To a couple of consecutive non-degenerate critical points one associates a relative attaching map. There is a second attaching map corresponding to the same pair of critical points but with respect to the function $-f$. What is the relation between these two maps ?

This question has been considered by Franks [7] who showed that in the Morse-Smale case and if the manifold is stably parallelizable, then the stable classes of these maps coincide up to sign.

It turns out that a better understanding of this problem passes through a study of a more general situation than the Morse one.

Here we look to a class of relatively general, isolated, critical points that we call reasonable. Passage through such a critical point corresponds to the attachment of a cone over a space generally more complicated than a sphere. A relative attaching map is again associated to a pair of consecutive reasonable critical points.

In this context, let k and k' be the homotopy classes of the attaching maps associated to f and, respectively, to $-f$ for the same pair of reasonable consecutive critical points. We prove that there is a space P and a self-homotopy equivalence u of P such that (some suspension of) k factors through P as a composition $k_1 \circ k_2$ and k' and $k_1 \circ u \circ k_2$ are Spanier-Whitehead duals.

The map u induces the identity in homology (with $\mathbf{Z}/2$ coefficients) and depends only on the normal structure coming from the embedding of our manifold in a high-dimensional sphere. The construction of u is explicit and is shown to be related to the $\mathbf{J}$-homomorphism.

This result is new even in the Morse case. Of course, if the manifold is stably parallelizable the two attaching maps are Spanier-Whitehead duals. This implies the result of Franks and shows that it is valid for Morse functions (and not only Morse-Smale ones).

The homological consequence of this duality is a special case of the result obtained by McCord [10] in the general context of the Conley index theory.

One key difference with respect to the Morse case is that, for a reasonable critical point, the stable and unstable "manifolds" can be quite complicated sets. Because of this, to extend the results form the Morse case, one has to use instead certain thickenings (called cyllindrical neighborhoods) of the union of these two sets. This also forces the proof of the duality result to be different from an extension of the method used by Franks. On the other hand, Spanier-Whitehead duality appears in a natural way in this setting.

There are two main ideas that underline this paper. Firstly, from the homotopy angle, many results that seem specific to the non-degenerate context actually hold for a much larger class of critical points.

Secondly, much of the homotopical information one usually extracts from gradient flows gains in significance after passage to the stable category.

Here is the plan of the paper.

Section 2 starts by recalling the definition and some basic properties of reasonable critical points. We then construct a special type of neighborhood of the invariant set formed by a pair of consecutive critical points together with the flow lines that join them. The setting here is that of the Conley index theory [2]. This is used to define a connection map for a couple of reasonable, consecutive critical points in terms of local information on the flow. This map coincides in relevant cases with the homotopy class of the attaching map mentioned before but it has meaning even when such a natural identification is not possible.

The first result of section 3 concerns a product formula. It implies that the effect of adding a quadratic function to f is to suspend the connection map. Section 3 ends with the proof of the duality. This proof is based on the explicit constructions of the second section and on the product formula.

Acknowledgement. I am indebted to Mark Mahowald who helped me understand the significance of the normal structure.

2. The connection map.

From now on let M be a smooth, riemannian, compact, n-dimensional manifold with possibly non-void boundary. We fix a few basic notions below.

Reasonable critical points and the A-index. An isolated critical point P of a smooth function $f : M \longrightarrow \mathbf{R}$ is *reasonable* [3] if for a sufficiently small, open disk B around P the pair $((f^{-1}(f(P)) - P) \bigcap B, P)$ is a Whitney stratification. Critical points that are reasonable are abundant: an isolated critical point P of a function analytic with respect to some chart around P is in this class. The homotopy type of the set $f^{-1}(-\infty, f(P)] \bigcap \partial B$ is called the A-index of P with respect to f, $A_f(P)$. All finite CW-complexes can appear as the A-index of some reasonable critical point. The A-index of a non-degenerate critical point of index k is S^{k-1}.

The direct and inverse flows. The smooth function $f : M \longrightarrow \mathbf{R}$ induces two flows on M: the *direct* flow, induced by the vector field $-\nabla f$ and the *inverse* flow, induced by ∇f. We denote by $\gamma_t(x)$ the direct flow.

If P is a reasonable critical point of f, then the A-indexes of P with respect to f and to $-f$ are clearly represented by complements in S^{n-1}. In particular, when connected, they are Spanier-Whitehead duals [14,15].

Cylindrical neighborhoods. Suppose P is a reasonable critical point of $f : M \longrightarrow \mathbf{R}$ and let $a = f(P)$. Denote by $B(P, \epsilon)$ the closed disk of radius ϵ around P.

For $\epsilon, \delta, \delta' \in \mathbf{R}^+$ sufficiently small let $K(P; \epsilon; \delta, \delta') \subset M$ consist of those points x that satisfy:

1. $a - \delta \leq f(x) \leq a + \delta'$.
2. One of the following three possibilities holds:
   (a) there is a $t \in \mathbf{R}$ such that $\gamma_t(x) \in f^{-1}(a) \bigcap B(P, \epsilon)$.
   (b) $lim_{t \to \infty} \gamma_t(x) = P$.
   (c) $lim_{t \to -\infty} \gamma_t(x) = P$.

Variants of this type of neighborhoods have been used often [3,4,5,6,11,16].

The Whitney stratification condition in the definition of a reasonable critical point implies [3] that for τ smaller than some constant the intersection of $f^{-1}(a)$ and $S(P, \tau) = \partial B(P, \tau)$ is transverse and by consequence $f^{-1}(a) \bigcap B(P, \tau)$ is the cone over its boundary. This shows that, when $\epsilon, \delta, \delta'$ are smaller than some fixed constants, $K(P; \epsilon; \delta, \delta')$ is a contractible neighborhood of P, all these neighborhoods are homeomorphic, each neighborhood of P contains one of this type, $\partial(K(P; \epsilon; \delta, \delta')) \simeq S^{n-1}$ ($\simeq$ denotes homotopy equivalence). One also sees that the sets $A_K = K(P; \epsilon; \delta, \delta') \bigcap f^{-1}(a - \delta)$ are all homeomorphic. We also denote $B_K = K(P; \epsilon; \delta, \delta') \bigcap f^{-1}(a + \delta')$. These sets have similar properties to the A_K's. Let W_K be the "vertical" walls of $K(P; \epsilon; \delta, \delta')$: it consists of those points of $K(P; \epsilon; \delta, \delta')$ situated on a flow line that passes through $S(P, \epsilon) \bigcap f^{-1}(a)$. Notice that $A_K \bigcap W_K = \partial A_K$, $B_K \bigcap W_K = \partial B_K$ and W_K is homeomorphic to $\partial A_K \times [-1, 1]$ (when it is non-empty). We will always consider that the -1 end of this cobordism is in A_K and the $+1$ end in B_K.

The homotopy type of A_K coincides with the A-index of P relative to f, $A_f(P) \simeq f^{-1}(-\infty, a] \bigcap S(P, \tau)$ (and similarly $B_K \simeq A_{-f}(P)$). Indeed, the manifold $(A_K \bigcup W_K) \bigcap f^{-1}(-\infty, a]$ is invertibly cobordant (mod boundary) to $f^{-1}(-\infty, a] \bigcap S(P, \tau)$. In particular they are homotopy equivalent [8]. On the other hand $(A_K \bigcup W_K) \bigcap f^{-1}(-\infty, a] \simeq A_K$.

The pair $(K(P; \epsilon; \delta, \delta'), A_K)$ is an index pair (in the sense of Conley index theory [2,12]) for P viewed as an invariant set for the direct flow. Hence, the homotopy type of the quotient $K(P; \epsilon; \delta, \delta')/A_K \simeq \Sigma A_K$ coincides with the Conley index of P.

Cone attachments. Assume that $(W; V_0, V_1)$ is a smooth compact cobordism and that $f : W \longrightarrow \mathbf{R}$ is a smooth function that is minimal, regular and equal to $a - \delta$ on V_0 and maximal, regular and equal to $a + \delta'$ on V_1. Assume also that f has exactly one critical point P which is reasonable and of critical value a. Then, one sees immediately that W admits as a deformation retract $V_0 \bigcup K(P; \epsilon; \delta, \delta')$ for ϵ, passage through a reasonable critical point corresponds to the attachment of a cone (over $A_f(P)$).

Consecutive critical points. Two critical points P and Q of $f : M \longrightarrow \mathbf{R}$ are called *consecutive* if $f(P) < f(Q)$, all the flow lines (of the direct flow) connecting Q to P are not broken and there are no flow lines connecting Q to P that pass through the boundary of M.

REMARK 1. When does the existence of a cobordism (W^{n+1}, V_0, V_1) for which there is a homotopy cofibration sequence $Z \longrightarrow V_0 \longrightarrow W$ imply the existence of a smooth function $f : W \longrightarrow \mathbf{R}$ maximal, regular and constant on V_1, minimal, constant and regular on V_0 and having a single critical point ? If the dimension of Z as a CW-complex is strictly smaller than half the dimension of V_0 and if Z and V_0 are both simply connected, then one can show [3] following a skeletal decomposition of Z that there is a thickening, Z' of Z in V_0 and a disk D^{n+1} in W^{n+1} such that $D^{n+1} \bigcap V_0 = Z'$, $\partial D^{n+1} - Int(Z')$ is a neat submanifold of W and the topological manifold $W - (Int(D^{n+1}) \bigcup Int(Z'))$ is homeomorphic to $V_1 \times [0,1]$. Applying now the construction of Takens [16] we obtain the required function (a slight modification of his construction allows one to assume that the critical point constructed in this way is also reasonable).

Assume X is a 2-connected, k-dimensional CW-complex having Lusternik-Schnirelmann category m. Applying the argument above inductively following a cone decomposition of X one obtains that for any smooth, compact manifold M of the homotopy type of X, having simply connected boundary and of dimension $\geq 2k$ there is a smooth function $f : M \longrightarrow \mathbf{R}$ constant, maximal and regular on ∂M with at most $m+2$ critical points.

The classical inequality of Lusternik and Schnirelmann shows that any function as above has at least $m+1$ critical points. The relation between the L.S.-category and the attachment of cones is given by the fact that a space of L.S.-category m has the homotopy type of an $m+1$ cone. If X is two connected one can even assume that the spaces over which the cones are attached are simply connected.

2.1. A special index pair for consecutive critical points. From now on let $f : M \longrightarrow \mathbf{R}$ be a fixed smooth function. If P is a reasonable critical point of f the pairs $(K(P;\epsilon;\delta,\delta'), A_K)$ are very useful index pairs (in the terminology of the Conley index theory) one advantage being that the set $K(P;\epsilon;\delta,\delta')$ is homotopically simple. We describe below a similar construction for the case of a couple of reasonable critical points.

Given the critical points P, Q we denote by $I(P,Q)$ the invariant set of the direct flow formed by P and Q together with all the points situated on the flow lines that join them.

PROPOSITION 2.1. *For any pair P, Q of consecutive, reasonable critical points there exists an index pair (N,E) of $I(P,Q)$ with N being homeomorphic to a union $K_1 \bigcup_{Z\times\{0\}} \bigcup Z \times [0,1] \bigcup_{Z\times\{1\}} K'$ with K_1 a cylindrical neighborhood of P, K' a cylindrical neighborhood of Q and Z a compact topological manifold. In particular, N is homotopy equivalent to a (non-reduced) suspension or to S^0.*

PROOF. It is natural to try to construct N as the union of three sets: one of the type of $K(Q;---)$ the second of the type of $K(P;---)$ and the third consisting of the points situated on the flow lines that join the first two. The construction is not so simple because we have to control better the properties of the third set. In fact, we will write it as $Z \times [0,1]$ such that for all $p \in Z$, $p \times [0,1]$ is a segment of a flow line with the 0 end in $B_{K(P;..)}$ and the 1 end in $A_{K(Q;..)}$; with Z a compact, topological manifold with boundary $\partial Z = \partial_1(Z) \bigcup \partial_2(Z)$ satisfying $\partial_1(Z) \times \{1\} \subset \partial A_{K(Q;---)}$ and $\partial_2(Z) \times \{0\} \subset \partial B_{K(P;---)}$.

First, notice that if there are no flow lines connecting Q to P the disjoint union of one index pair for Q and one for P (both of type $K(-;-;-,-)$) gives the wanted index pair ($Z = \emptyset$ here). In particular, this covers the cases when P is a local maximum or Q a local minimum.

Supose that there are connecting flow lines.

To fix ideas we will first assume that P is not a local minimum.

Let $K = K(P;\epsilon;\delta,\delta')$ and $K' = K(Q;\eta;\mu,\mu')$ such that $f(P)+\delta' < f(Q)-\mu$.

Consider the set $E = \{x \in A_{K'} :$ there is a $t(x) \in \mathbf{R}^+$ such that $\gamma_{t(x)}(x) \in B_K\}$. The set E embeds in B_K by $x \longrightarrow \gamma_{t(x)}(x)$. The key fact is that E is closed when ϵ and η are small enough.

Indeed, suppose that x_n is a sequence in E that converges to some y. We intend to show $y \in E$. Consider $x'_n = \gamma_{t(x_n)}(x_n)$. For compacity reasons we may assume that x'_n converges also to some point y'.

At this moment it is useful to assume that M is included in a second manifold M' wich is just M extended by a collar of its boundary and f is defined on all of M' (it is easy to see that this is not restrictive as P and Q are consecutive with respect to the restriction of f to $M - \{$ a sufficiently small collar of $\partial M\}$, in case $\partial M = \emptyset$ this argument is not necessary). As B_K is closed the following three possibilities are open to y: first, for some T, $\gamma_T(y) \notin M$; secondly $lim_{t\to\infty}\gamma_t(y) = R$ for some critical point R; thirdly, $y \in E$ and $\gamma_{t(y)}(y) = y'$. Notice that the first can be eliminated. Indeed, if y exits M (before reaching the $f(P)+\delta'$ critical level) there is an entire neighborhood of it that does the same. This contradicts the fact that x_n converges to y.

Hence, we are left with the second case. There is a critical point R, different from P and Q, and such that $\lim_{t\to\infty}\gamma_t(y) = R$. For a very small ν assume that $x''_i \in f^{-1}(f(R)-\nu)$ are the images of x_i by the flow induced embedding. We may assume that x''_i converges to some point y''. It is easy to see that y'' is situated on a flow line that originates at R. Reapplying the same reasoning for the sequence x''_n and y'' we see that the points y and y' are connected by a broken flow line. Now, if for arbitrarily small neighborhoods K and K' we can find pairs of points $y \in K'$, $y' \in K$ that are connected by a broken flow line, then, the same type of arguments as above show that, Q and P are connected by a broken flow line. This is not possible as P and Q are consecutive.

We assume in the following that ϵ and η are as small as required by the above argument. Moreover, because of the requirement relative to ∂M in the definition of consecutive critical points, we may assume that ϵ, η are small enough such that no point in $A_{K'}$ is connected to a point in B_K by a flow line that passes through the boundary of M.

Let $S = E \bigcap \partial A_{K'}$. Denote by S' the image of S by the embedding $e : x \longrightarrow \gamma_{t(x)}(x)$. Consider a neighborhood of S in $\partial A_{K'}$, S_1. If this neighborhood is close enough to S it is embedded in $f^{-1}(f(P)+\delta')$ by e as a codimension 1 submanifold S''. We now consider another neighborhood $K_1 = K(P;\epsilon';\delta,\delta')$ with $\epsilon' < \epsilon$. By

We will refer in the following to this transversality condition as *condition* Ψ.

Let $Z = \{x \in A_{K'} : \text{there is a } t(x) \in \mathbf{R}^+ \text{ such that } \gamma_{t(x)}(x) \in B_{K_1}\}$. Similarly to E this is also closed. Let $\partial_1(Z) = Z \bigcap \partial A_{K'}$. Of course, Z and $\partial_1(Z)$ embedd in B_{K_1} (in the same way as E and S do). We denote the image of these embeddings by Z' and respectively $\partial_1(Z')$. Notice that Z' is a topological manifold with boundary. In fact, $\partial Z'$ is equal to the union of $\partial_1(Z')$ and $\partial_2(Z')$ over I with $\partial_2(Z') = \partial B_{K_1} \bigcap Z'$. Indeed, suppose that $x \in (Z - \partial A_{K'})$ and is not an interior point of Z. The fact that there are no flow lines connecting points in $A_{K'}$ to points in B_{K_1} by passing through ∂M implies that there is a neighborhood of x in $A_{K'}$ that is embedded by the flow γ in $f^{-1}(f(P) + \delta')$. The image of any such neighborhood by this embedding is not contained in B_{K_1}. Hence, $x \in \partial B_{K_1}$. We denote by $\partial_2(Z)$ the image of $\partial_2(Z')$ by the inverse embedding.

Consider the set $W' = \{x \in M : f(P) + \delta' \leq f(x) \leq f(Q) - \mu; \text{ there is a } t \in \mathbf{R} \text{ such that } \gamma_t(x) \in Z'\}$. It is homeomorphic to $Z \times [0,1]$ (with Z' being identified to $Z \times \{0\}$). Using this parametrization let $\partial_2(W') = \partial_2(Z') \times [0,1] \subset W'$. We now let $N = K_1 \bigcup W' \bigcup K'$. The exit set is $E = A_{K_1} \bigcup W_{K_1} \bigcup \partial_2(W') \bigcup (A_{K'} - Z)$.

If P is a local minimum the same construction works without any modification except for the definition of the exit set which now becomes $E = A_{K'} - Z$. This definition makes sense as in this case Z is a union of connected components of $A_{K'}$. Indeed, it is easy to see that Z is both open and closed in $A_{K'}$. □

REMARK 2. There is an index pair of the type described inside any neighborhood of $I(P,Q)$.

Properties of N. For further use we collect the properties of the neighborhood N constructed above : $N = K_1 \bigcup W' \bigcup K'$; $W' = Z \times [0,1]$, $Z \times \{0\} \subset B_{K_1}$, $Z \times \{1\} \subset A_{K'}$; for all $z \in Z$, $z \times [0,1]$ is a segment of a flow line. Moreover, Z is a compact, topological manifold whose boundary is the union of two submanifolds $\partial_1(Z)$ and $\partial_2(Z)$ such that ∂Z has corners along $\partial_1(Z) \bigcap \partial_2(Z) = \partial(\partial_1(Z)) = \partial(\partial_2(Z))$ and $\partial_1(Z) \times \{1\} \subset \partial A_{K'}$, $\partial_2(Z) \times \{0\} \subset \partial B_{K_1}$.

We call the set Z as above the *incidence* set of K' and K_1.

2.2. Construction of the connection map. To define the connection map we use the explicit form of the index pair (N, E) of $I(P,Q)$ described in Proposition 2.1. This map is not defined if Q is a local minimum or P is a local maximum.

The idea of the construction is simple. We consider two neighborhoods $K' = K(Q; \eta; \mu, \mu')$ and $K_1 = K(P; \epsilon'; \delta, \delta')$ as in 2.1 and consider the map that sends a point of $A_{K'}$ that flows to a point in B_{K_1} to its impact point viewed as an element of $K'/(A_{K_1} \bigcup W_{K_1})$; a point that misses B_{K_1} is sent to the basis point in $K_1/(A_{K_1} \bigcup W_{K_1})$.

We proceed with a formal definition below. We assume that condition Ψ is satified.

We denote by $k(K', K_1)$ the homotopy class of the map $A_{K'} \longrightarrow *$ if Z is void and, if this is not the case, the homotopy class of the map $A_{K'} \longrightarrow \Sigma A_{K_1}$ that is given by the composition: $A_{K'} \longrightarrow A_K/\overline{(A_K - Z)} = Z/\partial_2(Z) \longrightarrow Z'/\partial_2(Z') \longrightarrow$ $\longrightarrow K_1/(A_{K_1} \bigcup W_{K_1}) \longrightarrow K_1/A_{K_1} \longrightarrow \Sigma A_{K_1}$. Here, $Z' = Z \times \{0\}$, the first map is the collapsing, the second map is the flow induced homeomorphism, the third is induced by the inclusion $(Z', \partial_2(Z')) \hookrightarrow (K_1, (A_{K_1} \bigcup W_{K_1}))$, the fourth map is induced by a homeomorphism that pushes W_{K_1} to a collar of the boundary in

A_{K_1}. Notice that as K_1 is contractible the homotopy type of the quotient K_1/A_{K_1} is that of ΣA_{K_1}. The fifth map is the homotopy inverse of the map induced by $(CA_{K_1}, A_{K_1}) \longrightarrow (K_1, A_{K_1})$ (which restricts to the identity on A_{K_1}; CX is the cone over the space X; $C\emptyset \simeq *$; $\Sigma\emptyset \simeq S^0$).

We now want to compare the different maps $k(K', K_1)$ constructed as above.
Assume R is a reasonable critical point of f. For $\epsilon < \eta$, $\delta < \tau$, $\delta' < \tau'$ there is a canonical inclusion (and homotopy equivalence)

$$Y = K(R; \epsilon; \delta, \delta') \hookrightarrow Y' = K(R; \eta; \tau, \tau')$$

Moreover, there is a homeomorphism $i(Y, Y') : Y \longrightarrow Y'$ which is homotopic to this inclusion. It is defined by noting that $K(R; \eta; \tau, \tau') - Int(K(R; \epsilon; \delta, \delta')$ is homeomorphic to $\partial(K(R; \epsilon; \delta, \delta')) \times [0, 1]$. We then use a small interior collar of $\partial(K(R; \epsilon; \delta, \delta'))$ that we can "inflate" as to cover $K(R; \eta; \tau, \tau') - Int(K(R; \epsilon; \delta, \delta')$. We can do this such that the homeomorphism $i(Y, Y')$ restricts to a homeomorphism of the boundaries of the two neigborhoods and moreover to the A_-, B_- and W_- parts of these boundaries.

PROPOSITION 2.2. *Assume that P and Q are two consecutive, reasonable, critical points with P different from a local extremum and Q from a local minimum. We consider two neighborhoods $K = K(Q; \epsilon; \delta, \delta')$, $K' = K(P; \mu; \tau, \tau')$ that satisfy the transversality condition Ψ. For any other pair $L = K(Q; \epsilon_1; \delta_1, \delta_1')$, $L' = K(P; \mu_1; \tau_1, \tau_1')$ that also satisfies condition Ψ and with the parameters for the first neighborhoods strictly bigger than the corresponding ones for L and L', we have the following homotopy commutative diagram:*

$$\begin{array}{ccc} A_L & \xrightarrow{k(L,L')} & \Sigma A_{L'} \\ \Big\downarrow{\scriptstyle i(L,K)} & & \Big\downarrow{\scriptstyle i(L',K')} \\ A_K & \xrightarrow{k(K,K')} & \Sigma A_{K'} \end{array}$$

PROOF. We denote by $Z = \{x \in A_K :$ there is a $t(x) \in \mathbf{R}^+$ such that $\gamma_{t(x)}(x) \in B_{K'}\}$, Z' is the image of Z in $B_{K'}$ by the flow induced homeomorphism; $\partial Z' = \partial_1(Z') \bigcup \partial_2(Z')$ with $\partial_2(Z') \subset \partial B_{K'}$, $\partial_1(Z')$ being the image of $\partial_1(Z)$ by the flow induced homeomorphism (see the proof of Proposition 2.1). The corresponding sets for the pair L, L' will be indicated by the $_1$ subscript. By using the flow induced deformations we may assume that $\delta = \delta_1$, $\delta' = \delta_1'$, $\tau = \tau_1$, $\tau' = \tau_1'$. Then we have $Z_1 \subset Z$ and $Z_1' \subset Z'$ and these inclusions commute with the flow induced homeomorphisms. We may also assume that K and L' satisfy condition Ψ. We will indicate the sets constructed as above for the pair K, L' by a $_2$ subscript. Notice first the relations : $A_L \subset A_K$, $Z_1 \subset A_L$, $Z_2 \subset A_K$, $Z_1 = Z_2 \bigcap A_L$, $Z \subset A_K$, $Z_2 \subset$

$$\begin{array}{ccccc}
A_L & \longrightarrow & A_K & \stackrel{id}{\longleftarrow} & A_K \\
\downarrow & & \downarrow & & \downarrow \\
Z_1/\partial_2(Z_1) & \longrightarrow & Z_2/\partial_2(Z_2) & \longleftarrow & Z/\partial_2(Z) \\
\downarrow & & \downarrow & & \downarrow \\
L'/(A_{L'} \bigcup W_{L'}) & \stackrel{id}{\rightarrow} & K'/((K' - Int(L')) \bigcup A_{K'}) & \leftarrow & K'/(A_{K'} \bigcup W_{K'})
\end{array}$$

The proof ends by remarking that the following diagram is homotopy commutative (the horizontal maps are the number four maps appearing in the composition defining the connection maps).

$$\begin{array}{ccc}
K'/(A_{K'} \bigcup W_{K'}) & \longrightarrow & K'/A_{K'} \\
\downarrow & & \uparrow \\
K'/((K' - Int(L')) \bigcup A_{K'}) & & \\
\downarrow{\scriptstyle id} & & \\
L'/(A_{L'} \bigcup W_{L'}) & \longrightarrow & L'/A_{L'}
\end{array}$$

□

The homotopy class of any of the maps $A_f(Q) \longrightarrow \Sigma A_f(P)$ constructed above will be denoted by $k_f(Q,P)$ (or $k(Q,P)$ when there is no ambiguity) and will be called the *connection map of Q and P with respect to f*.

REMARK 3. 1. It is obvious that $\Sigma(k(Q,P))$ is the usual connection map of an attractor-repeller pair that appears in Conley index theory.

2. Assume that (P,Q) and (Q,R) are two pairs of consecutive, reasonable, critical points of f and suppose $\partial M = \emptyset$. Let v_1, v_2, v_3, v_4 be four regular values of f that satisfy $v_1 < f(P)$, $f(P) < v_2 < f(Q)$, $f(Q) < v_3 < f(R)$, $v_4 > f(R)$. Assume also that there are no other critical points of f whose corresponding critical value belongs to the interval $[v_1, v_4]$. Let $W_i = f^{-1}(-\infty, v_i]$. We have homotopy cofibration sequences $A_f(P) \longrightarrow W_1 \longrightarrow W_2$, $A_f(Q) \longrightarrow W_2 \longrightarrow W_3$, $A_f(R) \longrightarrow W_3 \longrightarrow W_4$. It is immediate that $k(Q,P)$ and $k(R,Q)$ are identified with the resulting attaching maps: $A_f(Q) \longrightarrow \Sigma A_f(P)$ and $A_f(R) \longrightarrow \Sigma A_f(Q)$. Because of this we also have $\Sigma k(Q,P) \circ k(R,Q) \simeq *$.

Hence, if (P,Q), (Q,R) are two pairs of consecutive, reasonable critical points and $\Sigma k(Q,P) \circ k(R,Q)$ is not trivial, then there are some other critical points in $f^{-1}([f(P), f(R)])$ besides P, Q and R.

3. When f is Morse-Smale it is easy to see that our description of the connection map is close to that of Franks [7]. When $index(P) + 1 = index(Q)$ it implies the usual description given in terms of counting (with sign) the flow lines connecting Q to P (for example, see the proof given to this fact in [13]). In this case we can

choose the neighborhoods K' and K_1 from the definition of our map in such a way that each such flow line will correspond to one component of Z, each component of this type being a disk. It is then easy to identify $\partial_2(Z)$ and $\partial_1(Z)$ and the result follows.

3. Two properties of the connection map.

In this section we study the connection map of a product flow and the relation between the connection maps of the direct and reverse flows.

These two questions are interdependent: to compare the direct and reverse connection maps one has to pass in the stable category; to understand what suspending means in this context one needs a product formula.

We continue to discuss only the case of functions with reasonable, isolated critical points. However, it seems likely that both results admit generalizations to much more general invariant sets.

3.1. A product formula. Suspension. As before, we assume that $f : M \longrightarrow \mathbf{R}$ is smooth, M compact and that P and Q are consecutive reasonable, critical points of f. Assume also that $g : D^m \longrightarrow \mathbf{R}$ has just one critical point, $0 \in D^m$, which is reasonable. We consider $f \oplus g : M \times D^m \longrightarrow \mathbf{R}$, $f \oplus g(x,y) = f(x) + g(y)$. We assume that P is not a local maximum and that Q is not a local minimum.

THEOREM 3.1. *The points $(P,0)$ and $(Q,0)$ are consecutive, reasonable critical points of $f \oplus g$ and the respective connection map $k_{f\oplus g}((Q,0),(P,0))$ can be identified with $k_f(Q,P) * \mathbf{1}_{A_g(0)}$.*

REMARK 4. 1. The join $A * B$ is defined as the push out of the two inclusions $A \times B \longrightarrow A \times CB$ and $A \times B \longrightarrow CA \times B$ where CA, CB are the cones over A and B respectively (by a slight abuse we use $\emptyset * B \simeq B$). This construction is functorial. If $u : A \longrightarrow G$ and $t : B \longrightarrow D$ are two maps their join is denoted $u * t : A * B \longrightarrow G * D$.

2. At the beginning of the proof below we will make explicit the identifications of $A_f(P) * A_g(0)$ with $A_{f\oplus g}((P,0))$ and the corresponding one for Q that were refered to in the statement.

PROOF. The strategy of the proof is to replace in this case the usual connection map with one defined in terms of product neighborhoods, identify the incidence set in this case and then point out the needed join decompositions.

The facts that $(P,0)$, $(Q,0)$ are reasonable and consecutive are immediate. The space level part of the statement is Corollary 2.7 in [3] (and will be recalled below). We first assume that 0 is not a local minimum of g.

Products of cylindrical neighborhoods of Q and of 0. Consider two neighborhoods $K = K(Q;\epsilon;\delta,\delta)$ and $N = K(0;\epsilon;\delta,\delta)$. By modifying conveniently the different constants involved one can find inside any neighborhood $T = K((Q,0);\alpha;\beta,\beta')$ a product of type $K \times N$ and inside any such product a neighborhood like T. This allows one to use an invertible cobordism argument [3] to show that the set $P(K,N) = \{z \in \partial(K \times N) : (f \oplus g)(z) \leq 0\}$ has the same homotopy type as $A_{f\oplus g}((Q,0))$. On the other hand, it has the same homotopy type as $A_K * A_N$. Indeed, the join of A_K and A_N is given by the union $L(K,N) = A_K \times N \bigcup K \times A_N$

$J(K,N) = B_K \times N \bigcup K \times B_N$ in $\partial(K \times N)$. Moreover, on $L(K,N)$ the vector field $-\nabla(f \oplus g)$ points out of $K \times N$ and it points in on $J(K,N)$.

The connection map in terms of product neighborhoods. Suppose that T is such that it includes the product $K \times N$. We have two flow induced maps that are homotopy equivalences: $i : L(K,N) \longrightarrow A_T$ and $h : \Sigma A_T = T/(A_T \bigcup W_T) \longrightarrow (K \times N)/C(K,N)$. The first is given by just leting points flow down and stop once we reach A_T; the second does not move points belonging to the interior of $K \times N$ as for the exterior points, the ones that reach in forward time $Int(J(K,N))$ have as image the impact point, the others as well as $C(K,N)$ are sent to the base point.

Suppose now that K' is a neighborhood of P similar to K and N' one of 0 that is similar to N and that contains N. We are going to use the above notations for the product $K' \times N'$.

There is a connection map $k : L(K,N) \longrightarrow (K' \times N')/C(K',N')$: all the points from $L(K,N)$ that flow down to some points of $Int(J(K',N'))$ have the impact point as image, all the others are sent to the base point. Assume T' is a neighborhood of $(Q,0)$ similar to T and that it contains $K' \times N'$.

Obviously, the diagram below is homotopy commutative:

$$\begin{array}{ccc} L(K,N) & \xrightarrow{k} & (K' \times N')/C(K',N') \\ \Big\downarrow{\scriptstyle i} & & \Big\uparrow{\scriptstyle h} \\ A_T & \xrightarrow{k(T,T')} & \Sigma A_{T'} \end{array}$$

Therefore, what we need to show is $k \simeq k(K,K') * \mathbf{1}_{A_N}$ (obviously, i and h are both homotopy equivalences).

It is clear that, in fact, we do not need to use for the definition of the map k above the same constants ϵ, δ, δ for both K and N (and similarly for K' and N').

The incidence set for the product neighborhoods. Choose K and K' satisfying condition Ψ and let $Z \subset A_K$ be their incidence set as in Proposition 2.1. We pick N and N' such that the time needed for all the points in Z to reach $B_{K'}$ (under the flow γ induced by $-\nabla f$) is smaller than the time needed for any of the points in N to exit N' (under the flow γ^1 induced by $-\nabla g$). We may also choose these neighborhoods such that the time nedeed for all the points in A_N to reach $A_{N'}$ is strictly smaller than that nedded for any of the points in B_K to reach $B_{K'}$ (this is possible by choosing ϵ in the definition of K sufficiently small).

Denote the flow induced by $-\nabla(f \oplus g)$ by γ^2.

Let $Z_p = \{(x,y) \in L(K,N) : \text{there exists } T \in \mathbf{R}^+, \gamma^2_T(x,y) \in J(K',N')\}$. As in the proof of proposition 2.1 we see that Z_p is closed. The points in $L(K,N)$ can not reach (in forward time) $K' \times B_{N'}$ so $Z_p = \{(x,y) \in L(K,N) : \text{there are } t,k \geq 0, \gamma_t(x) \in Z \subset A_K, \gamma_{t+k}(x) \in B_{K'} \text{ and for } 0 \leq \tau \leq t+k, \gamma^1_\tau(y) \in N'\}$. We see that $Z \times N \subset Z_p$. In fact, Z_p is identified with $(Z \times N) \bigcup (Z \times A_N) \times [0,1]$ where we consider $Z \times \{0\} = Z \subset A_K \subset K$. With this identification $(Z \times A_N) \times \{1\}$ consists exactly of those points (x,y) of $L(K,N)$ such that for some $T \in \mathbf{R}^+$ we have $\gamma^2_T(x,y) \in B_{K'} \times \partial(N')$. There is a deformation of Z_p to $Z \times N$ given by $(x,y) \longrightarrow (\gamma_r(x), y)$ where $r \in \mathbf{R}^+$.

We now consider $\partial_2(Z_p) = (\partial_2(Z) \times N) \bigcup (\partial_2(Z) \times A_N \times [0,1]) \bigcup (Z \times A_N \times \{1\})$. Notice that $\partial_2(Z_p)$ consists exactly of the elements of Z_p that flow by γ^2 to points in $\partial(J(K',N'))$. This is a consequence of the facts that γ sends exactly the points of $\partial_2(Z)$ in $\partial(B_{K'})$ and that no other points of Z_p besides those of $Z \times A_N \times \{1\}$ reach (in forward time) $B_{K'} \times \partial(N')$. Moreover, $\partial Z_p \bigcap Int(L(K,N)) \subset \partial_2(Z_p)$, $\partial_2(Z_p) \subset \partial Z_p$, $\partial_2(Z_p) \bigcap \partial L(K,N) = \partial(\partial_2(Z_p))$.

Let us denote by $\partial_1(Z_p)$ the closure of the complement of $\partial_2(Z_p)$ in ∂Z_p. It is identificd with $((\partial_1(Z) \times N \bigcup Z \times \partial(N)) - Int(Z \times A_N)) \bigcup (\partial_1(Z) \times A_N \times [0,1]) \bigcup Z \times \partial(A_N) \times [0,1]$. Note that $\partial_1(Z_p)$ coincides with the intersection of $\partial(Z_p)$ with $\partial(L(K,N))$. Indeed, as $\partial_1(Z) \subset \partial(A_K)$ we have $(\partial_1(Z) \times N \bigcup Z \times \partial(N)) \subset (\partial(A_K) \times N \bigcup A_K \times \partial(N))$. Also, $\partial_1(Z) \times [0,1] \subset W_K$ so $\partial_1(Z) \times A_N \times [0,1] \subset \partial(K) \times A_N$, similarly $Z \times [0,1] \subset K$ so, $Z \times \partial(A_N) \times [0,1] \subset K \times \partial A_N$.

Identification of the joins. The connection map k is given by the composition $w \circ v$ with

$$v : L(K,N) \longrightarrow L(K,N)/\overline{(L(K,N) - Z_p)} \longrightarrow Z_p/\partial_2(Z_p)$$

where the first map is the collapsing map the second being the obvious homeomorphism and $w : Z_p/\partial_2(Z_p) \longrightarrow (K' \times N')/C(K',N')$ is induced by the inclusions.

There is a homeomorphism of Z_p with $Z \times N$ which extends to a homeomorphism of $L(K,N)$. We use it to write $L(K,N)$ as a union of two components the first, U, homeomorphic to $A_K \times N$ the second ,V, homeomorphic to $K \times A_N$ with $U \bigcap V$ homeomorphic to $A_K \times A_N$ and such that $Z_p \subset U$. We now identify via this homeomorphism U to $A_K \times N$, V to $K \times A_N$, Z_p to $Z \times N$ and $\partial_2(Z_p)$ with the subset of $\partial(Z \times N)$ given by $\partial_2(Z) \times N \bigcup Z \times A_N$. Under this identification $Z \times A_N \times \{1\} \subset Z_p$ corresponds to $Z \times A_N$.

We need the following general facts. Assume A, B are two spaces $U \subset A$.

1. We have the homeomorphisms: $(A \times CB)/((U \times CB) \bigcup (A \times B)) \longrightarrow ((A \times CB)/(U \times CB))/((A \times B)/(U \times B)) \longrightarrow ((A/U \times CB)/(* \times CB))/((A/U \times B)/(* \times B))$.

2. The collapsing map $(A/U) * B = (A/U \times CB) \bigcup (C(A/U) \times B) \longrightarrow ((A/U \times CB)/(* \times CB))/((A/U \times B)/(* \times B))$ gives a natural homotopy equivalence.

It follows that we have homeomorphisms: $Z_p/\partial_2(Z_p) = (Z \times N)/(\partial_2(Z) \times N \bigcup Z \times A_N) \longrightarrow ((Z \times N)/(\partial_2(Z) \times N))/((Z \times A_N)/(\partial_2(Z) \times A_N)) \longrightarrow ((Z/\partial_2(Z) \times N)/(* \times N))/((Z/\partial_2(Z) \times A_N)/(* \times A_N))$. This last space is naturally homotopy equivalent with $(Z/\partial_2(Z)) * A_N$.

To end the case $0 \neq min$ we need to remark that v is identified to the join of $v' : A_K \longrightarrow A_K/\overline{(A_K - Z)} \longrightarrow Z/\partial_2(Z)$ with the identity of A_N and that w is identified to the join of $Z/\partial_2(Z) \longrightarrow K'/(A_{K'} \bigcup W_{K'})$ with the flow induced homotopy equivalence $A_N \longrightarrow A_{N'}$.

For v notice that it equals the composition: $(A_K \times N) \bigcup (K \times A_N) \longrightarrow (A_K \times N)/(b \times N \bigcup A_K \times A_N) \xrightarrow{v''} (A_K \times N)/(\overline{(A_K - Z)} \times N \bigcup A_K \times A_N) \longrightarrow (Z \times N)/(\partial_2(Z) \times N \bigcup Z \times A_N) = Z_p/\partial_2(Z_p)$ where the first two maps are the obvious collapsing maps, the first being also a homotopy equivalence, the last map is a homeomorphism (because of 1. above and $Z/\partial_2(Z) = A_K/\overline{A_K - Z}$) and b is a point in $\overline{(A_K - Z)}$. In its turn v'' is naturally identified with the map $((A_K \times N)/(b \times N))/((A_K \times A_N)/(b \times A_N)) \longrightarrow (((A_K/\overline{(A_K - Z)}) \times N)/(* \times N))/(((A_K/\overline{(A_K - Z)}) \times A_N)/(* \times A_N))$ which is induced by the product of the

For w we write $C(K',N') = ((A_{K'} \bigcup W_{K'}) \times N') \bigcup (K' \times (A_{N'} \bigcup W_{N'}))$ and then use the same type of identifications as in the case of v.

If 0 is a local minimum of g we have $L(K,N) = A_K \times N$ and the same arguments as above imply the result in this case also. □

COROLLARY 3.2. *If g is a quadratic form of index i, then* $k_{f\oplus g}((Q,0),(P,0)) \simeq \Sigma^i k_f(Q,P)$

REMARK 5. 1. All the contructions in the proof above can be made in any given neighborhood of $I((P,0),(Q,0))$. In fact, the analogue of the index pair in Proposition 2.1 is obtained here by taking the union of $K \times N$ with $K' \times N'$ and $Z_p \times [0,1]$ using the inclusions of Z_p in $L(K,N)$ and in $J(K',N')$. We denote such a neighborhood by $T(K,N;K',N')$.

2. The space level of the join formula above is an extension of a formula of King [9] which concerns only the case when g is a nondegenerate quadratic form.

3.2. Duality. Here we compare the connection maps corresponding to the direct and inverse flows.

Spanier-Whitehead duality. We recall a few basic facts [14], [15].

Two pointed, connected CW-complexes X, X' are Spanier-Whitehead dual if there is a map $\mu : X \wedge X' \longrightarrow S^m$ such that the slant product $\mu^*(i_m)^*/ : H_q(X') \longrightarrow H^{m-q}(X)$ gives an isomorphism. The map μ is called an m-duality. If X and X' are m-duals with duality map μ, then $\Sigma^k X$ and $\Sigma^q X'$ are $m+k+q$-duals with respect to the obvious map induced by μ. There is a good notion of duality for maps also (all maps and spaces are pointed). Assume X, X' are duals by an m-duality, μ, and Y, Y' are duals by an m-duality ν. Two maps $f : X \longrightarrow Y$, $g : Y' \longrightarrow X'$ are duals if for some (possibly) large k,k' the following diagram homotopy commutes:

$$\begin{array}{ccc} \Sigma^k X \wedge \Sigma^{k'} Y' & \xrightarrow{f \wedge 1} & \Sigma^k Y \wedge \Sigma^{k'} Y' \\ \Big\downarrow{\scriptstyle 1\wedge g} & & \Big\downarrow{\scriptstyle \nu} \\ \Sigma^k X \wedge \Sigma^{k'} X' & \xrightarrow[\mu]{} & S^{m+k+k'} \end{array}$$

The reason why suspending is necessary is that only by allowing k and k' to be sufficiently large this notion of duality corresponds to a duality isomorphism between (stable) classes of maps $D_m(\mu,\nu) : \{X,Y\} \approx \{Y',X'\}$.

We need now to indicate one way to obtain duality maps [14]. Assume $A \subset S^{m+1}$, $B \subset S^{m+1}$ such that the inclusion induces an isomorphism $H_*(B) \longrightarrow H_*(S^{m+1} - A)$, and A, B are connected and disjoint. Then there is a duality map $\mu : A \wedge B \longrightarrow S^m$. This map is defined in the following way: we fix a point $p \in S^{m+1} - (A \bigcup B)$ and identify $S^{m+1} - \{p\}$ with $\mathbf{R}^{m+1}$. We then define $v : A \times B \longrightarrow S^m$ by $v(x,y) = (x-y)/||x-y||$. Connectivity implies that this map is null-homotopic when restricted to $A \vee B$. It defines a map $A \wedge B \longrightarrow S^m$ which is a duality. There is also a way to identify dual maps in this setting. For example if A', B' is another pair of subspaces of S^{m+1} that satisfies the conditions above and if $A \xrightarrow{i} A'$ and $B' \xrightarrow{i'} B$ are inclusions, then i and i' are dual with respect to the duality maps constructed as above.

Duality mod u. Suppose (X,X') and (Y,Y') are two pairs of m-dual spaces. We say that the two maps $f : X \longrightarrow Y$ and $g : Y' \longrightarrow X'$ are *Spanier-Whitehead duals mod u* if there is a $k \in \mathbf{N}$ and a space H together with a self homotopy equivalence

$u : H \longrightarrow H$ that induces the identity in homology (with $\mathbf{Z}/\mathbf{2}$ coefficients in general, or $\mathbf{Z}$ coefficients in an orientable context) with the property that that $\Sigma^k f$ factorizes as $\Sigma^k X \xrightarrow{r} H \xrightarrow{l} \Sigma^k Y$ and g and $l \circ u \circ r$ are Spanier-Whitehead duals.

Some useful homotopy equivalences. We need to fix another homotopical construction.

For a CW-complex Y let $Eq(Y)$ be the group of self-homotopy equivalences of Y. We consider the map: $s_Y : [Y * S^{k-1}, S^k] \xrightarrow{+id_{S^k}} [\Sigma(Y \times S^{k-1}), S^k] \xrightarrow{ad} [Y \times S^{k-1}, \Omega S^k] \xrightarrow{id_Y \times} [Y \times S^{k-1}, Y \times \Omega S^k] \xrightarrow{\Sigma} [\Sigma(Y \times S^{k-1}), \Sigma(Y \times \Omega S^k)] \xrightarrow{p} Eq(Y * S^{k-1})$. Here ad is the adjunction; for the first map we use the formula $\Sigma(Y \times Z) = \Sigma Y \vee \Sigma(Y \wedge Z) \vee \Sigma Z$. To define p we use the obvious retraction $r : \Sigma(Y \times \Omega S^k) \longrightarrow \Sigma(Y \times S^{k-1})$ of the inclusion $\Sigma(Y \times S^{k-1}) \xrightarrow{\Sigma(id_Y \times e)} \Sigma(Y \times \Omega S^k)$ ($e : S^{k-1} \longrightarrow \Omega S^k$ being the adjoint of the identity; r is induced by the retraction $r' : \Sigma\Omega S^k \longrightarrow \Sigma S^{k-1}$ of Σe). If $x \in [\Sigma(Y \times S^{k-1}), \Sigma(Y \times \Omega S^k)]$, then $p(x)$ is the restriction of $r \circ x$ to $\Sigma(Y \wedge S^{k-1})$, projected on $\Sigma(Y \wedge S^{k-1})$ (recall that $\Sigma(Y \wedge Z) = Y * Z$).

Consider an inclusion $A \xrightarrow{i} X$. We will use the map $s_{X,A} : [X * S^{k-1}, S^k] \longrightarrow Eq((X/A) * S^{k-1})$ obtained by taking $s_{X,A}(x)$ to be the self-homotopy equivalence of $(X/A) * S^{k-1}$ induced by $s_X(x)$ and $s_A(x \circ i)$ from the homotopy cofibration sequence $A * S^{k-1} \longrightarrow X * S^{k-1} \longrightarrow X/A * S^{k-1}$

We return now to our familiar setting $f : M \longrightarrow \mathbf{R}$ is smooth, M is compact, riemannian of dimension n and P, Q are two consecutive, reasonable critical points of f. We assume that all spaces $A_f(P)$, $A_{-f}(P)$, $A_f(Q)$, $A_{-f}(Q)$ are connected. Moreover, we assume that the dimension of M is different from four. Then all neighborhoods $K(P; \epsilon; \delta, \delta')$, $K(Q; \epsilon; \delta, \delta')$ are homeomorphic to disks. Both these assumptions are not too restrictive as they can be realized by adding a suitable quadratic form to the initial function. The previous section shows that the effect of such an addition is just suspending the homotopical data.

We are also going to assume that M is embedded in a high dimensional sphere S^{k+n} with $k > n$ and that its nomal bundle is classified by $\chi \in [M, BO(k)]$.

THEOREM 3.3. *The maps $k_f(Q, P)$ and $k_{-f}(P, Q)$ are Spanier-Whitehead duals mod u with u described as follows.*

There is a triple of subspaces of M, (T_1, T, T') with $T_1 \simeq \Sigma T$ and a map $Q : [M, BO(k)] \xrightarrow{i} [\Sigma T, BO(k)] \xrightarrow{ad} [T, O(k)] \xrightarrow{J} [T * S^{k-1}, S^k] \xrightarrow{s_{T,T'}}$ $\longrightarrow Eq((T/T') * S^{k-1})$ such that $u = Q(\chi)$.

Here i is restriction and J is the generalized J-homomorphism. The composition of the last three maps is a morphism of groups.

PROOF. We consider cylindrical neighborhoods K' of Q and K_1 of P as in the proof of Proposition 2.1. Let $N = K_1 \bigcup W' \bigcup K'$ the neighborhood of $I(P, Q)$ constructed there. As mentioned above we assume that all the spaces $A_{K'}$, A_{K_1}, $B_{K'}$, B_{K_1} are connected and that K' and K_1 are both homeomorphic to D^n. Hence, it is clear that $A_{K'}, B_{K'}$ and A_{K_1}, B_{K_1} are pairs of $(n-2)$-dual spaces. This implies that $A_{K'}, \Sigma B_{K'}$ and $\Sigma A_{K_1}, B_{K_1}$ are pairs of $(n-1)$-dual spaces. It is useful to know

$A'_{K_1} = A_{K_1} \bigcup W_{K_1}$. We embed first $K_1 \subset S^n$ then (as K_1 is a disk) the complement of K_1 in S^n is the cone over the boundary of K_1. We consider the union U of K_1 with the subset of this cone that gives the cone over A'_{K_1}. It is easy to see that $S^n - U$ retracts to $Int(B_{K_1})$ and that U represents ΣA_{K_1}. The duality map that we need is induced by this embedding. We denote this duality map by ν. Moreover, let $Z \hookrightarrow B_{K_1}$ be the incidence set appearing in 2.1 (we identify here Z and $Z \times \{0\}$). Recall that $\partial Z = \partial_2(Z) \bigcup \partial_1(Z)$ with $\partial_2(Z) \subset \partial B_{K_1} \subset W_{K_1}$. Hence, we may consider the inclusion $i : Z/\partial_2(Z) \longrightarrow \Sigma A'_{K_1} = K_1/A'_{K_1}$. It is again useful to think about it in S^n. It is the inclusion of the union, T, of Z and the part of the cone on ∂K_1 that gives the cone over $\partial_2(Z)$. The complement of T in S^n has as a deformation retract the union of K_1 and a copy of Z that is pasted to K_1 along $\partial_1(Z)$ (this type of argument is very classical in nature see [1]). Clearly, there is a (nontrivial) inclusion j of a copy of B_{K_1} in this complement. The conclusion is that $Z/\partial_2(Z), Z/\partial_1(Z)$ are also $n-1$-dual by a duality map η such that with respect to ν and η the two inclusions i and j are dual.

We can proceed in a similar fashion for the pair $A_{K'}, \Sigma B_{K'}$. We denote by ν' the duality map obtained for this pair. Clearly $Z \subset A_{K'}$ and $\partial_1(Z) \subset W_{K'}$. As above we obtain a duality map η' for the pair $Z/\partial_2(Z), Z/\partial_1(Z)$. We denote by i' the inclusion $Z/\partial_1(Z) \longrightarrow \Sigma B_{K'}$. It is dual (with respect to η' and ν') to the inclusion $j' : A_{K'} \longrightarrow Z/\partial_2(Z)$.

We use the notations $k = k(K', K_1)$ and $k' = k(K_1, K')$. The definitions of the connection maps imply that k and respectively k' factor as
$A_{K'} \xrightarrow{p} Z/\partial_2(Z) \xrightarrow{i} \Sigma A_{K_1}$ and $B_{K_1} \xrightarrow{p'} Z/\partial_1(Z) \xrightarrow{i'} \Sigma B_{K'}$.

We obtain that the following diagram $(**)$ commutes up to homotopy:

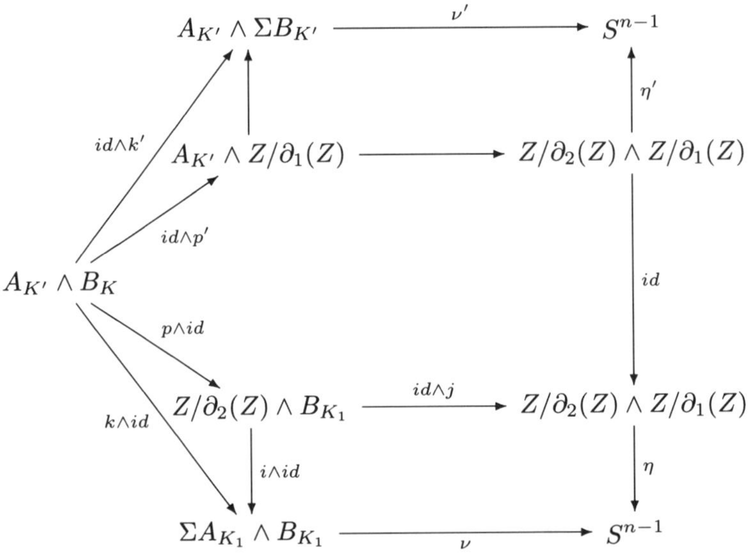

If η and η' are the same we obtain that the two connection maps are Spanier-Whitehead duals. In turn, this can be assumed if all the constructions above are done with respect to a neighborhood of type N that is embedded in a sphere of dimension n $(= dim(M))$.

The stably parallelizable case. Assume now that $i(\chi) = 0$. Consider a non-degenerate quadratic form $g : D^k \longrightarrow \mathbf{R}$. On a tubular neighborhood of N, $U =$

$N \times D^k$ we can define the function $f \oplus g : N \times D^k \longrightarrow \mathbf{R}$. For the invariant set $I((P,0),(Q,0))$ there is a neighborhood of type $T(K,N;K'',N'')$ (as in the last remark of subsection 3.1). All the above argument can be done with respect to this neighborhood by replacing K' with $K \times N$, K_1 with $K'' \times N''$, Z with Z_p, $\partial_1(Z)$ with $\partial_1(Z_p)$ and $\partial_2(Z)$ with $\partial_2(Z_p)$.

By the product formula we know that the connection maps constructed in this way (for $f \oplus g$) are iterated suspensions of the corresponding maps for f. Hence, in the stably parallelizable case we obtain that the two connection maps are dual.

The general case. We come now to the general case. Denote again by U a tubular neighborhood of N in S^{n+k}. As $N = K_1 \bigcup W' \bigcup K'$ with K_1 and K' contractible and $W' = Z \times [0,1]$, U is homeomorphic to the push out of the two inclusions $Z \times D^k \longrightarrow K_1 \times D^k$, $Z \times D^k \longrightarrow K' \times D^k$. The identification between $Z \times D^k \subset K' \times D^k$ and $Z \times D^k \subset K_1 \times D^k$ is given by a map $t : Z \times D^k \longrightarrow Z \times D^k$ with $t(x,a) = (x, t_x a)$, $t_x \in O(k)$, clearly we have $t_x \simeq (ad \circ i)(\chi)(x)$. We denote $\chi' = (ad \circ i)(\chi)$. We then have $t(x,a) = (x, \chi'(x)(a))$

Let $g : U \longrightarrow \mathbf{R}$ be the negative of the square of a function measuring the distance from M. Under the identifications above we may assume that the restriction of g at each of the sets $K_1 \times D^k$ and respectively $K' \times D^k$ coincides with the composition of $g'(a) = -a_1^2 - a_2^2 - \dots - a_k^2$ with the projection on D^k.

Consider the function $f + g$. It has two reasonable consecutive critical points $(P,0)$ and $(Q,0)$. The stably parallelizable case discussed before shows that $k_{f+g}((Q,0),(P,0))$ and $k_{-(f+g)}((P,0),(Q,0))$ are Spanier-Whitehead duals. We need to compare these two connection maps with $k_f(Q,P)$ and $k_{-f}(P,Q)$ respectively.

On each of the two sets $K_1 \times D^k$ and $K' \times D^k$ the behaviour of $k_{f+g}((Q,0),(P,0))$ is described in Theorem 3.1. However, because of the twisting given by χ' the connection map $k_{f+g}((Q,0),(P,0))$ cannot be evaluated by directly applying that theorem. Instead, using the notations of the proof of the theorem, one sees that this connection map is given by the composition $w \circ \chi'' \circ v$ (we recall $v : L(K,N) \longrightarrow L(K,N)/\overline{(L(K,N) - Z_p)} \longrightarrow Z_p/\partial_2(Z_p)$, $w : Z_p/\partial_2(Z_p) \longrightarrow (K' \times N')/C(K',N')$).

Here $\chi'' : Z_p/\partial_2(Z_p) \longrightarrow Z_p/\partial_2(Z_p)$ is derived from χ' in the following way: recall the identification of Z_p with $Z \times N$ and that of $\partial_2(Z_p)$ with $\partial_2(Z) \times N \bigcup Z \times A_N$. In our case $A_N = S^{k-1}$ and $N = D^k$. The map χ'' is induced by t and by the restriction of t to $\partial_2(Z_p)$.

Of course the same argument applies also to the function $-(f+g)$. In that case, one needs the function $\chi_2 : Z_p/\partial_1(Z_p) \longrightarrow Z_p/\partial_1(Z_p)$. Here $\partial_1(Z_p)$ is the complement of $\partial_2(Z_p)$ in ∂Z_p and χ_2 is induced by $t'(x,a) = (x, (\chi'(x))^{-1}(a))$. In this case we have $B_N = \emptyset$ and we immediately obtain that $\partial_1(Z_p) = \partial_1(Z) \times D^k$. Hence, $\chi_2 \simeq id$. Thus, $k_{-(f+g)}((P,0),(Q,0)) \simeq k_{-f}(P,Q)$.

Identification of u. To end the proof we take $T_1 = N$, $T = Z_p$, $T' = \partial_2(Z_p)$ and we still have to understand better the map χ''. In particular, we want to show $\chi'' \simeq Q(\chi)$. By identifying Z_p to $Z \times D^k$ and $\partial_2(Z_p)$ to $\partial_2(Z) \times D^k \bigcup Z \times S^{k-1}$ we write $Z_p/\partial_2(Z_p)$ as $((Z \times D^k)/(Z \times S^{k-1}))/((\partial_2(Z) \times D^k)/(\partial_2(Z) \times S^{k-1}))$. The map χ'' coincides with the map induced to the cofibre of the inclusion $(\partial_2(Z) \times D^k)/(\partial_2(Z) \times S^{k-1}) \longrightarrow (Z \times D^k)/(Z \times S^{k-1})$ by the maps $t' : (\partial_2(Z) \times D^k)/(\partial_2(Z) \times S^{k-1}) \longrightarrow$

Notice that $(Z \times D^k)/(Z \times S^{k-1}) \simeq (Z * S^{k-1}) \vee S^k$ and $(\partial_2(Z) \times D^k)/(\partial_2(Z) \times S^{k-1}) \simeq (\partial_2(Z) * S^{k-1}) \vee S^k$.

The restriction of both t' and t'' to S^k is just the inclusion of S^k into the wedge. Let t_1'' be the composition $Z * S^{k-1} \longrightarrow (Z * S^{k-1}) \vee S^k \xrightarrow{t''} (Z * S^{k-1}) \vee S^k \longrightarrow Z * S^{k-1}$ and define similarly t_1' (here the first map is the inclusion and the last one collapses S^k to a point).

Clearly, χ'' is induced on the cofibre of $(\partial_2(Z) * S^{k-1}) \vee S^k \longrightarrow (Z * S^{k-1}) \vee S^k$ by t_1'' and t_1'.

We now show that $t_1'' = (s_Z \circ J)(\chi')$. Let $b : Z \times S^{k-1} \longrightarrow Z \times S^{k-1}$ be given by the restriction of t. Notice that t_1'' is, up to homotopy, the composition $Z * S^{k-1} \hookrightarrow \Sigma(Z \times S^{k-1}) \xrightarrow{\Sigma b} \Sigma(Z \times S^{k-1}) \longrightarrow Z * S^{k-1}$ (the last map being, as before, the collapsing and the first the inclusion). On the other hand $J(\chi')$ is, by definition, the composition $Z * S^{k-1} \hookrightarrow \Sigma(Z \times S^{k-1}) \xrightarrow{\Sigma(p_2 \circ b)} S^k$ with $p_2 : Z \times S^{k-1} \longrightarrow S^{k-1}$ being the projection. Because $dim(Z) < k$ the restriction of Σb to ΣZ coincides up to homotopy with the inclusion $\Sigma Z \hookrightarrow \Sigma(Z \times S^{k-1})$. Hence, we can write $\Sigma(p_2 \circ b) = J(\chi') + id_{S^k}$. We take the adjoint, and get $ad(J(\chi') + id_{S^k}) = e \circ p_2 \circ b$ where $e : S^{k-1} \longrightarrow \Omega S^k$ is the adjoint of the identity. We take the product of the map $ad(J(\chi') + id_{S^k}) : Z \times S^{k-1} \longrightarrow \Omega S^k$ with the identity of Z and then suspend. We precompose the resulting map with $Z * S^{k-1} \hookrightarrow \Sigma(Z \times S^{k-1})$ and compose it with the retraction $\Sigma(Z \times \Omega S^{k-1}) \longrightarrow Z * S^{k-1}$.

It is easy to see that the map obtained in this way coincides with t_1'' and with $s_Z(J(\chi'))$.

The same argument applies to t_1' and gives the wanted description for χ''.

Notice also that t_1'' (as well as t_1') induces the identity in homology with $\mathbf{Z}/2$ coefficients in general and with integral coefficients if χ is orientable.

The fact that the composition of the last three maps in the statement is a homomorphism follows immediately from the description of $t_1' = (s_Z \circ J)(\chi')$ and of t_1'' in terms of t. □

REMARK 6. As pointed out to me by Mark Mahowald examples where the two connection maps are not dual are easy to produce. A simple one is provided by the Morse function with three critical points on $\mathbf{RP}^2$. Indeed, if one considers the critical points of index 0 and 1 the corresponding attaching map is null. The reverse attaching map however is $S^1 \xrightarrow{2} S^1$. Our theorem implies that for an orientable manifold this type of example when the two critical points are of indexes of difference one and the direct and reverse attaching maps are not Spanier-Whitehead duals can not occur.

References

1. M.F.Atiyah, *Thom Complexes*, Proc. London Math. Soc. **11** (1961), 291–310
2. Ch.Conley, *Isolated invariant sets and the Morse index*, CBMS Regional Conf.Ser. in Math 38, Amer. Math. Soc. , Providence, R. I. , 1976
3. O.Cornea, *Cone-decompositions and degenerate critical points*, To appear, Proc. London Math.Soc.
4. E.N.Dancer, *Degenerate critical points, homotopy indeces and Morse inequalities*, J. Reine Ang. Math. **350** (1984), 1–22
5. E.N.Dancer, *Degenerate critical points, homotopy indeces and Morse inequalities II*, J. Reine Ang. Math. **382** (1987), 145–164
6. E.N.Dancer, *Degenerate critical points, homotopy indeces and Morse inequalities III*, Bull. Austral. Math. Soc. **40** (1989), 97–108

7. J. Franks, *Morse-Smale flows and homotopy theory*, Topology **18** (1979), 199–215
8. H.King, *Real analytic germs and their varieties at isolated singularities*, Invent. Math. **37** (1976), 193–199
9. H.King, *The number of critical points in Morse approximations*, Comp. Math. **34** (1977), 285–288
10. Ch. McCord, *Poincare Lefschetz duality for the homology Conley index*, Trans. of the A.M.S. **329** (1992) 233–252
11. E.Rothe, *A relation between the type numbers of a critical point and the index of the corresponding field of gradient vectors*, Math. Nach. **4** (1951), 13–27
12. D.Salamon, *Connected simple systems and the Conley index of isolated invariant sets*, Trans. of the A.M.S. **291** (1985) 1–41
13. D.Salamon, *Morse theory, the Conley index and Floer homology*, J.Bull. Lond. Math. Soc. **22** (1990), 113–140
14. E.H.Spanier, J.H.C. Whitehead, *Duality in homotopy theory*, Mathematika **2** (1959), 56–80
15. E.H.Spanier, *Function spaces and duality*, Annals of Math. **70** (1959), 338–378
16. F.Takens, *The minimal number of critical points of a function on a compact manifold and the Lusternik-Schnirelmann category*, Invent. Math. **6** (1968), 197–244

Université de Lille 1, U.F.R. de Mathématiques & U.R.A. 751 au CNRS, 59655 Villeneuve D'Ascq, France

E-mail address: `cornea@gat.univ-lille1.fr`

URL: `http://www-gat.univ-lille1.fr/~cornea/octav.html`

Contemporary Mathematics
Volume **220**, 1998

Comparing Completions of a Space at a Prime

Paul Goerss[1]

Abstract: In this paper we compare the Bousfield-Kan p-completion of a space with the p-profinite completion of a space due to Morel. In particular, we develop tools for calculating with the p-profinite completion, and we we give the Bousfield-Kan completion the model category theoretic foundation enjoyed by the p-profinite completion.

There are two practical ways to complete an abelian group A at a prime p. One could complete A with respect to the neighborhood base of zero given by the subgroups $p^n A$. This completion, called p-completion, is $\lim A/p^n A$. Or one could complete A with respect to the neighborhood base of zero given by subgroups $B \subseteq A$ of finite p-power index. This completion, called p-pro-finite completion, is $\lim A_\alpha$ where A_α runs over the finite p-group quotients of A. They agree for finitely generated groups, but not in general: if V is an $\mathbb{F}_p$ vector space it is p-complete, but its p-pro-finite completion is the double dual V^{**}. Also, the former, p-completion, is easier to define, but it is neither left nor right exact and the category of p-complete groups is not abelian – it is not closed under cokernels. The latter, p-pro-finite completion, is initially less tractable, but it is right exact and the category of p-pro-finite abelian groups is an abelian category.

---

[1]The author was partially supported by the National Science Foundation.

AMS Subject Classification (1991): 55P00 55T15

There are two corresponding completions for topological spaces. The analog of p-completion is Bousfield-Kan completion [3] and the analog of p-pro-finite completion is related to Sullivan's pro-finite completion and has recently been given a homotopical definition by Morel [13]. The purpose of this note is to compare these two completions; in addition, we seek to give Morel's completion the same sort of computational footing that the Bousfield-Kan completion enjoys.

To underscore the similarities and differences, let me make some remarks on how these completions of spaces are constructed.

If X is a space, the mod p homology $H_*X = H_*(X, \mathbb{F}_p)$ is a graded coalgebra over $\mathbb{F}_p$ and, as such, there is an isomorphism $\operatorname{colim}_\alpha C_\alpha \cong H_*X$, where $C_\alpha \subseteq H_*X$ runs over the filtered system of sub-coalgebras which are finite dimensional in each degree. Thus there is an isomorphism in cohomology, $H^*X \cong \lim_\alpha C_\alpha^*$, where C_α^* is the algebra dual to C_α. In short, H^*X is a pro-finite algebra—which is to say the inverse limit equips H^*X with a topology—and if $X \to Y$ is a map of spaces, $H^*Y \to H^*X$ is a continuous map of algebras.

The Bousfield-Kan completion of X may be obtained by the following program. Define a tower of fibrations

$$\cdots \to X_s \xrightarrow{q_s} X_{s-1} \to \cdots \to X_1 \xrightarrow{q_1} X_0$$

equipped with maps $f_s : X \to X_s$ so that $q_s f_s = f_{s-1}$ and X_s is defined recursively. First X_0 is a generalized mod p Eilenberg-MacLane space (i.e., with trivial k-invariants) so that $\pi_* X_0$ is a graded $\mathbb{F}_p$ vector space and the continuous map $f_0^* : H^*X_0 \to H^*X$ is surjective. Having defined X_{s-1}, choose a generalized Eilenberg-MacLane space K_s with $\pi_* K_s$ an $\mathbb{F}_p$-vector space and a map $k_s : X_{s-1} \to K_s$ so that

$$H^*K_s \xrightarrow{k_s^*} H^*X_{s-1} \xrightarrow{f_{s-1}^*} H^*X$$

is an exact sequence of pro-finite algebras. Note that the universal property of Eilenberg-MacLane spaces implies that one may specify the continuous map k_s^* first; this then specifies k_s up to homotopy. Define $q_s : X_s \to X_{s-1}$ to be the pull-back, via k_s, of the path-fibration over K_s and f_s to be any lift of f_{s-1}. In the end, one may identify $\lim_s X_s$ with the Bousfield-Kan completion of X, and the induced map $X \to \lim_s X_s$ as the completion map. (See [3, III §6]).

The p-pro-finite completion of Morel arises when one runs this program after neglecting the topology on H^*X. To do this one needs a setting where non-topological cohomology arises naturally. This is provided by the continuous cohomology of a simplicial pro-finite set. As this sentence indicates this paper is written simplicially, so "space" means "simplicial set". The details of continuous cohomology and Morel's results are given in section 1. We linger there a bit to prove that the closed model category on simplicial pro-finite sets constructed by Morel in [13] is actually part of a simplicial model category structure. This helps with later constructions. Section 2 gives a similar homotopical foundation to the Bousfield-Kan completion. This enlarges on the idea, originally due to Dror-Farjoun [4], that one might regard the Bousfield-Kan completion as a pro-object.

The later sections provide, among other things, a computational foundation for p-pro-finite completion. Section 3 gives a description of p-pro-finite completion as the total space of a cosimplicial space and describes the resulting Bousfield-Kan spectral sequence. We prove also that if H_*X is finite in each degree, then the Bousfield-Kan completion and the p-pro-finite completion of X are weakly equivalent. This is certainly plausible, given the constructions above. Section 5 computes the homotopy groups of the p-pro-finite completion of a connected nilpotent spaces X in terms of the homotopy groups of X. As a preliminary section 4 discusses the derived functors of p-pro-finite completion of abelian groups. Unlike the p-completion functor needed for Bousfield-Kan completion, p-pro-finite completion is right exact and has straightforward homological algebra. Finally section 6 approaches the problem of computing the homology of the p-pro-finite completion of X in terms of H_*X. This is not simple. Even in the case of the Bousfield-Kan completion one must compute $H_*(\lim_s X_s)$, where the X_s are as above, by no means a straightforward operation. See [3, VI §5]. In the case of p-pro-finite completion, one is faced with a spectral sequence based on the right derived functors of the completion of H_*X in the category of coalgebras. Some examples are given.

1 Pro-finite spaces and p-pro-finite completion.

This section is devoted to recapitulating and interpreting the result of Morel

A pro-finite set X is a filtered inverse limit $X = \{X_\alpha\}$ of finite sets X_α, and a morphism of pro-finite sets $X = \{X_\alpha\} \to Y = \{Y_\alpha\}$ is a morphism in the pro-category—that is, an element in

$$\mathrm{Hom}_{\mathrm{pro}}(X,Y) = \lim_\beta \operatorname*{colim}_\alpha \mathrm{Hom}_{\mathrm{sets}}(X_\alpha, Y_\beta).$$

For X a pro-finite set, the inverse limit $\lim_\alpha X_\alpha$ acquires a topology by giving each X_α the discrete topology and requiring that

$$\lim_\alpha X_\alpha \subseteq \prod_\alpha X_\alpha$$

be a subspace, where the product has the product topology. Thus $\lim_\alpha X_\alpha$ is compact and totally disconnected. Let $\widehat{\mathcal{E}}$ be the category of pro-finite sets and $\mathcal{F}$ the category of compact totally disconnected topological spaces and continuous maps. For the following compare [14].

LEMMA 1.1. *The functor* lim : $\widehat{\mathcal{E}} \to \mathcal{F}$ *is an equivalence of categories.*

PROOF: Let $Z \in \mathcal{F}$. Define $\mathcal{R}(Z)$ to be the set of equivalence relations $R \subseteq Z\times Z$ so that R is open in $Z\times Z$. Then $\{Z/R\}_{\mathcal{R}(Z)}$ is a pro-finite set and $Z \to \lim_{\mathcal{R}(Z)} Z/R$ is an isomorphism. Also if $X \in \widehat{\mathcal{E}}$, then the relations defined by the projections $X \to X_\alpha$ are contained in $\mathcal{R}(\lim_\alpha X_\alpha)$, so X is pro-isomorphic to $\{\lim_\alpha X_\alpha/R\}_{\mathcal{R}(\lim_\alpha X_\alpha)}$. Finally the assignment $Z \to \{Z/R\}_{\mathcal{R}(Z)}$ is a functor, for if $f : Z \to Z'$ is continuous, $(f\times f)^{-1}$ defines a map $\mathcal{R}(Z') \to \mathcal{R}(Z)$ and hence a pro-map $\{Z/\mathcal{R}\}_{\mathcal{R}(Z)} \to \{Z'/R\}_{\mathcal{R}(Z')}$. ∎

Because of this result one often confuses $X \in \widehat{\mathcal{E}}$ with $\lim_\alpha X_\alpha \in \mathcal{F}$. I hope the context makes clear which I'm referring to.

Now let $\widehat{\mathbf{S}}$ be the category of simplicial pro-finite sets; one can identify this with the category of simplicial compact totally disconnected topological spaces. There is a closed-model category structure on $\widehat{\mathbf{S}}$, but before giving the details, let us remark that $\widehat{\mathbf{S}}$ is a simplicial category in the sense of Quillen [15, II §2]. To see this, first note that the "forgetful functor" to simplicial sets

(1.2.1) $$|\cdot| : \widehat{\mathbf{S}} \to \mathbf{S}$$

which sends X to $\lim_\alpha X_\alpha$ without the topology has a left adjoint given by completion:

(1.2.2) $$\widehat{Y} = \{Y_\alpha\}$$

where Y_α runs over all the quotients of Y that are level-wise finite in the sense that for all n the set of n simplices is finite.

To define the simplicial structure on $\widehat{\mathbf{S}}$, let $X \in \widehat{\mathbf{S}}$ and $K \in \mathbf{S}$ and define

$$X \otimes K = X \times \widehat{K} \tag{1.3.1}$$

in $\widehat{\mathbf{S}}$. For fixed K, the functor $X \mapsto X \otimes K$ from $\widehat{\mathbf{S}}$ to itself has right adjoint $X \mapsto \hom(K, X)$. If K has finitely many non-degenerate simplices,

$$\hom(K, X) = \{\mathrm{map}(K, X_\alpha)\}. \tag{1.3.2}$$

For more general K, one must proceed as in [15, II §2]. Finally, there is an external mapping space functor

$$\mathrm{map}(\cdot,\cdot) : \widehat{\mathbf{S}}^{\mathrm{op}} \times \widehat{\mathbf{S}} \to \mathbf{S}$$

with n-simplices given by the formula

$$\mathrm{map}(X, Y)_n = \mathrm{Hom}_{\widehat{\mathbf{S}}}(X \otimes \Delta^n, Y), \tag{1.3.3}$$

where Δ^n is the usual n-simplex.

REMARK 1.4: Let $* \in \widehat{\mathbf{S}}$ be the terminal object, the one-point space, and let $X \in \widehat{\mathbf{S}}$. Then, as always, one should distinguish between $\hom(*, X) \cong X$ and

$$\mathrm{map}(*, X) \cong |X| \in \mathbf{S}.$$

As above $|X| = \lim_\alpha X_\alpha$, neglecting the topology.

We next turn to the model category structure on $\widehat{\mathbf{S}}$. For $X \in \widehat{\mathbf{S}}$, define the continuous cohomology with $\mathbb{F}_p$ coefficients by

$$H^*_{\mathrm{cont}}(X, \mathbb{F}_p) = \operatorname*{colim}_\alpha H^*(X_\alpha, \mathbb{F}_p) \tag{1.5}$$

This has a description as the cohomology of continuous cochains as follows: regard X as a simplicial compact, totally disconnected space and set

$$C^n_{\mathrm{cont}}(X) = \mathrm{Hom}_{\mathrm{cont}}(X_n, \mathbb{F}_p).$$

Then $C^*_{\mathrm{cont}}(X)$ is a cosimplicial vector space and

If $X \in S$, then

$$C^n(X) = \mathrm{Hom}_{\mathrm{sets}}(X_n, \mathbb{F}_p) \cong \mathrm{Hom}_{\mathrm{cont}}(\widehat{X}, \mathbb{F}_p),$$

whence

$$H^* X \cong H^*_{\mathrm{cont}}(\widehat{X}). \tag{1.6}$$

Now define a morphism $f : X \to Y$ in $\widehat{\mathbf{S}}$ to be

1.7.1) a weak equivalence if $H^* f : H^*_{\mathrm{cont}} Y \to H^*_{\mathrm{cont}} X$ is an isomorphism;

1.7.2) a cofibration if it is a level-wise injection—that is, a level-wise injection of simplicial topological spaces; and

1.7.3) a fibration if it has the right lifting property with respect to all trivial cofibrations.

As always, a trivial cofibration is a cofibration which is also a weak equivalence. Also a morphism $f : X \to Y$ has the right lifting property with respect to a morphism $g : A \to B$ if any lifting problem

$$\begin{array}{ccc} A & \longrightarrow & X \\ {\scriptstyle g}\downarrow & \nearrow & \downarrow{\scriptstyle f} \\ B & \longrightarrow & Y \end{array}$$

can be solved in such a way that both triangles commute.

Morel's result is:

Theorem 1.8. *With these definitions of weak equivalence, cofibration, and fibration, $\widehat{\mathbf{S}}$ becomes a simplicial model category.*

Proof: The fact that $\widehat{\mathbf{S}}$ is a closed model category is Théorème 1 of [13]. This leaves only Axiom SM7, relating the simplicial structure to the model category structures. But this follows easily from SM7b [15, II §2], which is a triviality in this case. ∎

We will devote some space below to clarifying what the fibrations and fibrant objects of $\widehat{\mathbf{S}}$ are. But let us now describe the homotopical properties of the completion functor.

LEMMA 1.9. *1) The completion functor* $(\cdot)^{\wedge} : \mathbf{S} \to \widehat{\mathbf{S}}$ *preserves cofibrations and sends* $H_*(\cdot;\mathbb{F}_p)$ *isomorphisms to weak equivalences.*

2) The forgetful functor $|\cdot| : \widehat{\mathbf{S}} \to \mathbf{S}$ *preserves fibrations and weak equivalences among fibrant objects.*

PROOF: The second statement is a formal consequence of the first. To the first, note that $(\cdot)^{\wedge}$ preserves injections, since every inclusion of sets is split. Since cofibrations are level-wise injections in both categories, completion preserves cofibrations. The statement about $H_*(\cdot;\mathbb{F}_p)$ isomorphisms follows from 1.6. ∎

An easy consequence of Quillen's Theorem [15, I §4] and Lemma 1.9 is:

COROLLARY 1.10. *The functors* $(\cdot)^{\wedge}$ *and* $|\cdot|$ *induce an adjoint pair of total derived functors*

$$\mathbf{L}(\cdot)^{\wedge} : \mathbf{Ho}(\mathbf{S}) \rightleftarrows \mathbf{Ho}(\widehat{\mathbf{S}}) : \mathbf{R}|\cdot|. \qquad ∎$$

REMARKS 1.11: 1) Since every object of $\mathbf{S}$ is cofibrant, $\mathbf{L}(X)^{\wedge} \simeq \widehat{X}$. However, not every object of $\widehat{\mathbf{S}}$ is fibrant. To obtain a model for $\widehat{\mathbf{S}}$, choose a $H^*_{\text{cont}}(\cdot)$ isomorphism $X \to Y$ with Y fibrant; then $|Y| \in \mathbf{S}$ is a model for $\mathbf{R}|X|$.

2) By [1] there is a simplicial model category structure on $\mathbf{S}$ where $f : X \to Y$ is a cofibration if it is a level-wise inclusion and a weak equivalence if H_*f is an isomorphism. The fibrant objects are the Bousfield-local spaces. Hence Lemma 1.9.1 implies $|X|$ is Bousfield local if $X \in \widehat{\mathbf{S}}$ is fibrant.

We now define the p-pro-finite completion of a space X.

DEFINITION 1.12. *Let* $X \in \mathbf{S}$. *Then the* p*-pro-finite completion of* X *is an* $H^*_{\text{cont}}(\cdot)$ *isomorphism* $\widehat{X} \to X_p$ *with* X_p *fibrant in* $\widehat{\mathbf{S}}$.

Of course, X_p is uniquely defined up to homotopy equivalence, $|X_p|$ is a model for $\mathbf{R}|\widehat{X}|$ and the map

$$\eta : X \to |X_p|$$

in $\mathbf{S}$ can be called the completion map. On the level of homotopy categories the morphism η is a model for the unit of the adjunction $X \to \mathbf{R}|\mathbf{L}(X)^{\wedge}|$.

In section 3 we will point out that the choice $\widehat{X} \to X_p$ can be made functorially; alternatively, one could see this fact by noting that Morel actually proves that $\widehat{\mathbf{S}}$

We now turn to a discussion of fibrant objects. We begin by extending the definition of continuous cohomology. Let M be a pro-finite abelian group. Then if $X \in \widehat{\mathbf{S}}$, the continuous cochains $C^*_{\text{cont}}(X, M)$ are defined by

$$C^n_{\text{cont}}(X, M) = \text{Hom}_{\widehat{\mathcal{E}}}(X_n, M) \tag{1.13.1}$$

and the continuous cohomology of X with coefficients in M are given by

$$H^*_{\text{cont}}(X, M) = H^* C^*_{\text{cont}}(X, M). \tag{1.13.2}$$

Now, if $X = \{X_\alpha\}$ and $M = \{M_\beta\}$, one has

$$\begin{aligned} C^n_{\text{cont}}(X, M) &= \lim_\beta \operatorname{colim}_\alpha \text{Hom}_{\mathcal{E}}((X_\alpha)_n, , M_\beta) \\ & \lim_\beta C^n_{\text{cont}}(X, M_\beta) \end{aligned}$$

so one gets a spectral sequence

$$\lim_\beta{}^{(p)} H^q_{\text{cont}}(X, M_\beta) \Rightarrow H^{p+q}_{\text{cont}}(X, M) \tag{1.14}$$

and this has the following consequences:

LEMMA 1.15. *1) If $X \in \mathbf{S}$ is a space, there is a spectral sequence*

$$\lim_\beta{}^{(p)} H^q(X, M_\beta) \Rightarrow H^{p+q}_{\text{cont}}(\widehat{X}, M)$$

2) If $X \in \mathbf{S}$ has only finitely many simplices in each degree,

$$\lim_\beta H^q(X, M_\beta) \cong H^p_{\text{cont}}(\widehat{X}, M).$$

PROOF: In the first clause, one has $H^q(X, M_\beta) = H^q_{\text{cont}}(\widehat{X}, M_\beta)$. In the second, $\lim_\beta{}^{(p)} H^q(X, M_\beta) = 0$ for $p > 0$ by [11, 1.1], since $H^q(X, M_\beta)$ is a finite abelian group for all q and β. ∎

The next observation is that (normalized) continuous cochains and continuous cohomology are representable functors. The proofs are the usual ones.

Briefly, let $s\mathcal{A}b$ be the category of simplicial abelian groups and $c\mathcal{A}b$ the category of chain complexes of abelian abelian groups. The normalization functor $N : s\mathcal{A}b \to c\mathcal{A}b$ is an equivalence of categories with inverse K. For an abelian group A, let $L(A, n) \in s\mathcal{A}b$ be the object with

$$NL(A, n)_k = \begin{cases} A, & k = n, n+1 \\ 0, & k \neq n \end{cases}$$

and $\partial = 1 : NL(A,n)_{n+1} \to NL(A,n)_n$. Then for $X \in \mathbf{S}$, there is a natural isomorphism

$$NC^n(X,A) \cong \mathrm{Hom}_{\mathbf{S}}(X, L(A,n))$$

where $NC^n(X,A)$ denotes the normalized cochains. If $K(A,n) \in s\mathcal{A}b$ has

$$NK(A,n) \cong A$$

concentrated in degree n, then

$$Z^n(X,A) \cong \mathrm{Hom}_{\mathbf{S}}(X, K(A,n))$$

where $Z^n(X,A) \subseteq NC^n(X,A)$ are the cocycles. Finally, interpreting chain homotopies as simplicial homotopies yields

$$[X, K(A,n)] \cong H^n(X,A).$$

If M is a pro-finite abelian group, $L(M,n)$ and $K(M,n)$ are simplicial pro-finite abelian groups and the analogous statements hold.

LEMMA 1.16. *For $X \in \widehat{\mathbf{S}}$ and a pro-finite abelian group M, there are natural isomorphisms*

$$\mathrm{Hom}_{\widehat{\mathbf{S}}}(X, L(M,n)) \cong NC^n_{\mathrm{cont}}(X,M)$$
$$\mathrm{Hom}_{\widehat{\mathbf{S}}}(X, K(M,n)) \cong Z^n_{\mathrm{cont}}(X,M)$$
$$[X, K(M,n)]_{\widehat{\mathbf{S}}} \cong H^n_{\mathrm{cont}}(X,M).$$

PROOF: Only the statement about homotopy classes requires clarification. Again, one interprets a (continuous) simplicial homotopy as a continuous chain homotopy. ■

Strictly speaking, one shouldn't calculate $[X, K(M,n)]_{\widehat{\mathbf{S}}}$ unless $K(M,n)$ is fibrant. We will see below that $K(M,n)$ is fibrant if M is a pro-finite abelian p-group. For more general M, a slight modification of the model category structure (so $f : X \to Y$ is a weak equivalence if $H^*_{\mathrm{cont}} f : H^*_{\mathrm{cont}}(Y, \mathbb{Z}/\ell\mathbb{Z}) \to H^*_{\mathrm{cont}}(X, \mathbb{Z}/\ell\mathbb{Z})$ is an isomorphism for all primes ℓ) makes $K(M,n)$ fibrant. See [13], after Théorème

LEMMA 1.17. *Let M be a pro-finite abelian group. Then*

$$\pi_*|K(M,n)| \cong \lim_\alpha M_\alpha$$

concentrated in degree n.

PROOF: Note that $|K(M,n)| = \lim_\alpha K(M_\alpha, n)$ without the topology is a simplicial group. Hence for $t > 0$

$$\begin{aligned}\pi_t|K(M,n)| = [S^t, |K(M,n)|]_S &\cong [S^t, K(M,n)]_{\widehat{S}} \\ &\cong H^n_{\mathrm{cont}}(S^t, M)\end{aligned}$$

since S^t has only finitely many simplices in each degree. Similarly, for $t = 0$, $\pi_t|K(M,n)| = H^n_{\mathrm{cont}}(*, M)$. Now apply Lemma 1.15.2. ∎

That said, we can use these objects to give some feel for the fibrations in $\widehat{\mathbf{S}}$. Note that the coboundary

$$\partial : NC^n_{\mathrm{cont}}(X, M) \to Z^{n+1}_{\mathrm{cont}}(X, M)$$

defines a map of pro-finite simplicial abelian groups

$$L(M,n) \to K(M, n+1).$$

The fiber is $K(M,n)$.

LEMMA 1.18. *Let M be a pro-finite abelian p-group. The for all $n \geq 0$, $K(M,n)$ is fibrant and $L(M,n) \to K(M,n+1)$ is a fibration.*

PROOF: This is a variant of [13], §1.4, Lemma 2. The proof is the same, since normalization is exact. ∎

As mentioned above, the restriction to p-groups is necessary since we defined weak equivalences using only mod p cohomology.

All fibrations may be characterized as follows. Define a class of morphisms to be *saturated* if it is closed under isomorphisms, products, pull-backs, retracts, and countable inverse limits. This last means that given a tower of morphisms

$$\cdots \to X_s \to X_{s-1} \to \cdots \to X_1 \to X_0$$

in the class, then $\lim X_s \to X_0$ is in the class. The class of fibrations in a closed model category is saturated. The saturation of a class of morphisms is the smallest saturated class containing those morphisms.

LEMMA 1.19. *The fibrations in* $\widehat{\mathbf{S}}$ *are at once*

1) *the saturation of the class of morphisms* $L(M,n) \to K(M,n+1)$ $n \geq 0$ *and, with* M *a pro-finite abelian p-group;*
2) *the saturation of the set of morphisms,* $L(\mathbb{Z}/p,n) \to K(\mathbb{Z}/p,n+1)$, $n \geq 0$.

PROOF: This is implicit in the construction of fibrations given in [13] §1.4, Proposition 1. ∎

We close with a discussion of how p-pro-finite completion behaves with respect to coproducts. This requires a very short discussion of connectedness in $\widehat{\mathbf{S}}$.

A pro-finite space $X \in \widehat{\mathbf{S}}$ is connected if any isomorphism $X_1 \coprod X_2 \cong X$ in $\widehat{\mathbf{S}}$ implies $X_1 = \emptyset$ or $X_2 = \emptyset$. A connected component $X \subseteq Y$ is a maximal connected sub-object in $\widehat{\mathbf{S}}$. Let $X \in \widehat{\mathbf{S}}$ be arbitrary and $I = \{X_i\}$ be the set of connected components of X. Then the usual arguments imply $\coprod_I X_i \cong X$ and there is an isomorphism

$$I \mapsto \mathrm{Hom}_{\mathbb{F}_p\text{-alg}}(H^*_{\text{cont}}X, \mathbb{F}_p)$$

sending X_i to

$$H^0_{\text{cont}}X \to H^0_{\text{cont}}X_i \cong \mathbb{F}_p.$$

Thus if $f : X \to Y$ is any trivial cofibration in $\widehat{\mathbf{S}}$, f induces an isomorphism from the set of connected components of X to the connected components of Y and we have:

LEMMA 1.20. *A morphism in* $\widehat{\mathbf{S}}$ *is a fibration if and only if it has the right lifting property with respect to all trivial cofibrations* $X \to Y$ *with* X *connected.* ∎

We know that Bousfield-Kan completion commutes with all disjoint unions. The following is the best we can do for p-pro-finite completion.

PROPOSITION 1.21. *Let* X_i, $1 \leq i \leq n$, *be a finite set of spaces. Then there is a weak equivalence*

n n

Proof: Let $\widehat{X}_i \to Y_i$ be a set of weak equivalences with $Y_i \in \widehat{\mathbf{S}}$ fibrant. Then the natural map

$$H^*(\coprod_{i=1}^{n} X_i) \leftarrow H^*_{\text{cont}}(\coprod_{i=1}^{n} Y_i)$$

is an isomorphism, since the disjoint union is finite. Thus we need only show $\coprod_{i=1}^{n} Y_i$ is fibrant. But this follows from Lemma 1.20. ∎

2 The homotopical background of Bousfield-Kan completion.

As an exercise in using new thoughts to understand old ones, we show how the ideas of the previous section can be specialized to show how the Bousfield-Kan p-completion of a space can be obtained as the inverse limit of a fibrant object in a suitable model category structure on towers of simplicial sets. This has the aesthetic advantage of making the important tower lemmas of [3, §III.6] a straightforward consequence of standard model category facts. For the sake of brevity, we will only sketch certain proofs.

One way to interpret this section is as follows: it is a useful idea of Dror-Farjoun's [4] that one can regard the Bousfield-Kan completion as a pro-object. What we do here provides a homotopical foundation for this observation.

Let **Tow** be the category of towers $X = \{X_n\}$ of simplicial sets with morphisms the pro-maps

$$\text{Hom}(X, Y) = \lim_n \operatorname{colim}_k \mathbf{S}(X_k, Y_n).$$

Thus a morphism $X \to Y$ is an equivalence class of tower maps, meaning a "commutative ladder" of the form

$$\begin{array}{ccccccccc} \cdots \longrightarrow & X_{k_{s+1}} & \longrightarrow & X_{k_s} & \longrightarrow \cdots & X_{k_2} & \longrightarrow & X_{k_1} \\ & \downarrow & & \downarrow & & \downarrow & & \downarrow \\ \cdots \longrightarrow & Y_{n_{s+1}} & \longrightarrow & Y_{n_s} & \longrightarrow \cdots & Y_{n_2} & \longrightarrow & Y_{n_1} \end{array}$$

where the horizontal maps are induced from the tower projections of X and Y and $\lim k_s = \infty = \lim n_s$.

The first thing to notice is that **Tow** is a simplicial category. Indeed if $K \in \mathbf{S}$ and $X \in \mathbf{Tow}$, then the functors

(2.1.1) $$X \mapsto X \otimes K = \{X_n \times K\}$$

and

(2.1.2) $$X \mapsto X^K = \{\mathrm{map}_{\mathbf{S}}(K, X_n)\}$$

are adjoint and the mapping space functor **map** : **Tow** $\times$ $\mathbf{Tow}^{\mathrm{op}} \to \mathbf{S}$ with N simplices

(2.1.3) $$\mathbf{map}(X, Y)_n = \mathrm{Hom}_{\mathbf{Tow}}(X \otimes \Delta^n, Y)$$

completes the structure.

There is also a notion of continuous cohomology with $\mathbb{F}_p$ coefficients. If $X \in$ **Tow**, define

(2.2) $$H^*_{\mathrm{cont}} X = H^*_{\mathrm{cont}}(X, \mathbb{F}_p) = \operatorname*{colim}_n H^*(X_n, \mathbb{F}_p).$$

The notation is justified by the observation that if $\lim X_n$ is topologized with the inverse limit topology (i.e., $\lim X_n$ is a subspace of $\prod X_n$, where each X_n has the discrete topology), thus $H^*_{\mathrm{cont}} X$ can be computed using continuous cochains on $\lim X_n$.

Now define a morphism $f : X \to Y$ in **Tow** to be

2.3.1) a weak equivalence if $H^*_{\mathrm{cont}} f : H^*_{\mathrm{cont}} Y \to H^*_{\mathrm{cont}} X$ is an isomorphism;
2.3.2) a cofibration if $\lim X_n \to \lim Y_n$ is an inclusion;
2.3.3) a fibration if it has the right lifting with respect to all trivial fibrations.

THEOREM 2.4. *With the simplicial structure of (2.1) and the notion of weak equivalence, cofibrations, and fibration of (2.3), the category* **Tow** *becomes a simplicial model category.*

The proof is essentially the same as that given by Morel for $\widehat{\mathbf{S}}$. The relevant observations are, first, that $H^k_{\mathrm{cont}}(\cdot)$ is a representable homotopy functor and, second, that if $V = \{V_n\}$ is any tower of $\mathbb{F}_p$ vector spaces, then the map $L(V, k) \to K(V, k)$ in *Tow* is a fibration. The representing object for $H^k_{\mathrm{cont}}(\cdot)$ is $K(\mathbb{Z}/p\mathbb{Z}, k)$ regarded as a constant tower.

Note that if X is a constant tower $H^*_{\mathrm{cont}} X = H^* X$.

DEFINITION 2.5. *Let $X \in$ **S** be a space. Then we may regard X as a constant tower in* **Tow**. *Define the Bousfield-Kan p-completion $(\mathbb{F}_p)_\infty X$ of X to be a fibrant*

Thus $(\mathbb{F}_p)_\infty X$ is well-defined up to weak equivalence in **Tow**. As with the p-pro-finite completion one sometimes confuses the difference between $(\mathbb{F}_p)_\infty X \in \mathbf{Tow}$ and its inverse limit; in fact, I've already done so, as Bousfield and Kan define $(\mathbb{F}_p)_\infty X$ to as an inverse limit. I hope that, in context, this confusion will cause no difficulty.

We now reprove one of Bousfield and Kan's tower lemmas, using Theorem 2.4.

PROPOSITION 2.6. ($\mathbb{F}_p$-nilpotent tower lemma, [3], p. 88). *Let $X \in \mathbf{S}$ be regarded as a constant tower and $X \to Y$ a map of towers so that*

1) *$H_*X \to H_*Y = \{H_*Y_n\}$ is a pro-isomorphism;*
2) *for all n, Y_n is an $\mathbb{F}_p$-nilpotent space.*

Then Y is weakly equivalent to the Bousfield-Kan p-completion of X.

PROOF: The first statement is equivalent to the assertion that

$$H^*X \cong H^*_{\text{cont}}X \cong H^*_{\text{cont}}Y,$$

so we need only show that the second statement implies Y is fibrant. Fix n and let $Y\langle n\rangle$ be the tower with

$$Y\langle n\rangle_k = \begin{cases} Y_n & k \geq n \\ * & k < n. \end{cases}$$

Then, by Proposition 5.3iii of [3, p. 83] (a refined Postnikov tower for $\mathbb{F}_p$ nilpotent spaces given at the beginning of section 5) and the fact that $L(V,k) \to K(V,k)$ is a fibration in **Tow** for all towers of vector spaces V, the tower $Y\langle n\rangle$ is fibrant in **Tow**. Now $Y = \lim_n Y\langle n\rangle$, and since a directed inverse limit of fibrant objects is fibrant, Y is fibrant in **Tow**. ■

3 A cosimplicial description of p-pro-finite completion.

In this section, we interpret a construction of Morel's to describe X_p as the total space of a cosimplicial p-pro-finite space. We describe the E_2 terms of the resulting homotopy spectral sequence and use this to compare Bousfield-Kan p-completion to p-pro-finite completion.

Let $X \in \widehat{\mathbf{S}}$ be a simplicial pro-finite set. Define $\mathbb{F}_p X$ by the equation

$$\mathbb{F}_p X = \{\mathbb{F}_p X_\alpha\} \tag{3.1}$$

where $\mathbb{F}_p X_\alpha$ is the simplicial vector space on X_α.

Note that $\mathbb{F}_p X$ is a simplicial pro-finite vector space and the functor $X \to \mathbb{F}_p X$ is left adjoint to the forgetful functor from $\widehat{\mathbf{S}}$ to the category of simplicial pro-finite $\mathbb{F}_p$ vector spaces. We can also forget the vector space structure and obtain a functor

$$\overline{\mathbb{F}}_p(\cdot) : \widehat{\mathbf{S}} \to \widehat{\mathbf{S}}.$$

Thus, by the usual properties of adjoint functors, is the functor of a triple on $\widehat{\mathbf{S}}$; hence for each $X \in \widehat{\mathbf{S}}$ we obtain an augmented cosimplicial object in $\widehat{\mathbf{S}}$

$$X \to \overline{\mathbb{F}}_p^\bullet X \tag{3.2}$$

where $(\overline{\mathbb{F}}_p^\bullet X)^s = \overline{\mathbb{F}}_p \circ \cdots \circ \overline{\mathbb{F}}_p X$, the composition taken $s+1$ times. The cosimplicial object $\overline{\mathbb{F}}_p^\bullet X$ is a resolution of X in the sense that, applying functor $\mathbb{F}_p$ degree-wise yields an augmented cosimplicial pro-finite vector space

$$\mathbb{F}_p X \to \mathbb{F}_p(\overline{\mathbb{F}}_p^\bullet X)$$

which comes equipped with a canonical contraction.

Morel's next observation is that $\overline{\mathbb{F}}_p^\bullet X$ can be used to build the p-pro-finite completion on X. To state the result, note that $\widehat{\mathbf{S}}$ is a simplicial model category, it has homotopy colimits in the style of Bousfield and Kan. In particular, if $Z^\bullet$ is a cosimplicial object in $\widehat{\mathbf{S}}$, the coequalizer diagram in $\widehat{\mathbf{S}}$ one can form the total space $\mathrm{Tot}(Z^\bullet)$ defined by the coequalizer diagram in $\widehat{\mathbf{S}}$

$$\mathrm{Tot}(Z^\bullet) \to \prod_{[n]} \hom(\Delta^n, Z^n) \rightrightarrows \prod_{[n]\to[m]} \hom(\Delta^n, Z^m) \tag{3.3}$$

where $\Delta^n \in \mathbf{S}$ is the canonical n-simplex and $\hom(\Delta^n, Z^n)$ is the internal mapping space functor

$$\hom(\Delta^n, Z^n) = \{\hom(\Delta^n, Z^n_\alpha)\}.$$

The second product is over all morphisms in the ordinal number category Δ.

To further analyze the structure of $\mathrm{Tot}(Z^\bullet)$ as an object of $\widehat{\mathbf{S}}$, let $\Delta^\bullet$ be the standard cosimplicial space which is Δ^k in cosimplicial degree k and let sk_n be the nth skeleton functor. If $Z^\bullet$ is a cosimplicial object in $\widehat{\mathbf{S}}$, then $Z^\bullet = \{Z^\bullet/R\}_{R\in\mathcal{R}(Z^\bullet)}$, using the notation of Lemma 1.1. For all $n < \infty$, let

Since $sk_n\Delta^\bullet$ is generated, as a cosimplicial space, by the canonical n-simplex $\iota_n \in (\Delta^n)_n$, $\mathrm{Tot}_n(Z^\bullet)$ is an object in $\widehat{\mathbf{S}}$ and there is an isomorphism in $\widehat{\mathbf{S}}$

$$\mathrm{Tot}(Z^\bullet) \cong \lim_n \mathrm{Tot}_n(Z^\bullet). \tag{3.4}$$

To obtain the p-pro-finite completion, we have

Proposition 3.5. *Let $X \in \widehat{\mathbf{S}}$ be a simplicial pro-finite set. Then the induced map $X \to \mathrm{Tot}(\overline{\mathbb{F}}_p^\bullet X)$ is weak equivalence and $\mathrm{Tot}(\overline{\mathbb{F}}_p^\bullet X)$ is fibrant. If $X = \widehat{Y}$, where Y is a space, then*

$$Y \mapsto \mathrm{Tot}(\overline{\mathbb{F}}_p^\bullet \widehat{Y})$$

is a model for the p-pro-finite completion of Y.

Proof: The second clause follows from the first and the definition p-pro-finite completion. To see $X \to \mathrm{Tot}(\overline{\mathbb{F}}_p^\bullet X)$ is a cohomology isomorphism note that

$$\begin{aligned} H^*_{\text{cont}} X &\cong \operatorname*{colim}_\alpha H^* X_\alpha \cong \operatorname*{colim}_\alpha \operatorname*{colim}_n H^* \mathrm{Tot}_n \overline{\mathbb{F}}_p^\bullet X_\alpha \\ &\cong H^*_{\text{cont}} \mathrm{Tot}(\overline{\mathbb{F}}_p^\bullet X). \end{aligned}$$

Thus one need only show $\mathrm{Tot}(\overline{\mathbb{F}}_p^\bullet X)$ is fibrant. For this it is sufficient to show $\overline{\mathbb{F}}_p^\bullet X$ is fibrant in the Reedy model category structure on cosimplicial objects in $\widehat{\mathbf{S}}$. But the standard argument [3, p. 276] shows that the canonical map to the nth matching object

$$s : \overline{\mathbb{F}}_p^n X \to M_n \overline{\mathbb{F}}_p^\bullet X$$

is a split surjection of simplicial pro-finite vector spaces. The result now follows from Lemma 1.18. ∎

This resolution is a key tool in the investigation of p-pro-finite completion. We begin with the following preliminary result. Let $|\cdot| : \widehat{\mathbf{S}} \to \mathbf{S}$ be the forgetful functor.

Lemma 3.6. *Let $Z^\bullet$ be a cosimplicial object in $\widehat{\mathbf{S}}$. Then there is a natural isomorphism*

$$|\mathrm{Tot}(Z^\bullet)| \cong \mathrm{Tot}(|Z^\bullet|).$$

Proof: If $K \in \mathbf{S}$ has only finitely many simplices, then for all $X \in \widehat{\mathbf{S}}$,

$$\begin{aligned} |\hom(K, X)| &\cong \lim_{\mathcal{R}(X)} \operatorname{map}(K, X/R) \\ &\cong \operatorname{map}(K, |X|). \end{aligned}$$

Since $|\cdot|$ is a right adjoint it preserves products and equalizers. The result now follows by applying $|\cdot|$ to the diagram of (3.3). ■

An object $X \in \widehat{\mathbf{S}}$ is *pointed* if it comes equipped with a map $* \to X$ in $\widehat{\mathbf{S}}$ from the one-point space; this amounts to choosing a basepoint for $|X|$. A cosimplicial object $Z^\bullet$ is pointed if there is a map $* \to Z^\bullet$ from the constant one-point cosimplicial object to $Z^\bullet$. This amounts to choosing a basepoint for $|aZ^\bullet|$ where $aZ^\bullet$ is the equalizer of $d^0, d^1 : Z^0 \to Z^1$. If $Z^\bullet$ is pointed, there is a homotopy spectral sequence

$$\pi^s\pi_t|Z^\bullet| \Rightarrow \pi_{t-s}\mathrm{Tot}(|Z^\bullet|) \cong \pi_{t-s}|\mathrm{Tot}(Z^\bullet)|. \tag{3.7}$$

For the spectral sequence to make homotopical sense, $|Z^\bullet|$ must be fibrant in the sense of [3], X §4. Thus we have the extremely technical:

Lemma 3.8. *Let $Z^\bullet$ be a cosimplicial object in $\widehat{\mathbf{S}}$. If $Z^\bullet$ is fibrant in the Reedy model category structure on cosimplicial objects in $\widehat{\mathbf{S}}$, then $|Z^\bullet|$ is fibrant as a cosimplicial space.*

Proof: One needs

$$|s| : |Z^n| \to M_n|Z^\bullet|$$

to be a fibration. By hypothesis $s : Z^n \to M_nZ^\bullet$ is a fibration in $\widehat{\mathbf{S}}$, and Lemma 1.9.2 says $|\cdot|$ preserves fibrations. Now one need only observe that $M_n|Z^\bullet| \cong |M_nZ^\bullet|$ since the matching space is an inverse limit. ■

This lemma applies, in particular, to $\overline{\mathbb{F}}_p^\bullet X$, $X \in \widehat{\mathbf{S}}$. See the proof of Proposition 3.5. We next interpret the E_2 term $\pi^s\pi_t\overline{\mathbb{F}}_p^\bullet X$.

Lemma 3.9. *Let $W = \{W_\alpha\}$ be a simplicial pro-finite vector space. Then*

$$\pi_*|W| \cong \lim_\alpha \pi_*|W_\alpha|.$$

Proof: Every short exact sequence of pro-finite vector spaces is split. Hence there are pro-finite vector space U_n, $n \geq 1$, V_n, $n \geq 0$ and a (non-natural) isomorphism

$$W \cong \prod_n L(U_n, n) \times \prod_n K(V_n, n).$$

In this result, we chose $0 \in |W|$ as the basepoint; however, because $|W|$ is simplicial group, the result is equally true for any choice of basepoint.

We now use Lemma 3.9 to compute $\pi_t \overline{\mathbb{F}}_p X$, $X \in \widehat{\mathbf{S}}$. The result is similar to the non-pro-finite case (see, for example [13] §1 for a lucid exposition); therefore, I will be brief.

Let $\mathcal{K}$ be the category of unstable algebras over the Steenrod algebra. If $X \in \widehat{\mathbf{S}}$, then $H^*_{\text{cont}} X \in \mathcal{K}$. Let $\mathcal{K}_*$ be the category of augmented objects in $\mathcal{K}$, i.e., objects $H \in \mathcal{K}$ equipped with a morphism $\epsilon : H \to \mathbb{F}_p$. If $X \in \widehat{\mathbf{S}}$ is pointed, then $H^*_{\text{cont}} X \in \mathcal{K}_*$. The augmentation ideal functor from $\mathcal{K}_*$ to graded vector spaces has a left adjoint G, and we let $\overline{G} : \mathcal{K}_* \to \mathcal{K}_*$ be the composite functor. Then $\overline{G}$ is the functor of a cotriple on $\mathcal{K}_*$. If $X \in \widehat{\mathbf{S}}$ is pointed, then the unit $X \to \overline{\mathbb{F}}_p X$ gives $\overline{\mathbb{F}}_p X$ a basepoint.

LEMMA 3.10. *Let $X \in \widehat{\mathbf{S}}$ be pointed. Then with the induced basepoint on $\overline{\mathbb{F}}_p X$, there is a natural isomorphism*

$$\pi_t \overline{\mathbb{F}}_p X \cong \mathrm{Hom}_{\mathcal{K}_*}(\overline{G} H^*_{\text{cont}} X, H^* S^t).$$

PROOF: Applying Lemma 3.9 yields

$$\pi_t \overline{\mathbb{F}}_p X \cong \pi_t \mathbb{F}_p X \cong \lim_{\mathcal{R}(X)} \pi_t \mathbb{F}_p (X/R).$$

Then we have natural isomorphisms

$$\begin{aligned} \lim_{\mathcal{R}(X)} \pi_t \mathbb{F}_p (X/R) &\cong \lim_{\mathcal{R}(X)} \mathrm{Hom}_{\mathcal{K}_*}(\overline{G} H^*(X/R), H^* S^t) \\ &\cong \mathrm{Hom}_{\mathcal{K}_*}(\overline{G} H^*_{\text{cont}} X, H^* S^t) \end{aligned}$$

since $\overline{G}$ commutes with filtered colimits. ■

Now, if $H \in \mathcal{K}_*$, let $\overline{G}_\bullet H \to H$ be the cotriple resolution H induced by $\overline{G}$ and define

$$\mathrm{Ext}^s_{\mathcal{K}_*}(H, H^* S^t) = \pi^s \mathrm{Hom}_{\mathcal{K}_*}(\overline{G}_\bullet H, H^* S^t).$$

If $t = 0$, this is defined only for $s = 0$.

PROPOSITION 3.11. *Let $X \in \widehat{\mathbf{S}}$ be pointed. Then the homotopy spectral sequence of the cosimplicial space $|\overline{\mathbb{F}}_p^\bullet X|$ is*

$$\mathrm{Ext}^s_{\mathcal{K}_*}(H^*_{\text{cont}} X, H^* S^t) \Rightarrow \pi_{t-s} |\mathrm{Tot}(\overline{\mathbb{F}}_p^\bullet X)|.$$

PROOF: Combine the spectral sequence of 3.7 with $Z^\bullet = \mathbb{F}_p^\bullet X$ with Lemma 3.10.■

COROLLARY 3.12. *Let $Y \in \mathbf{S}$ be a pointed space. Then there is a homotopy spectral sequence*

$$\mathrm{Ext}^s_{\mathcal{K}_*}(H^*Y, H^*S^t) \Rightarrow \pi_{t-s}|Y_p|.$$

PROOF: One has $H^*Y = H^*_{\mathrm{cont}}\widehat{Y}$ and $|\mathrm{Tot}(\mathbb{F}^\bullet_p \widehat{Y})| = |Y_p|$ by Proposition 3.5. ■

One can use these tools to give a comparison between the Bousfield-Kan and p-pro-finite completions of a space Y. For $Y \in \mathbf{S}$, let $\mathbb{F}_p Y$ be the simplicial vector space on Y, and let $\overline{\mathbb{F}}_p : \mathbf{S} \to \mathbf{S}$ be the resulting triple. Then the augmented cosimplicial space $Y \to \overline{\mathbb{F}}^\bullet_p Y$ is fibrant and the induced map $Y \to \mathrm{Tot}(\overline{\mathbb{F}}^\bullet_p Y) = (\mathbb{F}_p)_\infty Y$ is the Bousfield-Kan p-completion of Y by [3], p. 88. By the same reasoning that arrived at Proposition 3.11, the homotopy spectral sequence of $\overline{\mathbb{F}}^\bullet_p Y$ is the Bousfield Bousfield-Kan spectral sequence

$$\mathrm{Ext}^s_{\mathcal{CA}_*}(H_*S^t, H_*Y) \Rightarrow \pi_{t-s}\mathrm{Tot}(\overline{\mathbb{F}}^\bullet_p Y). \tag{3.13}$$

Here $\mathcal{CA}_*$ is the category of augmented unstable coalgebras over the Steenrod algebra. The claim is that one can compare this spectral sequence to the one of Corollary 3.12.

Dualization induces a map

$$\mathrm{Hom}_{\mathcal{CA}_*}(H_*S^t, H_*Y) \to \mathrm{Hom}_{\mathcal{K}_*}(H^*Y, H^*S^t).$$

This extends to a map of derived functors

$$\mathrm{Ext}^s_{\mathcal{CA}_*}(H_*S^t, H_*Y) \to \mathrm{Ext}^s_{\mathcal{K}_*}(H^*Y, H^*S^t). \tag{3.14}$$

This can be realized homotopically as follows. Let $\widehat{Y} = \{Y_\alpha\}$ be the completion of Y. Then the induced maps $\mathbb{F}_p Y \to \mathbb{F}_p Y_\alpha$ define a map of cosimplicial spaces $\overline{\mathbb{F}}^\bullet_p Y \to |\overline{\mathbb{F}}^\bullet_p \widehat{Y}|$ and one gets:

PROPOSITION 3.15. *Let $Y \in \mathbf{S}$ be pointed. Then the map $\overline{\mathbb{F}}^\bullet_p Y \to |\overline{\mathbb{F}}^\bullet_p \widehat{Y}|$ induces a map of homotopy spectral sequences*

$$\begin{array}{ccc} \mathrm{Ext}^s_{\mathcal{CA}_*}(H_*S^t, H_*Y) & \Longrightarrow & \pi_{t-s}(\mathbb{F}_p)_\infty Y \\ \downarrow & & \downarrow \end{array}$$

A space Y is of finite type if H_kY is finite for all k. If Y is of finite type, so is $\mathbb{F}_pY$ and this implies that

$$\mathrm{Ext}^s_{\mathcal{CA}_*}(H_*S^t, H_*Y) \cong \mathrm{Ext}^s_{\mathcal{K}_*}(H^*Y, H^*S^t).$$

COROLLARY 3.16. *Let $Y \in \mathcal{S}$ be of finite type. Then the natural map of completions*

$$(\mathbb{F}_p)_\infty Y \to |Y_p|$$

is a weak equivalence.

PROOF: If Y is connected, this follows from Proposition 3.15 and the mapping lemma of [3], p. 285. In general, Y will have only finitely many components, so one can apply Proposition 1.21 and [3] §I.7.1. ∎

REMARK 3.17: One of the main points of this section was to discuss the comparison of spectral sequences of Proposition 3.15, and Corollary 3.16 was a by-product. A shorter non-functorial route to this last result is as follows: for Y of finite type, define a tower of fibrations

$$\cdots \to Y_s \xrightarrow{q_s} Y_{s-1} \cdots \to Y_2 \xrightarrow{q_2} Y_1 \xrightarrow{q_1} Y_0$$

and a map $f_s : Y \to Y_s$ so that

1) $q_s f_s = f_{s-1}$,
2) each Y_s is of finite type, Y_0 is a product of Eilenberg-MacLane spaces of type $K(\mathbb{Z}/p, n)$, and there is a homotopy pull-back square for $s \leq 0$,

$$\begin{array}{ccc} Y_{s+1} & \longrightarrow & * \\ \downarrow & & \downarrow \\ Y_s & \xrightarrow{\iota_s} & K_s \end{array}$$

where K_s is also a product of Eilenberg-MacLane space of type $K(\mathbb{Z}/p, n)$ and of finite type and
3) $H^*K_s \to H^*Y_s \to H^*Y$ is exact in the sense that

$$\mathbb{F}_p \otimes_{H^*K_s} H^*Y_s \cong H^*Y.$$

Then $\mathrm{colim} H^*Y_s \cong H^*Y$ and Lemma 1.19 implies $\lim Y_s$ is weakly equivalent to $|Y_p|$. On the other hand, the tower lemma Proposition 2.6 implies $\lim Y_s \simeq (\mathbb{F}_p)_\infty Y$.

4 The p-pro-finite completion of abelian groups.

In this section we discuss the derived functors of p-pro-finite completion.

Let $\mathcal{A}b$ be the category of abelian groups and $\mathcal{A}b_p$ the category of pro-finite abelian p-groups. This category is equivalent, via inverse limit, to the category of compact, totally disconnected abelian groups all of whose open subgroups have p-power index. There is also an extremely useful Pontrjagin duality. Let $T\mathcal{A}b_p$ be the category of p-torsion abelian groups. Then if $A \in T\mathcal{A}b_p$, the dual

$$A^* = \mathrm{Hom}_{\mathcal{A}b}(A, \mathbb{Z}/p^\infty) \tag{4.1.1}$$

is naturally an object in $\mathcal{A}b_p$, since every torsion group is the filtered colimit of its finite subgroups. Conversely, if $B = \{B_\alpha\} \in \mathcal{A}b_p$, then the continuous dual

$$B^\# = \operatorname*{colim}_\alpha \mathrm{Hom}_{\mathcal{A}b}(B_\alpha, Z/p^\infty) \tag{4.1.2}$$

is naturally object in $T\mathcal{A}b_p$ and the functors $(\cdot)^*$ and $(\cdot)^\#$ induce an equivalence of categories.

The forgetful functor $|\cdot| : \mathcal{A}b_p \to \mathcal{A}b$ given by $|B| = \lim_\alpha B_\alpha$ (without topology) has left adjoint $\Phi(A) = \{A_\alpha\}$ where the filtered system has objects all surjections $A \to A_\alpha$ with A_α a finite abelian p-group. We call this functor p-pro-finite completion. If we regard $\Phi(A)$ as a compact totally disconnected topological group—that is, take $\Phi(A) = \lim_\alpha A_\alpha$ with the topology—then the unit of the adjunction $A \to |\Phi(A)|$ really is a completion map $A \to \lim_\alpha A_\alpha$.

It is possible to give a simple description of the Pontrjagin dual of $\Phi(A)$. If A is an abelian group, let

$$T\mathrm{Hom}(A, \mathbb{Z}/p^\infty) = \operatorname*{colim}_n \mathrm{Hom}(A, \mathbb{Z}/p^n\mathbb{Z}).$$

This is the p-torsion in $\mathrm{Hom}(A, \mathbb{Z}/p^\infty)$.

LEMMA 4.2. *Let A be an abelian group. Then there is a natural isomorphism*

PROOF: For if we write $\Phi(A) = \{A_\alpha\}$ where A_α is a finite abelian p-group

$$\begin{aligned} T\mathrm{Hom}(A, \mathbb{Z}p^\infty) &\cong \operatorname*{colim}_n \mathrm{Hom}(A, \mathbb{Z}/p^n\mathbb{Z}) \\ &\cong \operatorname*{colim}_n \mathrm{Hom}_{\mathcal{A}b_p}(\Phi(A), \mathbb{Z}/p^n\mathbb{Z}) \\ &\cong \operatorname*{colim}_n \operatorname*{colim}_\alpha \mathrm{Hom}(A_\alpha, \mathbb{Z}/p^n\mathbb{Z}) \\ &\cong \operatorname*{colim}_\alpha \mathrm{Hom}(A_\alpha, \mathbb{Z}/p^\infty\mathbb{Z}) = \Phi(A)^\#. \end{aligned}$$ ∎

It follows that, in contrast to p-completion, p-pro-finite completion is right exact: indeed, since $\mathbb{Z}_{p^\infty}$ is an injective abelian group, $T\mathrm{Hom}(A, \mathbb{Z}/p^\infty)$ is left exact.

Now let $L_s\Phi : \mathcal{A}b \to \mathcal{A}b_p$ be the left derived functors of Φ. Since Φ is right exact, $L_0\Phi \cong \Phi$ and, of course, $L_s\Phi = 0$ for $s > 1$. If we let $T : \mathcal{A}b \to \mathcal{A}b$ be the functor which assigns to an abelian group its torsion sub-group, T is left exact and has right derived functors R^sT; again, $R^0T \cong T$ and $R^sT = 0$ for $s > 1$.

PROPOSITION 4.3. *For all $A \in \mathcal{A}b$, there is a natural isomorphism*

$$(L_s\Phi(A))^\# \cong (R^sT)\mathrm{Hom}(A, \mathbb{Z}/p^\infty).$$

PROOF: The case $s = 0$ is Lemma 4.2. Also, if A is free abelian, $\mathrm{Hom}(A, \mathbb{Z}_p^\infty)$ is an injective abelian group, so one has

$$L_s\Phi(A) = 0 = (R^sT)\mathrm{Hom}(A, \mathbb{Z}_{p^\infty})$$

for $s > 0$. The result follows from general homological algebra. ∎

EXAMPLE 4.4: If $A = \mathbb{Z}/p^\infty$, then $\mathrm{Hom}(A, \mathbb{Z}/p^\infty) \cong \mathbb{Z}_p$, the p-adic integers. The short exact sequence

$$0 \to \mathbb{Z}_p \to \mathbb{Q}_p \to \mathbb{Z}/p^\infty \to 0$$

show $(R^1T)\mathrm{Hom}(A, \mathbb{Z}/p^\infty) = \mathbb{Z}/p^\infty$ and $T\mathrm{Hom}(A, \mathbb{Z}/p^\infty) = 0$. Hence $\Phi(\mathbb{Z}/p^\infty) = 0$ and $L_1\Phi(\mathbb{Z}/p^\infty) \cong \mathbb{Z}_p$.

We would now like to compare p-pro-finite completion to the p-completion functor

$$\Psi(A) = \lim_n A/p^nA \cong \lim_n(\mathbb{Z}/p^nZ \otimes A).$$

Because lim is only left exact and tensor product is only right exact Ψ is neither left nor right exact; nonetheless, Ψ still has left derived functors $L_s\Psi$. While $L_s\Psi = 0$ for $s > 1$, one can only claim there is a natural surjection $L_0\Psi \to \Psi$. In fact, one has

LEMMA 4.5. *There is a natural exact sequence*

$$0 \to \lim^1 \mathrm{TOR}(\mathbb{Z}/p^n\mathbb{Z}, A) \to L_0\Psi(A) \to \lim(\mathbb{Z}/p^n\mathbb{Z} \otimes A) \to 0$$

and a natural isomorphism

$$\lim \mathrm{TOR}(\mathbb{Z}/p^n\mathbb{Z}, A) \cong L_1\Psi(A).$$

PROOF: See [8]. If A is free abelian group, $\lim^1(\mathbb{Z}/p^n\mathbb{Z} \otimes A) = 0$. This is a consequence of the exact sequence in $\lim^i$ and the short exact sequence of towers

$$\{0 \to A \xrightarrow{p^n} A \to \mathbb{Z}/p^n\mathbb{Z} \otimes A \to 0\}.$$

Let $F_* \to A$ be a projective resolution. Then there is a short exact sequence of chain complexes

$$0 \to \lim(\mathbb{Z}/p^n \otimes F_*) \to \prod_n (\mathbb{Z}/p^n\mathbb{Z} \otimes F_*) \xrightarrow{\partial} \prod (\mathbb{Z}/p^n\mathbb{Z} \otimes F_*) \to 0.$$

Analyzing the long exact sequence in homology completes the proof. ■

As a consequence one has the description of $L_s\Psi$ familiar to topologists (see [3], p. 166).

COROLLARY 4.6. *There are natural isomorphisms*

$$L_0\Psi(A) \cong \mathrm{EXT}(\mathbb{Z}/p^\infty\mathbb{Z}, A)$$
$$L_1\Psi(A) \cong \mathrm{Hom}(\mathbb{Z}/p^\infty\mathbb{Z}, A).$$

PROOF: There are natural isomorphisms $\mathrm{TOR}(\mathbb{Z}/p^n\mathbb{Z}, A) \cong \mathrm{Hom}(\mathbb{Z}/p^n\mathbb{Z}, A)$, so, by Lemma 4.5

$$L_1\Psi(A) \cong \lim \mathrm{Hom}(\mathbb{Z}/p^n\mathbb{Z}, A) \cong \mathrm{Hom}(\mathbb{Z}/p^\infty\mathbb{Z}, A).$$

Since $L_0\Psi(A) = 0$ if A is injective, the result follows. ■

Note that Corollary 4.6 also implies Lemma 4.5 using the composite functor spectral sequence

$$\lim^{(p)} \mathrm{EXT}^q(\mathbb{Z}/p^n\mathbb{Z}, A) \Rightarrow \mathrm{EXT}^{p+q}(\mathbb{Z}/p^\infty\mathbb{Z}, A)$$

To compare p-completion $\Psi(\cdot)$ to p-pro-finite completion Φ, note that any map $A \to A_\alpha$ from A to a finite abelian p-group factors through $\mathbb{Z}/p^n\mathbb{Z} \otimes A \to A_\alpha$ for some n. This yields a map of groups

$$\theta : \Psi(A) = \lim(\mathbb{Z}/p^n\mathbb{Z} \otimes A) \to |\Phi(A)| = \lim_\alpha A_\alpha. \tag{4.7}$$

One can include the topology: $\Psi(A)$, with the inverse limit topology, is a Hausdorff abelian group and the natural map of topological abelian groups in (4.7) is continuous.

This can be made more explicit: by Lemma 4.2,

$$\begin{aligned}\Phi(A) &= (T\mathrm{Hom}(A, \mathbb{Z}/p^\infty))^* \\ &= \lim_n \mathrm{Hom}(A, \mathbb{Z}/p^n\mathbb{Z})^*\end{aligned}$$

where $*$ is $\mathbb{Z}/p^\infty$ duality. Evaluation defines a homomorphism

$$\epsilon : \mathbb{Z}/p^n\mathbb{Z} \otimes A \to \mathrm{Hom}(A, \mathbb{Z}/p^n\mathbb{Z})^* \tag{4.8.1}$$

and the morphism θ of (4.7) is isomorphic to

$$\epsilon : \lim_n(\mathbb{Z}/p^n\mathbb{Z} \otimes A) \to \lim_n \mathrm{Hom}(A, \mathbb{Z}/p^n\mathbb{Z})^*. \tag{4.8.2}$$

This also makes it possible to describe behavior on derived functors. Since $(\cdot)^*$ is exact, evaluation (4.8.1) defines morphism

$$\mathrm{Tor}(\mathbb{Z}/p^n\mathbb{Z}, A) \to \mathrm{Ext}(A, \mathbb{Z}/p^n\mathbb{Z})^* \tag{4.8.3}$$

and one has

LEMMA 4.9. *The natural maps of derived functors $L_s\Psi(A) \to L_s\Phi(A)$ are isomorphic to*

$$\lim_n \mathrm{Tor}(\mathbb{Z}/p^n\mathbb{Z}, A) \to \lim_n \mathrm{Ext}(A, \mathbb{Z}/p^n\mathbb{Z})^* \tag{$s=1$}$$

and

$$L_0\Psi(A) \to \lim_n(\mathbb{Z}/p^n\mathbb{Z} \otimes A) \to \lim_n \mathrm{Hom}(A, \mathbb{Z}/p^n\mathbb{Z})^*. \tag{$s=0$}$$

PROOF: This follows from the formulas of 4.8, the proof of Lemma 4.5, and the fact that $\lim^1 \mathrm{Hom}(A, \mathbb{Z}/p^n\mathbb{Z})^* = 0$ since $\mathrm{Hom}(A, \mathbb{Z}/p^n\mathbb{Z})^*$ is a pro-finite group. ■

PROPOSITION 4.10. *Let A be an abelian group with the property that $\mathbb{Z}/p\mathbb{Z} \otimes A$ and $\mathrm{Tor}(\mathbb{Z}/p\mathbb{Z}, A)$ are finitely generated. Then $L_0\Psi(A) \cong \lim_n(\mathbb{Z}/p^n\mathbb{Z} \otimes A)$ and the natural maps*

$$L_s\Psi(A) \to L_s\Phi(A)$$

are isomorphisms.

PROOF: By induction on n, $\mathrm{Tor}(\mathbb{Z}/p^n\mathbb{Z}, A)$ is finite, hence $\lim^1 \mathrm{Tor}(\mathbb{Z}/p^n\mathbb{Z}, A) = 0$. So $L_0\Psi(A) \cong \lim_n(\mathbb{Z}/p^n\mathbb{Z} \otimes A)$ by Lemma 4.5. Also by induction on n, the maps

$$\epsilon : \mathrm{Tor}^s(\mathbb{Z}/p^n\mathbb{Z}, A) \to \mathrm{Ext}^s(\mathbb{Z}/p^n\mathbb{Z}, A)^*$$

are isomorphisms for all s. Hence the result follows from Lemma 4.9. ■

Examples of groups satisfying the hypothesis of Proposition 4.10 include finitely generated groups, their p-completions, and finite direct sums of the standard injectives $\mathbb{Q}, \mathbb{Q}/\mathbb{Z}, \mathbb{Z}_{p^\infty}, \mathbb{Q}_p$, etc.

The p-complete abelian groups do not form an abelian sub-category of the category of all abelian groups. The smallest abelian sub-category containing the p-complete groups is the category of Ext-p-complete groups (or p-cotorsion groups); that is, those abelian groups A so that $\mathrm{Ext}(\mathbb{Z}\left[\frac{1}{p}\right], A) = 0 = \mathrm{Hom}(\mathbb{Z}\left[\frac{1}{p}\right], A)$. A clear explanation of this difficulty is in [10, §4].

As a final note we record

LEMMA 4.11. *Let A be an abelian group.*

1) *The p-pro-finite completion ΦA of A is Ext-p-complete.*
2) *The group $L_1\Phi A$ is torsion free.*

PROOF: For 1) write $\Phi A = \lim_\alpha A_\alpha$ where A_α are finite abelian p-groups. Then there is a spectral sequence

$$\lim_\alpha^p \mathrm{Ext}^q(\mathbb{Z}\left[\frac{1}{p}\right], A_\alpha) \Rightarrow \mathrm{Ext}^{p+q}(\mathbb{Z}\left[\frac{1}{p}\right], \Phi A).$$

But the E_2 term is zero.

For 2) Proposition 4.[illegible] implies $L_1\Phi(A)^{\#}$ is divisible. Hence $L_1\Phi(A)$ is torsion

5 The homotopy groups of the p-pro-finite completion of a nilpotent space.

Let X be a pointed, connected nilpotent space. Then there is a tower of principal fibrations

$$\begin{array}{ccccccc} \cdots \longrightarrow X_s \longrightarrow X_{s-1} & & \cdots \longrightarrow X_2 \longrightarrow X_1 = K(A_1, n_1) \\ \uparrow & & \uparrow \\ K(A_s, n_s) & & K(A_2, n_2) \end{array}$$

equipped with a weak equivalence $X \to \lim_s X_s$ and so that $\lim_{s\to\infty} n_s = \infty$. Using this fact and the results of the previous section, we give a formula for calculating $\pi_*|X_p|$.

We begin with the case of an Eilenberg-MacLane space. Let $L_s\Phi$ denote the derived functors of p-pro-finite completion of abelian groups.

LEMMA 5.1. *Let A be an abelian group and $n \geq 1$. then there are natural isomorphisms*

$$\pi_n|K(A,n)_p| \cong \Phi(A)$$

$$\pi_{n+1}|K(A,n)_p| \cong L_1\Phi(A)$$

and $\pi_k|K(A,n)_p| = 0$ if $k \neq n$ or $n+1$.

PROOF: Recall that $K(A,n)$ is the simplicial abelian group whose normalization $NK(A,n) \cong A$ in degree n. Define a simplicial abelian group $K_c(A,n)$ (c for cofibrant) and a weak equivalence $K_c(A,n) \to K(A,n)$ as follows. Choose a free resolution

$$0 \to F_1 \to F_0 \to A \to 0$$

of A as an abelian group and let $K_c(A,n)$ be the simplicial abelian group with normalization on the chain complex $F_1 \to F_0$ with F_0 in degree n. Then there is a weak equivalence $K_c(A,n) \to K(A,n)$. Hence $K_c(A,n)_p \to K(A,n)_p$ is a weak equivalence. Let $\Phi K_c(A,n)$ be the simplicial p-pro-finite obtained by applying the p-pro-finite completion functor level-wise. Since normalization is exact, this result follows from the next. ■

LEMMA 5.2. *The simplicial p-pro-finite abelian group $\Phi K_c(A,n)$ is a model for the p-pro-finite completion of $K_c(A,n)$.*

PROOF: By Lemmas 1.18 and 1.19, $\Phi K_c(A,n)$ is fibrant. Thus we need only show that the map $K_c(A,n)^\wedge \to \Phi K_c(A,n)$ adjoint to $K_c(A,n) \to |\Phi K_c(A,n)|$ induces an isomorphism $H^* K_c(A,n) \cong H^*_{\text{cont}} \Phi K_c(A,n)$. To do this we introduce a spectral sequence.

If V is a cosimplicial vector space and $\Lambda(\cdot)$ is the exterior algebra functor, there is a functor $G_0(\cdot)$ on graded vector spaces so that

$$G_0(\pi^* V) \cong \pi^* \Lambda(V).$$

This is a result of Dold. Since Λ commutes with filtered colimits so does G_0. If B is a simplicial abelian group which is free in each degree, there is a spectral sequence

$$G_0(\pi^* \text{Hom}(B, \mathbb{Z}/p\mathbb{Z})) \Rightarrow H^* \overline{W} B. \tag{5.3}$$

Here $\overline{W} B$ is the "suspension functor" on simplicial abelian groups; on normalizations $(N\overline{W}B)_k = B_{k-1}$. In particular $\overline{W} K_c(A, n-1) \cong K_c(A,n)$. We will construct the spectral sequence below. Assuming this to be the case, consider the map $B^\wedge \to \Phi B$ adjoint to $B \to |\Phi B|$. Then, since $\mathbb{Z}/p\mathbb{Z}$ is a finite and, hence, p-pro-finite abelian group

$$\text{Hom}(B, \mathbb{Z}/p\mathbb{Z}) \cong \text{Hom}_{\text{cont}}(\Phi B, \mathbb{Z}/p\mathbb{Z}) \cong \operatorname*{colim}_\alpha \text{Hom}(B_\alpha, \mathbb{Z}/p\mathbb{Z}).$$

Hence one has a diagram of spectral sequences

$$\begin{array}{ccc} G_0(\pi^* \text{Hom}(B, \mathbb{Z}/p\mathbb{Z})) & \Longrightarrow & H^* \overline{W} B \\ \downarrow & & \downarrow \\ \operatorname*{colim}_\alpha G_0(\pi^* \text{Hom}(B_\alpha, \mathbb{Z}/p\mathbb{Z})) & \Longrightarrow & \operatorname*{colim}_\alpha H^* \overline{W} B_\alpha. \end{array}$$

Since G_0 commutes with filtered colimits, one has an isomorphism on E_2 terms. Since normalization is exact, $\Phi \overline{W} B \cong \overline{W} \Phi B$ and we conclude

$$H^* \overline{W} B \cong H^*_{\text{cont}} \Phi \overline{W} B.$$

Setting $B = K_c(A, n-1)$ proves the result.

To produce the spectral sequence (5.3), form the bisimplicial group $\overline{W}_\cdot B$ with the cochains functor Hom $=$ $\mathrm{Hom}_{\mathrm{sets}}(\cdot, \mathbb{Z}/p\mathbb{Z})$ to $\overline{W}_\bullet B$ and filtering degree in s yields a spectral sequence

$$\pi^s\pi^t\mathrm{Hom}(\overline{W}_\bullet B) \Rightarrow H^{s+t}\overline{W}B.$$

For fixed s, $\pi^*\mathrm{Hom}(\overline{W}B_s) \cong H^*(B_s) \cong \Lambda(\mathrm{Hom}(B_s, \mathbb{Z}/p\mathbb{Z}))$ since B_s is free. The spectral sequence follows. ∎

We now come to the analog of Bousfield and Kan's nilpotent fiber lemma.

PROPOSITION 5.4. *Let $F \to E \xrightarrow{q} B$ be a nilpotent fiber sequence of connected spaces. Then there is a homotopy fiber sequence*

$$F_p \to E_p \to B_p.$$

PROOF: One can argue exactly as in [3], Chap. II. Here is another way. By [3] II.4.7, the fibration $q : E \to B$ has a refined Postnikov tower

$$\begin{array}{ccccccccc} \cdots \longrightarrow & E_s & \xrightarrow{q_s} E_{s-1} \longrightarrow \cdots \longrightarrow & E_2 & \xrightarrow{q_2} & E_1 & \xrightarrow{q_1} E_0 = B \\ & \uparrow & & \uparrow & & \uparrow & \\ & K(A_s, n_s) & & K(A_2, n_2) & & K(A_1, n_1) & \end{array}$$

so that $E \simeq \lim E_s$ and $q \simeq \lim q_s$, each q_s is a principal fibration and $\lim\limits_{s\to\infty} n_s = \infty$. If F_s is the fiber of $E_s \to B_s$, one gets an induced tower

$$\begin{array}{ccccc} \cdots \longrightarrow & F_s & \xrightarrow{q_s} F_{s-1} \longrightarrow \cdots \longrightarrow & F_2 & \longrightarrow F_1 = K(A_1, n_1) \\ & \uparrow & & \uparrow & \\ & K(A_s, n_s) & & K(A_2, n_2) & \end{array}$$

with $\lim F_s \cong F$.

Choose a weak equivalence $B^\wedge \to B_p$ with B_p fibrant and factor $\widehat{E} \to \widehat{B} \to B_p$ as $\widehat{E} \xrightarrow{j} E_p \xrightarrow{f} B_p$ where j is a weak equivalence and f is a fibration. As the notation indicates E_p and B_p are models for the p-pro-finite completions. Let F' be the fiber of f. Then F' is automatically fibrant, so we need only show the induced map $\widehat{F} \to F'$ induces an isomorphism $H^*F \cong H^*_{\mathrm{cont}}F'$. Suppose we knew this to be true in the case where q is principal with fiber $K(A,n) = F$. Then, we can show inductively that there is a fibration sequence $(F_s)_p \to (E_s)_p \to B_p$. Indeed, with proper choices of fibrant models, one has a diagram with columns and top and bottom rows fibration sequences.

$$\begin{array}{ccccc}
K(A_s, n_s)_p & \xrightarrow{\simeq} & K(A_s, n_s)_p & \longrightarrow & * \\
\downarrow & & \downarrow & & \downarrow \\
(F_s)_p & \longrightarrow & (E_s)_p & \longrightarrow & B_p \\
\downarrow & & \downarrow & & \downarrow{=} \\
(F_{s-1})_p & \longrightarrow & (E_{s-1})_p & \longrightarrow & B_p
\end{array}$$

It follows that the middle row is also a homotopy fibration sequence. To see this, let $(F_s)'$ be the fiber of $(E_s)_p \to B_p$. Then there is a map of fibration sequences

$$\begin{array}{ccccc}
K(A_s, n_s)_p & \longrightarrow & (F_s)_p & \longrightarrow & (F_{s-1})_p \\
\downarrow{=} & & \downarrow & & \downarrow{=} \\
K(A_s, n_s)_p & \longrightarrow & (F_s)' & \longrightarrow & (F_{s-1})_p
\end{array}$$

whence, by the Serre Spectral Sequence in $\widehat{\mathbf{S}}$ (cf. [13], the final remark), we have that $H^*_{\text{cont}}(F_s)_p \cong H^*_{\text{cont}}(F'_s)$. Since both are fibrant $(F_s)_p \simeq (F_s)'$.

Finally, note that $\lim(E_s)_p$ is a model for E_p. This is because $\lim\limits_s(E_s)_p$ is fibrant and (because $\lim\limits_{s\to\infty} n_s = \infty$)

$$H^*E \cong \operatorname*{colim}_s H^*E_s \cong \operatorname*{colim}_s H^*_{\text{cont}}(E_s)_p \cong H^*_{\text{cont}}(\lim(E_s))_p.$$

Now the fiber of $\lim_s(E_s)_p \to B_p$ is $\lim(F_s)_p$. But $\lim(F_s)_p$ is a model for F_p, by the same reasoning and we are left only with the following lemma:

LEMMA 5.5. *Let $K(A,n) \to E \xrightarrow{q} B$ be a principal fibration. Then*

$$K(A,n)_p \to E_p \to B_p$$

is a fibration sequence in $\widehat{\mathbf{S}}$.

PROOF: Consider the principal fibration

$$K(A,n) \to WK(A,n) \to \overline{W}K(A,n).$$

Since $WK(A,n)$ is contractible Lemma 5.2 implies that the result holds in this simplicial p-pro-finite abelian groups. Let $B \to \overline{W}K(A,n)$ classify q, and define E' by the homotopy pull-back diagram in $\widehat{\mathbf{S}}$.

$$\begin{array}{ccc} E' & \longrightarrow & WK(A,n)_p \\ \downarrow & & \downarrow \\ B_p & \longrightarrow & \overline{W}K(A,n)_p. \end{array}$$

We need only show the induced map $\widehat{E} \to E'$ induces an isomorphism $H^*\widehat{E} \cong H^*_{\text{cont}}E'$. Then E' is a model for E_p and since the fiber of $E' \to B_p$ is $K(A,n)_p$, we will be done.

To finish the argument, consider the diagram

$$\begin{array}{ccccc} K(A,n)^\wedge & \longrightarrow & \widehat{E} & \longrightarrow & \widehat{B} \\ \downarrow & & \downarrow & & \downarrow \\ K(A,n)_p & \longrightarrow & E' & \longrightarrow & B_p. \end{array}$$

This induces a map of Serre Spectral Sequences

$$\begin{array}{ccc} H^*(B, H^*K(A,n)) & \Longrightarrow & H^*E \\ \downarrow & & \downarrow \\ H^*_{\text{cont}}(B_p, H^*_{\text{cont}}K(A,n)_p) & \Longrightarrow & H_{\text{cont}}E'. \end{array}$$

Since the E_2 terms are isomorphic, the result follows. ∎

With this result in hand, we can make use of the following definition.

DEFINITION 5.6. *Let G be a group. Define the derived functors of p-pro-finite completion by the formula*

$$|L_s\Phi(G)| = \pi_{s+1}|K(G,1)_p|.$$

By Lemma 5.1, this agrees with the usual definition if G is abelian.

Here is a model for $K(G,1)_p$. Choose a free simplicial resolution $X_\bullet \to G$ in the category of groups and let $\Phi X_\bullet$ denote the level-wise p-pro-finite completion of X. Now there is a weak equivalence $\overline{W}X_\bullet \simeq K(G,1)$. As a consequence of Proposition 2, §1.5 of [13], $\overline{W}\Phi X_\bullet$ is a fibrant. See Corollairé 1, §1.5 of [13]. Hence it is only a matter of showing that $H^*\overline{W}X_\bullet \cong H^*_{\text{cont}}\overline{W}\Phi X_\bullet$. Now, one has, by Kan [12], that

$$H^{n+1}\overline{W}X_\bullet \cong \pi^n\text{Hom}(X_\bullet, \mathbb{Z}/p\mathbb{Z}).$$

Hence

$$\begin{aligned} H^{n+1}\overline{W}X_\bullet &\cong \pi^n \mathrm{Hom}_{\mathrm{cont}}(\Phi X_\bullet, \mathbb{Z}/p\mathbb{Z}) \\ &\cong \operatorname*{colim}_\alpha \pi^n \mathrm{Hom}((X_\bullet)_\alpha, \mathbb{Z}/p\mathbb{Z})) \\ &\cong H^*_{\mathrm{cont}} \overline{W}\Phi X. \end{aligned}$$

Note that because Φ is a left adjoint and lim is exact on pro-finite groups,

$$L_0\Phi(G) = \pi_1|\overline{W}\Phi X_\bullet| \cong |\pi_0\Phi X_\bullet| \cong |\Phi G|. \tag{5.7}$$

LEMMA 5.8. *1) If G is a nilpotent group, $L_s\Phi(G) = 0$ for $s > 1$.*
2) If $1 \to K \to G \to H \to 1$ is a short exact sequence of nilpotent groups, there is a long exact sequence

$$0 \to L_1\Phi K \to L_1\Phi G \to L_1\Phi H \to \Phi K \to \Phi G \to \Phi H \to 1.$$

PROOF: First consider the short exact sequence of the second statement. Then

$$K(K,1) \to K(G,1) \to K(H,1)$$

is a nilpotent fibration sequence. So there is a long exact sequence in $L_*\Phi(\cdot)$. If G is nilpotent, one can use this long exact sequence to induct over the lower central series to show $L_s\Phi(G) = 0$ for $s > 2$. ■

PROPOSITION 5.9. *Let X be a connected nilpotent space. Then there is a splittable short exact sequence for all $n \geq 1$*

$$0 \to \Phi(\pi_n X) \to \pi_n|X_p| \to L_1\Phi(\pi_{n-1}X) \to 0.$$

PROOF: The existence of the short exact sequence follows from the refined Postnikov tower for X, and Lemmas 5.1 and 5.8, and the nilpotent fiber lemma Proposition 5.4. The fact that sequence splits follows from [9, p. 370] because $\Phi(\pi_n X)$ is EXT-p-complete and $L_1\Phi(\pi_{n-1}X)$ is torsion free. See Lemma 3.11. Note that if G is nilpotent $L_1\Phi(G)$ is torsion free by Lemma 5.8.2 and Lemma 3.11. ■

If X is connected and nilpotent, and $(\mathbb{F}_p)_\infty X$ is its Bousfield-Kan completion, there is a splittable short exact sequence ([3], p. 183) where $L_s\Psi$ are the derived functors of p-completion. It is an easy exercise to show that the induced map $\pi_*(\mathbb{F}_p)_\infty X \to \pi_n|X_p|$ fits into a map of short exact sequence 5.10 to 5.9 with the natural maps $L_s\Psi(\pi_n X) \to L_s\Phi(\pi_n X)$ on the ends.

6 The homology of the p-pro-finite completion.

Let X be a space. One would like the mod p homology $H_*|X_p|$ as a functor of H_*X, at least if X is connected and nilpotent. Since $H^*_{\text{cont}}X_p \cong H^*X$, one can subsume this question into the larger problem of computing $H_*|Y|$ as a functor of $H^*_{\text{cont}}Y$. This seems to be a difficult problem. This section is devoted to what I know, and is mostly an outgrowth of previous work on the homology of homotopy inverse limits [7].

We begin with some algebra. A *pro-finite graded* (cocommutative) *coalgebra* $C = \{C_\alpha\}$ is a filtered system of graded coalgebras C_α which are finite in each degree. The category $\mathcal{CA}_{pf}$ of pro-finite graded coalgebras has these coalgebras as objects and pro-morphisms (as in 1.1) as morphisms. If $X = \{X_\alpha\} \in \widehat{\mathbf{S}}$, then $H_*X = \{H_*X_\alpha\} \in \mathcal{CA}_{pf}$.

The category $\mathcal{CA}_{pf}$ is equivalent, via a Pontrjagin style duality, to the category $\mathcal{A}$ of graded commutative algebras over $\mathbb{F}_p$. Indeed, if $C = \{C_\alpha\} \in \mathcal{CA}_{pf}$, then

$$(6.1.1) \qquad C^{\#} = \mathrm{Hom}_{\text{cont}}(C, \mathbb{F}_p) = \operatorname*{colim}_\alpha C^*_\alpha$$

is in $\mathcal{A}$, and if $A \in \mathcal{A}$, then

$$(6.1.2) \qquad A^* = \{A^*_\alpha\} \in \mathcal{CA}_{pf}$$

where $A_\alpha \subseteq A$ runs over the sub-algebras of finite type. These two dualization functors define the equivalence of categories. Note that this implies that if $X \in \widehat{\mathbf{S}}$, then $H_*X \in \mathcal{CA}_{pf}$ is determined, up to isomorphism in $\mathcal{CA}_{pf}$, by the algebra

$$H^*_{\text{cont}}X \cong (H_*X)^{\#}$$

in $\mathcal{A}$. Thus we may recast the question that opened this section as follows: if $Y \in \widehat{\mathbf{S}}$ is fibrant, can one compute $H_*|Y|$ as a functor of $H_*Y \in \mathcal{CA}_{pf}$? The answer we give is an attenuated affirmative one.

Incidentally, the fibrancy condition of Y is to guarantee that the weak equivalence class of $|Y|$, and hence the coalgebra $H_*|Y|$, depends only on the weak equivalence class of Y in $\widehat{\mathbf{S}}$.

To begin with, let $C = \{C_\alpha\} \in \mathcal{CA}_{pf}$. Then we can form the limit $\lim\limits_{\alpha} C_\alpha$ in the category $\mathcal{CA}$ of graded commutative coalgebras. This limit is only a sub-object of the limit $\lim\limits_{\alpha}{}^{\mathbb{F}_p} C_\alpha$ of graded vector spaces; the inverse limit functor need not be exact, only left exact. See Example 6.12 below. If $Y \in \widehat{\mathbf{S}}$, then there is a natural map

$$\eta : H_*|Y| = H_* \lim Y_\alpha \to \lim H_* Y_\alpha \tag{6.2}$$

and one might hope that this is an edge homomorphism in a spectral sequence whose E_2 term depends on right derived functors of the limit functor. This is what I wish to explain.

The first point to be made is that the limit functor $\lim : \mathcal{CA}_{pf} \to \mathcal{CA}$ has suitably defined right derived functors. Since $\mathcal{CA}_{pf}$ is not an abelian category, one would expect non-abelian derived functors defined with the aid of a cosimplicial resolution. To facilitate this one has the next result. Let $c\mathcal{CA}_{pf}$ be the category of cosimplicial objects in $\mathcal{CA}_{pf}$.

LEMMA 6.3. *The category $c\mathcal{CA}_{pf}$ has the structure of a simplicial model category where a morphism $f : C^\bullet \to D^\bullet$ is*

1) *a weak equivalence if $\pi^* f : \pi^* C^\bullet \to \pi^* D^\bullet$ is an isomorphism, and*
2) *a cofibration if the normalized cochain complex $Nf : NC^\bullet \to ND^\bullet$ is an injection above cochain degree 0.*

The simplicial structure is determined by

$$\hom(K, C^\bullet)^n \cong \prod_{x \in K^n} C^n$$

where $K \in \mathbf{S}$ and $C^\bullet \in \mathcal{CA}_{pf}$, and the induced coface and codegeneracy maps.

PROOF: The opposite category of $\mathcal{CA}_{pf}$ is equivalent to the category $\mathcal{A}$ of graded commutative algebras over $\mathbb{F}_p$. The result follows by interpreting Quillen's standard simplicial model category structure on the category $s\mathcal{A}$ of simplicial objects in $\mathcal{A}$.

DEFINITION 6.4. *Let $C \in \mathcal{CA}_{pf}$. The total right derived functor* $\mathbf{R}\lim$ *is defined by the formula*

$$\mathbf{R}\lim C = \lim D^{\bullet}$$

where $C \to D^{\bullet}$ is a weak equivalence in $c\mathcal{CA}_{pf}$ to a fibrant object, and lim *is applied in each cosimplicial degree. Thus* $\mathbf{R}\lim C$ *is a cosimplicial coalgebra and the derived functors are defined by*

$$R^s \lim C = \pi^s \mathbf{R}\lim C. \qquad \blacksquare$$

As usual $R^* \lim C$ is well-defined up to isomorphism. Also, since $\lim D^{\bullet}$ is a cosimplicial coalgebra, $R^* \lim C$ is a bigraded, cocommutative coalgebra.

To interpret this object in homotopy theory one needs:

LEMMA 6.5. *Let $V = \{V_\alpha\}$ be a simplicial pro-finite vector space. Then the natural map*

$$\eta : H_*|V| \to \lim H_* V_\alpha$$

is an isomorphism.

PROOF: This is a consequence of Lemma 3.9 and the fact that if W is any simplicial vector space, then the coalgebra H_*W is a functor of π_*W. ■

The main result of this section is the following. No assertion is made here about convergence. This question will be addressed in the remark following.

PROPOSITION 6.6. *Let $Y \in \widehat{\mathbf{S}}$ be fibrant. Then there is a second quadrant homology spectral sequence*

$$R^* \lim H_*Y \Rightarrow H_*|Y|.$$

PROOF: Let $Y \to \overline{\mathbb{F}}_p^{\bullet} Y$ be the cosimplicial resolution of section 2. See 2.2. Then there is a second-quadrant homology spectral sequence

$$\pi^s H_t|\overline{\mathbb{F}}_p^{\bullet} Y| \Rightarrow H_{t-s}\mathrm{Tot}|\overline{\mathbb{F}}_p^{\bullet} Y|.$$

We need only interpret the E_2-term and the abutment. For the latter, we have, by Proposition 3.5, that $|Y| \to \mathrm{Tot}|\overline{\mathbb{F}}_p^{\bullet} Y| \cong |\mathrm{Tot}(\overline{\mathbb{F}}_p^{\bullet} Y)|$ is a weak equivalence. For the former, we have that

$$H_*|\overline{\mathbb{F}}_p^{\bullet} Y| \cong \lim H_*\overline{\mathbb{F}}_p^{\bullet} Y,$$

by the previous result; hence, we need only assert that $H_*\overline{\mathbb{F}}_p^\bullet Y$ is fibrant in $c\mathcal{CA}_{pf}$. If $(\cdot)^\#$ is the duality functor of 6.1.1, then one need only assert that

$$(H_*\overline{\mathbb{F}}_p^\bullet Y)^\# \cong H^*_{\text{cont}}\overline{\mathbb{F}}_p^\bullet Y$$

is cofibrant in the category of simplicial commutative algebras. This is well-known, and essentially follows from the fact that the cohomology of an Eilenberg-MacLane space is a free commutative algebra. (See [6], among many sources.) ■

REMARK 6.7: Convergence is always a problem with this type of spectral sequence. Since we are especially interested in the case of non-finite type spaces, the best source is probably [2, §3]. In particular, each of the spaces in the cosimplicial space $|\mathbb{F}_p^\bullet Y|$ is p-nilpotent, so if

$$[R^s \lim H_*X]_t = 0$$

for $t-s \leq 1$, and for each s there are only finitely many k so that $[R^s \lim H_*X]_{s+k} = 0$, the spectral sequence will converge strongly.

EXAMPLE 6.8: There are times when $R^s \lim C = 0$ for $s > 0$. For example suppose $C = \{C_\alpha\} \in \mathcal{CA}_{pf}$ is a diagram of coalgebras obtained from a diagram of connected, bicommutative Hopf algebras by forgetting the algebra structure. Then the techniques of [7] and the fact that limit is exact on pro-finite groups implies $R^s \lim C = 0$ $s \geq 1$. It is always true that $R^0 \lim C \cong \lim C$.

This applies to the spectral sequence of Proposition 6.6 as follows. Let $Y = \{Y_\alpha\} \in \widehat{\mathbf{S}}$ be pointed and fibrant and suppose each Y_α is 2-connected. Then $\text{map}_*(S^2, Y) = \{\text{map}_*(S^2, Y_\alpha)\}$ is also fibrant and $H_*\text{map}_*(S^2, Y_\alpha) \in \mathcal{CA}_{pf}$ is a diagram underlying a diagram of connected bicommutative Hopf algebras. The spectral sequence will converge if $[\lim H_*\text{map}_*(S^2, Y)]_1 = 0$, by Bousfield's result. Actually, it should converge without the last assumption, but I've not looked very hard for a proof.

If $X \in \mathbf{S}$ is 2-connected, then there exists a model for $X_p = \{X_\alpha\} \in \widehat{\mathbf{S}}$ so that each X_α is 2-connected. See the construction of fibrations given by Morel in [13], for example.

We now interpret Proposition 6.6 in the case where $Y = X_p$ is the p-pro-finite completion of a space X. This amounts to providing an interpretation of the pro-we may define $\Phi C \in \mathcal{CA}_{pf}$ to be the diagram $\{C_\alpha\}$ obtained from taking finite type coalgebras under C. Thus one considers the category with objects surjective coalgebra maps $C \to C_\alpha$ and morphisms commutative triangles and obtains ΦC by sending $C \to C_\alpha$ to C_α. There is a natural map $C \to \lim \Phi C$; this may be regarded as a pro-finite completion of C in the category of coalgebras.

The description of $(H^*X)^*$ given in (6.1.2) implies

$$(H^*X)^* \cong \Phi H_*X$$

and Proposition 6.6 implies there is a spectral sequence

$$R^* \lim \Phi H_*X \Rightarrow H_*|X_p|. \tag{6.9}$$

EXAMPLE 6.10: Suppose $Z \in \mathbf{S}$ is pointed, 2-connected, and fibrant. Then the space $\text{map}_*(S^2, Z_p)$ is a model for $\text{map}_*(S^2, Z)_p$. We know $\text{map}_*(S^2, Z_p)$ is fibrant and

$$\text{map}_*(S^2, Z) \to \text{map}_*(S^2, Z_p)$$

induces an isomorphism $H^*_{\text{cont}}\text{map}_*(S^2, Z_p) \cong H^*\text{map}_*(S^2, Z)$, by the Serre Spectral Sequence. By Example 6.7,

$$R^s \lim H_*\text{map}_*(S^2, Z_p) = 0$$

for $s > 0$. Also $\lim H_*\text{map}_*(S^2, Z_p) = \lim \Phi H_*\text{map}_*(S^2, Z_p)$. Thus, if Z is 3-connected (see Example 6.7 for this hypothesis),

$$H_*\text{map}_*(S^2, Z_p) \cong \Phi H_*\text{map}_*(S^2, Z_p). \qquad \blacksquare$$

EXAMPLE 6.11: Some sort of hypothesis is needed to guarantee convergence. If X is finite type, then ΦH_*X has a final object—namely, H_*X itself and one easily checks that this implies $R^s \lim \Phi H_*X = 0$ for $s > 0$ and $\lim \Phi H_*X = H_*X$. But one cannot conclude from (6.9) that $H_*|X_p| \cong H_*X$, without some further hypothesis.. By the results of section 2, $|X_p|$ is weakly equivalent to Bousfield-Kan completion $(\mathbb{F}_p)_\infty X$ and there are spaces X for which $H_*X \not\cong H_*(\mathbb{F}_p)_\infty X$. ■

EXAMPLE 6.12: We show, by example, that $\lim : \mathcal{CA}_{pf} \to \mathcal{CA}$ is not right exact. Let C_n, $n \geq 1$, be the connected coalgebra so $(C_n)_k = 0$, $(C_n)_2 \cong \mathbb{F}_p$ with generator

z, $(C_n)_1 \cong \mathbb{F}_p^{2n}$ with generators $x_1, x_2, \ldots, x_n, y_1, y_2 \ldots, y_n$, and the only non-trivial diagonal given by

$$\Delta z = z \otimes 1 + 1 \otimes z + \sum_{i=1}^{n} x_i \otimes y_i + y_i \otimes x_i.$$

Thus C_n is the homology of an n-holed torus. There are quotient maps $C_n \to C_{n-1}$ sending z to z and x_i, y_i to the like-named object, except x_n, y_n go to zero. Let $C = \{C_n\} \in \mathcal{CA}_{pf}$. It is a simple calculation to show the inverse limit $(\lim C_n)_k = 0$ for $k \geq 2$. In particular if $D = H_*S^2$ and $C_n \to H_*S^2$ is the non-trivial map, then $C \to D$ is surjective in $\mathcal{CA}_{pf}$, but

$$\lim C \to \lim D \cong H_*S^2$$

is not.

References

1. A.K. Bousfield, "The localization of spaces with respect to homology," *Topology* 14 (1975), pp. 133-150.

2. A.K. Bousfield, "On the homology spectral sequence of a cosimplicial space", *Amer. J. of Math.* 109 (1987), pp. 361-394.

3. A.K. Bousfield and D.M. Kan, *Homotopy Limits, Completions and Localizations*, Lecture Notes in Mathematics 304, Springer-Verlag 1972.

4. E. Dror, "Pro-nilpotent representations of homology types", *Proc. AMS* 38 (1973), pp. 657-660.

5. W.G. Dwyer and D.M. Kan, *Model Categories and General Abstract Homotopy Theory*, in preparation.

6. P.G. Goerss, "André-Quillen cohomology and the Bousfield-Kan spectral sequence," *Astérisque* 191 (1990), pp. 109-209.

7. P.G. Goerss, "On the homology of homotopy inverse limits," to appear in *J.P.A.A.*.

8. J.P.C Greenlees and J.P. May, "Derived functors of I-adic completion and local cohomology", *J. of Algebra* 149 (1992) no. 2, pp. 438-453.

9. D.K. Harrison, "Infinite abelian groups and homological methods", Annals of Math. 69 (1956), pp. 366-391.

10. U. Jannsen, "Continuous étale cohomology", *Math. Ann.* 280 (1988), pp. 207-245.

11. C.U. Jensen, "On the vanishing of $\lim^{(i)}$", *J. of Algebra* 15 (1970), pp. 151-166.

13. F. Morel, "Ensembles profinis simpliciaux et interprétation géométrique du foncteur T", *Bulletin de la SMF* 124 (1996) fac. 2, pp. 347-373.

14. D.G. Quillen, "An application of profinite simplicial groups", *Comm. Math. Helv.* 44 (1969), pp. 45-60.

15. D.G. Quillen, *Homotopical Algebra*, Lecture Notes in Mathematics 43, Springer-Verlag, Berlin 1967.

Contemporary Mathematics
Volume **220**, 1998

Toward the homotopy groups of the higher real K-theory EO_2

Vassily Gorbounov and Peter Symonds

ABSTRACT. In this note we compute the initial term of the spectral sequence which converges the homotopy groups of the higher real K-theory EO_2 at the prime 3. We also describe precisely the way in which the integral modular forms are embedded in this initial term.

1. Introduction

Our main result is the calculation of the initial term of the spectral sequence $H^*(G,E)^{\mathrm{Gal}} \Rightarrow \pi_* EO_2$, which converges to the homotopy groups of the higher real K-theory EO_2 at the prime 3. Here G is the group $\mathbb{Z}/3 \rtimes \mathbb{Z}/4$ and E is an infinitely generated module for G over $\mathbb{Z}_3$, which arises from the theory of formal groups. We also show how the integral modular forms of Deligne appear naturally in this initial term. This calculation was originally sketched by Hopkins and Miller [**5**], but the details were never published, so we have proceeded by our own methods.

For each prime p and natural number n there is a p-adic Lie group $\mathbb{S}_n$, which is the group of absolute endomorphisms of the Honda formal group law $\bar{F}$ over $\mathbb{F}_p$. There is also a ring E_n^0, which can be considered as the moduli space of lifts of $\bar{F}$ to a formal group law of height n over a p-local ring. Thus $\mathbb{S}_n$ acts on E_n^0. This action is extended to an action on $E_n = E_n^0[u, u^{-1}]$. According to the Morava Change of Rings Theorem, there is a spectral sequence $H_c^*(\mathbb{S}_n, E_n)^{\mathrm{Gal}} \Rightarrow \pi_* L_{K(n)}$, where $L_{K(n)}$ is the localization of the sphere spectrum at the nth Morava K-theory $K(n)$ [**7**].

Morava [**10**] introduced a spectrum $\mathbb{E}_n$ such that $\pi_*\mathbb{E}_n = E_n$. Later, Hopkins and Miller [**5**] sketched how to construct a nice action of $\mathbb{S}_n$ on $\mathbb{E}_n$ that induces the original action on E_n.

1991 *Mathematics Subject Classification.* Primary 55N20, 55S25, 55Q10, 55T15.

Key words and phrases. Fixed Point Spectral Sequence, Morava Stabilizer Algebra, Homotopy Groups, Modular Forms.

It is known that if $n = p-1$, then $\mathbb{S}_n$ has a unique maximal finite subgroup G_n, up to conjugation. When $p = 3$ and $n = 2$, this is the group G of order 12 mentioned above. Hopkins and Miller define EO_n to be the homotopy fixed point spectrum of the action of G_n on $\mathbb{E}_n$, so there is a spectral sequence $H^*(G_n, E_n)^{\mathrm{Gal}} \Rightarrow \pi_* EO_n$. When $p = 2$ and $n = 1$, EO_1 turns out to be KO, ordinary real K-theory, whilst $K(1)$ is ordinary complex K-theory. This justifies the name "higher real K-theories". When $n = 2$ this construction can be generalized to give an integral version of elliptic cohomology, which agrees with the usual elliptic cohomology when 6 is inverted [**6**]. The spectrum which represents this cohomology theory is called the spectrum of topological modular forms, TMF.

The calculation presented here is for the case $p = 3$ and $n = 2$, and can be viewed as a computation of the initial term of the spectral sequence which converges to $\pi_* TMF_{(3)}$. It has the interesting property that $(E^G)^{\mathrm{Gal}}$, which is part of the initial term, is related to the ring of integral modular forms of Deligne [**1**]. A problem here is that the Morava Change of Rings Theorem concerns a specific formal group of height n over the field $\mathbb{F}_p$, the Honda formal group. When $n = 2$, a supersingular elliptic curve yields another formal group, which is naturally connected to the integral modular forms. These two formal groups only become isomorphic after a field extension.

The authors wish to thank Mark Mahowald for helpful discussions.

2. The Moduli Space

As a general reference for this section we recommend [**3**]. Let F_{u_1} be the formal group law over $\mathbb{Z}_3$ which is obtained from the usual universal commutative 3-typical formal group law over $\mathbb{Z}_3[v_1, v_2, \ldots]$ by setting $v_1 = u_1$, $v_2 = 1$, $v_i = 0$, $i \geq 3$.

$$F_{u_1}(x, y) = x + y - u_1(x^2y + xy^2) - (x^6y^3 + x^3y^6) + \ldots.$$

Its reduction modulo 3 and u_1 is

$$\bar{F}(x, y) = x + y - (x^6y^3 + x^3y^6) + \ldots.$$

Now $\mathbb{S}_2 \subset \mathbb{F}_9[[x]]$ is the ring of absolute endomorphisms of $\bar{F}$: it also can be characterized as the group of units in the maximal order O of a certain division algebra. This O, in turn, can be described as follows. Let W be the ring of Witt vectors over $\mathbb{F}_9$, ie the integers of the unramified extension of degree 2 of $\mathbb{Q}_3$. Then $\mathbb{S}_2$ contains an element S such that, additively, $O \cong W \oplus SW$, and the multiplication is described by $S^2 = 3$, $wS = S\bar{w}$, $w \in W$. Thus $\mathbb{S}_2$ contains a finite group $G \cong \mathbb{Z}/3 \rtimes \mathbb{Z}/4$ generated by $a = \dfrac{\zeta S - 1}{2}$, of order 3, and $b = \zeta^2$, of order 4, where $\zeta \in W$ is a primitive 8th root of unity.

Now we must identify these elements in $\mathbb{F}_9[[x]]$. Any $w \in W$ corresponds to $f^{-1}(wf(x))$, where $f(x)$ is the logarithm of $\bar{F}$,

$$f(x) = x + \tfrac{1}{3}x^9 + \tfrac{1}{9}x^{81} + \ldots.$$

Therefore $f(\zeta x) = \zeta f(x)$, and so multiplication by ζ corresponds to ζx and b corresponds to $\zeta^2 x$. Now S corresponds to x^3 and $\zeta S - 1$ has series

$$\zeta x^3 +_{\bar{F}} (-x) = -x + \zeta x^3 + \zeta^3 x^{15} + \ldots.$$

Also $[2]_{\bar{F}} = x +_{\bar{F}} x = -x + x^9 + \ldots$, so $[\frac{1}{2}]_{\bar{F}} = -x - x^9 - x^{81} - \ldots$ and thus a has series $x - \zeta x^3 + x^9 - \zeta^3 x^{15} + \ldots$.

A theorem of Lubin and Tate [**9, 3**] shows that the group S_2 acts continuously on $E_0 = W[[u_1]]$ according to the formula

$$F_{u_1}(\hat{g}\alpha x, \hat{g}\alpha y) = \hat{g}\alpha F_{g(u_1)}(x, y),$$

where $g \in S_2 \subset \mathbb{F}_9[[x]]$, $\hat{g}$ is a lift of g to $E_0[[x]]$, and $\alpha \in E_0[[x]]$ is such that its reduction to $\mathbb{F}_9[[x]]$ is the identity. Here $g(u_1) \in E_0$. Both $g(u_1)$ and $\hat{g}\alpha$ are uniquely determined by this equation, which we shall simplify by writing $\tilde{g} = \hat{g}\alpha$ so that

$$F_{u_1}(\tilde{g}x, \tilde{g}y) = \tilde{g}F_{g(u_1)}(x, y).$$

The action can be extended to $E = E_0[u, u^{-1}]$ by the formula $g(u) = (\tilde{g}'(0))u$.

PROPOSITION 2.1. *Suppose that* $g \in S_2$, $\tilde{g} = \gamma_1 x + \gamma_2 x^2 + \gamma_3 x^3 + \ldots$, $\gamma_i \in E_0$. *Then* $g(u_1) = \gamma_1^2 u_1 + 3\gamma_3\gamma_1^{-1}$ *and* $\gamma_2 = 0$ *in* E_0.

PROOF. Notice that

$$F_{u_1}(\tilde{g}x, \tilde{g}y) = \tilde{g}x + \tilde{g}y - u_1\gamma_1^3(x^2y + xy^2)$$

modulo degree 4 in x and y and

$$\tilde{g}F_{g(u_1)}(x, y) = \gamma_1(x + y - g(u_1)(x^2y + xy^2)) +$$
$$\gamma_2(x^2 + y^2 + 2xy) + \gamma_3(x^3 + y^3 + 3(x^2y + xy^2))$$

modulo degree 4 in x and y. Examing the coefficients of x^2y and xy yields the result. □

COROLLARY 2.2. *All of* $\mathbb{S}_2$ *fixes* $u^{-2}u_1 \in \bar{E}$.

3. The Action of G on E

Let $\bar{E} = \mathbb{F}_9 \otimes_W E = \mathbb{F}_9[[u_1]][u, u^{-1}]$.

PROPOSITION 3.1. *In* $\bar{E}$ *we have:*

$$a(u_1) = u_1 + \zeta^3 u_1^2 \mod u_1^3$$

and

$$a(u) = u(1 - \zeta^3 u_1) \mod u_1^2.$$

PROOF. Let $\tilde{a}(x) = \alpha_1 x + \alpha_2 x^2 + \ldots$, $\alpha_i \in \bar{E}_0$. Then $\alpha_1 = 1$, $\alpha_3 = -\zeta$, $\alpha_9 = 1$ modulo u_1, and the other α_i, $i \leq 9$ satisfy $\alpha_i = 0$ modulo u_1. We work in $\bar{E}$ throughout.

$$F_{u_1}(\tilde{a}(x), \tilde{a}(y)) = \tilde{a}x + \tilde{a}y - u_1[(x - \zeta x^3)^2(y - \zeta y^3) + (x - \zeta x^3)(y - \zeta y^3)^2] -$$
$$\alpha_1^9(x^6y^3 + x^3y^6)$$

modulo u_1^2 and degree 4 in x and y.

The coefficient of x^6y^3 is $-1 + \zeta^3 u_1$ modulo u_1^2. Now

$$\tilde{a}F_{a(u_1)}(x, y) = \tilde{a}[x + y - a(u_1)(x^2y + xy^2)] - \alpha_1(x^6y^3 + x^3y^6)$$

powers 1, 3 and 9. It is easy to check that these do not infact do so, hence the coefficient of x^6y^3 is $-\alpha_1$. We have $-\alpha_1 = -1+\zeta^3 u_1 \bmod u_1^2$ and so $\alpha_1^2 = 1+\zeta^3 u_1 \bmod u_1^2$. The result follows from Proposition 2.1 and the definition of the action on u. □

PROPOSITION 3.2. *We have* $b(u_1) = -u_1$, $b(u) = \zeta^2 u$ *in* E.

PROOF. $\zeta F_{\zeta^2 u_1}(x,y) = F_{u_1}(\zeta x, \zeta y)$ by construction. So $\tilde{b}(x) = \zeta^2 x$. □

4. The Module Structure of $\bar{E}$

Let $k = \mathbb{F}_9$ and consider the group algebra of $\langle a\rangle$. Let $A = a-1 \in k\langle a\rangle$. Then $k\langle a\rangle = k[A]/(A^3)$. We know that $a(u_1) = u_1 + \zeta^3 u_1^2 + Xu_1^3$, $X \in \bar{E}_0$, so $Au_1 = \zeta^3 u_1^2 + Xu_1^3$, and $A^2 u_1 = -\zeta^2 u_1^3 \bmod \deg 4$.

Let $y = A^2 u_1 \neq 0$. Since y is invariant under a, so is $k[[y]] \subset \bar{E}_0$. Let F be the $k\langle a\rangle$-submodule generated by u_1, which has k-basis u_1, Au_1, A^2u_1. Then as F has dimension 3 it must be free and $\bar{E}_0 = k \oplus (F \otimes k[[y]])$. Note that $b(y) = -y$, and that F is easily checked to be a kG-module.

It will be convenient for us to discuss the ring $\mathbb{L}E = E[\frac{1}{u_1}]$, and similarly $\mathbb{L}\bar{E}$. Note that $\mathbb{L}\bar{E}_0 = F \otimes k[[y]][\frac{1}{y}]$.

By Proposition 2.1, $u^2\mathbb{L}\bar{E}_0$ contains an invariant element $u^2u_1^{-1}$, and multiplication by $u^2u_1^{-1}$ yields an isomorphism of kG-modules $\mathbb{L}\bar{E}_0 \to u^2\mathbb{L}\bar{E}_0$. The invariants of $u^2\mathbb{L}\bar{E}_0$ under $\langle a\rangle$ are therefore $u^2u_1^{-1}k[[y]][\frac{1}{y}]$ so occur in u_1-filtration -1 modulo 3. Now if $w \in u\mathbb{L}\bar{E}_0$ is invariant under a and is an eigenvector for b, then so is w^2, and so w must have u_1 filtration 1 modulo 3. After multiplying by a suitable power of y we may assume that w is in filtration 1, so its leading term is uu_1 and the eigenvalue under the action of b must be $-\zeta^2$, ie $b(w) = -\zeta^2 w$. Multiplication by w^r yields an isomorphism of $k\langle a\rangle$-modules $\mathbb{L}\bar{E}_0 \to u^r\mathbb{L}\bar{E}_0$.

Finally set $x = y^{-1}w^3$, which has u_1-filtration 0. To sum up we have:
$A = a-1$, $y = A^2u_1$, $Aw = Ax = Ay = 0$, and $bu_1 = -u_1$, $bw = -\zeta^2 w$, $bx = -\zeta^2 x$, $by = -y$.

This is represented pictorially in figure 1. The blocks in unbroken lines represent indecomposable summands of $\bar{E}$ as a kG-module, and each block has a k-basis with u_1-filtration and u-grading corresponding to the points in the block with integral rectangular coordinates. The dotted lines indicate the blocks of $\mathbb{L}\bar{E}$. Each complete block of dimension 3 is projective.

The picture has vertical periodicity 3 as an $\langle a\rangle$-module (given by multiplication by x) and periodicity 12 as a G-module (multiplication by x^4).

This decomposition into blocks also lifts to E and to $\mathbb{L}E$. This is because there is a one-to-one correspondence between the finite dimensional indecomposable modules for G over W and over k. It can be seen explicitly as follows: Let m be the left-most basis element in of any block of $\mathbb{L}\bar{E}$. Lift m to an element $\tilde{m}$ of $\mathbb{L}E$ that is an eigenvector under the action of b. Then the WG-lattice generated by $\tilde{m}$ is a direct summand of $\mathbb{L}E$ that is equal to the original block modulo 3.

Let B denote the submodule of $\bar{E}$ with basis $\{1, wy^{-1}Au_1, w\}$. In other words B consists of the two incomplete blocks in u-grading 0 and 1. Then $\bar{E} = B \otimes k[x, \frac{1}{x}] \oplus \Pi(\text{projective})$.

Now $x \in \bar{E}$ lifts to $\tilde{x} \in E$ such that $a\tilde{x} = \tilde{x}$, $b\tilde{x} = -\zeta^2\tilde{x}$, and B lifts to $\tilde{B}$. Thus $E = (\tilde{B} \otimes \mathbb{Z}_3[\tilde{x}, \frac{1}{\tilde{x}}]) \oplus \Pi(\text{projective})$.

REMARK 4.1. *We have a product of projectives because E involves power series in u_1. Over k a product of projectives is also projective, but over W this is not true, since since it is not even projective over W (According to* **[8]**, *$\Pi_\Lambda \mathbb{Z}_p$ is not a free $\mathbb{Z}_p$-module when Λ is infinite). However Π(projective) is at least cohomologically trivial.*

5. Cohomology

In Tate cohomology, $\hat{H}^*(\langle a \rangle, E) = \hat{H}^*(\langle a \rangle, B) \otimes \mathbb{Z}_3[\tilde{x}, \tilde{x}^{-1}]$, where we identify $\tilde{x}$ with its image in $\hat{H}^0(\langle a \rangle, E)$.

The cohomology of $\tilde{B}$ is easy to calculate: One part is $\hat{H}^*(\langle a \rangle, W) = k[z, z^{-1}]$, deg $z = 2$, and $b(z) = -z$, as is known from the calculation of the cohomology of the dihedral group of order 6. The cohomology of the other part is calculated using dimension shifting in the block of $\mathbb{L}E$ that contains it to reduce to the case of the 1-dimensional WG-lattice on $wy^{-1}u_1$. The result is a free $k[z, z^{-1}]$-module on a generator t of degree 1, and $bt = -\zeta^2 t$.

Hence $\hat{H}^*(\langle a \rangle, E) = k[\tilde{x}, \tilde{x}^{-1}, z, z^{-1}] \otimes E(t)$.

Now $\hat{H}^*(G, E) = \hat{H}^*(G, E)^{(b)}$, and calculating the invariants yields:

THEOREM 5.1. $\hat{H}^*(G, E) = \mathbb{F}_9[\tilde{x}^4, \tilde{x}^{-4}, \tilde{x}^2 z, (\tilde{x}^2 z)^{-1}] \otimes E(\tilde{x}^{-1}t)$

There is an action of $\mathrm{Gal}(W/\mathbb{Z}_3) = \langle \sigma \rangle \cong \mathbb{Z}/2$ on G by conjugation with S. There is also an action on E via the action on the coefficients of the series. But conjugation by S has the same effect on a as conjugation by b, so does not affect $\hat{H}^*(G, \mathbb{Z}_3)$, and thus $\mathrm{Gal}(W/\mathbb{Z}_3)$ acts on $\hat{H}^*(G, E)$ via its action on E only. Now $\hat{H}^*(G, W) = \hat{H}^*(G, \mathbb{Z}_3) \otimes W$, so z could have been chosen invariant under the action. Consider the action of σ on $\tilde{x}$. We seek a replacement $\hat{x}$ for x that is invariant under σ. Now $\tilde{x} = (\frac{\sigma+1}{2})x + (\frac{\sigma-1}{2})x$, so at least one of the two terms on the right hand side, taken modulo 3, has u_1-filtration 0. If it is the first, take $\hat{x} = (\frac{\sigma+1}{2})x$. If it is the second, take $\hat{x} = \zeta^2(\sigma - 12)x$. It is easy to check that $\hat{x}^4$ is still invariant under G. A similar argument applies to $\hat{x}^{-1}t$ and we obtain

THEOREM 5.2. *We have:*

$$\hat{H}^*(G, E)^{Gal} = \mathbb{F}_3[\hat{x}^4, \hat{x}^{-4}, \hat{x}^2 z, (\hat{x}^2 z)^{-1}] \otimes E(\hat{x}^{-1}t),$$

$$H^*(G, E)^{Gal} = \mathbb{F}_3[\hat{x}^4, \hat{x}^{-4}, \hat{x}^2 z] \otimes E(\hat{x}^{-1}t), \qquad * > 0.$$

6. Invariants

In $\bar{E}$, y is not invariant under G, but y^2 is. Also $r = u^{-2}u_1$ is invariant. Together with x, these generate the invariants (see figure 2).

THEOREM 6.1. $\bar{E}^G = \mathbb{F}_9[[y^2]][x^4, x^{-4}, r]/(r^6 = x^4 y^2)$

There are no invariant elements with odd u-grading.

When we lift this to W, each invariant element lifts to an invariant element in the corresponding block of $\mathbb{L}E$, but its u_1-filtration may increase by 1 or 2. In the other hand $p = r^2$ lifts to $\tilde{p}$ and $q = r^3$ lifts to $\tilde{q} = \tilde{x}^{-2}\tilde{y}$. Thus the image of E^G in $\bar{E}^G$ misses precisely the part with basis the monomials rx^i (this could also be seen from the long exact sequence in cohomology for $3E \to E \to \bar{E}$). By construction, $\tilde{p}^3 = \tilde{q}^2 \pmod 3$.

THEOREM 6.2. *$E^G = W[[\tilde{y}^2]][\tilde{x}^4, \tilde{x}^{-4}, \tilde{p}, \tilde{q}]$, where $\tilde{q}^2 = \tilde{x}^{-4}\tilde{y}^2$, and $\tilde{p}^3 - \tilde{q}^2 = 3\delta$ for some $\delta \in E^G$.*

(see figure 3). As before,

THEOREM 6.3. *$(E^G)^{Gal} = \mathbb{Z}_3[[\hat{y}^2]][\hat{x}^4, \hat{x}^{-4}, \hat{p}, \hat{q}]$, where $\hat{q}^2 = \hat{x}^4\hat{y}^2$, and $\hat{p}^3 - \hat{q}^2 = 3\delta$ for some $\delta \in (E^G)^{Gal}$.*

7. The Ring Structure of the Invariants

THEOREM 7.1. *The invariant elements p and q can be chosen in such a way that $p^3 - q^2$ is divisible by 3^3. This is the best possible: they can not be chosen so as to make $p^3 - q^2$ divisible by 3^4. In fact, if $\delta = (p^3 - q^2)/3^3 \in E$, then the leading term of δ is a unit times u^{-12}.*

PROOF. The first part will follow from section 8 using modular forms, but we give an elementary proof here. Chose $p \in E_0 u^{-4}$ to be any element invariant under both G and Gal that reduces to r^2 modulo 3. Normalize it so that the coefficient of $u_1^2 u^{-4}$ is 1.

Then $p = (u_1^2 + 3s)u_1^{-4}$ for some $s \in E_0$ which is an even function of u_1, and so $p^3 = (u_1^6 + 3u_1^4 s)u^{-12} \bmod 3^3$.
Let $\tilde{q} = (u_1^3 + \frac{3^2}{2}u_1 s)u^{-1}$, so that $p^3 - \tilde{q}^2 = 0 \pmod{3^3}$.

Now since p is invariant, so is $\tilde{q}^2 \pmod{3^3}$. In other words $a(\tilde{q}^2) - \tilde{q}^2 = 0 \pmod{3^3}$ or $3^3 | (a(\tilde{q}) - \tilde{q})(a(\tilde{q}) + \tilde{q})$. But $a(\tilde{q}) + \tilde{q} = (2u_1^3 + \ldots)u^{-6} \pmod 3$, so we must have $3^3 | a(\tilde{q}) - \tilde{q}$.

Now $\tilde{q} \in E_0 u^{-6}$, and from the decomposition of E as a sum of indecomposables we can see that $H^1(G, E_0 u^{-6}) = 0$. Let $H = G \rtimes \langle\sigma\rangle$, so that $H^1(H, E_0 u^{-6}) = 0$.

The long exact sequence for cohomology becomes

$$0 \to (E_0 u^{-6})^H \to (E_0 u^{-6})^H \to (E_0 u^{-6}/3^3)^H \to 0,$$

and so $\tilde{q}$ lifts to an invariant element $q \in E_0 u^{-6}$ such that $q = \tilde{q} \pmod{3^3}$ and $p^3 - q^2 = 0 \pmod{3^3}$.

To show that this is the best possible power of 3, note that p must be an even function of u_1 and that q must be odd, so it suffices to prove that the constant term (in u_1) of p is not divisible by 3^2.

Let $\mathfrak{m}$ denote the maximal ideal of E_0: it is generated over W by 3 and u_1. Then $p = (c + u_1^2)u^{-4} \pmod{\mathfrak{m}^3}$ for some constant c, and an argument similar to the proof of 3.1, but working modulo $\mathfrak{m}^2$ instead of 3, shows us that $a(u_1) = -3\zeta + u_1 \pmod{\mathfrak{m}^2}$, so $a(u_1) = -3\zeta + u_1 + e$, say, where the constant term of e is divisible by 3^2.

Also $a(u) = \alpha_1 u \pmod{\mathfrak{m}}$, where $\alpha_1 = 1 \pmod{\mathfrak{m}}$. Thus

$$a(p) = (c + (-3\zeta + u_1 + e)^2)\alpha_1^{-4}u^{-4} \quad \bmod \mathfrak{m}^3.$$

But since p is invariant this must be equal to $(c + u_1^2)u^{-4} \pmod{\mathfrak{m}^3}$, and in particular the constant terms should be equal $\pmod{3^3}$. Using the subscript 0 to denote the constant term, this yields

$$(c+(-3\zeta+e_0)^2)(\alpha_1)_0^{-4}=c \quad (\mathrm{mod} 3^3)$$

and hence

$$c((\alpha_1)_0^4-1)=(-3\zeta+e_0)^2 \quad (\mathrm{mod} 3^3).$$

Now $3^2|e_0$ and $3|((\alpha_1)_0^4-1)$ so we see that if $3^2|c$ then $3|\zeta$, a contradiction. □

According to 7.1, if $p^3-q^2=3^3\delta$ then the leading term of δ is u^{-12}, up to a unit. This enables us to express $\hat{x}^{-4}$ in terms of δ and y^2, hence:

THEOREM 7.2.

$$(E^G)^{Gal}=\mathbb{Z}_3[[q^2\delta^{-1}]][\delta,\delta^{-1},p,q]/(p^3-q^2=3^3\delta)$$

8. Modular Forms

In this section we describe a connection between the ring of integral modular forms computed by Deligne **[1]** and the ring of invariants E^G.

We recall some standard material about elliptic curves and modular forms. An elliptic curve over a scheme S is a smooth morphism $p\colon C\to S$, whose geometric fibers are connected curves of genus one, together with a section $e\colon S\to C$. We denote by $\underline{\omega}_{C/S}$ the invertible sheaf $p_*(\Omega^1_{C/S})$ on S. The following definition is taken from **[1]**:

DEFINITION 8.1. *An integral modular form of weight $k\in\mathbb{Z}$ and level one is a rule which assigns to any elliptic curve over any scheme S a section $f(C/S)$ of $(\underline{\omega}_{C/S})^{\otimes k}$ over S such that the following two conditions are satisfied.*
1. $f(C/S)$ depends only on the S-isomorphism class of the elliptic curve C/S.
2. The formation of $f(C/S)$ commutes with arbitrary change of base $g\colon S'\to S$ (meaning that $f(C'/S')=g^*f(C/S)$.

We are interested in the situation in which the base S is an affine scheme, $S=\mathrm{Spec}(R)$, R is a ring, and $\underline{\omega}_{C/R}$ is a free R-module of rank one with basis ω. Then any section of $\underline{\omega}^k_{C/R}$ can be written as $f(C/R,\omega)\cdot\omega^{\otimes k}$, where f is a rule which assigns to every pair $(C/R,\omega)$ consisting of an elliptic curve C over a ring R and a nowhere vanishing section of $\Omega^1_{C/R}$ on C, an element $f(C/R,\omega)\in R$, such that the following three conditions are satisfied:
1. $f(C/R,\omega)$ depends only on the R-isomorphism class of the pair $(C/R,\omega)$.
2. f is homogeneous of degree $-k$ in the second variable; for any $\lambda\in R^\times$, $f(C/R,\lambda\omega)=\lambda^{-k}f(C/R,\omega)$.
3. The formation of $f(C/R,\omega)$ commutes with arbitrary extension of scalars $g:R\to R'$ (meaning that $f(C_{R'/R'},\omega')=g(f(C/R,\omega)))$.

Let $(C/R,\omega)$ be a pair of the type considered above. Choose coordinate functions x and y such that the image of y in $\underline{\omega}^{-1}_{C/R}\otimes\underline{\omega}^{-1}_{C/R}\otimes\underline{\omega}^{-1}_{C/R}$ is $\omega^{-1}\otimes\omega^{-1}\otimes\omega^{-1}$ and the image of x in $\underline{\omega}^{-1}_{C/R}\otimes\underline{\omega}^{-1}_{C/R}$ is $\omega^{-1}\otimes\omega^{-1}$. Consider the corresponding Weierstrass equation

$$y^2+a_1xy+a_3y=x^3+a_2x^2+a_4x+a_6$$

of the curve C. The symbols c_4, c_6 and Δ denote the usual polynomials in the a_i's **[1]**. An elementary computation shows that the rules which assign to the pair $(C/R,\omega)$ the differential forms $c_4\omega^4$, $c_6\omega^6$ and $\Delta\omega^{12}$ are indeed modular forms

THEOREM 8.2. *(Deligne)* $\mathfrak{M} = \mathbb{Z}[c_4, c_6, \Delta]/(c_4^3 - c_6^2 = 12^3\Delta)$

Let us also recall the Serre-Tate Theorem. A good reference is [**2**]. Let R be a ring and I an ideal such that the ideal (p, I) is nilpotent.

THEOREM 8.3. *The following categories are equivalent:*
a. The category $\mathfrak{A}$ *of elliptic curves over* R *and* R*-homomorphisms.*
b. The category $\mathfrak{B}$ *of triples* (F, C_0, i) *where* F *is a* p*-divisible group over* R*,* C_0 *is an elliptic curve over* $R/(p, I)$*,* i *is an isomorphism between the reduction of* F *modulo* (p, I) *and* F_{C_0}*. Morphisms are pairs* (r_1, r_2)*, where* r_1 *is an* R*-homomorphism of* p*-divisible groups and* r_2 *is an* $R/(p, I)$*-homomorphism of elliptic curves which make the obvious diagram commutative.*

Denote by FG and Ell the functors which provide the equivalence of the above categories.

The argument we will present below works for all primes, but for the purposes of this paper we restrict ourselves to the case when the prime p is equal to 3. Denote by $\mathbb{F}_{sep}$ the separable closure of $\mathbb{F}_3$ and by S the ring of integers of the maximal unramified extension of $\mathbb{Z}_3$, (which has residue class field $\mathbb{F}_{sep}$). From now on take R to be the ring $S[[u_1]]$. Note that R is the inverse limit of rings R_n to which the Serre-Tate Theorem applies. Take for example R_n to be $R/(3, u_1)^n$. Denote by r_n and p_n the canonical homomorphisms $R_n \to R_{n-1}$ and $R \to R_n$ respectively. The ideal $(3, u_1)$ is invariant under G, so G also acts on R_n.

Take C_0 to be a supersingular elliptic curve over $\mathbb{F}_{sep}$. There is only one such up to isomorphism. Recall [**1**] that $\mathrm{Aut}(C_0)$ is isomorphic to $G \cong \mathbb{Z}/3 \rtimes \mathbb{Z}/4$, and denote by F_0 the formal group of C_0.

First we construct a map from $\mathfrak{M}$ to $R_n[u^{-1}, u]^G$ for each n. According to the classification of formal groups over $\mathbb{F}_{sep}$, see for example [**3**], F_0 is isomorphic to $\bar{F}$. Let i be such an isomorphism, and consider the triple (F_{u_1}, F_0, i). Denote by T_n the triple $(p_n^* F_{u_1}, F_0, i)$. Let $r = (r_1, r_2) : T_n \to FG(Ell(T_n))$ be part of the natural isomorphism $id_{R_n} \to FG \circ Ell$. Let $C_n = Ell(T_n)$.

LEMMA 8.4. *There is a choice of coordinates on the elliptic curve* C_n *over* R_n *such that the corresponding formal group law* F_n *is strictly isomorphic to* $p_n^* F_{u_1}$*. Furthermore, this can be done in such a way that* F_n *has the form*

$$F_n(x, y) = x + y - u_1(x^2y + xy^2) \quad (mod\ deg\ 4)$$

modulo 3.

PROOF. Choose coordinates x and y on C_n. Then r_1 becomes a series over R_n such that $r_1'(0)$ is an invertible element of R_n. Then

$$x' = (r_1'(0))^2 x,$$
$$y' = (r_1'(0))^3 y$$

provide the required coordinates.

Next we want to show that we can make a choice of coordinates such that the strict isomorphism between $p_n^* F_{u_1}$ and F_n has the form $r_1(z) = z$ mod deg 3 and modulo 3. Then F_n has to be of the form given in the statement of the lemma. Suppose that the strict isomorphism had the form $r_1(z) = z + \alpha z^2$ mod deg 3 and modulo 3. An elementary computation shows that the change of coordinates $x' = x$ and $y' = y - \alpha x$ on C_n yields the change of the coordinate on F_n given by the series $Z(z) = z - \alpha z^2$ mod deg 3 and modulo 3. The resulting formal group law is strictly isomorphic to $p_n^* F_{u_1}$ and the strict isomorphism is $Z(r(z))$, which is equal to z mod deg 3 and modulo 3. □

Let $g \in G$. We denote by the same letter the element of $\mathrm{Aut}R$ associated to it in Section 2. Let $g(x)$ be the representative of g in $\mathbb{F}_{sep}[[x]]$ and denote by $\tilde{g}(x)$ the Lubin-Tate lift of $g(x)$. Recall that it has the following property

$$\tilde{g}^{-1} F_{u_1}(\tilde{g}(x), \tilde{g}(y)) = g^* F_{(u_1)}(x, y).$$

Denote by $\tilde{g}_n(x)$ the series $\tilde{g}(x)$ with the coefficients reduced modulo the ideal $(3, u_1)^n$. The following proposition is an immediate consequence of the Serre-Tate Theorem. We are using standard notation from the theory of elliptic curves (see [**1**]).

PROPOSITION 8.5. *For every $g \in G$ the elliptic curves C_n and $g^* C_n = C_n'$ are isomorphic. In terms of the Weierstrass equations of C_n and C_n' such an isomorphism is given by a substitution $l_n(g)$:*

$$x = a^2 x' + b$$

$$y = a^3 y' + a^2 c x' + d, \quad a, b, c, d \in R_n.$$

With coordinates on C_n chosen as in 8.4 and corresponding coordinates on C_n', the substitution $l_n(g)$ which provides an isomorphism between C_n' and C_n has the property $\tilde{g}_n'(0) = a^{-1}$.

PROOF. If $FG(C_n) = (F_1, E_1, i_1)$, then by definition of FG we have $FG(g^* C_n) = (g^* F_1, E_1, i_1)$. There is the following diagram, all the maps of which are isomorphisms in the category $\mathfrak{B}$ which corresponds to the ring R_n:

$$\begin{array}{ccc} T_n & \xleftarrow[(\tilde{g}_n, g)]{} & g^* T_n \\ {\scriptstyle (r_1, r_2)} \big\downarrow & & \big\downarrow {\scriptstyle g^*(r_1, r_2)} \\ (F_1, E_1, i_1) & & (g^* F_1, E_1, i_1). \end{array} \tag{8.6}$$

This shows that $FG(C_n)$ and $FG(g^* C_n)$ are isomorphic in $\mathfrak{B}$, therefore we obtain the required R_n-isomorphism $l_n(g)$ of C_n and $g_* C_n$ from Serre-Tate Theorem and we have a commutative diagram

$$\begin{array}{ccc} T_n & \xleftarrow[(\tilde{g}, g)]{} & g^* T_n \\ {\scriptstyle (r_1, r_2)} \big\downarrow & & \big\downarrow {\scriptstyle g^*(r_1, r_2)} \\ (F_1, E_1, i_1) & \xleftarrow[(FG(l_n(g)), g)]{} & (g^* F_1, E_1, i_1). \end{array} \tag{8.7}$$

By 8.4 we can choose coordinates on C_n such that the series r_1 provides a strict isomorphism of the corresponding formal group laws. For this choice of coordinates $FG(l_n(g))$ becomes a series such that its derivative at zero is equal to a^{-1}, as follows from the definition of the functor FG. □

We now construct a ring homomorphism $j_n : \mathfrak{M} \to R_n[u, u^{-1}]$. Let C_n be the elliptic curve defined above. Choose coordinates on C_n as in 8.4 and denote by ω the corresponding section of ω. Let f be a modular form of weight k. Define $j_n(f)$

PROOF. Indeed, taking into account the properties of modular forms and the formula for the action of $l_n(g)$ on the invariant differential from [1], we obtain:

$$g(f(C_n,\omega)u^k) = f(g^*C_n,\omega)g(u^k) = f(C_n, l_n(g)^{-1}\omega)a^{-k}u^k =$$

$$f(C_n, a^{-1}\omega)a^{-k}u^k = f(C_n,\omega)u^k.$$

□

It is hard to identify the images of the generators of $\mathfrak{M}$ in $R_n[u,u^{-1}]$ directly, because we know very little about the curve C_n. We can however say something in this direction if we change the base to $\bar{R}_n = R_n \otimes \mathbb{F}_3$. First we need to recall the structure of the ring of modular forms in characteristic 3 as computed in [1].

THEOREM 8.9. *The ring $\mathfrak{M}_3$ of modular forms in characteristic 3 is isomorphic to*

$$\mathbb{F}_3[b_2, \Delta]$$

and $c_4 = b_2^2$, $c_6 = -b_2^3$.

An argument like the one presented above provides us with a map

$$j_n^{(3)} : \mathfrak{M}_3 \to \bar{R}_n[u,u^{-1}]^G.$$

PROPOSITION 8.10. $j_3^{(3)}(b_2) = u_1u^2$

PROOF. From Proposition 8.4 we know that there is a choice of coordinates over R_n for C_n such that its formal group law has the form

$$F_n'(x,y) = x + y - u_1(x^2y + xy^2) \pmod{\deg 4}$$

modulo 3. On the other hand according to a computation presented in [11], p115, this means that the element b_2 of the Weierstrass equation which corresponds to this choice of coordinates is equal to u_1. □

Finally we show that the maps j_{n-1} and r_nj_n are equal for all n and, therefore, that we have constructed a map $j : \mathfrak{M} \to R[u^{-1},u]^G$. Indeed, since the triples T_{n-1} and $r_n^*T_n$ coincide, the elliptic curves C_{n-1} and $r_n^*C_n$ are isomorphic over R_{n-1} by the Serre-Tate theorem. Furthermore, since this isomorphism induces a strict isomorphism of the formal groups laws of C_{n-1} and C_n (with respect to the coordinates chosen), the maps j_{n-1} and r_nj_n coincide by construction. The same argument shows that the maps $j_n^{(3)}$ and $\bar{r}_nj_n^{(3)}$ are equal, where $\bar{r}_n$ is the reduction of r_n modulo 3.

This implies that $j(c_4)$ and $j(c_6)$ are suitable candidates for the elements p and q of Theorem 7.2, and since $(c_4^3 - c_6^2)/3^3 = 4^3\Delta$, Theorem 7.1 shows that $j(\Delta)$ is a suitable replacement for δ. Identifying $\mathfrak{M}$ with its image under j, we can rewrite 7.2 as:

COROLLARY 8.11. $(S \otimes E)^G = S[[c_6^2\Delta^{-1}]][\Delta, \Delta^{-1}, c_4, c_6]/(c_4^3 - c_6^2 = 12^3\Delta)$.

This can be expressed more succinctly as follows.

COROLLARY 8.12. *$(S \otimes E)^G$ is isomorphic as a ring to $\mathfrak{M}[\Delta^{-1}] \otimes S$ completed with respect to $c_6^2\Delta^{-1}$.*

References

[1] P. Deligne, *Courbes Elliptiques: Formulaire (d'après J. Tate)*, in Modular Functions of One Variable IV, Springer LNM 476, 53–74.

[2] V. Drinfeld, *Coverings of p-adic symmetric domains(Russian)*, Func. Anal. i Prilozen, **10**, no. 2, (1976), 29-40.

[3] M. Hazewinkel, *Formal Groups and Applications*, Academic Press (1978).

[4] M. J. Hopkins, B. H. Gross, *The rigid analytic period mapping, Lubin-Tate space, and stable homotopy theory*, Bull. Amer. Math. Soc. **30** (1994), 76-86.

[5] M. Hopkins, H. Miller, Midwest Topology Seminar, Fall 1994.

[6] M. Hopkins, H. Miller, AMS Summer Research Institute, Seattle, 1996.

[7] M. Hopkins, M. Mahowald, H. Sadofsky, *Constructions of elements in Picard Groups, in Topology and Representation Theory*, Contemporary Mathematics, **158**, (1994).

[8] Kaplansky, *Infinite Abelian Groups*, Univ. Michigan Press 1954.

[9] J. Lubin, J. Tate, *Formal moduli for one parameter formal Lie groups*, Bull. Soc. Math. France, **94** (1966), 49-60.

[10] J. Morava, *Noetherian localisation of categories of cobordism comodules*, Ann. Math., **121** (1985), 1-39.

[11] J. H. Silverman, *The Arithmetic of Elliptic Curves*, Springer Graduate Texts in Mathematics, **106**, 1986.

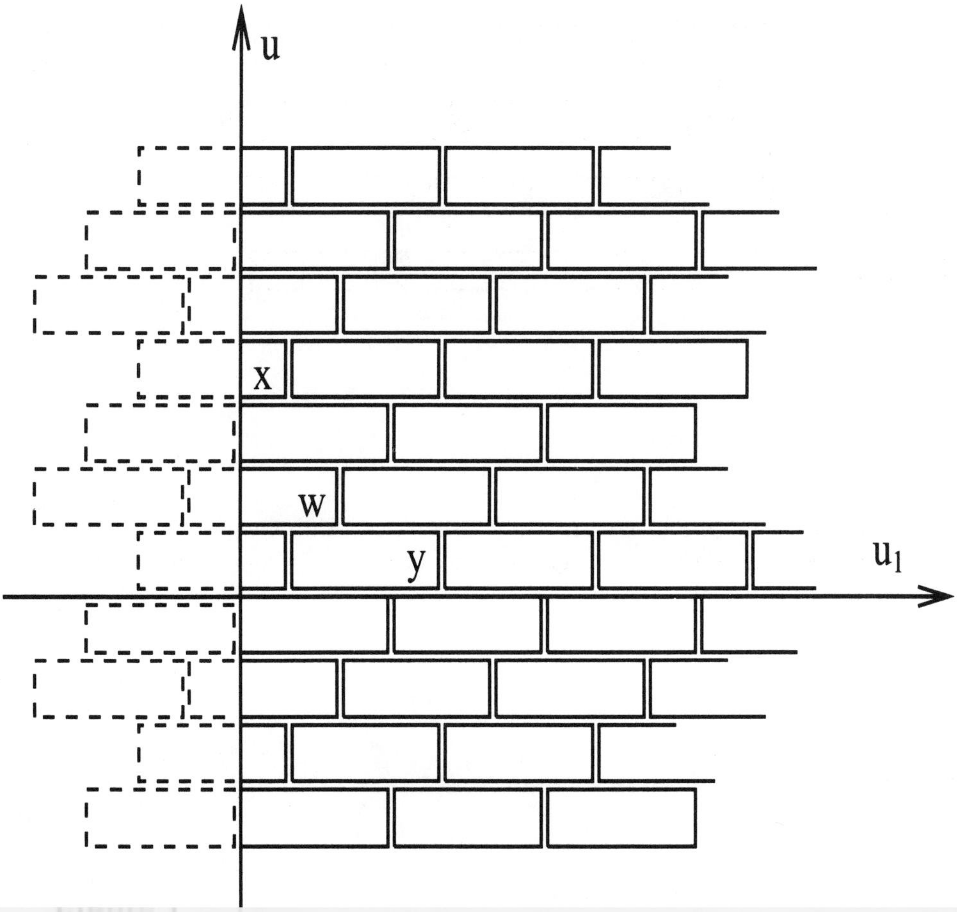

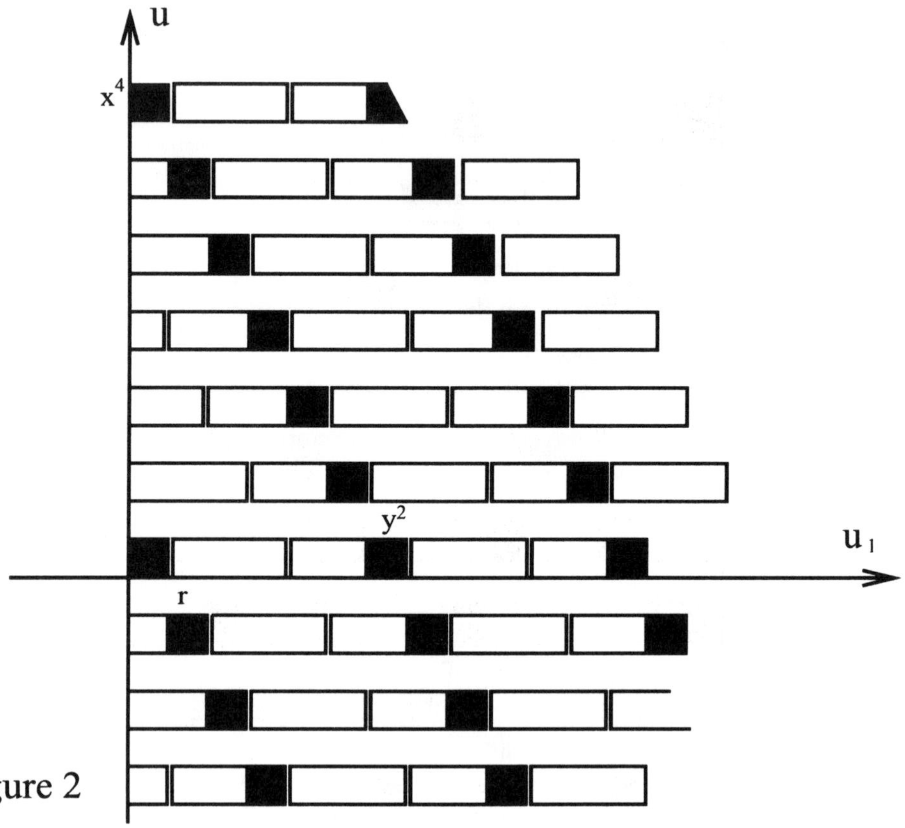

Figure 2

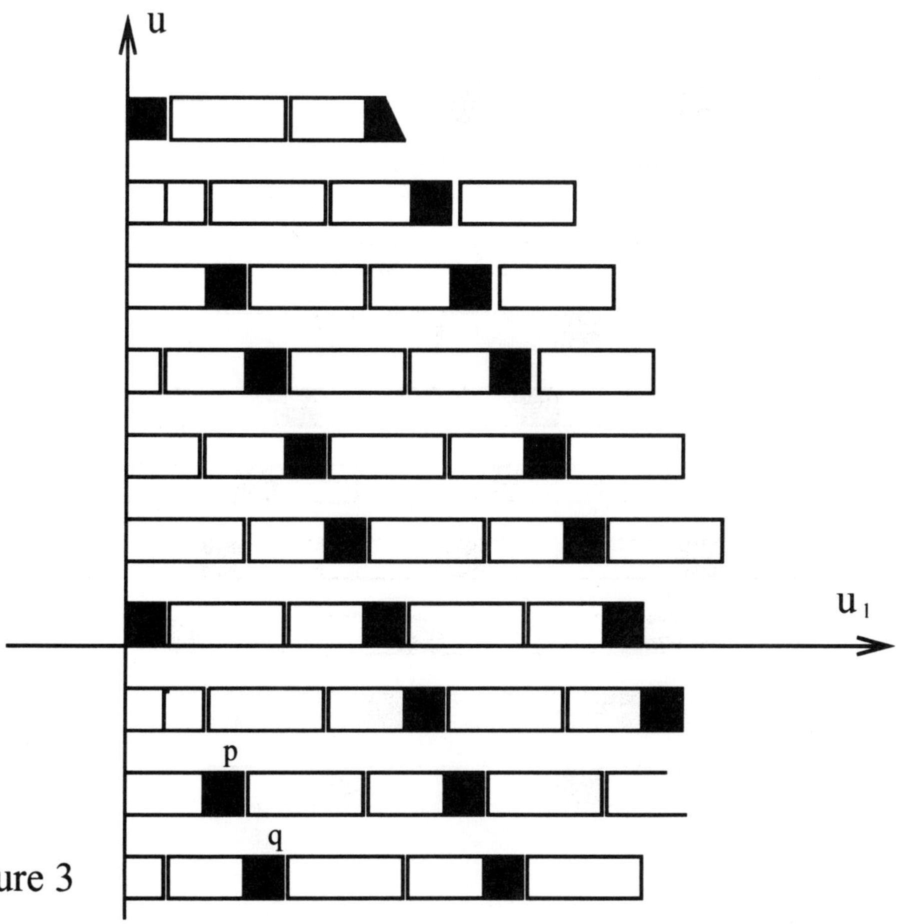

Figure 3

Department of Mathematics, University of Kentucky, Lexington, KY 40506-0027, USA

Department of Mathematics, UMIST, P O Box 88, Manchester M60 1QD, England
E-mail address: vgorb@ms.uky.edu, symonds@ms.uky.edu

Contemporary Mathematics
Volume **220**, 1998

Skew Symmetric Bundle Maps

Daniel Henry Gottlieb

ABSTRACT. We study the "Lie Algebra" of the group of Gauge Transformations of Space-time. We obtain topological invariants arising from this Lie Algebra. Our methods give us fresh mathematical points of view on Lorentz Transformations, orientation conventions, the Doppler shift, Pauli matrices, Electro-Magnetic Duality Rotation, Poynting vectors, and the Energy Momentum Tensor T .

1. Introduction

Let M be a space-time and $T(M)$ its tangent bundle. Thus M is a 4-dimensional manifold with a nondegenerate inner product $\langle\ ,\rangle$ on $T(M)$ of index $-+++$. We study the space of bundle maps $F : T(M) \to T(M)$ which are skew symmetric with respect to the metric, i.e. $\langle Fv, v\rangle = 0$ for all $v \in T_x(M)$ and all $x \in M$.

A skew symmetric F has invariant planes and eigenvector lines in each $T_x(M)$. We give necessary and sufficient conditions as to when these plane systems and line systems form subbundles in Theorem 7.3. Also we determine the space of those F which give the same underlying structure. This is done by introducing the bundle map $T_F = F \circ F - \frac{1}{4}(tr\ F^2)I : T(M) \to T(M)$. Then the space of skew symmetric F which give rise to the same T is homeomorphic to $\mathrm{Map}(M, S^1)$, the space of maps of M into the circle S^1. (See Theorem 7.11.)

We also show that the space of skew symmetric F has a natural complexification. (see Propositions 2.2 and 2.3) This leads to an equivalence between the F and vector fields on the complexified tangent bundle $T(M) \otimes \mathbb{C}$. The complexified study leads to several beautiful relations which link our subject matter to Clifford Algebras and Quaternions. (See Corollaries 4.6 and 4.7 and Theorem 4.8.) We naturally find many points of contact with Physics, especially classical electromagnetism. These considerations frequently govern our choice of notation. The physical motivations and remarks will be explored in the Scholia; and the mathematical motivations and links will be found in the Remarks.

SCHOLIUM 1.1. PHYSICAL CONNECTIONS.

1991 *Mathematics Subject Classification.* 57R22, 58D30, 83C50.

Key words and phrases. Electro-magnetism, energy-momentum, vector bundles, Lorentz Transformations, Clifford Algebras, Doppler shift.

a) Each skew symmetric F corresponds to a two-form $\widehat{F}$. The electro-magnetic tensor is a two form. In the classical theory it satisfies Maxwell's equations. The symmetric bundle map T_F corresponds to the energy-momentum tensor of the electro-magnetic field. The homotopy invariants arising from the existence of subbundles must give physical information if there is any physical content in Classical Electro-Magnetism. We show that the invariants distinguish the two main cases; a classical free electron and a classical electron in a magnetic field.

b) We give formulas in terms of **E** and **B** for the eigenvectors of F. Changing observers gives the same eigenvector multiplied by a factor. For "radiative" F, this factor reduces to the Doppler shift. One wonders if the more general shift for non-radiative F has any physical meaning.

c) The space of skew symmetric F has a canonical splitting of space and time. It is mapped isomorphically onto $T(M) \otimes \mathbb{C}$ by a choice of a field of observers. Thus any complex tangent vector field corresponds to a skew-symmetric F. So, for example, if the solutions of the Dirac equation have any physical content, then the homotopy invariants of the corresponding F must havc physical import.

REMARK 1.2. MATHEMATICAL MOTIVATION.

The mathematical point of view of this work stems from the author's study of the space of bundle equivalences in [$\mathbf{G}_1$], [$\mathbf{G}_2$], [$\mathbf{G}_3$]. These bundle equivalences form spaces which later became popular known as groups of gauge transformations. The main result of these papers is that the classifying space of these groups of gauge transformations is the space of maps of the base space into the classifying space of the fibration in question.

This theorem has played an important role, at least in the mathematical part of of Gauge Theory. It entered into the theory via Proposition 2.4 of [**AB**]. But the point of view of these works concerned spaces of connections, instead of spaces of bundle equivalences. The original point of view was furthered in papers by Booth, Heath and Piccinini among others, see for example [**BP**].

In this present work, we study other types of bundle maps. The "Lie Algebras" of "Gauge Transformation Groups" seems to be a natural class to study. The skew-symmetric bundle maps of space-time are the "Lie Algebra" of the group of isometries on $T(M)$, i.e. bundle maps $Q : T(M) \to T(M)$ so that $\langle Qv, Qw\rangle = \langle v, w\rangle$.

SCHOLIUM 1.3. PHYSICAL POINT OF VIEW.

Galileo's famous quote that the Laws of Nature are written in the language of geometry should be revised in view of the development of Topology in this century. As topology underlies geometry, one would expect that some Laws of Nature would be expressed in terms of the elementary homotopy invariants of topology. Among these are the degrees of maps and the index of vector fields.

Our method for discovering these laws follows Galileo. To the argument that no one had seen an object travel at a constant velocity forever along a straight line, Galileo replies: Let us assume it is true, derive its mathematical consequences, and see if they relate to what is observed. Thus we begin by studying infinitesimal rigid motions F on space-time M, and observe connections with electromagnetism, etc. The idea of separating the physical from the mathematical arguments via Scholia is borrowed from Newton's Principia.

REMARK 1.4. LEVELS OF NOTATION.

We proceed by adding layers of notation to our space-time. We descend one level for every choice we estimate we make. We begin at Level –1 with the inner product and continue by choosing an orientation at Level –2. By Level –10 we have chosen an orthonormal basis for the tangent space of M. We eventually end at Level –16, which are the standard coordinates for Minkowski Space.

This approach permits us to understand that choosing an orientation is like taking a complex conjugate. It also allows a clear view of Lorentz Transformations at the various levels. The major technique of computation in this paper is given by a Level –10 block matrix which allows Level –10 calculations to produce Level -2 statements.

ACKNOWLEDGEMENTS.

I have had very productive conversations with Barrett O'Neill, Stephen Parrot, and Solomon Gartenhaus. Gartenhaus gave me a key example which forced me to think more deeply at the beginning of this work. Barrett O'Neill gave me many ideas. The best one is the definition of the complexification map c. Barrett O'Neill's book Semi-Riemannian Manifolds [**O**] exposes space-time in a rigorous mathematical manner. Stephen Parrott's book [**P**] provided great stimulation and guidance.

2. Notation and Preliminaries

A *space-time* M is a smooth 4-dimensional orientable manifold with a Lorentzian metric $\langle\ ,\ \rangle$ defined on the tangent bundle $T(M)$ and a nonzero future pointing time-like vector field. If $x \in M$, then $T_x = T_x(M)$ will denote the 4-dimensional tangent space over x. The space of vectors orthogonal to a vector $u \in T_x$ will be denoted by T_x^u.

A skew symmetric bundle map is a map $F : T(M) \to T(M)$ which covers the identity on the base, is a vector bundle map, and is skew symmetric that is,

$$F(\alpha v_x + \beta w_x) = \alpha F(v_x) + \beta F(w_x) \in T_x \tag{1}$$

$$\text{and } \langle F(v_x), w_x\rangle = \langle v_x, -F(w_x)\rangle. \tag{2}$$

Let ℓ be the vector bundle over M whose fibre ℓ_x is the vector space of skew symmetric linear transformation $F_x : T_x \to T_x$. Then the space of cross-sections $\Gamma(\ell)$ to ℓ corresponds to the space of bundle maps in the usual manner. Let $\Lambda^2(M)$ be the bundle of two forms over M. Thus the fibre $\Lambda^2(M)_x$ are bilinear antisymmetric maps $\widehat{F}_x : T_x \times T_x \to \mathbb{R}$. Any two-form is a cross-section to $\Lambda^2(M)$.

Now ℓ is bundle equivalent to $\Lambda^2(M)$. Let $\rho : \ell \to \Lambda^2(M)$ so that $\rho(F_x) = \widehat{F}_x$ where

$$\widehat{F}_x(v_x, w_x) = \langle v_x, F_x(w_x)\rangle. \tag{3}$$

The non-degeneracy of $\langle\ ,\ \rangle$ implies that ρ is an isomorphism on each fibre, thus ℓ sets up a bijection between two-forms and bundle maps.

LEVEL –1. LORENTZ INNER PRODUCT.

Notation plays an important role in Mathematics and Physics. It is a powerful choices of frame fields and coordinates we descend eventually to Level -16, which is the canonical coordinates of Minkowski Space-time. The number describing the Levels approximates the number of choices made to introduce the notation. We have already made one choice in Level -1 by assuming that $\langle\ ,\ \rangle$ has signature $-+++$, we could have assumed signature $+---$. Level 0 then has innerproduct $\epsilon\langle\ ,\ \rangle$ where ϵ is ± 1. The geometry does not change with the change of ϵ. The geodesics remain the same and skew-symmetric bundle maps remain the same so the choice $-+++$ does not affect our work. But in comparing our results with other authors, be aware that the electro-dynamicists usually choose $+---$. Thus S. Parrott [**8**] chooses $+---$ where as O'Neill [**9**] chooses $-+++$.

LEVEL -2. ORIENTATION.

Since M is orientable, there is a volume form $\Omega \in \Lambda^4(M)$. There are two choices consistent with the metric, $\pm\Omega$. We choose Ω as the orientation. We could have chosen $-\Omega$. Now the Hodge dual is an isomorphism defined on $\Lambda^2(M)$, satisfying $*(*\eta) = -\eta$ for $\eta \in \Lambda^2(M)$. Under $\rho : \ell \to \Lambda^2(M)$ the Hodge dual corresponds to an operator $*$ on $\Gamma(\ell)$. It satisfies

$$(aF)^* = aF^* \quad \textit{and} \quad (F+G)^* = F^* + G^* \quad \textit{and} \quad F^{**} = -F. \tag{4}$$

Let $u \in T_x(M)$ be an *observer*. That is u is a future pointing time-like vector such that $\langle u, u\rangle = -1$. Then we define

$$\mathbf{E}_u = Fu \quad \textit{and} \quad \mathbf{B}_u = -F^*u. \tag{5}$$

Note that $\mathbf{E}_u$ and $\mathbf{B}_u \in T^u$. If we change the orientation, we obtain a new $*'$. This is related to the old $*$ by $F^{*'} = -F^*$. Thus for change of orientation, $\mathbf{E}_u$ remains the same, but $\mathbf{B}_u$ becomes $-\mathbf{B}_u$.

If v and w are space-like vectors in T_x, they span a space-like plane if and only they are linearly independent and

$$v^2w^2 - \langle \mathbf{v}, \mathbf{w}\rangle^2 > 0. \tag{6a}$$

If

$$v^2w^2 - \langle \mathbf{v}, \mathbf{w}\rangle^2 = 0, \tag{6b}$$

they span a light-like plane and if

$$v^2w^2 - \langle \mathbf{v}, \mathbf{w}\rangle^2 < 0 \tag{6c}$$

they span a space-like plane.

Let u be an observer. We define the dot product and cross product on T^u_m.

DEFINITION 2.1. We define the dot product and the cross product in the space T^u_m. Let $\mathbf{v}$ and $\mathbf{w} \in T^u_m$. Define the dot product by

$$\mathbf{v}\cdot\mathbf{w} = \mathbf{v}\cdot_u \mathbf{w} = \langle \mathbf{v}, \mathbf{w}\rangle \tag{7}$$

Then $v^2 = \mathbf{v}\cdot\mathbf{v}$ and $\mathbf{v}\cdot\mathbf{w} = vw\cos\theta$ where θ is defined to be the angle between $\mathbf{v}$ and $\mathbf{w}$. Define the cross product in T^u_m by $\mathbf{v}\times\mathbf{w} = \mathbf{v}\times_u\mathbf{w} =$ the unique vector orthogonal to $\mathbf{v}$ and $\mathbf{w}$ in T^u_m of length $|vw\sin\theta|$ so that $\Omega(\mathbf{u}, \mathbf{v}, \mathbf{w}, \mathbf{v}\times\mathbf{w}) > 0$

This cross product satisfies the usual relations:

$$\begin{aligned}
&\mathbf{v}\times\mathbf{w} = -\mathbf{w}\times\mathbf{v}\\
&\mathbf{v}\times(\alpha\mathbf{w}+\beta\mathbf{x}) = \alpha(\mathbf{v}\times\mathbf{w})+\beta(\mathbf{v}\times\mathbf{x})\\
&\mathbf{v}\times\mathbf{w} = \mathbf{0} \text{ if and only if } \alpha\mathbf{v} = \beta\mathbf{w}\\
&(\mathbf{u}\times\mathbf{v})\times\mathbf{w} = (\mathbf{w}\cdot\mathbf{u})\mathbf{v} - (\mathbf{w}\cdot\mathbf{v})\mathbf{u}\\
&\mathbf{u}\cdot(\mathbf{v}\times\mathbf{w}) = -\mathbf{v}\cdot(\mathbf{u}\times\mathbf{w})\\
&\mathbf{v}\cdot(\mathbf{v}\times\mathbf{w}) = 0
\end{aligned}\tag{8}$$

We use F^* to impose a complex structure on ℓ. Define

$$e^{i\theta}F = \cos\theta F + \sin\theta F^*. \tag{9}$$

PROPOSITION 2.2. *The action* $e^{i\theta}$ *on* $\Gamma(\ell)_x$ *induces a complex structure.*

PROOF. Any complex number $z = ae^{i\theta}$, so $z\cdot F = e^{i\theta}(aF)$. We check that $e^{i\theta'}(e^{i\theta}F) = e^{i(\theta+\theta')}F$ and $e^{i\theta}\cdot(F+F') = e^{i\theta}\cdot F + e^{i\theta}\cdot F'$. □

Consider $T(M)\otimes\mathbb{C}$. We define the innerproduct $\langle\ ,\ \rangle_{\mathbb{C}}$ on $T(M)\otimes\mathbb{C}$ by

$$\langle iu,v\rangle_{\mathbb{C}} = \langle u,iv\rangle_{\mathbb{C}} = i\langle u,v\rangle \quad when \quad u,v\in T_x(M). \tag{10}$$

If $\mathbf{a}$ and $\mathbf{b}$, $\mathbf{c}$ and $\mathbf{d}$ are in T_x^u, we define

$$(\mathbf{a}+i\mathbf{b})\times(\mathbf{c}+i\mathbf{d}) = (\mathbf{a}\times\mathbf{c}-\mathbf{b}\times\mathbf{d})+i(\mathbf{b}\times\mathbf{c}+\mathbf{a}\times\mathbf{d}). \tag{11}$$

Let $\ell_{\mathbb{C}}$ be the bundle of linear maps $\mathbb{F}: T_x\otimes\mathbb{C}\to T_x\otimes\mathbb{C}$ skew symmetric with respect to $\langle\ ,\ \rangle_{\mathbb{C}}$. Let $F_x\in\ell_x$ act on $T_x\otimes\mathbb{C}$ by

$$F(\mathbf{a}+i\mathbf{b}) = F(\mathbf{a})+iF(\mathbf{b}). \tag{12}$$

Define $c:\ell\to\ell_{\mathbb{C}}$ and $\bar{c}:\ell\to\ell_{\mathbb{C}}$ by

$$cF = F - iF^* \quad\text{and}\quad \bar{c}F = F + iF^*. \tag{13}$$

Note that changing the orientation means replacing F^* by $F^{*'} := -F^*$. Hence the complex structure is changed so that c becomes $\bar{c} = c'$.

PROPOSITION 2.3. *c is a complex bundle map.*

PROOF. cF_x is skew symmetric on $T_x\otimes\mathbb{C}$. Also c commutes with addition and multiplication. It is complex because

$$c(e^{i\theta}\cdot F) = e^{i\theta}(cF). \tag{14}$$

This follows because

$$\begin{aligned}
e^{i\theta}(cF) &= (\cos\theta + i\sin\theta)(F - iF^*)\\
&= \cos\theta F + \sin\theta F^* + i(\sin\theta F - \cos\theta F^*)\\
&= e^{i\theta}\cdot F - i(e^{i\theta}\cdot F^*) = e^{i\theta}\cdot F - i(e^{i\theta}F)^*\\
&= c(e^{i\theta}F).
\end{aligned}$$

SCHOLIUM 2.4. MAXWELL'S EQUATIONS AND LORENTZ' LAW.

a) We chose $\rho : \ell \to \Lambda^2$ to be given by (3), $\widehat{F}(v,w) = \langle v, F(w)\rangle$, in order to agree with the standard index conventions of tensor analysis. Parrott's otherwise careful book makes the opposite choice, $\widehat{F}(v,w) = \langle F(v), w\rangle$, and is thus inconsistent with his index conventions. This has little import for his book, since he deals mostly with forms, but it could cause confusion if one is using skew symmetric operators.

b) Electro-magnetic tensors are two-forms. Classically they satisfy Maxwell's equations:

$$d\widehat{F} = 0, \quad d * \widehat{F} = J. \tag{15}$$

We can write Maxwell's equations in terms of skew symmetric bundle maps as follows.

$$\text{div } F = j \ , \quad \text{div } F^* = 0 \tag{16}$$

where j is a one form. We may reduce this to one equation by extending div to the complex case by $\text{div}(iF) = i \text{ div}(F)$. Then F satisfies Maxwell's equation if and only if $\text{div}(cF)$ is real.

c) *The Lorentz Law:* Suppose a particle with charge q is moving in an electromagnetic field $\widehat{F}$ with 4–velocity u. Then its acceleration is $a = qFu$ where $\rho(F) = \widehat{F}$. This is the reason we chose the symbol $\mathbf{E}$ to equal Fu. The charge is motionless with respect to the u observer, hence its acceleration is given by the electric field $\mathbf{E}$ as seen by that observer. Also $\mathbf{B} = -F^*u$ corresponds to the magnetic field, as will be seen shortly.

LEVEL -9. ORTHONORMAL BASES. We may choose orthonormal vector fields e_0, e_1, e_2, e_3, so

$$\langle e_0, e_0\rangle = -1 \quad \textit{and} \quad \langle e_i, e_j\rangle = \delta_{ij}. \tag{17}$$

Already this notation restricts the topology of the M. It must be parallelizable for such a basis to exist. Fortunately we can find local regions which admits these orthogonal frame fields. Now $F(e_i) = \sum F_{ij} e_j$. So $\langle F(e_i), e_j\rangle = F_{ij}\langle e_j, e_j\rangle$. Hence F is skew symmetric if and only if $F_{ji}\langle e_i, e_i\rangle = -F_{ij}\langle e_j, e_j\rangle$. So we can represent F by a matrix of the form

$$F = \left(\begin{array}{c|ccc} 0 & E_1 & E_2 & E_3 \\ \hline E_1 & 0 & B_3 & -B_2 \\ E_2 & -B_3 & 0 & B_1 \\ E_3 & B_2 & -B_1 & 0 \end{array}\right) \quad \begin{array}{l} \text{where } F_{0i} = F_{i0} = E_i \\ \text{and } F_{ij} = -F_{ji} = B_k. \end{array} \tag{18}$$

We find it convenient to partition this matrix into blocks. So

$$F = \left(\begin{array}{c|c} 0 & \mathbf{E}^T \\ \hline \mathbf{E} & \times\mathbf{B} \end{array}\right). \tag{19}$$

where $E = \begin{pmatrix} E_1 \\ E_2 \\ E_3 \end{pmatrix}$ and $\mathbf{B} = \begin{pmatrix} B_1 \\ B_2 \\ B_3 \end{pmatrix}$. Here the notation $\times\mathbf{B}$ means

$$(\times\mathbf{B}) \begin{pmatrix} v_1 \\ v_2 \\ v_3 \end{pmatrix} = (v_1\mathbf{e}_1 + v_2\mathbf{e}_2 + v_3\mathbf{e}_3) \times (B_1 e_1 + B_2 e_2 + B_3 e_3),$$

or

$$(\times\mathbf{B})\mathbf{v} = \mathbf{v} \times \mathbf{B} \tag{20}$$

for short. This assumes that $e_1 \times e_2 = e_3$. If $e_1 \times e_2 = -e_3$, then $(\times\mathbf{B})\mathbf{v} = \mathbf{B} \times \mathbf{v}$.

LEVEL -10. ORIENTED ORTHONORMAL BASES. Same as in Level –9, but here we require $e_1 \times e_2 = e_3$.

Now in Level –9 we have

$$F = \begin{pmatrix} 0 & \mathbf{E}^T \\ \mathbf{E} & \times\mathbf{B} \end{pmatrix} \quad \text{and} \quad F^* = \begin{pmatrix} 0 & -\mathbf{B}^T \\ -\mathbf{B} & \times\mathbf{E} \end{pmatrix}. \tag{21}$$

Then

$$cF = \begin{pmatrix} 0 & \mathbf{A}^T \\ \mathbf{A} & \times(-i\mathbf{A}) \end{pmatrix} \quad \text{where} \quad \mathbf{A} = \mathbf{E} + i\mathbf{B} \tag{22}$$

Note that any matrix of the form $\begin{pmatrix} 0 & \mathbf{E}^T \\ \mathbf{E} & \times\mathbf{B} \end{pmatrix}$ represents a skew symmetric linear map.

SCHOLIUM 2.5. LORENTZ TRANSFORMATION AT LEVEL -2.
Let u and u' be observers. Then

$$u' = \frac{1}{\sqrt{1-w^2}}(u + \mathbf{w}) \tag{23}$$

where $\mathbf{w}$ is space-like in T^u_x. We call $\mathbf{w}$ the velocity of u' relative to u. There is a symmetric formula

$$u = \frac{1}{\sqrt{1-w'^2}}(u' + \mathbf{w}')$$

But note that $\mathbf{w}'$ does not lie in the same subspace as $\mathbf{w}$. However $w = w'$ and $\mathbf{w}$ and $\mathbf{w}'$ both lie in the u, u' plane. Now if a particle moves along u' as seen by u, then

$$\mathbf{a} = qFu' = \frac{q}{\sqrt{1-w^2}} \begin{pmatrix} 0 & \mathbf{E}^T \\ \mathbf{E} & \times\mathbf{B} \end{pmatrix} \begin{pmatrix} 1 \\ \mathbf{w} \end{pmatrix} = q\left[(\mathbf{E}\cdot\mathbf{w})u + \mathbf{E} + \mathbf{w}\times\mathbf{B}\right]/\sqrt{1-w^2}. \tag{24}$$

This is a more familiar form of the Lorentz Law.

The block matrix of Level –10 gives a very effective way of discovering facts about F. Most of the time we will use Level –2 proofs or Level –10 proofs. But what are definitely superior are Level –2 statements.

Now from the block matrices of Level –10 we quickly find several facts.

PROPOSITION 2.6. *a)* $\dim_{\mathbb{R}} \ell_x = 6$, *so* $\dim_{\mathbb{C}} \ell_x = 3$.

b) For a given observer field u, *there is an* F *for every pair of vector fields* $\mathbf{E}$ *and* $\mathbf{B}$ *in* T^u.

c) The map $c : \ell \to \ell_{\mathbb{C}}$ *is injective, since the map* $\phi_u : \ell \to T^u \otimes \mathbb{C}$ *is a vector bundle equivalence where*

$$\phi_u(F) = cFu = Fu - iF^*u = \mathbf{E} + i\mathbf{B} \tag{25}$$

3. Key Relations

LEMMA 3.1. *Let the commutator be denoted by* $[x, y] = xy - yx$.
a) $(\times\mathbf{A})(\times\mathbf{B}) = \mathbf{A}\mathbf{B}^T - (\mathbf{A}\cdot\mathbf{B})I$
where I *is the* 3×3 *identity matrix and all vectors are column vectors.*
b) $[(\times\mathbf{B}'), (\times\mathbf{B})] = \times(\mathbf{B}'\times\mathbf{B})$.
c) $[\mathbf{E}'\mathbf{E}^T, \mathbf{E}\mathbf{E}'^T] = \times(\mathbf{E}'\times\mathbf{E})$.
d) $\times(\mathbf{B}+\mathbf{B}') = \times\mathbf{B} + \times\mathbf{B}'$.
e) $(\times\mathbf{B})^T = \times(-\mathbf{B}) = -(\times\mathbf{B})$
f) $\mathbf{v}^T(\times\mathbf{B}) = (\mathbf{B}\times\mathbf{v})^T$

PROOF OF F).

$$\begin{aligned}\mathbf{v}^T(\times\mathbf{B}) &= [(\times\mathbf{B})^T\mathbf{v}]^T = [\times(-\mathbf{B})\mathbf{v}]^T \\ &= [\mathbf{v}\times(-\mathbf{B})]^T = [\mathbf{B}\times\mathbf{v}]^T.\end{aligned}$$

□

A key result is the following

THEOREM 3.2.

$$FF^* = F^*F = -(\mathbf{E}\cdot\mathbf{B})I.$$

PROOF. Use (21) and multiply out using Lemma 3.1a.

COROLLARY 3.3. $\langle Fv, F^*v\rangle = (\mathbf{E}\cdot\mathbf{B})\langle v, v\rangle$ *for any* $v \in T(M)$. *Hence* $\mathbf{E}_u\cdot\mathbf{B}_u = \mathbf{E}_{u'}\cdot\mathbf{B}_{u'}$ *for any two observers.*

PROOF.

$$\begin{aligned}\langle Fv, F^*v\rangle &= -\langle F^*Fv, v\rangle = -\langle -(\mathbf{E}\cdot\mathbf{B})v, v\rangle \\ &= \mathbf{E}\cdot\mathbf{B}\langle v, v\rangle.\end{aligned}$$

Thus $\mathbf{E}_{u'}\cdot(-\mathbf{B}_{u'}) = \mathbf{E}\cdot\mathbf{B}(-1)$. □

COROLLARY 3.4. $-\mathbf{E}\cdot\mathbf{B} = \lambda_F\lambda_{F^*}$ *where* λ_F *is the eigenvalue for an eigenvector* s *of* F *and* λ_{F^*} *is the eigenvalue of* s *for* F^*.

PROOF. Since F and F^* commute, they have a common eigenvector s. Then

$$\lambda_{F^*}\lambda_F s = F^*Fs = -(\mathbf{E}\cdot\mathbf{B})s.$$

□

COROLLARY 3.5. $F^2 - F^{*2} = (E^2 - B^2)I$.

PROOF. Apply Theorem 3.2 to $(F+F^*)(F+F^*)^*$. So

$$\begin{aligned}(F+F^*)(F+F^*)^* &= -\langle (F+F^*)u, -(F+F^*)^*u\rangle I \\ -(F^2 - F^{*2}) &= -(\mathbf{E}-\mathbf{B})\cdot(\mathbf{B}+\mathbf{E})I \\ F^2 - F^{*2} &= (E^2 - B^2)I.\end{aligned}$$

The second equation follows from (4) and the definition of $\mathbf{E}$ and $\mathbf{B}$. □

COROLLARY 3.6. $E_u^2 - B_u^2 = E_{u'}^2 - B_{u'}^2$.

COROLLARY 3.7. $\lambda_F^2 - \lambda_{F^*}^2 = E^2 - B^2$.

DEFINITION 3.8. Let $T_F = \frac{1}{2}(F^2 + F^{*2})$. Thus T_F is a bundle map which is symmetric with respect to $\langle \ , \ \rangle$.

PROPOSITION 3.9. $T_F = F^2 - \frac{(E^2-B^2)}{2} I$.

PROOF. Usc Corollary 3.5.

PROPOSITION 3.10.

$$T_F = \left[\begin{array}{c|c} \frac{E^2+B^2}{2} & -(\mathbf{E} \times \mathbf{B})^T \\ \hline \mathbf{E} \times \mathbf{B} & \mathbf{E}\mathbf{E}^T + \mathbf{B}\mathbf{B}^T - \frac{E^2+B^2}{2} I \end{array} \right].$$

PROOF. Compare [**P**], p.117, equation (28). Use equations Lemma 3.1a and Proposition 3.9. □

COROLLARY 3.11. Trace $(T_F) = 0$.

COROLLARY 3.12. Trace $(F^2) = 2(E^2 - B^2)$, *hence*

$$T_F = F^2 - \frac{1}{4} \operatorname{tr}(F^2) I.$$

PROOF. Use Corollary 3.11 and Proposition 3.9.

SCHOLIUM 3.13. ENERGY-MOMENTUM TENSOR.

a) Physically T_F is a multiple of the energy-momentum tensor. See [**P**], p.116, equation (20).

b) The Poynting 4-vector as seen by observer u is

$$Tu = \frac{E^2 + B^2}{2} u + \mathbf{E} \times \mathbf{B}. \tag{26}$$

Thus $\frac{E^2+B^2}{2}$ is interpreted as the energy of the electromagnetic field F, and $\mathbf{E} \times \mathbf{B}$ is interpreted as the 3-momentum per unit volume of the field F.

4. The Complex Structure and Commutators

Using the commutator relations Lemma 3.1b and c and matrix multiplication, we obtain the following key result for commutators $[F_1, F_2] = F_1F_2 - F_2F_1$.

THEOREM 4.1.

$$[F_1, F_2] = \left[\begin{array}{c|c} 0 & (-\mathbf{E}_1 \times \mathbf{B}_2 - \mathbf{B}_1 \times \mathbf{E}_2)^T \\ \hline (-\mathbf{E}_1 \times \mathbf{B}_2 - \mathbf{B}_1 \times \mathbf{E}_2) & \times(\mathbf{E}_1 \times \mathbf{E}_2 - \mathbf{B}_1 \times \mathbf{B}_2) \end{array} \right]$$

In other words

$$[F', F]u = -\mathbf{E}' \times \mathbf{B} - \mathbf{B}' \times \mathbf{E}$$
$$-[F', F]^* u = \mathbf{E}' \times \mathbf{E} - \mathbf{B}' \times \mathbf{B}.$$

COROLLARY 4.2.

$$[F_1, F_2]^* = [F_1, F_2^*] = [F_1^*, F_2]. \tag{27}$$

Hence

$$e^{i(\theta+\phi)} \cdot [F_1, F_2] = [e^{i\theta} \cdot F_1 \quad , \quad e^{i\phi} \cdot F_2]. \tag{28}$$

PROOF. (27) follows from (Theorem 4.1) and (28) follows from (27).

Hence the complexification of $\Gamma(\ell)$ commutes with the Lie algebra structure of $\Gamma(\ell)$.

THEOREM 4.3.

$$[cF, cG] = 2c([F, G]).$$

PROOF.

$$\begin{aligned}
[cF, cG] &= (F - iF^*)(G - iG^*) - (G - iG^*)(F - iF^*) \\
&= FG - F^*G^* - i(FG^* + F^*G) - GF + G^*F - i(-GF^* - G^*F) \\
&= [F, G] + [G^*, F^*] - i([F, G^*] + [F^*, G]) \\
&= [F, G] + [G, F]^{**} - i([F, G] + [F, G])^* \\
&= 2([F, G] - i[F, G]^*) = 2c[F, G].
\end{aligned}$$

where the last equality comes from the definition of c, and the previous two equalities come from (27) and (4). □

COROLLARY 4.4.

$$(c[F_1, F_2])u = i(\mathbf{E}_1 + i\mathbf{B}_1) \times (\mathbf{E}_2 + i\mathbf{B}_2)$$

for observer u.

PROOF. $c[F_1, F_2] = \frac{1}{2}[cF_1, cF_2]$ by Theorem 4.3. Now $cF_1 = \begin{pmatrix} 0 & \mathbf{A}_1^T \\ \mathbf{A}_1 & \times(-i\mathbf{A}_1) \end{pmatrix}$ where $A_1 = \mathbf{E}_1 + i\mathbf{B}_1$ and similarly for cF_2. Now by Theorem 4.1 for complex F, we have

$$\begin{aligned}
[cF_1, cF_2]u &= -\mathbf{A}_1 \times (-i\mathbf{A}_2) - (-i(\mathbf{A}_1) \times \mathbf{A}_2) \\
&= 2i\mathbf{A}_1 \times \mathbf{A}_2 = 2i(\mathbf{E}_1 + i\mathbf{B}_1) \times (\mathbf{E}_2 + i\mathbf{B}_2).
\end{aligned}$$

□

THEOREM 4.5. *Let* $\mathbb{F} = \begin{pmatrix} 0 & \mathbf{A}^T \\ \mathbf{A} & \times\mathbf{C} \end{pmatrix}$ *where* $\mathbf{A}$ *and* $\mathbf{C}$ *are complex 3-vectors. Then* $\mathbb{F}^2 = kI$ *if and only if* $k = \mathbf{A} \cdot \mathbf{A}$ *and* $\mathbf{C} = \pm i\mathbf{A}$.

PROOF. Assume $\mathbb{F}^2 = kI$. Then $\mathbb{F}^2 w = kw$ for any vector w. For $w = \begin{pmatrix} 1 \\ 0 \\ 0 \\ 0 \end{pmatrix}$ we get $\mathbf{A} \cdot \mathbf{A} = k$ and $\mathbf{A} \times \mathbf{C} = \mathbf{0}$. Thus $\mathbf{C} = s\mathbf{A}$ for some $s \in \mathbb{C}$. Hence $\mathbb{F} = \begin{pmatrix} 0 & \mathbf{A}^T \\ \mathbf{A} & \times(s\mathbf{A}) \end{pmatrix}$ and $\mathbb{F}^2 = (\mathbf{A} \cdot \mathbf{A})I$. Apply $\begin{pmatrix} 0 \\ \mathbf{v} \end{pmatrix}$ to this last equation. We obtain

$$(\mathbf{v} \cdot \mathbf{A})\mathbf{A} + (\mathbf{v} \times s\mathbf{A}) \times (s\mathbf{A}) = (\mathbf{A} \cdot \mathbf{A})\mathbf{v}.$$

Using the third and fourth equations of (8) and rearranging terms we get

$$(1+s^2)(\mathbf{A}\cdot\mathbf{v})\mathbf{A} = (1+s^2)(\mathbf{A}\cdot\mathbf{A})\mathbf{v}$$

for arbitrary $\mathbf{v}$. Thus $1+s^2=0$. Hence $s=\pm i$.

For the converse, first suppose $s=-i$, so $\mathbb{F} = \begin{pmatrix} 0 & \mathbf{A}^T \\ \mathbf{A} & \times(-i\mathbf{A}) \end{pmatrix}$. Then $\mathbb{F} = F - iF^* = cF$ where $cFu = \mathbf{E}+i\mathbf{B} = \mathbf{A}$. Then

$$\begin{aligned}\mathbb{F}^2 &= (cF)^2 = (F-iF^*)^2 = F^2 - F^{*2} - i(FF^* + F^*F)\\ &= (E^2-B^2)I - 2i(-\mathbf{E}\cdot\mathbf{B})I\\ &= (E^2 - B^2 + 2\mathbf{E}\cdot\mathbf{B})i)I\\ &= (\mathbf{E}+i\mathbf{B})\cdot(\mathbf{E}+i\mathbf{B})I = (\mathbf{A}\cdot\mathbf{A})I.\end{aligned}$$

Similarly if $s=i$, $\mathbb{F}^2 = (\mathbf{A}\cdot\mathbf{A})I$ implies $\mathbb{F}^2 = (\mathbf{A}\cdot\mathbf{A})I$. □

COROLLARY 4.6.

$$\begin{aligned}(cF)^2 &= (\mathbf{A}\cdot\mathbf{A})I = \lambda_{cF}^2 I\\ (\bar{c}F)^2 &= \lambda_{\bar{c}F}^2 I.\end{aligned}$$

PROOF. The inner equality of the first line was proved above. Apply this equation to an eigenvector s to get the last equality of the first line. The second equality follows from complex conjugation.

COROLLARY 4.7. $cF_1cF_2 + cF_2cF_1 = 2(\mathbf{A}_1\cdot\mathbf{A}_2)I$.

PROOF.

$$\begin{aligned}cF_1cF_2 + cF_2cF_1 &= (cF_1+cF_2)^2 - cF_1^2 - cF_2^2\\ &= [(\mathbf{A}_1+\mathbf{A}_2)\cdot(\mathbf{A}_1+\mathbf{A}_2) - (\mathbf{A}_1\cdot\mathbf{A}_1 - \mathbf{A}_2\cdot\mathbf{A}_2]I = 2(\mathbf{A}_1\cdot\mathbf{A}_2)I.\end{aligned}$$

□

THEOREM 4.8. $cF_1\overline{cF}_2 = \overline{cF}_2cF_1$.

PROOF. We must show that $[cF_1, \overline{cF}_2] = 0$. Apply theorem 4.1 where $\mathbf{E}_j = \mathbf{A}_j$ and $\mathbf{B}_j = (-1)^j i\mathbf{A}_j$ for $j=1,2$. Then all cross products must be zero in Theorem 4.1 and we obtain the desired result. □

REMARK 4.9. CLIFFORD ALGEBRAS.

According to the first proposition of [**LM**], Corollary 4.6 is a clue that $c : \ell_x \to \mathbb{C}(4)$ involves representations of Clifford modules. Here $\mathbb{C}(4)$ is the space of linear maps on $T_x\otimes\mathbb{C}$, a 16 dimensional space which is a complex Clifford Algebra. The image of c in $\mathbb{C}(4)$ generates the Quaternions tensored with $\mathbb{C}$. The complex conjugate $\bar{c}$ generates another complex representation of the Quaternions in $\mathbb{C}(4)$. The two representations commute, and they generate all of $\mathbb{C}(4)$ under composition.

SCHOLIUM 4.10. PAULI MATRICES.

The Pauli matrices of physics play an important role in quantum mechanics. The relations among their products are their key features, the actual form of the matrices is not important. Thus we have $\boldsymbol{\sigma}_x$, $\boldsymbol{\sigma}_y$, $\boldsymbol{\sigma}_z$ so that $\{\boldsymbol{\sigma}_i, \boldsymbol{\sigma}_j\} = 2\delta_{ij}I$ and $[\boldsymbol{\sigma}_x, \boldsymbol{\sigma}_y] = 2i\boldsymbol{\sigma}_z$, **[F, III, 11-4]**. We get the same relations using cF as follows. Let E_x, E_y, E_z be the F with zero $\mathbf{B}$ field and with unit $\mathbf{E}$ fields pointing along the x, y, z axes, respectively, of Minkowski space. So for example $E_x = \begin{pmatrix} 0 & \mathbf{e}_x \\ \mathbf{e}_x & 0 \end{pmatrix}$. Denote $\boldsymbol{\sigma}_x = cE_x$, $\boldsymbol{\sigma}_y = cE_y$ and $\boldsymbol{\sigma}_z = cE_z$. Then $\boldsymbol{\sigma}_x$, $\boldsymbol{\sigma}_y$, $\boldsymbol{\sigma}_z$ satisfy the Pauli matrix relations. In addition, $\overline{\boldsymbol{\sigma}}_x$, $\overline{\boldsymbol{\sigma}}_y$, $\overline{\boldsymbol{\sigma}}_z$ commute with the $\boldsymbol{\sigma}$'s and satisfy the Pauli relations among themselves except that $\overline{\boldsymbol{\sigma}}_x\overline{\boldsymbol{\sigma}}_y = -i\overline{\boldsymbol{\sigma}}_z$. Also $\boldsymbol{\sigma}_x$, $\boldsymbol{\sigma}_y$, $\overline{\boldsymbol{\sigma}}_x$, $\overline{\boldsymbol{\sigma}}_y$ generate the Clifford algebra $\mathbb{C}(4)$. This can be shown by brute force.

5. Eigenvectors

Recall our notation in which $\bar{c}F = \overline{cF}$ and $\lambda_{\bar{c}F} = \overline{\lambda_{cF}}$.

PROPOSITION 5.1. *$cF \circ \overline{cF} = 2T_F$. Hence $\lambda_{cF}\overline{\lambda_{cF}} = 2\lambda_T$.*

PROOF. $cF \circ \bar{c}F = (F - iF^*)(F + iF^*) = F^2 + F^{*2}$ since $FF^* = F^*F$. Now apply the definition of T_F, (Definition 3.8).

COROLLARY 5.2.

$$T_{e^{i\theta}\cdot F} = T_F.$$

PROOF.

$$\begin{aligned} T_{e^{i\theta}\cdot F} &= \frac{1}{2}c(e^{i\theta} \cdot F) \circ \overline{c(e^{i\theta} \cdot F)} \\ &= \frac{1}{2}(e^{i\theta}cF) \circ e^{-i\theta}\overline{cF} = \frac{1}{2}cF\bar{c}F = T_F. \end{aligned}$$

□

COROLLARY 5.3. *$T^2 = \lambda_T^2 I$ where λ_T is an eigenvalue of T.*

PROOF. $T^2 = \frac{1}{4}((cF)(\overline{cF}))^2 = \frac{1}{4}(cF)^2(\overline{cF})^2 = \frac{1}{4}\lambda_{cF}^2\overline{\lambda}_{cF}^2 I$ by Theorem 4.8. So $T^2 = \lambda_T^2 I$.

THEOREM 5.4. *Let $F \in \Gamma(\ell)$ and let λ_F be an eigenvalue of F and λ_T be an eigenvalue of T_F.*

a) $\lambda_T = \sqrt{(\frac{E^2 - B^2}{2})^2 + (\mathbf{E} \cdot \mathbf{B})^2}$

b) $\lambda_F = \pm\sqrt{\lambda_T + \frac{(E^2 - B^2)}{2}}$, $\quad \lambda_{F^*} = \pm\sqrt{\lambda_T - \frac{(E^2 - B^2)}{2}}$.

c) $\lambda^4 - (E^2 - B^2)\lambda^2 - (\mathbf{E} \cdot \mathbf{B})^2$ *, or equivalently,*
$\lambda^4 - (\lambda_F^2 - \lambda_{F^*}^2)\lambda^2 - (\lambda_F\lambda_{F^*})^2$*, is the characteristic polynomial of F.*

PROOF. Corollaries 3.4 and 3.7 gives the equations $\lambda_F\lambda_{F^*} = -\mathbf{E}\cdot\mathbf{B}$ (Corollary 3.4) and $\lambda_F^2 - \lambda_{F^*}^2 = E^2 - B^2$ (Corollary 3.7). Eliminating λ_{F^*} from (Corollary 3.4) and (Corollary 3.7) gives $\lambda_F^4 - (E^2 - B^2)\lambda_F^2 - (\mathbf{E} \cdot \mathbf{B})^2 = 0$. Solving gives b). Then a) follows from $\lambda_T = \lambda_F^2 - \frac{(E^2 - B^2)}{2}$ which follows from Corollary 3.7. To be absolutely certain that c) is the characteristic polynomial, one must calculate $\det(F - \lambda I)$ for F represented as a matrix in (18). □

PROPOSITION 5.5. *If s is an eigenvector of $F \in \ell_x$, then $\lambda_F\langle s,s\rangle = 0$. So if $\lambda_F \neq 0$, then s is a null vector. Both λ_F and λ_{F^*} are zero if and only if $\lambda_T = 0$. In that case s is a multiple of $\frac{E^2+B^2}{2}\, u + \mathbf{E}\times\mathbf{B}$, which is null.*

PROOF. $\lambda_F\langle s,s\rangle = \langle\lambda_F s,s\rangle = \langle Fs,s\rangle = -\langle s,F_s\rangle = -\lambda_F\langle s,s\rangle$. The same argument holds for the complex cF, so $\lambda_{cF}\langle s,s\rangle = 0$. Since $\lambda_T = \frac{1}{2}\lambda_{cF}\overline{\lambda_{cF}}$, we get the second sentence. Now $\lambda_T = 0$ if and only if $E = B$ and $\mathbf{E}\cdot\mathbf{B} = 0$. Under those conditions, use (19) to show that $\frac{E^2+B^2}{2}\, u + \mathbf{E}\times\mathbf{B}$ is an eigenvector and is a null vector. □

SCHOLIUM 5.6. THE NULL AND NON NULL CASES.

The null and non-null cases are when $\lambda_T = 0$ and $\lambda_T \neq 0$ respectively. If $\lambda_T = 0$ then $E = B$ and $\mathbf{E}\cdot\mathbf{B} = 0$. This is called the null case mathematically. Physicists identify an electro-magnetic field with $E = B$ and $\mathbf{E}\cdot\mathbf{B} = 0$ as the *radiative* or *wave-like* case. In the null case $\lambda_F = \lambda_{F^*} = \lambda_T = \lambda_{cF} = 0$. The characteristic polynomial is λ^4, $T = F^2$, F^2u is the eigenvector of F^2.

PROOF. $F^2(F^2u) = F^4u = T^2u = 0$. So $s = F^2u = Tu = \frac{E^2+B^2}{2}\, u + \mathbf{E}\times\mathbf{B}$. (The Poynting 4-vector). Now s is null, i.e. $\langle s,s\rangle = 0$. Since $\langle s,s\rangle = \langle Ts,Ts\rangle = \langle T^2s,s\rangle = \langle 0,s\rangle = 0$. So image $(T^2) =$ span s. Then dim ker $T = 3$. □

Now consider the non-null case. Then $\lambda_T \neq 0$. Hence $\lambda_{cF} \neq 0$, so one of λ_F or λ_{F^*} is not zero. Hence there are two real null eigenvectors of cF, s for λ_{cF} and s_- for $-\lambda_{cF}$. Both s and s_- are linearly independent. Since $T = \frac{1}{2}\, cF\overline{cF}$, s and s_- are both eigenvectors of T with eigenvalue $\lambda_T > 0$.

Let Π_+ be the space of eigenvectors of T in $T_x(M)$ corresponding to λ_T and let Π_- be the space of eigenvectors corresponding to $-\lambda_T$. Then $\Pi_+ =$ image (Φ_+) and $\Pi_- =$ image (Φ_-) where $\Phi_+ = \lambda_T I + T$ and $\Phi_- = -\lambda_T I + T$. Now $\Phi_\pm$ are symmetric with respect to $\langle\ ,\ \rangle$. Note that $\Phi_\pm^2 = \pm 2\lambda_T\Phi_\pm$ and $\Phi_+\Phi_- = \Phi_-\Phi_+ = 0$, all because of the fact that $T^2 = \lambda_T^2 I$. From this we obtain:

PROPOSITION 5.7. *Let F be non null.*
a) Π_+ is orthogonal to Π_-.
b) Π_+ is time-like and Π_- is space like.
c) $\dim\Pi_+ = \dim\Pi_- = 2$.
d) $F(\Pi_\pm) \subset (\Pi_\pm)$, i.e. $\Pi_\pm$ are invariant subspaces of F.

PROOF. The following two lemmas prove a), b) and c). And d) follows since for $v \in \Pi_\pm$ we have $\pm\lambda_T F(v) = F(\pm\lambda_T v) = F(T(v)) = T(F(v))$. So $F(v) \in \Pi_\pm$.□

LEMMA 5.8. *Suppose $Q : T_x \to T_x$ is symmetric with respect to $\langle\ ,\ \rangle$. If Q has a time-like eigenvector, then Q has an orthonormal frame of eigenvectors.*

PROOF. Let u be a time-like eigenvector of Q. We may assume that $\langle u,u\rangle = -1$. Consider T_x^u, the space of vectors orthogonal to u. Then $Q : T_x^u \to T_x^u$ since $\langle u,Qv\rangle = \langle Qu,v\rangle = \lambda_Q\langle u,v\rangle = 0$ if $v \in T_x^u$. Hence $Q : T^u \to T^u$. But T^u is space-like and $\langle\ ,\rangle$ on T^u is positive definite and Q is symmetric. Hence there is an

LEMMA 5.9. *Let $Q : T_x \to T_x$ be a linear map which is*

a) symmetric with respect to $\langle\ ,\ \rangle$, i.e. $\langle Qv, w\rangle = \langle v, Qw\rangle$.

b) $Q^2 = \lambda^2 I$.

c) Trace $(Q) = 0$.

d) $\langle u, Qu\rangle < 0$ for some future timelike u.

Then if $\lambda = 0$, there is a null eigenvector s so that image (Q) = span s. If $\lambda \neq 0$, then the set of all eigenvectors corresponding to $\pm\lambda$ form two 2 dimensional subspaces $\Pi_\pm$, and Π_+ is orthogonal to Π_-, and Π_+ is time-like and Π_- is space like.

PROOF. Suppose $\lambda = 0$. Then $\langle Qv, Qv\rangle = \langle Q^2 v, v\rangle = 0$ for all $v \in T_x$. So the image of Q consists of null-vectors. Since $\langle u, Qu\rangle < 0$ for some time-like u, we see that $Qu \neq 0$ and that $Q(Qu) = 0$. So Qu is the desired s.

Suppose $\lambda \neq 0$. Let $\langle u, Qu\rangle < 0$ for observer u. Consider $\lambda u + Qu$. Then $\langle \lambda u + Qu, \lambda u + Qu\rangle = -2\lambda^2 + 2\lambda\langle u, Qu\rangle < 0$. So $\lambda u + Qu$ is time like. But $Q(\lambda u + Qu) = \lambda Qu + Q^2 u = \lambda(\lambda u + Qu)$. So $\lambda u + Qu$ is a time-like eigenvector. Thus by Lemma 5.8, there is an orthonormal eigenvector frame. Since trace $(Q) = 0$, two of the vectors of the frame correspond to λ and generate a time-like plane Π_+ and the orthogonal two generate Π_- and are space-like. □

COROLLARY 5.10. *If $Q : T_x \to T_x$ is as in the theorem above, there is an antisymmetric $F : T_x \to T_x$ so that $T_F = Q$.*

PROOF. If $\lambda \neq 0$, Π_+ intersects the light cone in two null-subspaces generated by, say, s_+ and s_- respectively. Let $\lambda_F = \sqrt{2\lambda}$. Define $Fs_+ = \lambda_F s_+$ and $Fs_- = -\lambda_F s_-$. Let $F(v) = 0$ for all v in Π_- so we are defining $\lambda_{F^*} = 0$. Then there is a unique linear map which satisfies these conditions and $Q = F^2 - \lambda_F^2 I$. Note F is antisymmetric on Π_- since it is trivial and on Π_+ since

$$\langle \alpha s_+ + \beta s_-, F(\alpha s_+ + \beta s_-)\rangle = \langle \alpha s_+ + \beta s_-, \alpha\lambda_F s_+ - \beta\lambda_F s_-\rangle = 0,$$

since s_+ and s_- are null.

If $\lambda = 0$, choose observer u and let $s = Qu$. Choose $\mathbf{E}$ and $\mathbf{B} \in T^u$ so that $s, \mathbf{E}, \mathbf{B}$ are in the kernel of Q and are mutually orthogonal and of sufficient length so that $s = E^2 u + \mathbf{E} \times \mathbf{E}$ where $B = E$. Then let $Fu = \mathbf{E}$, $F(\mathbf{B}) = 0$, $F(s) = 0$ and $F(\mathbf{E}) = s$. Then F is determined and $F^2 = Q$. □

REMARK 5.11.

The question is, given Q over TM, does there exist an F so that $Q = T_F$?

6. Complex Eigenvectors

Let $\phi_+ = \lambda_{cF} I + cF$ and $\phi_- = -\lambda_{cF} I + cF$. Let $\overline{\phi}_+ = \overline{\lambda}_{cF} I + c\overline{F}$ and $\overline{\phi}_- = -\overline{\lambda}_{cF} I + c\overline{F}$. Since $cF^2 = \lambda_{cF}^2 I$ and $cFc\overline{F} = c\overline{F}cF$, we obtain the following facts.

THEOREM 6.1. *Let $cF : T_x \otimes \mathbb{C} \to T_x \otimes \mathbb{C}$ and $cF \neq 0$.*

a) The image of $(\phi_\pm)$ equals the $\pm\lambda_{cF}$ eigenspace of cF and the image of $(\overline{\phi}_\pm)$ equals the $\pm\overline{\lambda}_{cF}$ eigenspace of $c\overline{F}$.

b) The kernel of $(\phi_\pm)$ equals the $\mp\lambda_{cF}$ eigenspace of cF, and the kernel of $(\overline{\phi}_\pm)$ equals the $\mp\overline{\lambda}_{cF}$ eigenspace of $c\overline{F}$.

c) *The eigenspaces of* cF *and* $\overline{cF}$ *consist of null vectors.*
d) *The eigenspaces of* cF *and* $\overline{cF}$ *have dimension 2.*

PROOF.
We easily see that

$$cF\phi_\pm = \pm\lambda_{cF}\,\phi_\pm, \quad \overline{cF}\,\overline{\phi_\pm} = \pm\overline{\lambda_{cF}}\,\overline{\phi_\pm} \tag{29}$$

$$\phi_\pm\phi_\mp = 0, \ \overline{\phi_\pm}\,\overline{\phi_\mp} = 0 \tag{30}$$

$$\langle\phi_\pm v, w\rangle = \langle v, \phi_\mp w\rangle \tag{31}$$

a) follows from (29)
b) follows from (30) and a)
c) follows from a) and (31).
d) For an observer u, the vectors $\phi_+\overline{\phi}_+u$ and $\phi_+\overline{\phi}_-u$ are eigenvectors of cF by (29).

Now $\phi_+\overline{\phi}_+u$ is an eigenvector of cF corresponding to λ_{cF} as well as an eigenvector of $\overline{cF}$ corresponding to $\overline{\lambda}_{cF}$. On the other hand $\phi_+\overline{\phi}_-u$ is an eigenvector of cF corresponding to λ_{cF} and also an eigenvector of $\overline{cF}$ corresponding to $-\overline{\lambda}_{cF}$. If $\phi_+\overline{\phi}_+u$ is linear dependent on $\phi_+\overline{\phi}_-u$, then $-\overline{\lambda}_{cF} = \overline{\lambda}_{cF}$, hence $\lambda_{cF} = 0$, hence F is null. Thus if F is nonnull, $\phi_+\overline{\phi}_+u$ and $\phi_+\overline{\phi}_-u$ are linearly independent eigenvectors. If F is null, then $F^2u = E^2u + \mathbf{E}\times\mathbf{B}$ and $cFu = \mathbf{E} + i\mathbf{B}$ are linearly independent eigenvectors of the eigenspace. Hence $\dim(\text{image}(\phi_+)) \geq 2$ and similarly $\dim(\ker(\phi_+)) \geq 2$. Therefore d) is proved. □

LEMMA 6.2. *Suppose* a *and* b *are real vectors in* T_x. *Then* $a + ib \in T_x \otimes \mathbb{C}$ *is null if and only if either* a *and* b *are linear dependent null vectors, or* a *and* b *are both space–like and have the same length and are orthogonal.*

PROOF. Let $v = a + ib$. Now $\langle v, v\rangle = 0$ if and only if $\langle a, a\rangle = \langle b, b\rangle$ and $\langle a, b\rangle = 0$. If a or b is null, so is the other. Since they are orthogonal null vectors, they must be linearly dependent.

On the other hand, if one of a or b is space–like, so is the other and they have equal lengths and are orthogonal. Neither a or b can be time–like, since if one were, they both would be. But no two time–like vectors are orthogonal. □

LEMMA 6.3. *Let* $\mathbf{a}$ *and* $\mathbf{b}$ *be space–like in* T_x. *Then* $\mathbf{a}$ *and* $\mathbf{b}$ *span a space–like plane if and only if* $a^2b^2 - \langle\mathbf{a}, \mathbf{b}\rangle^2 > 0$. *Thus if* $\mathbf{a}$ *and* $\mathbf{b}$ *are orthogonal and space–like, they span a space–like plane.*

PROOF. $\langle\alpha\mathbf{a}+\beta\mathbf{b},\ \alpha\mathbf{b}+\beta\mathbf{b}\rangle$ is greater than zero if and only if the determinant of

$$\begin{pmatrix} a^2 & \langle a, b\rangle \\ \langle a, b\rangle & b^2 \end{pmatrix}$$

is greater than zero.

LEMMA 6.4. *Any null subspace of* $T_x \otimes \mathbb{C}$ *has a degenerate inner product. That is any two vectors in a subspace of null vectors are orthogonal.*

REMARK 6.5. NULL PLANES.

Suppose $\mathbf{E} + i\mathbf{B} \in T_x^u \otimes \mathbb{C}$ is a null vector in the rest space of an observer u. Then $s = E^2u + \mathbf{E} \times_u \mathbf{B}$ is a real null vector. Now s and $\mathbf{E}+i\mathbf{B}$ span a null plane V, which is the image of cF where F is a null skew symmetric operator with $Fu = \mathbf{E}$ and $F^*u = -\mathbf{B}$.

Also $s_- = E^2u - \mathbf{E} \times \mathbf{B}$ is a real null vector. Again s_- and $\mathbf{E}+i\mathbf{B}$ span a null plane V', which is the image of $\overline{c}G$ for a null G so that $Gu = \mathbf{E}$ and $G^*u = \mathbf{B}$.

Thus we have two kinds of null planes, those which are the images of null cF and those which are the images of null $\overline{cF}$.

We can think of these null planes from a geometric point of view. Suppose $\mathbf{E}+i\mathbf{B}$ is a space–like null vector. Then $\mathbf{E}$ and $\mathbf{B}$ span a space–like plane $\Pi_s \subset T_x$, by Lemma 6.3. Let Π_t be the time–like plane orthogonal to Π_s. Then Π_t intersects the light cone in two one–dimensional null lines. One of these real null lines and $\mathbf{E}+i\mathbf{B}$ spans a null plane and the other line and $\mathbf{E}+i\mathbf{B}$ spans the "conjugate" null plane containing $\mathbf{E}+i\mathbf{B}$.

Thus given a space–like null vector v, there are exactly two null planes containing v. We say these two planes are *$*$–conjugate with respect to v*. If V is a null plane and contains a light–like null vector v, then we say that $\overline{V}$ *is $*$–conjugate* to V with respect to v. The planes which are the image of a null cF are called *$*$–consistent* null planes and those which are the image of a null $\overline{c}F$ are called *$*$–inconsistent.*

LEMMA 6.6. *In $T_x \otimes \mathbb{C}$*

a) *Every null plane contains a real null vector*
b) *The eigenspaces of cF are $*$–consistent planes. The eigenspaces of $\overline{cF}$ are $*$–inconsistent planes.*
c) *The intersection of a $*$–consistent and a $*$–inconsistent plane is one dimensional.*

PROOF. a) Choose an appropriate basis and use analysis to obtain the conditions for a null–plane.

b) By continuity and connectivity of $\ell_x \oplus \mathbb{C}$.
c) Let V be the null plane spanned by $\mathbf{A} = \mathbf{E}+i\mathbf{B}$ and $s = E^2u + \mathbf{E} \times \mathbf{B}$ for u an observer orthogonal to $\mathbf{E}$ and $\mathbf{B}$. Then V is both the image and kernel of a null cF such that $cFu = \mathbf{A}$, since $cF^2 = 0$. Let $\overline{W}$ be a $*$–inconsistent null plane. It is the image of some null $\overline{c}G$. Now $\overline{W} \neq V$ since $\overline{W}$ is $*$–inconsistent, so $cF(\overline{W}) \neq 0$. Since cF and $\overline{c}G$ commute by Theorem 4.8, we see that $\overline{W} \cap V \neq 0$. So $W \cap V$ is one dimensional. □

THEOREM 6.7. *Let F and G be skew symmetric bundle maps. Let $\phi = \lambda_{cF}I + cF$ and $\overline{\gamma} = \overline{\lambda}_{cG}I + c\overline{G}$. Note that the choice of which of the two eigenvectors $\pm\lambda_{cF}$ is not reflected in the notation.*

a) *$\phi\overline{\gamma} = \overline{\gamma}\phi$.*
b) *The image of $(\phi\overline{\gamma})$ is one dimensional and is generated by a null vector which is an eigenvector of both cF and $c\overline{G}$ with associated eigenvalues λ_{cF} and $\overline{\lambda}_{cG}$ respectively.*
c) *The image of $(\phi\overline{\phi})$ is generated by a real null vector s which is an eigenvector of cF corresponding to λ_{cF}.*

PROOF.

b) From (a), the image of $\phi\overline{\gamma}$ is the one dimensional sub space (image (ϕ)) $\cap$ (image $(\overline{\gamma})$).

c) Let $\gamma = \phi$ and apply (b).

COROLLARY 6.8. *The eigenvector s for a skew symmetric bundle map satisfies the following equation in terms of $\mathbf{E}_u$ and $\mathbf{B}_u$,*

$$s = 2(\lambda_T u + \frac{E^2 + B^2}{2} u + \mathbf{E} \times \mathbf{B} + \lambda_F \mathbf{E} - \lambda_{F^*} \mathbf{B}).$$

PROOF. Recall $s = \phi\overline{\phi}u$ where $\phi\overline{\gamma} = \overline{\gamma}\phi$. Expand that equation and use equations (5), (13), and (26) and Proposition 5.1. □

Define $\Psi : \ell \oplus \mathbb{C} \oplus T(M) \longrightarrow T(M) \otimes \mathbb{C}$ by $\Psi(F, \alpha, v) = (\alpha I + cF)v$. Let $\Psi_v : \ell \oplus \mathbb{C} \to T(M) \otimes \mathbb{C}$ be defined by

$$\Psi_v(F, \alpha) = \Psi(F, \alpha, v).$$

THEOREM 6.9. *$\Psi_v : \ell \oplus \mathbb{C} \to T(M) \otimes \mathbb{C}$ is a bundle equivalence if v is a non null vector field.*

PROOF. Both bundles are 4 dimensional and Ψ_v is a bundle map, so we only need to show that Ψ_v has zero kernel. So assume $(\alpha I + cF)v = 0$. Then v is an eigenvector of cF, hence by Theorem 6.1c we have, in contradiction to the hypothesis, that v is a null vector.

7. Eigenbundles

Given a skew symmetric $F \in \Gamma(\ell)$, we define a map $\psi_F : M \longrightarrow \mathbb{C}$ by setting

$$\psi_F(m) = \lambda^2_{cF_m} = (E^2 - B^2) + 2i(\mathbf{E} \cdot \mathbf{B}) \tag{32}$$

evaluated at m.

We define a sequence of open submanifolds $M \supset M_0 \supset M_1$ based on the given F.

$$M_0 = \{m \in M \mid F_m \text{ is defined and not identically zero}\} \tag{33}$$
$$M_1 = \{m \in M \mid F_m \text{ is not null}\} \tag{34}$$

Since F_m is null if and only if $\lambda_{cF} = 0$, we see that

$$\psi_F^{-1}(\mathbb{C} - 0) = M_1. \tag{35}$$

DEFINITION 7.1. We define the degree of F, denoted deg F, to be the degree of $\psi_F : M_1 \longrightarrow \mathbb{C} - 0$. We define the degree of $\psi : M_1 \longrightarrow \mathbb{C} - 0$ to be the integer which corresponds to the generator of the subgroup (image (ψ)) $\subset H_1(\mathbb{C} - 0) \cong \mathbb{Z}$.

REMARK 7.2. The degree of ψ in Definition 7.1 is related to the usual Brouwer degree of Algebraic Topology. This can be seen in [$\mathbf{G}_4$]. Note, the definition of deg ψ yields a non–negative integer, in contrast to the usual Brouwer degree.

THEOREM 7.3. *The following are equivalent:*

a) $\deg \psi$ is even.

b) There is a line bundle of eigenvectors of F over M_1.

PROOF. Consider $\widetilde{M_1}$, the set of pairs (m, α) where $m \in M_1$, and $\alpha \in \mathbb{C} - 0$ is equal to either one of the two eigenvalues $\pm\lambda_{cF_m}$. Then $\widetilde{M_1}$ is a double covering space of M_1. If $\widetilde{M_1}$ is not connected, then it is possible to choose one α at each m in a continuous way over M_1. The choice of the eigenvector corresponding to $\alpha(m)$ gives the line bundle of eigenvectors. Conversely, a line bundle of eigenvectors over M_1 will select a continuous choice of corresponding eigenvalues, so $\widetilde{M_1}$ will be disconnected. Now we have a commutative diagram

$$\begin{array}{ccc} \widetilde{M} & \xrightarrow{\ \widetilde{\psi}\ } & \mathbb{C} - 0 \\ \Big\downarrow p & & \Big\downarrow \text{sq} \\ M & \xrightarrow[\ \psi\ ]{} & \mathbb{C} - 0 \end{array}$$

where $p(m, \alpha) = m$ and $\widetilde{\psi}(m, \alpha) = \alpha$ and $\text{sq}(z) = z^2$. If $\widetilde{M}$ is not connected, there is a cross-section s to p. Then deg ψ = deg $(\text{sq} \circ \widetilde{\psi} \circ s)$ is even since the degree of sq is 2. This proves that (a) and (b) are equivalent.

For (c), the plane bundle Π_+ of time-like invariant planes of F is also the eigenbundle of T_F corresponding to $\lambda_T > 0$. Now we can always choose a nonzero time-like vector field u over M. Then $\Phi_+(u) = \lambda_T u + Tu$ is a non zero vector field of eigenvectors of T. Hence there is a trivial line sub-bundle ε in Π_+. Hence $\Pi_+ = \varepsilon \oplus \nu$, where ν is the orthogonal line bundle. If Π_+ were orientable, then ν would be trivial and we could use the direction in ν to choose at each m one of the two null eigenvector subspaces in $(\Pi_+)_m$. Hence we would get a line bundle η of eigenvectors of F. Conversely, if the eigenbundle η existed, then $\Pi_+ = \varepsilon \oplus \eta$. Since η is a trivial line bundle (because M is time-oriented) this implies that Π_+ is orientable.

Now d) is equivalent to b) because we assumed that M was time orientable. Thus the line bundle over M_0 must be trivial and hence gives a nonzero vector field over M_1. This vector field obviously extends continuously over M_0. In fact, the equation of Corollary 6.8 gives the vector field, the possible ambiguity of the choice of eigenvectors being eleminated by the fact that the bundle in b) is trivial. The converse, d) implies b), is obvious. □

SCHOLIUM 7.4. THE PHASE OF ψ.

We may write $\psi_F(m) = \lambda^2_{cF_m} = 2\lambda_{T_m} e^{i\alpha}$ for some angle α, which we will call the *phase* of ψ. Suppose we have two paths in space-time from A to B which do not pass over radiation. If we measure the difference of the phase after having traveled from point A to point B along the two paths, we will find that they differ by a multiple n of 2π. If n is not zero, then the two paths linked wave-like regions. If n is even, then the continuous extension of the same eigenvector at A along the two paths result in the same eigenvector at B.

COROLLARY 7.5. *Let F be a skew symmetric bundle map. There is a plane sub-eigen-bundle η of $T(M_0) \otimes \mathbb{C}$ if and only if* $\deg \psi_F$ *is even.*

PROOF. If deg ψ_F is even, then we can choose continuously one λ_{cF_m} out of the two possible. Thus $\phi = \lambda_{cF} I + cF$ is a well-defined bundle map since there is no ambiguity with λ_{cF}. Now over M_0, the image of (ϕ_m) is always a two plane by Theorem 6.1d. The unambiguous choice of λ_F gives a bundle map ϕ whose image is a plane bundle η. Conversely, if η is a plane eigenbundle, it selects the eigenvalue λ_{cF_m} at each m which correspond to the plane η_m.

SCHOLIUM 7.6. A NEW ELECTRO-MAGNETIC INVARIANT.
First note that from Corollary 6.8, that for any observer the vectors of the form

$$\mathbf{E} \times \mathbf{B} + \lambda_F \mathbf{E} - \lambda_{F^*} \mathbf{B} \tag{36}$$

can never be zero as long as F is defined and not identically equal to zero. Now if $\deg \psi_F$ is even, then (36) gives rise to a vector field which is nonzero over M_0. The index of that vector field on any closed compact space-like manifold whose boundry is contained in M_0 must be an invariant of the field, independent of the observer field. The index is defined in [$\mathbf{G}_5$], for example.

COROLLARY 7.7. *Let* $F \in \Gamma(\ell)$ *and suppose that* $\mathbf{E} \cdot \mathbf{B} = 0$ *for all* $m \in M_0$. *Then* $\deg \psi_F = 0$ *and* $\ker(F)$ *is a plane sub-bundle of* $T(M_0)$.

COROLLARY 7.8. *If* $0 \in \mathbb{C}$ *is a regular value of* ψ_F, *then* $\deg \psi_F = 1$.

PROOF. If 0 is a regular value of ψ_F we can find a small circle about 0 which lifts to M_1. Thus $1 \in \mathbb{Z} \cong H_1(\mathbb{C} - 0;\ \mathbb{Z})$ is the image of ψ_*.

SCHOLIUM 7.9. ELECTRONS.

a) A classical free electron at rest in Minkowski space M can be represented by an F such that $\mathbf{E}(\mathbf{r}, t) = -\frac{\mathbf{r}}{r^3}$ and $\mathbf{B}(\mathbf{r}, t) = 0$. Thus $M_0 = M_1 = M-$ (the time axis). The deg of the free electron is zero by Corollary 7.7.
b) A classical electron at rest in a constant magnetic field will be represented by an F such that $\mathbf{E}(\mathbf{r}, t) = -\frac{\mathbf{r}}{r^3}$ and $\mathbf{B} = \mathbf{e}_x$. Then $M_0 = M-$ (the time axis) and $M_1 = M_0 - S$ where S in each space slice is a circle of radius 1 in the yz plane centered on the electron. The deg of the election in a constant magnetic field is 1 by Corollary 7.8.
c) An assembly of point charges at rest in a constant magnetic field will have odd degree if the number of charged points is odd. This follows from the considerations of Scholium 7.6: If the degree were even, then (36) gives a nonzero space-like vector field everywhere except at the charged points. The vector field near a charged point points outward if the point has positive charge and inward if the point has negative charge, as can be seen by the equations of Theorem 5.4 where E is much larger than B. Thus the index of each singularity contributes a positive or negative 1 to the index of the vector field, the sign of the 1 depending upon the sign of the charge times the sign of the eigenvalue λ_F. Since the number of charges is odd, the index of the vector field cannot be zero by the summation equation (8) of [$\mathbf{G}_5$]. Hence the index is not zero by hypothesis. On the other hand the index of the vector field must be zero (by the existance of defects, property (8) of [$\mathbf{G}_5$]), since far away from the points it looks like the constant B field (seen by using theorem 5.4 for B much larger than E). Thus the vector field cannot exist, so the degree must be odd.

skew symmetric bundle map $\varphi \cdot F$ defined by $(\varphi \cdot F)_m = \varphi(m) F_m$ gives rise to the same T as does F.

On the other hand, suppose $F' = \varphi F$. Then $\psi_{F'} = \varphi^2 \psi_F$. So $\psi'_* = (2\varphi_* + \psi_*)$ on the first homology groups. Thus deg $\psi' =$ deg $\psi + 2k$ for some k. So deg (φF) has the same parity as deg (F).

THEOREM 7.11. *The space of skew symmetric operators over M_0 which gives rise to the same T is homeomorphic to the space of maps $\varphi : M_0 \longrightarrow S^1$. The path components of map (M_0, S^1) correspond to the elements of $H^1(M_0; \mathbb{Z})$.*

PROOF. We only need show that given F'_m and F_m with the same T, there is a θ such that $F'_m = e^{i\theta} F_m$. Now F'_m and F_m must have the same invariant planes and the same eigenvectors. Also $\lambda_F^2 + \lambda_{F^*}^2 = 2\lambda_T$ and the same holds for F'. So we can rotate F until $\lambda_F = \lambda_{F'}$ and $\lambda_{F^*} = \lambda_{F'^*}$. So F and F' agree on Π_+. Similarly they agree on Π_-. For null F and F', the $\mathbf{E}$ and $\mathbf{B}$ must have the same length and same $\mathbf{E} \times \mathbf{B}$, so one can rotate into the other.

SCHOLIUM 7.12. ELECTRON "STATES" FOR THE SAME ENERGY MOMENTUM.

a) For the free electron F of 7.10, $H^1(M^0; \mathbb{Z}) \cong 0$. So all the "states" $e^{i\theta} F$ are homotopic to one another. All of them have eigenvector bundles.
b) For the electron in a constant magnetic field, F, there are infinitely many homotopy classes of "states" giving rise to the same energy momentum T_F. Since $H^1(M_1; \mathbb{Z}) \cong \mathbb{Z}$, these states correspond to the integers. Each state has odd degree. Thus there is no eigenvector line bundle over M_1 for any state.

8. Lorentz Transformations

Lorentz Transformations play an important role in Physics. They are an artifact of Level -16, the standard coordinates of Minkowski space. As we move up through the levels of notation they seem to dwindle in importance. That is because one of their main functions, relating different choices of systems of notation, is eliminated as the choices are eliminated. What remains are two things, changes of observers in Level -2 as mentioned in Scholium 2.5, and the Gauge group of bundle isometries of Level 0. At these levels we obtain a fresh perception of the Lorentz Transformation.

LEVEL -16. MINKOWSKI SPACE–TIME.

At Level -10 we have coordinates for the tangent space, but not for the manifold. A choice of 4 functions coordinatizes M. We need to tie in our bases of the tangent bundle with the gradients of the coordinate functions. We can use the gradients as a basis, but usually they will not be orthonormal. Or, we can use the Gram Schmidt process on them to get a more complicated orthonormal basis. The best thing would be to find coordinates whose gradients are orthonormal. That is what is done for Minkowski space.

So let $M = \mathbb{R}^4$. Put coordinates t, x, y, z on M with orthonormal gradients $\frac{\partial}{\partial t}, \frac{\partial}{\partial x}, \frac{\partial}{\partial y}, \frac{\partial}{\partial z}$. We could choose another such coordinate system t', x', y', z'. The formulas relating them are called the Lorentz transformation. See [F], page I - 15 - 3.

SCHOLIUM 8.1. LORENTZ TRANSFORMATIONS OF ELECTRO-MAGNETISM.

Feynman in [F],(Vol. II, Table 26.4), carries out the Lorentz transformation in Level -16, and then tries to express the results in notation at Level -2. Calling $\mathbf{E}'$ and $\mathbf{B}'$ the transformed version of the original $\mathbf{E}$ and $\mathbf{B}$, he relates them by the formula

$$\begin{aligned} \mathbf{E}'_{\|} &= \mathbf{E}_{\|} & \mathbf{B}'_{\|} &= \mathbf{B}_{\|} \\ \mathbf{E}'_{\perp} &= \frac{(\mathbf{E}+\mathbf{w}\times\mathbf{B})_{\perp}}{\sqrt{1-w^2}} & \mathbf{B}'_{\perp} &= \frac{(\mathbf{B}-\mathbf{w}\times\mathbf{E})_{\perp}}{\sqrt{1-w^2}}. \end{aligned} \tag{37}$$

Here $\mathbf{E}'_{\|}$ means the component of $\mathbf{E}'$ parallel to the relative velocity $\mathbf{w}$ of the two coordinate frames and $E'_{\perp}$ means the component of $\mathbf{E}'$ orthogonal to $\mathbf{w}$. This formula is both correct and meaningless.

Let us give a Level -2 derivation of (37). Let $\mathbf{E} = Fu$ and $\mathbf{B} = -F^*u$. Let $u' = \dfrac{1}{\sqrt{1-w^2}}(u+\mathbf{w})$. Then $\mathbf{E}' = Fu'$ and $\mathbf{B}' = -F^*u'$. Substituting (21) and (23) into these formulas results in

$$\mathbf{E}' = \frac{\mathbf{E}\cdot\mathbf{w}}{\sqrt{1-w^2}}\left(u+\frac{\mathbf{w}}{w^2}\right) + \frac{1}{\sqrt{1-w^2}}\left(\mathbf{E} - \frac{\mathbf{E}\cdot\mathbf{w}}{w^2}\mathbf{w} + \mathbf{w}\times_u\mathbf{B}\right) \tag{38}$$

$$\mathbf{B}' = \frac{\mathbf{B}\cdot\mathbf{w}}{\sqrt{1-w^2}}\left(u+\frac{\mathbf{w}}{w^2}\right) + \frac{1}{\sqrt{1-w^2}}\left(\mathbf{B} - \frac{\mathbf{B}\cdot\mathbf{w}}{w^2}\mathbf{w} + \mathbf{w}\times_u\mathbf{E}\right) \tag{39}$$

Each of the four terms on the right hand sides of the above equations are orthogonal to u' and hence lie in $T^{u'}$. The last terms in each equation are orthogonal to $\mathbf{w}$ and lies in a plane orthogonal to the u, u' plane. These are $\mathbf{E}'_{\perp}$ and $\mathbf{B}'_{\perp}$ respectively. The first terms in each equation are the parallel components

$$\mathbf{E}'_{\|} = \frac{(\mathbf{E}\cdot\mathbf{w})}{\sqrt{1-w^2}}\left(u+\frac{\mathbf{w}}{w^2}\right) \text{ and } \mathbf{B}'_{\|} = \frac{(\mathbf{B}\cdot\mathbf{w})}{\sqrt{1-w^2}}\left(u+\frac{\mathbf{w}}{w^2}\right) \tag{40}$$

But $\mathbf{E}_{\|} = \dfrac{\mathbf{E}\cdot\mathbf{w}}{w}\mathbf{w}$ and $\mathbf{B}_{\|} = \dfrac{\mathbf{B}\cdot\mathbf{w}}{w}\mathbf{w}$. So $\mathbf{E}'_{\|} \neq \mathbf{E}_{\|}$ and $\mathbf{B}'_{\|} \neq \mathbf{B}_{\|}$ contrary to the assertion in (37). However they are both in the $u, \mathbf{w}$ plane and $E'_{\|} = E_{\|}$ and $B'_{\|} = B_{\|}$.

PROOF.

$$\begin{aligned} \mathbf{E}'_{\|}\cdot\mathbf{E}'_{\|} &= \frac{(\mathbf{E}\cdot\mathbf{w})^2}{1-w^2}\langle\left(u+\frac{\mathbf{w}}{w^2}\right), \left(u+\frac{\mathbf{w}}{w^2}\right)\rangle \\ &= \frac{(\mathbf{E}\cdot\mathbf{w})^2}{1-w^2}\left(-1+\frac{w^2}{w^4}\right) = \\ &= \left(\mathbf{E}\cdot\frac{\mathbf{w}}{w}\right)^2 = E_{\|}^2 \end{aligned}$$

Similarly for $B'_{\|} = B_{\|}$.

SCHOLIUM 8.2. THE DOPPLER SHIFT.

Let s_u be an eigenvector of F corresponding to λ_F as seen by an observer u.

is another observer. Then u' sees a different eigenvector $s_{u'}$. But $s_{u'}$ must be a multiple of s_u since they are eigenvectors. So the question is, what is the multiple in terms of $\mathbf{E}, \mathbf{B}$ and $\mathbf{w}$? The answer is:

$$s_{u'} = \frac{1}{\sqrt{1-w^2}}\left[1 + \frac{-(\mathbf{E}\times\mathbf{B})\cdot\mathbf{w} + \lambda_F\mathbf{E}\cdot\mathbf{w} - \lambda_{F^*}\mathbf{B}\cdot\mathbf{w}}{\lambda_T + \frac{E^2+B^2}{2}}\right] s_u. \tag{42}$$

PROOF. Define

$$\varphi(v) = \frac{\langle v, s_-\rangle}{\langle u, s_-\rangle} s_u \tag{43}$$

where s_- is an eigenvector corresponding to $-\lambda_F$. Then φ is a linear map whose image is the span of s_u and whose kernel is the space of vectors orthogonal to s_-. Now $\varphi(u) = s_u$.

Now $\Phi := (\lambda_{cF}I + cF)\circ(\overline{\lambda_{cF}I + cF})$ has the same properties and let $\Phi(u) := s_u$. Then $\Phi = \varphi$. Let $s_- = \Phi_-(u) = (-\lambda_{cF}I + cF)\circ(\overline{-\lambda_{cF}I + cF})u$. Now

$$s_u = 2\left(\lambda_T u + \frac{E^2+B^2}{2}u + \mathbf{E}\times\mathbf{B} + \lambda_F\mathbf{E} - \lambda_{F^*}\mathbf{B}\right) \tag{44}$$

and

$$s_- = 2\left(\lambda_T u + \frac{E^2+B^2}{2}u + \mathbf{E}\times\mathbf{B} - \lambda_F\mathbf{E} + \lambda_{F^*}\mathbf{B}\right) \tag{45}$$

from Corollary 6.8 and s_- is the same with the signs changed on λ_F and λ_{F^*}.

Now $s_{u'} = \varphi(u') = \dfrac{\langle u', s_-\rangle}{\langle u, s_-\rangle} s_u$. Substituting (41) into this equation yields

$$s_{u'} = \frac{1}{\sqrt{1-w^2}}\left(1 + \frac{\langle \mathbf{w}, s_-\rangle}{\langle u, s_-\rangle}\right) s_u. \tag{46}$$

Now

$$\langle u, s_-\rangle = -2\left(\lambda_T + \frac{E^2+B^2}{2}\right) \tag{47}$$

using (45). Then using (45) to calculate $\langle \mathbf{w}, s_-\rangle$ and substituting this into (46) we obtain (42). □

Now (42) holds for all $F \in \Gamma(\ell)$. If we restrict to null F we should see (42) reduce to a simpler form. In the null case $\lambda_F = \lambda_{F^*} = 0$ and $E = B$. So equation (42) reduces to

$$s_{u'} = \frac{1}{\sqrt{1-w^2}}\left(1 - \mathbf{w}\cdot\frac{(\mathbf{E}\times\mathbf{B})}{E^2}\right) s_u. \tag{48}$$

Now $\mathbf{w}\cdot\dfrac{(\mathbf{E}\times\mathbf{B})}{E^2}$ is the component along the $\mathbf{E}\times\mathbf{B}$ direction. If we assume that $\mathbf{w} = \mathbf{w}_r$, that is $\mathbf{w}$ is pointing in the radial direction, then

$$s_{u'} = \sqrt{\frac{1-w_r}{1+w_r}}\, s_u. \tag{49}$$

Here $\sqrt{\dfrac{1-w_r}{1+w_r}}$ is the Doppler shift ratio. This suggests that null F propagate along null geodesics by parallel translation.

SCHOLIUM 8.3. ELIMINATING $\mathbf{E} \times \mathbf{B}$.

In the non–null case there is a Lorentz transformation so that $\mathbf{E}' \times \mathbf{B}' = 0$. We may see this clearly using Level -2 methods. Suppose u' is an eigenvector of T_F. Then

$$T_F u' = \lambda_T u' = \frac{E_{u'}^2 + B_{u'}^2}{2} u' + \mathbf{E}_{u'} \times \mathbf{B}_{u'}. \tag{50}$$

The second equality shows that $\mathbf{E}_{u'} \times \mathbf{B}_{u'} = 0$ and $\dfrac{E_{u'}^2 + B_{u'}^2}{2} = \lambda_T$. Now we can always find an eigenvector u' by setting $u' = (\lambda_T I + T)u/k$. Here $k = \sqrt{2\lambda_T{}^2 + 2\lambda_T(E^2 + B^2)/2}$ where the k is the factor which makes u' an observer. Thus the relative velocity is

$$\mathbf{w} = (\mathbf{E}_u \times \mathbf{B}_u) \Big/ \left(\lambda_T + \frac{E_u^2 + B_u^2}{2} \right). \tag{51}$$

At Level -10, the Lorentz transformations become equations relating the choice of orthonormal bases e_0, e_1, e_2, e_3 and e_0', e_1', e_2', e_3'. In the block matrices formalism, the Lorentz transformation becomes an invertible matrix Λ so that

$$\begin{pmatrix} 0 & \mathbf{E}' \\ \mathbf{E}' & \times\mathbf{B}' \end{pmatrix} = \Lambda^{-1} \begin{pmatrix} 0 & \mathbf{E} \\ \mathbf{E} & \times\mathbf{B} \end{pmatrix} \Lambda. \tag{52}$$

Although we used many Level -10 arguments in this paper, our statements were usually Level -2. The only choices necessary were of different observers. The algebraic component of the Lorentz Transformations Λ becomes the bundle isometries of $T(M)$, that is the group of Gauge Transformations. These can be thought of at Level 0.

REMARK 8.4. THE EXPONENTIAL MAP e^F.

The exponential map maps the "Lie Algebra" $\Gamma(\ell)$ onto the group of bundle isometries $\mathcal{G}$ of $T(M)$. This exponential map is a diffeomorphism near the identity. It has a beautiful representation using the e^F notation.

$$e^F := I + F + \frac{1}{2!}F^2 + \frac{1}{3!}F^3 + \dots \tag{53}$$

where F^n means F composed with itself n–times. For F a bundle map, e^F satisfies several properties.

a) e^F is a well–defined bundle map
b) $(e^F)^{-1} = e^{-F}$ if F is skew symmetric
c) $\langle e^F v, w\rangle = \langle v, e^{-F} w\rangle$ if F is skew symmetric
d) $e^{F+F'} = e^F \circ e^{F'}$ if $FF' = F'F$
e) $\left.\dfrac{d}{dt} e^{tF}\right|_{t=0} = F$
f) Every isometry Q can be written as $Q = e^F$ for a skew symmetric F, at least locally.
g) If s is an eigenvector of F corresponding to λ_F, then s is an eigenvector of e^F corresponding to e^{λ_F}.
h) If F is skew symmetric and null, then since $F^3 = 0$ we have $e^F = I + F + \frac{1}{2}F^2$.

THEOREM 8.5. $e^{cF} = \cosh(\lambda_{cF})I + \dfrac{\sinh(\lambda_{cF})}{\lambda_{cF}}(cF)$
where $\cosh(x) = \dfrac{e^x + e^{-x}}{2}$
and $\sinh(x) = \dfrac{e^x - e^{-x}}{2}$.

COROLLARY 8.6.

$$e^F = \left(\cosh\left(\frac{\lambda}{2}\right)I + \frac{\sinh\left(\frac{\lambda}{2}\right)}{\lambda}cF\right) \circ \overline{\left(\cosh\left(\frac{\lambda}{2}\right)I + \frac{\sinh\left(\frac{\lambda}{2}\right)}{\lambda}cF\right)}$$

where $\lambda = \lambda_{cF}$.

PROOF. $e^F = e^{(cF+\overline{c}F)/2} = e^{cF/2}e^{\overline{c}F/2}$, this last by Remark 8.4d. Then apply Theorem 8.5.

We leave it as an exercise to the reader to expand Corollary 8.6 and obtain an equation involving only real quantities.

COROLLARY 8.7.

$$e^{-F}(cG)e^F = c(e^{-F}Ge^F) = e^{-cF/2}(cG)e^{cF/2}.$$

PROOF. First we note the following result.

$$(e^{-F}Ge^F)^* = e^{-F}G^*e^F. \tag{54}$$

This follows from the fact that $(e^{-F}Ge^F)^{-1} = e^{-F}G^{-1}e^F$ and that $G^{-1} = \dfrac{G^*}{-(\vec{E}\cdot\vec{B})}$ when $\vec{E}\cdot\vec{B} \neq 0$. The case for $\vec{E}\cdot\vec{B} = 0$ follows by continuity.

Now $c(e^{-F}Ge^F) = e^{-F}Ge^F - i(e^{-F}Ge^F)^*$

$$\begin{aligned} &= e^{-F}Ge^F - ie^{-F}G^*e^F \\ &= e^{-F}(cG)e^F \\ &= e^{-cF/2}(cG)e^{cF/2}. \end{aligned}$$

The last equality follows from subsituting $e^F = e^{(cF+\overline{c}F)/2} = e^{cF/2}e^{\overline{c}F/2}$ and the fact that $\overline{cF}$, and hence $e^{\overline{c}F/2}$, commutes with cG. □

References

[AB] M. F. Atiyah and R. Bott, *The Yangs-Mills Equations over Riemann Surfaces*, Phil.Trans. R. Soc. Lond. **A 308** (1982), 523-615.

[BP] P. Booth, P. Heath, C. Morgan, R. Piccinini, *Remarks on the homotopy type of groups of gauge transformations*, C. R. Math. Rep. Acad. Sci. Canad. **3** (1981), 3-6.

[F] R. Feynman, R. Leighton, and M. Sands, *The Feynman Lectures in Physics*, Addison-Weseley **Reading, Massachusetts** (1964).

[G1] Daniel H. Gottlieb, *On fibre spaces and the evaluation map*, Annals of Math **87** (1968), 42-55.

[G2] Daniel H. Gottlieb, *Correction to my paper "On fibre spaces and the evaluation map*, Annals of Math. **91** (1970), 640-642.

[G3] Daniel H. Gottlieb, *Applications of bundle map theory*, Trans. Amer. Math. Soc. **171** (1972), 23-50.

[G_4] Daniel H. Gottlieb, *The trace of an action and the degree of a map*, Trans. Amer. Math. Soc. **293** (1986), 381-410..

[G_5] Daniel H. Gottlieb and Geetha Samarayanake, *The Index of Discontinuous Vector Fields*, New York Journal of Mathematics **1** (1994-1995), 130-148.

[LM] H. B. Lawson and M. L. Michelson, *Spin Geometry*, Princeton University Press, Princeton (1989).

[ON] Barrett O'Neill, *Semi-Riemannian Geometry*, Academic Press, New York (1983).

[P] Stephen Parrott, *Relativistic Electrodynamics and Differential Geometry*, Springer-Verlag, New York (1987).

Department of Mathematics, Purdue University, West Lafayette, Indiana 47906

Current address: Department of Mathematics, Purdue University, West Lafayette, Indiana 47906

E-mail address: `gottlieb@math.purdue.edu`

Contemporary Mathematics
Volume **220**, 1998

The Adams spectral sequence for Minami's Theorem

Wen-Hsiung Lin and Mark Mahowald

1. Introduction

N. Minami proves in [16] an interesting result dealing with the Kervaire invariant elements $\theta_i \in {}_2\pi_*^s$ which, in the mod 2 Adams spectral sequence, have the name $h_i^2 \in Ext_A^{2,2^{i+1}}(\mathbb{Z}/2, \mathbb{Z}/2)$ where A denotes the mod 2 Steenrod algebra. He frames his result in the terms of the double transfer. Let P denote the infinite real projective space. There is a stable map $P \xrightarrow{\bar{t}} S^0$, the transfer map to the 2-fold covering $S^\infty \to P$, which is the adjoint of the composite $P \to SO \to QS^0$ where the first map is the standard embedding and the second map is the J-homomorphism ([3], [12]), The double transfer referred to is the composite

$$\bar{t} \wedge \bar{t}:\ P \wedge P \xrightarrow{id \wedge \bar{t}} P \wedge S^0 = P \xrightarrow{\bar{t}} S^0.$$

If $\theta_i:\ S^{2^{i+1}-2} \to S^0$ exists then, by Kahn-Priddy Theorem ([8]), it factors through the transfer $\bar{t}$. Minami proves the following result.

THEOREM 1.1 ([16]). *Suppose $\theta_i:\ S^{2^{i+1}-2} \to S^0$ exists. Then, for $i \geq 5$, θ_i does not factor through the double transfer $\bar{t} \wedge \bar{t}$.*

We remark that the Kervaire invariant problem is still open for most j. More precisely, θ_j for $j \leq 5$ are known to exist; the cases $0 \leq j \leq 3$ are classical and the cases $j = 4, 5$ are proved in [13], [4]. For $j \geq 6$ whether there exists a θ_j is still not known. Let e_k be the generator of $H_k(P) = \mathbb{Z}/2$, $1 \leq k < \infty$. Here, and for the rest of this paper, all homology or cohomology groups of a space or a spectrum

---

1991 *Mathematics Subject Classification.* 55Q45, 55T15.
The first author is supported by N.S.C. of Taiwan and the Foundation for Advancement of Outstanding Scholarship.

have $\mathbb{Z}/2$ coefficients. If θ_i exists and factors through the double transfer $\bar{t} \wedge \bar{t}$ then the class $e_{2^i-1} \otimes e_{2^i-1} \in H_{2^{i+1}-2}(P \wedge P)$ must be in the Hurewicz image. What Minami really proves in [16] for (1.1) is the following which is independent of the Kervaire invariant problem.

THEOREM 1.1′. *$e_{2^i-1} \otimes e_{2^i-1} \in H_{2^{i+1}-2}(P \wedge P)$ is not in the Hurewicz image for $i \geq 5$.*

In the terms of the Adams spectral sequence for $P \wedge P$ this means the corresponding class in

$$Ext_A^{0,2^{i+1}-2}(P \wedge P) = Ext_A^{0,2^{i+1}-2}(H^*(P \wedge P), \mathbb{Z}/2),$$

which is denoted by $\overline{e_{2^i-1} \otimes e_{2^i-1}}$, cannot detect homotopy elements in $\pi_*^s(P \wedge P)$ for $i \geq 5$. This is proved in [16] by BP homology theory. In this paper we give, in the setting of the mod 2 Adams spectral sequence, a proof of this fact. We will actually prove the following more concrete result where each $\overline{e_{2^j-1} \otimes e_{2^j-1}} \in Ext_A^{0,2^{j+1}-2}(P \wedge P)$ is as defined above.

THEOREM 1.2. *For $i \geq 5$,*

$$d_3\left(\overline{e_{2^i-1} \otimes e_{2^i-1}}\right) = \overline{e_{2^{i-1}-1} \otimes e_{2^{i-1}-1}} h_i h_0^2 \neq 0$$

in the Adams spectral sequence for $P \wedge P$.

For $1 \leq i \leq 3$, that $\overline{e_{2^i-1} \otimes e_{2^i-1}}$ is an infinite cycle is a trivial fact. In this paper we will also prove the following result for the case $i = 4$.

THEOREM 1.3. *$\overline{e_{15} \otimes e_{15}}$ is an infinite cycle in the Adams spectral sequence for $P \wedge P$.*

This implies that the θ_4, constructed in [13], does lift to the double transfer.

The proof of Theorem 1.2 will be given in two stages. The first stage is a proof of (1.2) based on the following assumption.

ASSUMPTION 1.4. *There exists a stable map $f : M = S^{2^i-2} \cup_{2\iota} e^{2^i-1} \to P$ such that $f^* : H^{2^i-1}(P) = \mathbb{Z}/2 \longrightarrow H^{2^i-1}(M) = \mathbb{Z}/2$ is non-zero, where $i \geq 5$.*

It is not difficult to see that (1.4) is equivalent to the assumption that there is a θ_{i-1} with $2\theta_{i-1} = 0$. This connection, however, will not be needed here. To describe the second stage, let $B(k)$ denote the 2-primary k^{th} Brown-Gitler spectrum ([6]). The top two cells of $\Sigma^{7\cdot 2^{i-3}} B\left(2^{i-4}\right)$ are just M. We will show that there exists a stable map

$$\Sigma^{7\cdot 2^{i-3}} B\left(2^{i-4}\right) \xrightarrow{f'} P$$

which on the top two cells behaves as f in (1.4). The second stage of the proof is a real proof of (1.2) which is a modification of that in stage one applied to f'. Our major work is the proof in the first stage, and this will be given in §4. Once this is done the second stage proof follows by parallelism without too much difficulties as will be seen. This will be given in §5. For the first stage proof we need results on some relevant Ext groups including those for $P, Y, P \wedge P$ and $Y \wedge Y$ where Y is the cofiber of the map f in (1.4). In §2 we state these results and describe our methods of calculations for obtaining them. In §3 we determine the differentials d_2 on some of the elements in these Ext groups, in relevant Adams spectral sequences, that we need for the first stage proof. The proof of Theorem 1.3, which deals with the case $i = 4$, will be given in §6.

2. Results on some Ext groups

For a left module N over the Steenrod algebra A we write $Ext_A^{s,t}(N)$ to denote $Ext_A^{s,t}(N, \mathbb{Z}/2)$, and for a space or a spectrum X we write $Ext_A^{s,t}(X)$ to denote $Ext_A^{s,t}(H^*(X))$. For $X = S^0$ we simply write $Ext_A^{s,t}$ to denote $Ext_A^{s,t}(S^0) = Ext_A^{s,t}(\mathbb{Z}/2, \mathbb{Z}/2)$. We will begin by recalling the structure of $Ext_A^{s,*}$ for $s \leq 4$ as this is the basis on which the Ext groups $Ext_A^{*,*}(X)$ for $X = P, Y, P \wedge P, Y \wedge Y$ that we need will be described and calculated.

Let Λ be the mod 2 lambda algebra ([5]). Recall that Λ is a bigraded differential algebra over $\mathbb{Z}/2$ generated by $\lambda_j \in \Lambda^{1,j}$, $j \geq 0$ with relations

$$(1) \quad \lambda_j \lambda_{2j+1+m} = \sum_{\nu \geq 0} \binom{m-\nu-1}{\nu} \lambda_{j+m-\nu} \lambda_{2j+1+\nu}$$

and the differential

$$(2) \quad \delta(\lambda_k) = \sum_{\nu \geq 0} \binom{k-\nu-1}{\nu+1} \lambda_{k-\nu-1} \lambda_\nu.$$

Then $H^{s,t}(\Lambda, \delta) = Ext_A^{s,s+t}$. From (1) we see $\{\lambda_{j_1} \cdots j_{j_s} \mid j_i \geq 2j_{i+1}\}$ is a $\mathbb{Z}/2$-base for Λ. Such monomial are said to be admissible. There is an operation $Sq^0 : \Lambda \to \Lambda$ given by

$$(3) \quad Sq^0(\lambda_{j_1} \cdots \lambda_{j_s}) = \lambda_{2j_1+1} \cdots \lambda_{2j_s+1}$$

where $\lambda_{j_1} \cdots \lambda_{j_s}$ is not necessarily admissible. This operation respects the relations (1) and commutes with the differential δ in (2). So it induces a map $Sq^0 : H^{s,t-s}(\Lambda) \to H^{s,2t-s}(\Lambda)$ which coincides with the first Steenrod operation $Sq^0 : Ext_A^{s,t} \to Ext_A^{s,2t}$ in [11]. There are the following classes in $Ext_A^{*,*}$ where $(Sq^0)^0 =$ identity map and $(Sq^0)^i = \underbrace{Sq^0 \cdots Sq^0}_{i}$ for $i > 0$.

(2.1) (1) $h_i = \{\lambda_{2^i-1}\} \in Ext_A^{1,2^i}$, $i \geq 0$,

(2) $c_i = \{\lambda_{2^{i+1}+2^i-1} \lambda_{2^{i+2}-1}^2\} \in Ext_A^{3,2^{i+3}+2^{i+1}+2^i}$, $i \geq 0$,

(3) $d_i = \left\{(Sq^0)^i \left(\lambda_6 \lambda_2 \lambda_3^2 + \lambda_4^2 \lambda_3^2 + \lambda_2 \lambda_4 \lambda_5 \lambda_3 + \lambda_1 \lambda_5 \lambda_1 \lambda_7\right)\right\}$
$\in Ext_A^{4,2^{i+4}+2^{i+1}}$, $i \geq 0$

(4) $e_i = \left\{(Sq^0)^i \left(\lambda_8 \lambda_3^3 + \lambda_4 \left(\lambda_5^2 \lambda_3 + \lambda_7 \lambda_3^2\right) + \lambda_2 \left(\lambda_3 \lambda_5 \lambda_7 + \lambda_1 \lambda_{11} \lambda_3\right)\right)\right\}$
$\in Ext_A^{4,2^{i+4}+2^{i+2}+2^i}$, $i \geq 0$

(5) $f_i = \left\{(Sq^0)^i \left(\lambda_4 \lambda_0 \lambda_7^2 + \lambda_3 \left(\lambda_9 \lambda_3^2 + \lambda_3 \lambda_5 \lambda_7\right) + \lambda_2^2 \lambda_7^2\right)\right\}$
$\in Ext_A^{4,2^{i+4}+2^{i+2}+2^{i+1}}$, $i \geq 0$

(6) $g_{i+1} = \left\{(Sq^0)^i \left(\lambda_6 \lambda_0 \lambda_7^2 + \lambda_5 \left(\lambda_9 \lambda_3^2 + \lambda_3 \lambda_5 \lambda_7\right) + \lambda_3 \left(\lambda_5 \lambda_9 \lambda_3 + \lambda_{11} \lambda_3^2\right)\right)\right\}$
$\in Ext_A^{2,2^{i+4}+2^{i+3}}$, $i \geq 0$

(7) $p_i = \left\{(Sq^0)^i\left(\lambda_{14}\lambda_5\lambda_7^2 + \lambda_{10}\lambda_9\lambda_7^2 + \lambda_6\lambda_9\lambda_{11}\lambda_7\right)\right\}$
$\in Ext_A^{4,2^{i+5}+2^{i+2}+2^i}, i \geq 0,$

(8) $D_3(i) =$
$\left\{(Sq^0)^i\left(\lambda_{22}\lambda_1\lambda_7\lambda_{31} + \lambda_{16}\lambda_7^2\lambda_{31} + \lambda_{14}\lambda_9\lambda_7\lambda_{31} + \lambda_{12}\lambda_{11}\lambda_7\lambda_{31}\right)\right\}$
$\in Ext_A^{4,2^{i+6}+2^i}, i \geq 0,$

(9) $p_i' = \left\{(Sq^0)^i\left(\lambda_0\lambda_{39}\lambda_{15}^2 + \lambda_0\lambda_{15}\lambda_{23}\lambda_{31}\right)\right\} =$

$$\left\{(Sq^0)^i\begin{pmatrix}\lambda_{38}\lambda_1\lambda_{15}^2 + \lambda_{30}\lambda_9\lambda_{15}^2 + \lambda_{28}\lambda_{11}\lambda_{15}^2 + \lambda_{22}\lambda_{17}\lambda_{15}^2 \\ + \lambda_{20}\lambda_{19}\lambda_{15}^2 + \lambda_{14}\lambda_1\lambda_{23}\lambda_{31} + \lambda_{12}\lambda_3\lambda_{23}\lambda_{31}\end{pmatrix}\right\}$$

$\in Ext_A^{4,2^{i+6}+2^{i+3}+2^i}, i \geq 0.$

THEOREM 2.2. *(1) Modulo indecomposable elements in* $Ext_A^{4,*}$*, the algebra* $Ext_A^{s,*}$ *for* $s \leq 4$ *is generated by* h_i *and* c_i *subject only to the following relations:*

$$h_ih_{i+1} = 0, \quad h_ih_{i+2}^2 = 0, \quad h_i^3 = h_{i-1}^2h_{i+1},$$
$$h_i^2h_{i+3}^2 = 0, \quad h_jc_i = 0 \quad \text{for} \quad j = i-1, i, i+2 \quad \text{and} \quad i+3.$$

(2) The classes in (2.1)(3) through (2.1)(9) form a $\mathbb{Z}/2$*-base for the indecomposable elements in* $Ext_A^{4,*}$*.*

The result (2.2)(1) for $s \leq 3$ is proved in [2], [18]. The relations $h_i^2h_{i+3}^2 = 0$ are due to May [15]. The result (2.2)(2) for $i = 0$ in (2.1)(3) through (2.1)(9) is due to Tangora [17]. The rest of the assertions in (2.2) are proved in [10]. In [10], the $\mathbb{Z}/2$-basis of the indecomposable elements in (2.2)(2) is proved by May spectral sequence rather than making calculations in the lambda algebra Λ. We exhibit cycle representations in Λ for these indecomposable elements as given in (2.1) because we want to use them to describe the corresponding elements in $Ext_A^{3,*}(P)$ that we will discuss next.

Let N be a locally finite left module over A and let N_* be its $\mathbb{Z}/2$-dual which is a right A-module by transposing the left A-module structure on N. Bigrade $N_* \otimes \Lambda$ by $(N_* \otimes \Lambda)^{s,t} = \underset{k}{\Sigma} N_k \otimes \Lambda^{s,t-s-k}$. For an arbitrary sequence $I = (i_1, \cdots, i_s)$ of non-negative integers we write λ_I to denote $\lambda_{i_1}\cdots\lambda_{i_s} \in \Lambda$. And we simply write $m_*\lambda_I$ to denote $m_* \otimes \lambda_I$ and m_* for $m_* \otimes 1$ in $N_* \otimes \Lambda$. $N_* \otimes \Lambda$ is a bigraded differential right Λ-module with differential δ given by

$$(4) \quad \delta(m_*\lambda_I) = m_*\delta(\lambda_I) + \sum_{j \geq 0} m_*Sq^{j+1}\lambda_j\lambda_I.$$

Then $Ext_A^{s,t}(N) = H^{s,t}(N_* \otimes \Lambda, \delta)$. The differential (5) induces a right action of $Ext_A^{*,*}$ on $Ext_A^{*,*}(N)$ which makes $Ext_A^{*,*}(N)$ a right $Ext_A^{*,*}$-module.

For $N_* = H_*(P)$, the right A-action is given by

$$(5) \qquad e_kSq^j = \binom{k-j}{j} e_{k-j}$$

where e_k is the generator of $H_k(P) = \mathbb{Z}/2$, $1 \leq k < \infty$. From (4), (5) we see the differential δ of $H_*(P) \otimes \Lambda$ on the generators e_k is given by

(6) $$\delta(e_k) = \sum_{\nu\geq 0} \binom{k-\nu-1}{\nu+1} e_{k-\nu-1}\lambda_\nu.$$

From (2), (4), (6) we see the Λ-map $\varphi : H_*(P)\otimes\Lambda \to \Lambda$ given on the generators e_k by $\varphi(e_k) = \lambda_k$ is a chain map. The induced map $\varphi_* : Ext_A^{s,t}(P) \to Ext_A^{s+1,t+1}(S^0)$ is precisely the induced map $\bar{t}_*$ of the transfer $P \xrightarrow{\bar{t}} S^0$ in E_2 terms of the Adams spectral sequences for P and S^0 (see [7], [9]).

In $H_*(P)\otimes\Lambda$ there is also an operation $Sq^0 : H_*(P)\otimes\Lambda \to H_*(P)\otimes\Lambda$ given by

$$Sq^0(e_k\lambda_{j_1}\cdots\lambda_{j_s}) = e_{2k+1}\lambda_{2j_1+1}\cdots\lambda_{2j_s+1}$$

which is a chain map. e_{2^i-1} for $i\geq 1$ and $e_{2^{i+1}+2^i-1}\lambda^2_{2^{i+2}-1}$ for $i\geq 0$ are cycles in $H_*(P)\otimes\Lambda$. Let $\bar{e}_{2^i-1} = \{e_{2^i-1}\} \in Ext_A^{0,2^i-1}(P)$ and $\hat{c}_i = \left\{e_{2^{i+1}+2^i-1}\lambda^2_{2^{i+2}-1}\right\} \in Ext_A^{2,2^{i+3}+2^{i+1}+2^i-1}(P)$. It is clear that $\bar{e}_{2^i-1}$ and $\hat{c}_i$ are mapped respectively to h_i and c_i under the induced map $\varphi_* = \bar{t}_* : Ext_A^{s,t}(P) \to Ext_A^{s+1,t+1}$. For $3\leq j\leq 9$ let z_j be the cycle in (2.1)(j) for $i=0$. Thus $z_3 = \lambda_6\lambda_2\lambda_3^2 + \lambda_4\lambda_4\lambda_3^2 + \lambda_2\lambda_4\lambda_5\lambda_3 + \lambda_1\lambda_5\lambda_1\lambda_7$, etc. Let $\hat{z}_j$ be the chain in $H_*(P)\otimes\Lambda$ obtained from z_j by replacing the leading λ_k in each term of the cycle z_j by e_k. Thus, for example, $\hat{z}_3 = e_6\lambda_2\lambda_3^2 + e_4\lambda_4\lambda_3^2 + e_2\lambda_4\lambda_5\lambda_3 + e_1\lambda_5\lambda_1\lambda_7$. Straightforward calculations show that each $\hat{z}_j$ is indeed a cycle. The sums $\hat{z}_{10} = e_2\lambda_0\lambda_7^2 + e_1(\lambda_9\lambda_3^2 + \lambda_3\lambda_5\lambda_7)$, $\hat{z}_{11} = e_2\lambda_5\lambda_7^2$ and

$$\begin{aligned}\hat{z}_{12} = {} & e_{12}\lambda_5\lambda_7^2 + e_{10}\lambda_7^3 + e_6\lambda_1^2\lambda_{23} + e_5(\lambda_8\lambda_3\lambda_{15} + \lambda_7\lambda_4\lambda_{15} + \lambda_{11}\lambda_0\lambda_{15}) \\ & + e_3(\lambda_0\lambda_5\lambda_{23} + \lambda_9\lambda_4\lambda_{15} + \lambda_{10}\lambda_3\lambda_{15})\end{aligned}$$

are also cycles in $H_*(P)\otimes\Lambda$. We thus have the following classes in $Ext_A^{*,*}(P)$.

(2.3)
(3) $\hat{d}_i = \left\{(Sq^0)^i(\hat{z}_3)\right\} \in Ext_A^{3,2^{i+4}+2^{i+1}-1}(P),\quad i\geq 0$

(4) $\hat{e}_i = \left\{(Sq^0)^i(\hat{z}_4)\right\} \in Ext_A^{3,2^{i+4}+2^{i+2}+2^i-1}(P),\quad i\geq 0$

(5) $\hat{f}_i = \left\{(Sq^0)^i(\hat{z}_5)\right\} \in Ext_A^{3,2^{i+4}+2^{i+2}+2^{i+1}-1}(P),\quad i\geq 0$

(6) $\hat{g}_{i+1} = \left\{(Sq^0)^i(\hat{z}_6)\right\} \in Ext_A^{3,2^{i+4}+2^{i+3}-1}(P),\quad i\geq 0$

(7) $\hat{p}_i = \left\{(Sq^0)^i(\hat{z}_7)\right\} \in Ext_A^{3,2^{i+5}+2^{i+2}+2^i-1}(P),\quad i\geq 0$

(8) $\hat{D}_3(i) = \left\{(Sq^0)^i(\hat{z}_8)\right\} \in Ext_A^{3,2^{i+6}+2^i-1}(P),\quad i\geq 0$

(9) $\hat{p}'_i = \left\{(Sq^0)^i(\hat{z}_9)\right\} \in Ext_A^{3,2^{i+6}+2^{i+3}+2^i-1}(P),\quad i\geq 0$

(10) $\alpha_{16}(i) = \left\{(Sq^0)^i(\hat{z}_{10})\right\} \in Ext_A^{3,2^{i+4}+2^{i+2}-1}(P),\quad i\geq 0$

(11) $\alpha_{21}(i) = \left\{(Sq^0)^i(\hat{z}_{11})\right\} \in Ext_A^{3,2^{i+4}+2^{i+3}+2^i-1}(P),\quad i\geq 0$

(12) $\alpha_{31}(i) = \left\{(Sq^0)^i(\hat{z}_{12})\right\} \in Ext_A^{3,2^{i+5}+2^{i+1}+2^i-1}(P),\quad i\geq 0$

It is clear that the class in (2.3)(j), $3\leq j\leq 9$, is mapped to the class in (2.1)(j) under $\bar{t}_*$.

$Ext_A^{*,*}$-module $Ext_A^{s,*}(P)$ for $s \leq 3$ is generated by $\bar{e}_{2^i-1}$ and $\hat{c}_i$ subject only to the following relations:

$$\begin{aligned}
&\bar{e}_{2^i-1}h_{i-1} = 0, && \bar{e}_{2^{i+2}-1}h_i^2 = \bar{e}_{2^{i+1}-1}h_{i+1}^2,\\
&\bar{e}_{2^{i+2}-1}h_{i+2}h_i = 0, && \bar{e}_{2^{i+3}-1}h_{i+3}h_i^2 = 0,\\
&\bar{e}_{2^{i+2}-1}h_{i+3}^2h_i = 0, && \bar{e}_{2^i-1}c_j = 0 \quad for \quad j = i-2, i-3,\\
&\hat{c}_ih_j = 0 \quad for \quad j = i-1, i, i+2, i+3, &&\\
&\hat{c}_ih_{i+1} = \bar{e}_{2^{i+1}-1}c_i \neq 0. &&
\end{aligned}$$

(2) The classes which are described in (2.3) form a $\mathbb{Z}/2$-base for the indecomposable elements in $Ext_A^{3,*}(P)$.

The result (2.4)(1) is proved in [7] and the result (2.4)(2) is new. In [7], (2.4)(1) is proved by making calculations with a spectral sequence which we recall as follows. Define a filtration $\{F(j)\}_{j\geq 1}$ of $H_*(P) \otimes \Lambda$ by $F(j) = \sum_{k\leq j} e_k \otimes \Lambda$. Then $F(j)/F(j-1) \cong \Sigma^j\Lambda$. This filtration gives rise to a spectral sequence $\{E_r^{j,s,t}\}_{r\geq 1}$ with $\sum_j E_\infty^{j,s,t} \cong Ext_A^{s,t}(P)$ as a $\mathbb{Z}/2$-module and

$$E_1^{j,s,t} = H^{s,t}\left(F(j)/F(j-1)\right) \cong Ext_A^{s,t-j}.$$

For each $r \geq 1$ the differential of the spectral sequence goes from $E_r^{j,s,t}$ to $E_r^{j-r,s+1,t}$. Since $Ext_A^{s,*}$ for $s \leq 4$ are now completely known (by (2.2)) the calculations with this spectral sequence for proving (2.4)(1) in [7] can be extended so as to obtain (2.4)(2). These extended calculations are very lengthy and will not be given here. We just remark that the classes α in (2.3) are represented by the following infinite cycles α' in the spectral sequence, indicated as $\alpha \leftrightarrow \alpha'$.

(2.3)′
(3) $\hat{d}_i \leftrightarrow e_{2^{i+2}+2^{i+1}+2^i-1}c_i, \quad i \geq 0,$
(4) $\hat{e}_i \leftrightarrow e_{2^{i+3}+2^i-1}h_{i+2}^3, \quad i \geq 0,$
(5) $\hat{f}_i \leftrightarrow e_{2^{i+2}+2^i-1}h_ih_{i+3}^2, \quad i \geq 0,$
(6) $\hat{g}_{i+1} \leftrightarrow e_{2^{i+2}+2^{i+1}+2^i-1}h_ih_{i+3}^2, \quad i \geq 0,$
(7) $\hat{p}_i \leftrightarrow e_{2^{i+4}-2^i-1}c_{i+1}, \quad i \geq 0,$
(8) $\hat{D}_3(i) \leftrightarrow e_{2^{i+4}+2^{i+3}-2^i-1}h_{i+1}h_{i+3}h_{i+5}, \quad i \geq 0,$
(9) $\hat{p}'_i \leftrightarrow e_{2^{i+5}+2^{i+3}-2^i-1}h_{i+1}h_{i+4}^2, \quad i \geq 0,$
(10) $\alpha_{16}(i) = e_{2^{i+1}+2^i-1}h_ih_{i+3}^2, \quad i \geq 0$
(11) $\alpha_{21}(i) = e_{2^{i+1}+2^i-1}c_{i+1}, \quad i \geq 0$
(12) $\zeta_{31}(i) = e_{2^{i+3}+2^{i+2}+2^i-1}c_{i+1}, \quad i \geq 0$

Let $M = S^{2^i-2} \cup_{2\iota} e^{2^i-1} \xrightarrow{f} P$ be as in Assumption 1.4 in §1 where $i \geq 5$. So $f_* \neq 0$ in dimension $2^i - 1$. It is easy to see $f_* = 0$ in dimension $2^i - 2$. Let Y be the cofiber of f. Then $H_k(Y) = \mathbb{Z}/2$ for $1 \leq k < \infty$ also. For $k \neq 2^i - 1$ the generator of $H_k(Y) = \mathbb{Z}/2$ is still denoted by e_k and that of $H_{2^i-1}(Y) = \mathbb{Z}/2$ is denoted by $\tilde{e}_{2^i-1}$. In the ASS for P, $d_2(\bar{e}_{2^i-1}) = \bar{e}_{2^{i-1}-1}h_{i-1}h_0 \neq 0$ ([7]). This implies the right A-action on $H_*(Y)$ is given by

(7) $$e_kSq^j = \binom{k-j}{j} e_{k-j} \quad \text{and} \quad \tilde{e}_{2^i-1}Sq^j = \begin{cases} e_{2^{i-1}-1} & j = 2^{i-1} \\ 0 & j \neq 2^{i-1} \end{cases}$$

$$\text{where} \quad e_{k-j} = 0 \quad \text{if} \quad k - j = 2^i - 1$$

From (4), (7) we see the differential δ of $H_*(Y) \otimes \Lambda$ on the generators e_k and $\tilde{e}_{2^i-1}$ is given by

$$(8) \quad \delta(e_k) = \sum_{\nu \geq 0} \binom{k-\nu-1}{\nu+1} e_{k-\nu-1}\lambda_\nu, \quad \delta(\tilde{e}_{2^i-1}) = e_{2^{i-1}-1}\lambda_{2^{i-1}-1}$$

$$\text{where} \quad e_{k-\nu-1} = 0 \quad \text{if} \quad k - \nu - 1 = 2^i - 1.$$

Consider the classes

$$\bar{e}_{2^\ell-1} = \{e_{2^\ell-1}\} \in Ext_A^{0,2^\ell-1}(Y) \ (\ell \neq i),$$

$$\bar{e}_{2^i+2^{i-1}-1} = \{e_{2^i+2^{i-1}-1}\} \in Ext_A^{0,2^i+2^{i-1}-1}(Y)$$

and

$$\tau = \{c\} \in Ext_A^{2,2^i+2^{i-1}}(Y)$$

where c is the cycle

(9) $c = e_{2^i+2^{i-2}-1}\lambda_{2^{i-2}-1}\lambda_0 + \tilde{e}_{2^i-1}\lambda_{2^{i-1}-1}\lambda_0 + e_{2^{i-1}+2^{i-2}-1}\lambda_{2^{i-1}+2^{i-2}-1}\lambda_0$
in $H_*(Y) \otimes \Lambda$.

PROPOSITION 2.5. *(1)* $\{\bar{e}_{2^\ell-1} | \ \ell \neq i\} \cup \{\bar{e}_{2^i+2^{i-1}-1}\}$ *is a* $\mathbb{Z}/2$*-base for* $Ext_A^{0,*}(Y)$.
(2) $Ext^{2,2^i+2^{i-1}}(Y) = \mathbb{Z}/2 \oplus \mathbb{Z}/2$ *on* $\bar{e}_{2^{i-1}-1}h_ih_0$ *and* τ.
(3) $Ext_A^{1,2^i}(Y) = 0$ *and* $Ext_A^{2,2^i}(Y) = 0$.
(4) $\bar{e}_{2^\ell-1}h_\ell \neq 0$ *in* $Ext_A^{1,2^{\ell+1}-1}(Y)$ *for all* i $(i \neq \ell)$.

(2.5)(1) is easily seen from (8). We have short exact sequences of A-modules

$$(10) \qquad 0 \to \Sigma^{2^i-1}\mathbb{Z}/2 \xrightarrow{j} H_*(P) \to \tilde{P} \to 0,$$

$$0 \to \tilde{P} \to H_*(Y) \xrightarrow{k} \Sigma^{2^i-1}\mathbb{Z}/2 \to 0$$

where j maps 1 to e_{2^i-1}, k maps $\tilde{e}_{2^i-1}$ to 1 and $\tilde{P} = coker j$. It is not difficult to see that the conclusions (2.5)(2), (2.5)(3) and (2.5)(4) follow from the long exact sequences in Ext groups corresponding to the two short exact sequences in (10) and from our knowledge of $Ext_A^{s,*}(P)$ for $s \leq 3$ in (2.4).

We proceed to describe $Ext_A^{s,t}(P \wedge P)$ and $Ext_A^{s,t}(Y \wedge Y)$ for some s and t that will be needed for the first stage proof of (1.2).

If K and L are locally finite left A-modules then $K \otimes L$ is also a locally finite left A-module via Cartan formula. The right A-action on $(K \otimes L)_* = K_* \otimes L_*$ also satisfies Cartan formula. Thus (4), when applied to $N_* \otimes \Lambda = (K \otimes L)_* \otimes \Lambda = K_* \otimes L_* \otimes \Lambda$, becomes

$$(11) \quad \delta((k_* \otimes \ell_*)\lambda_I) = (k_* \otimes \ell_*)\delta(\lambda_I) + \sum_{j \geq 0}\left(\sum_{\nu=0}^{j+1} k_*Sq^\nu \otimes \ell_*Sq^{j+1-\nu}\right)\lambda_j\lambda_I.$$

In the next two paragraphs we will consider $K_* \otimes L_* = H_*(P) \otimes H_*(P)$ and $K_* \otimes$

$K_* \otimes L_* \otimes \Lambda$ which give rise to various classes in the respective Ext groups and state some results about these Ext groups. Proofs of these results will be described afterwards.

It is easy to see, from (6) and (11), that $e_{2^\ell-1} \otimes e_{2^\ell-1}$ are cycles in $H_*(P) \otimes H_*(P) \otimes \Lambda$. Straightforward calculations show that $e_{2^\ell-2} \otimes e_1 + \sum_{1 \le j \le 2^\ell-3} e_j \otimes e_{2^\ell-j-1}$ are also cycles in $H_*(P) \otimes H_*(P) \otimes \Lambda$.

Let $\overline{e_{2^\ell-1} \otimes e_{2^\ell-1}} = \{e_{2^\ell-1} \otimes e_{2^\ell-1}\} \in Ext_A^{0,2^{\ell+2}-2}(P \wedge P)$ and $\overline{e_{2^\ell-2} \otimes e_1} = \left\{e_{2^\ell-2} \otimes e_1 + \sum_{1 \le j \le 2^\ell-3} e_j \otimes e_{2^\ell-j-1}\right\} \in Ext_A^{0,2^\ell-1}(P \wedge P)$. We claim the following.

(12) Let
$$z' = (e_{22} \otimes e_{31})\lambda_1\lambda_7 + (e_{14} \otimes e_{31})\lambda_9\lambda_7 + (e_{12} \otimes e_{31})\lambda_{11}\lambda_7 + (e_{16} \otimes e_{31})\lambda_7^2 + (e_8 \otimes e_{31})(\lambda_3\lambda_{19} + \lambda_{19}\lambda_3) + (e_8 \otimes e_{47})\lambda_3^2.$$
In $H_*(P) \otimes H_*(P) \otimes \Lambda$ there is a cycle z of the form $z = z' + z''$ where $z'' = \sum_\nu (e_4 \otimes e_{j_\nu})\lambda_{I_\nu} + \sum_\mu (e_3 \otimes e_{j_\mu})\lambda_{I_\mu} + \sum_m (e_2 \otimes e_{j_m})\lambda_{I_m} + \sum_n (e_1 \otimes e_{j_n})\lambda_{I_n}$.

To see this we note, by direct calculations, that

$$(*) \qquad \delta(z') = (e_4 \otimes e_{47})\lambda_3^3 + (e_4 \otimes e_{31})(\lambda_3^2\lambda_{19} + \lambda_3\lambda_{19}\lambda_3 + \lambda_{19}\lambda_3^2) \in (H_*(P) \otimes H_*(P) \otimes \Lambda)^{3,63}.$$

Since, by (2.4), $Ext_A^{3,t}(P) = 0$ for $t = 59, 60, 61, 62$ it follows that there exists a chain z'' of the form described in (12) such that $\delta(z'')$ is the sum in $(*)$. This proves (12). Let $\overline{(e_{22} \otimes e_{31})h_1h_3} \in Ext_A^{2,63}(P \wedge P)$ be the class represented by the cycle z in (12). There is also an operation $Sq^0 : H_*(P) \otimes H_*(P) \otimes \Lambda \to H_*(P) \otimes H_*(P) \otimes \Lambda$ given by

$$Sq^0((e_k \otimes e_\ell)\lambda_{i_1} \cdots \lambda_{i_s}) = (e_{2k+1} \otimes e_{2\ell+1})\lambda_{2i_1+1} \cdots \lambda_{2i_s+1}.$$

It is easy to see, from (3), (5) and (11), that Sq^0 is a chain map. Let

$$\overline{(e_{45} \otimes e_{63})h_2h_4} = Sq^0\left(\overline{(e_{22} \otimes e_{31})h_1h_3}\right) = \{Sq^0(z)\} \in Ext_A^{2,128}(P \wedge P).$$

PROPOSITION 2.6. *(1)* $Ext_A^{0,2^{\ell+1}-2}(P \wedge P) = \mathbb{Z}/2$ *on* $\overline{e_{2^\ell-1} \otimes e_{2^\ell-1}}$ *for* $\ell \ge 1$.

(2) $Ext_A^{1,2^{i+1}-2}(P \wedge P) = \mathbb{Z}/2$ *on* $\overline{e_{2^{i-1}-1} \otimes e_{2^{i-1}-1}}\,h_i$ *for* $i \ge 4$.

(3) $Ext_A^{1,2^{i+1}-1}(P \wedge P) = \mathbb{Z}/2$ *on* $\overline{e_{2^i-1} \otimes e_{2^i-1}}\,h_0$ *and* $\overline{e_{2^i-2} \otimes e_1}\,h_i$ *for* $i \ge 4$.

(4) $\overline{e_{2^i-2} \otimes e_1}\,h_{i-1} = 0$ *in* $Ext_A^{1,2^i+2^{i-1}-1}(P \wedge P)$ *and* $\overline{e_{2^{i-1}-1} \otimes e_{2^{i-1}-1}}\,h_{i-1}^2 = 0$ *in* $Ext_A^{2,2^{i+1}-2}(P \wedge P)$ *for* $i \ge 2$.

(5) $\overline{e_{2^{i-1}-1} \otimes e_{2^{i-1}-1}}\,h_ih_0^2 \ne 0$ *in* $Ext_A^{3,2^{i+1}}(P \wedge P)$ *for* $i \ge 5$.

(6) $Ext_A^{2,2^{i+1}-1}(P \wedge P)$
$$= \begin{cases} \mathbb{Z}/2 & \text{on } \overline{e_{2^{i-1}-1} \otimes e_{2^{i-1}-1}}\,h_ih_0 \quad \text{for } 4 \le i < \infty, i \ne 5. \\ \mathbb{Z}/2 \oplus \mathbb{Z}/2 & \text{on } \overline{e_{15} \otimes e_{15}}\,h_5h_0 \text{ and } \overline{(e_{22} \otimes e_{31})h_1h_3} \text{ for } i = 5. \end{cases}$$

(7) $Ext_A^{2,2^i}(P \wedge P) = \mathbb{Z}/2 \oplus \mathbb{Z}/2 \oplus \mathbb{Z}/2$ *on* $\overline{e_{2^{i-1}-1} \otimes e_{2^{i-1}-1}}\,h_0^2$,

$\overline{e_{2^{i-1}-2} \otimes e_1 h_0 h_{i-1}}$ *and* $\overline{e_{2^{i-2}-1} \otimes e_{2^{i-2}-1}\, h_1 h_{i-1}}$ *for* $i \geq 6$, $i \neq 7$. *For* $i = 7$, *we add one more* $\mathbb{Z}/2$ *on* $\overline{(e_{45} \otimes e_{63})\, h_2 h_4}$. *For* $i = 5$, *this group is* $\mathbb{Z}/2 \oplus \mathbb{Z}/2$ *on* $\overline{e_{15} \otimes e_{15}}\, h_0^2$ *and* $\overline{e_{14} \otimes e_1}\, h_0 h_4$.

(8) $\overline{e_{2^{i-1}-1} \otimes e_{2^{i-1}-1}}\, h_{i-1} h_0^2$, $\overline{e_{2^i-2} \otimes e_1}\, h_{i-2}^2 h_0$ *and* $\overline{e_{2^{i-2}-1} \otimes e_{2^{i-2}-1}}\, h_1 h_{i-1}^2$ *are linearly independent in* $Ext_A^{3,2^i+2^{i-1}}(P \wedge P)$ *for* $i \geq 6$.

(9) $\overline{e_{15} \otimes e_{15}}\, h_0^3 h_5 \neq 0$ *in* $Ext_A^{4,65}(P \wedge P)$.

Recall that Y is the cofiber of $M = S^{2^i-2} \cup_{2\iota} e^{2^i-1} \xrightarrow{f} P$ in (1.4) with $i \geq 5$. So we have the cofibrations $M \xrightarrow{f} P \xrightarrow{q} Y$ and $M \wedge P \xrightarrow{f \wedge id} P \wedge P \xrightarrow{q \wedge id} Y \wedge P$. The induced maps in chain complexes and in Ext groups of $P \wedge P \xrightarrow{q \wedge id} Y \wedge P$, $Y \wedge P \xrightarrow{id \wedge q} Y \wedge Y$ and $P \wedge P \xrightarrow{q \wedge q} Y \wedge Y$ both will be denoted by $(q \wedge id)_*$, $(id \wedge q)_*$ and $(q \wedge q)_*$ respectively. For the remainder of this paragraph, the integer i is the integer in $S^{2^i-2} \cup_{2\iota} e^{2^i-1} \xrightarrow{f} P$ with $i \geq 5$. The images of $\overline{e_{2^{i-1}-1} \otimes e_{2^{i-1}-1}} \in Ext_A^{0,2^i-2}(P \wedge P)$ (resp. $\overline{(e_{22} \otimes e_{31})\, h_1 h_3} \in Ext_A^{2,63}(P \wedge P)$) in $Ext_A^{0,2^i-2}(Y \wedge P)$ (resp. $Ext_A^{2,63}(Y \wedge P)$) and in $Ext_A^{0,2^i-2}(Y \wedge Y)$ (resp. $Ext_A^{2,63}(Y \wedge Y)$) under $(q \wedge id)_*$ and $(q \wedge q)_*$ respectively are also denoted by $\overline{e_{2^{i-1}-1} \otimes e_{2^{i-1}-1}}$ (resp. $\overline{(e_{22} \otimes e_{31})\, h_1 h_3}$). We should stress that when we consider $\overline{(e_{22} \otimes e_{31})\, h_1 h_3}$ in $Ext_A^{2,63}(Y \wedge P)$ or in $Ext_A^{2,63}(Y \wedge Y)$ then this Y is the Y corresponding to $i = 5$, that is, Y is the cofiber of $S^{30} \cup_{2\iota} e^{31} \xrightarrow{f} P$. Both images $\overline{e_{2^{i-1}-1} \otimes e_{2^{i-1}-1}}$ are also represented by $e_{2^{i-1}-1} \otimes e_{2^{i-1}-1}$ in $H_*(Y) \otimes H_*(P) \otimes \Lambda$ and in $H_*(Y) \otimes H_*(Y) \otimes \Lambda$ respectively. $\overline{(e_{22} \otimes e_{31})\, h_1 h_3} \in Ext_A^{2,63}(Y \wedge P)$ is represented by $(q \wedge id)_*(z)$ which has the same form as that for z given in (12). $\overline{(e_{22} \otimes e_{31})\, h_1 h_3} \in Ext_A^{2,63}(Y \wedge Y)$ is represented by $(q \wedge q)_*(z) = (e_8 \otimes e_{47})\, \lambda_3^2 + (q \wedge q)_*(z'')$ where z'' is as in (12), since $q_*\,(e_{31}) = 0$. In $H_*(Y) \otimes H_*(Y) \otimes \Lambda$ there are the cycles $e_{2^{i-1}-1} \otimes e_{2^i+2^{i-1}-1}$, $e_{2^i+2^{i-1}-1} \otimes e_{2^{i-1}-1}$ and the cycles

(13)
$$
\begin{aligned}
z_1 &= (e_{2^i+2^{i-2}-1} \otimes e_{2^{i-1}-1})\, \lambda_{2^{i-2}-1}\lambda_0 + (\tilde{e}_{2^i-1} \otimes e_{2^{i-1}-1})\, \lambda_{2^{i-1}-1}\lambda_0 \\
&\quad + (e_{2^{i-1}+2^{i-2}-1} \otimes e_{2^{i-1}-1})\, \lambda_{2^{i-1}+2^{i-2}-1}\lambda_0, \\
z_2 &= (e_{2^{i-1}-1} \otimes e_{2^i+2^{i-2}-1})\, \lambda_{2^{i-2}-1}\lambda_0 + (e_{2^{i-1}-1} \otimes \tilde{e}_{2^i-1})\, \lambda_{2^{i-1}-1}\lambda_0 \\
&\quad + (e_{2^{i-1}-1} \otimes e_{2^{i-1}+2^{i-2}-1})\, \lambda_{2^{i-1}+2^{i-2}-1}\lambda_0.
\end{aligned}
$$

Let

$$
\overline{e_{2^{i-1}-1} \otimes e_{2^i+2^{i-1}-1}} = \{e_{2^{i-1}-1} \otimes e_{2^i+2^{i-1}-1}\},
$$

$$
\overline{e_{2^i+2^{i-1}-1} \otimes e_{2^{i-1}-1}} = \{e_{2^i+2^{i-1}-1} \otimes e_{2^{i-1}-1}\} \in Ext_A^{0,2^{i+1}-2}(Y \wedge Y)
$$

and

$$
\alpha_j = \{z_j\} \in Ext_A^{2,2^{i+1}-1}(Y \wedge Y),\ j = 1, 2.
$$

z_1 in (13) when considered as a chain in $H_*(Y) \otimes H_*(P) \otimes \Lambda$ will be denoted by $\tilde{z}_1$. It is easy to see that $\tilde{z}_1$ is also a cycle. The sum

(14)
$$
\begin{aligned}
\tilde{z}_2 &= (\tilde{e}_{2^i-1} \otimes e_{2^{i-1}-1})\, \lambda_{2^{i-1}-1}\lambda_0 \\
&\quad + (e_{2^{i-1}-1} \otimes e_{2^i+2^{i-2}-1})\, \lambda_{2^{i-2}-1}\lambda_0 \\
&\quad + (e_{2^{i-1}-1} \otimes e_{2^{i-1}-1})\, \lambda_{2^i-1}\lambda_0
\end{aligned}
$$

is also a cycle in $H_*(Y)\otimes H_*(P)\otimes\Lambda$. Let $\tilde{\alpha}_j=\{\tilde{z}_j\}\in Ext_A^{2,2^{i+1}-1}(Y\wedge P)$, $j=1,2$.

PROPOSITION 2.7. *(1) α_1 and α_2 are linearly independent in $Ext_A^{2,2^{i+1}-1}(Y\wedge Y)$ for $i\geq 6$. For $i=5$, α_1, α_2 and $\overline{(e_{22}\otimes e_{31})\,h_1h_3}$ are linearly independent in $Ext_A^{2,63}(Y\wedge Y)$.*

(2) $\alpha_1+\alpha_2=\overline{e_{2^{i-1}-1}\otimes e_{2^{i-1}-1}}\,h_ih_0$ *in* $Ext_A^{2,2^{i+1}-1}(Y\wedge Y)$.

(3) $Ext_A^{0,2^{i+1}-2}(Y\wedge Y)=\mathbb{Z}/2\oplus\mathbb{Z}/2$ *on* $\overline{e_{2^{i-1}-1}\otimes e_{2^i+2^{i-1}-1}}$ *and* $\overline{e_{2^i+2^{i-1}-1}\otimes e_{2^{i-1}-1}}$.

PROPOSITION 2.8. *(1) $\tilde{\alpha}_1$ and $\tilde{\alpha}_2$ are linearly independent in $Ext_A^{2,2^{i+1}-1}(Y\wedge P)$ for $i\geq 6$. For $i=5$, $\tilde{\alpha}_1$, $\tilde{\alpha}_2$ and $\overline{(e_{22}\otimes e_{31})\,h_1h_3}$ are linearly independent in $Ext_A^{2,63}(Y\wedge P)$.*

(2) $\tilde{\alpha}_1+\tilde{\alpha}_2=\overline{e_{2^{i-1}-1}\otimes e_{2^{i-1}-1}}\,h_ih_0$ *in* $Ext_A^{2,2^{i+1}-1}(Y\wedge P)$.

Before proving (2.6), (2.7) and (2.8) we remark that only (2.6) and (2.7) will be needed in the first stage proof for (1.2). We will show in a moment that (2.8) implies (2.7)(1), (2). Why (2.7)(1), (2) are to be proved via (2.8) will be explained later.

(2.8) $\Rightarrow$ (2.7)(1), (2) is shown as follows. From the definitions of α_1 and $\tilde{\alpha}_1$ we have $(id\wedge q)_*(\tilde{\alpha}_1)=\alpha_1$. Since $q_*(e_{2^i-1})=0$, the cycle $\tilde{z}_2$ in (14) is mapped, under $(id\wedge q)_*$, to the cycle

$$\begin{aligned}z_2'=&(\tilde{e}_{2^i-1}\otimes e_{2^{i-1}-1})\,\lambda_{2^{i-1}-1}\lambda_0+(e_{2^{i-1}-1}\otimes e_{2^i+2^{i-2}-1})\,\lambda_{2^{i-2}-1}\lambda_0\\&+(e_{2^{i-1}-1}\otimes e_{2^{i-1}-1})\,\lambda_{2^i-1}\lambda_0+(e_{2^{i-1}-1}\otimes e_{2^{i-1}+2^{i-2}-1})\,\lambda_{2^{i-1}+2^{i-2}-1}\lambda_0\end{aligned}$$

in $H_*(Y)\otimes H_*(Y)\otimes\Lambda$. Straightforward calculations, by (7) and (11), show that, in $H_*(Y)\otimes H_*(Y)\otimes\Lambda$, $\delta\left((\tilde{e}_{2^i-1}\otimes\tilde{e}_{2^i-1})\,\lambda_0\right)=z_2+z_2'$ where z_2 is as in (13). So $(id\wedge q)_*(\tilde{\alpha}_2=\{\tilde{z}_2\})=\{z_2'\}=\{z_2\}=\alpha_2$. Again straightforward calculations, by (5), (7), (11), also show that, in $H_*(Y)\otimes H_*(P)\otimes\Lambda$,

$$\begin{aligned}&\delta\left((e_{2^i+2^{i-2}-1}\otimes e_{2^{i-1}+2^{i-2}-1})\,\lambda_0+(e_{2^{i-1}+2^{i-2}-1}\otimes e_{2^i+2^{i-2}-1})\,\lambda_0\right.\\&\qquad\left.+\,(e_{2^{i-1}-1}\otimes e_{2^i+2^{i-1}-1})\,\lambda_0\right)\\&=\tilde{z}_1+\tilde{z}_2+(e_{2^{i-1}-1}\otimes e_{2^{i-1}-1})\,\lambda_{2^i-1}\lambda_0\end{aligned}$$

where $\tilde{z}_1=z_1$ is as given in (13). So $\tilde{\alpha}_1+\tilde{\alpha}_2=\overline{e_{2^{i-1}-1}\otimes e_{2^{i-1}-1}}\,h_{i-1}h_0$ in $Ext_A^{2,2^{i+1}-1}(Y\wedge P)$. This proves (2.8)(2). Applying $(id\wedge q)_*$ we get $\alpha_1+\alpha_2=\overline{e_{2^{i-1}-1}\otimes e_{2^{i-1}-1}}\,h_{i-1}h_0$ in $Ext_A^{2,2^{i+1}-1}(Y\wedge Y)$. This proves (2.7)(2). Tensoring $H_*(Y)$ with the short exact sequences in (10) we obtain the short exact sequences

$$\begin{aligned}&0\to\Sigma^{2^i-1}H_*(Y)\to H_*(Y\wedge P)=H_*(Y)\otimes H_*(P)\to H_*(Y)\otimes\tilde{P}\to 0\\&0\to H_*(Y)\otimes\tilde{P}\to H_*(Y\wedge Y)=H_*(Y)\otimes H_*(Y)\to\Sigma^{2^i-1}H_*(Y)\to 0.\end{aligned}$$

From the long exact sequences in Ext groups corresponding to these short exact sequences and from (2.5)(2), (3) we see

$$Ext_A^{2,2^{i+1}-1}(Y\wedge P)\xrightarrow{(q\wedge id)_*}Ext_A^{2,2^{i+1}-1}(Y\wedge Y)\qquad\text{is monomorphic.}$$

This proves that (2.8)(1) implies (2.7)(1).

(2.6)(4) is proved as follows. By direct computations we find that

$$\delta\left(\sum_{1\le k\le 2^i+2^{i-1}-2} e_k\otimes e_{2^i+2^{i-1}-k-1}\right)=\left(\sum_{1\le j\le 2^i-2} e_j\otimes e_{2^i-j-1}\right)\lambda_{2^{i-1}-1}$$

and

$$\begin{aligned}&\delta\left((e_{2^i+2^{i-2}-1}\otimes e_{2^{i-1}-1})\lambda_{2^{i-2}-1}+(e_{2^{i-1}+2^{i-2}-1}\otimes e_{2^{i-1}-1})\lambda_{2^{i-1}+2^{i-2}-1}\right.\\&\qquad\left.+(e_{2^i-1}\otimes e_{2^{i-1}+2^{i-2}-1})\lambda_{2^{i-2}-1}\right)\\&=(e_{2^{i-1}-1}\otimes e_{2^{i-1}-1})\lambda^2_{2^{i-1}-1}\end{aligned}$$

in $H_*(P)\otimes H_*(P)\otimes\Lambda$. So $\overline{e_{2^i-2}\otimes e_1}\,h_{i-1}=0$ and $\overline{e_{2^{i-1}-1}\otimes e_{2^{i-1}-1}}\,h^2_{i-1}=0$. This proves (2.6)(4).

The rest of (2.6) and (2.7)(3), (2.8)(1) are proved by making calculations with a spectral sequence. This spectral sequence is described as follows.

Suppose K_* is a right A-module such that $K_j=0$ for $j\le 0$ and $K_j=\mathbb{Z}/2$ for $j\ge 1$, say, generated by e'_j. For example, $H_*(P)$ and $H_*(Y)$ have this property. Filter the complex $K_*\otimes L_*\otimes\Lambda$ by $\bar F(j)=\sum_{u\le j}K_u\otimes L_*\otimes\Lambda$. Then

$$\bar F(j)/\bar F(j-1)\cong K_j\otimes L_*\otimes\Lambda\cong\begin{cases}\Sigma^j L_*\otimes\Lambda & j\ge 1\\ 0 & j\le 0.\end{cases}$$

This filtration gives rise to a spectral sequence $\{\bar E_r^{j,s,t}=\bar E_r^{j,s,t}(K\otimes L)\}_{r\ge 1}$ with $\sum_j \bar E_\infty^{j,s,t}\cong Ext_A^{s,t}(K\otimes L)$ as a $\mathbb{Z}/2$-module and

$$\bar E_1^{j,s,t}=H^{s,t}\left(\bar F(j)/\bar F(j-1)\right)=\begin{cases}Ext_A^{s,t-j}(L)\text{ with base }e'_j & \text{for } j\ge 1\\ 0 & \text{for } j\le 0.\end{cases}$$

Thus, for $j\ge 1$, basis elements in $\bar E_1^{j,s,t}$ are of the form $e'_j\otimes\alpha$ with $\alpha\ne 0$ in $Ext_A^{s,t-j}(L)$. For each $r\ge 1$ the differential d_r of the spectral sequence goes from $\bar E_r^{j,s,t}$ to $\bar E_r^{j-r,s+1,t}$. If $d_r\left(e'_j\otimes\alpha\right)=e'_{j-r}\otimes\beta$ is a non-trivial differential then we write $e'_j\otimes\alpha\to e'_{j-r}\otimes\beta$ and if $e'_j\otimes\alpha$ is an infinite cycle, that is, $d_r\left(e'_j\otimes\alpha\right)=0$ for all r then we write $e'_j\otimes\alpha\to 0$. Via this spectral sequence if $Ext_A^{s,*}(L)$ are known for $s\le\bar s$ then one may (i) compute $Ext_A^{s,*}(K\otimes L)$ for $s\le\bar s-1$ and (ii) determine whether or not a finite family of specific known classes in $Ext_A^{\bar s,*}(K\otimes L)$ is linearly independent.

The results in (2.6) (except (2.6)(4)) and (2.7)(3), (2.8)(1) are proved this way since $Ext_A^{s,*}(P)$ for $s\le 3$ and $Ext_A^{0,*}(Y)$ (together with (2.5)(4), as will be seen) are known by (2.4) and (2.5)(1). We are not going to do here all the calculations in the spectral sequence for these. Rather, we illustrate in the next two paragraphs the use of this spectral sequence for proving four of these results, which are (2.6)(1)

For (2.6)(1) we have to consider $e_{2^{\ell+1}-2^k-1} \otimes \bar{e}_{2^k-1}$ for a fixed $\ell \geq 1$ and for $1 \leq k \leq \ell$ which form a $\mathbb{Z}/2$-base for $\bar{E}_1^{0,2^{\ell+1}-2}(P \wedge P)$. Here $\bar{E}_1^{s,t}(X \wedge P) = \sum_j \bar{E}_1^{j,s,t}(X \wedge P)$ for $X = P$ or Y. We have the differentials $e_{2^{\ell+1}-2^k-1} \otimes \bar{e}_{2^k-1} \to e_{2^{\ell+1}-2^{k+1}-1} \otimes \bar{e}_{2^k-1}h_k$ for $k \leq \ell - 1$ and $e_{2^\ell-1} \otimes \bar{e}_{2^\ell-1} \to 0$ for $k = \ell$. Thus $Ext_A^{0,2^{\ell+1}-2}(P \wedge P) = \mathbb{Z}/2$ on $\overline{e_{2^\ell-1} \otimes e_{2^\ell-1}}$ which is represented by the infinite cycle $e_{2^\ell-1} \otimes \bar{e}_{2^\ell-1}$. (2.7)(3) is similarly proved as follows. By (2.5)(1) we see that $e_{2^{i+1}-2^k-1} \otimes \bar{e}_{2^k-1}$ for $k \leq i-1$ together with $e_{2^{i-1}-1} \otimes \bar{e}_{2^i+2^{i-1}-1}$ form a $\mathbb{Z}/2$-base for $\bar{E}_1^{0,2^{i+1}-2}(Y \wedge Y)$. We have

$$e_{2^{i+1}-2^k-1} \otimes \bar{e}_{2^k-1} \longrightarrow e_{2^{i+1}-2^{k+1}-1} \otimes \bar{e}_{2^k-1}h_k \quad \text{for} \quad k \leq i-2,$$

$$e_{2^i+2^{i-1}-1} \otimes \bar{e}_{2^{i-1}-1} \longrightarrow 0 \quad \text{and} \quad e_{2^{i-1}-1} \otimes \bar{e}_{2^i+2^{i-1}-1} \longrightarrow 0$$

where $\bar{e}_{2^k-1}h_k \neq 0$ for $k \leq i-2$ by (2.5)(4). Thus $Ext_A^{0,2^{i+1}-2}(Y \wedge Y) = \mathbb{Z}/2 \oplus \mathbb{Z}/2$ on $\overline{e_{2^{i-1}-1} \otimes e_{2^i+2^{i-1}-1}}$ and $\overline{e_{2^i+2^{i-1}-1} \otimes e_{2^{i-1}-1}}$ respectively. This proves (2.6)(1) and (2.7)(3).

We proceed to prove (2.6)(3) and (2.8)(1). We begin with (2.8)(1). It is equivalent to proving for (2.8)(1) that $\tilde{\alpha}_1 + \tilde{\alpha}_2 = \overline{e_{2^{i-1}-1} \otimes e_{2^{i-1}-1}}\, h_i h_0$ (by (2.8)(2)), $\tilde{\alpha}_2$ and $\overline{(e_{22} \otimes e_{31})\, h_1 h_3}$ are linearly independent. For a class α in $Ext_A^{*,*}(P \wedge P)$ or in $Ext_A^{*,*}(Y \wedge P)$ if α is represented by a cycle of the form $\left(e_j \otimes \sum_\nu e_\nu\right)\lambda_{I_\nu} + x$ where $x \in \bar{F}(j-1)$ then we write $\alpha \equiv \left\{\left(e_j \otimes \sum_\nu e_\nu\right)\lambda_{I_\nu}\right\} \bmod \bar{F}(j-1)$. Recall $\overline{e_{2^{i-1}-1} \otimes e_{2^{i-1}-1}}\, h_0 h_i$ is represented by the cycle $(e_{2^{i-1}-1} \otimes e_{2^{i-1}-1})\lambda_0\lambda_{2^i-1}$. In $H_*(Y) \otimes H_*(P) \otimes \Lambda$,

$$\begin{aligned}
&\delta\left((e_{2^{i-1}} \otimes e_{2^{i-1}-1})\lambda_{2^i-1}\right)\\
&\equiv (e_{2^{i-1}-1} \otimes e_{2^{i-1}-1})\lambda_0\lambda_{2^i-1} + (e_{2^{i-1}-2} \otimes e_{2^{i-1}-1})\lambda_1\lambda_{2^i-1}\\
&\qquad\qquad \bmod \quad \bar{F}\left(2^{i-1}-3\right).
\end{aligned}$$

So $\overline{e_{2^{i-1}-1} \otimes e_{2^{i-1}-1}}\, h_0 h_i \equiv \{(e_{2^{i-1}-2} \otimes e_{2^{i-1}-1})\lambda_1\lambda_{2^i-1}\} \bmod \bar{F}\left(2^{i-1}-3\right)$. From (14) and (12) we see

$$\tilde{\alpha}_2 \equiv \{(\tilde{e}_{2^i-1} \otimes e_{2^{i-1}-1})\lambda_{2^{i-1}-1}\lambda_0\}\, mod \bar{F}\left(2^i-2\right)$$

and

$$\overline{(e_{22} \otimes e_{31})\, h_1 h_3} \equiv \{(e_{22} \otimes e_{31})\lambda_1\lambda_7\}\, mod \bar{F}(21).$$

These imply
$e_{2^{i-1}-2} \otimes \bar{e}_{2^{i-1}-1}h_1h_i$, $\tilde{e}_{2^i-1} \otimes \bar{e}_{2^{i-1}-1}h_{i-1}h_0$ and $e_{22} \otimes \bar{e}_{31}h_1h_3$ are infinite cycles in the spectral sequence $\bar{E}_r^{*,*,*}(Y \wedge P)$. If we can show they are not boundaries then we are done. The classes

(**) $\quad e_{2^{i+1}-2^\ell-2^j} \otimes \bar{e}_{2^\ell-1}h_j$ for $1 \leq 2^{i+1} - 2^\ell - 2^j \neq 2^i - 1$ with $j \neq \ell - 1$ and $\tilde{e}_{2^i-1} \otimes \bar{e}_{2^i-1}h_0$

form a $\mathbb{Z}/2$-base for $\bar{E}_1^{1,2^{i+1}-1}(Y \wedge P)$. We have the following differentials.

(2.8)(1)* (1) $e_{2^{i+1}-2^\ell-2^j} \otimes \bar{e}_{2^\ell-1}h_j \to e_{2^{i+1}-2^\ell-2^j-1} \otimes \bar{e}_{2^\ell-1}h_0h_j$, $2 \leq j < \ell - 1 < i$ or $3 \leq j = \ell \leq i-2$ or $2 \leq \ell < j \leq i$,

(2) $e_{2^{i+1}-2^\ell-2}\otimes \bar{e}_{2^\ell-1}h_1 \longrightarrow e_{2^{i+1}-2^{\ell+1}-2}\otimes \bar{e}_{2^\ell-1}h_\ell h_1,\ 4\le \ell\le i-1,$
(3) $e_{2^{i+1}-10}\otimes \bar{e}_7 h_1 \longrightarrow e_{2^{i+1}-24}\otimes \bar{e}_7 h_3^2,$
(4) $e_{2^i}\otimes \bar{e}_{2^{i-1}-1}h_{i-1}\longrightarrow e_{2^i-2}\otimes \bar{e}_{2^{i-1}-1}h_1 h_{i-1},$
(5) $e_{2^{i+1}-2^j-2}\otimes \bar{e}_1 h_j \longrightarrow e_{2^{i+1}-2^{j+1}-2}\otimes \bar{e}_1 h_j^2,\ 1\le j\le i-1,$
(6) $\tilde{e}_{2^i-1}\otimes \bar{e}_{2^i-1}h_0 \longrightarrow 0,$
(7) $e_{2^i-2}\otimes \bar{e}_{2^i-1}h_1 \longrightarrow 0,$
(8) $e_{2^i-2}\otimes \bar{e}_1 h_i \longrightarrow 0,$
(9) $e_{2^{i+1}-2^\ell-1}\otimes \bar{e}_{2^\ell-1}h_0 \longleftarrow e_{2^{i+1}-2^\ell}\otimes \bar{e}_{2^\ell-1},\ 2\le\ell\le i-1.$

From these differentials we see $e_{2^{i-1}-2}\otimes \bar{e}_{2^{i-1}-1}h_i h_1$, $\tilde{e}_{2^i-1}\otimes \bar{e}_{2^{i-1}-1}h_{i-1}h_0$ and $e_{22}\otimes \bar{e}_{31}h_1h_3$ are not boundaries; so they persist to $\bar{E}_\infty^{*,*,*}(Y\wedge P)$ representing respectively the classes $\tilde{\alpha}_1+\tilde{\alpha}_2=\overline{e_{2^{i-1}-1}\otimes e_{2^{i-1}-1}}h_0h_i$, $\tilde{\alpha}_2$ and $\overline{(e_{22}\otimes e_{31})\,h_1h_3}$. This proves (2.8)(1). At this point we would like to make the following remarks. We have shown earlier that (2.8)(1) implies (2.7)(1). The reason (2.7)(1) is proved via the above proof of (2.8)(1) is that $Ext_A^{s,*}(P)$ for $s\le 2$ are known by (2.4) so that we are able to determine the differentials in (2.8)(1)* to see $\tilde{\alpha}_1+\tilde{\alpha}_2$, $\tilde{\alpha}_2$ and $\overline{(e_{22}\otimes e_{31})\,h_1h_3}$ are linearly independent in $Ext_A^{2,2^{i+1}-1}(Y\wedge P)$. This concludes our remarks. Next we prove (2.6)(3). The classes in (**) with $\tilde{e}_{2^i-1}\otimes \bar{e}_{2^i-1}h_0$ replaced by $e_{2^i-1}\otimes \bar{e}_{2^i-1}h_0$ form a $\mathbb{Z}/2$-base for $\bar{E}_1^{1,2^{i+1}-1}(P\wedge P)$. All the differentials in (2.8)(1)* are also differentials in the spectral sequence $\bar{E}_r^{*,*,*}(P\wedge P)$ with $\tilde{e}_{2^i-1}\otimes \bar{e}_{2^i-1}h_0\longrightarrow 0$ in (2.8)(1)* (6) replaced by (2.8)(1)* (6)′ $e_{2^i-1}\otimes \bar{e}_{2^i-1}h_0 \longleftarrow e_{2^i}\otimes \bar{e}_{2^i-1}$. From these differentials we see only $e_{2^i-2}\otimes \bar{e}_{2^i-1}h_1$ and $e_{2^i-2}\otimes \bar{e}_1 h_i$ can possibly persist to $\bar{E}_\infty^{1,2^{i+1}-1}(P\wedge P)$. Since the classes $e_{2^{i+1}-2^\ell}\otimes \bar{e}_{2^\ell-1}$ for $1\le \ell\le i$ form a $\mathbb{Z}/2$-base for $\bar{E}_1^{0,2^{i+1}-1}(P\wedge P)$ and since $e_{2^{i+1}-2}\otimes \bar{e}_1\longrightarrow 0$, from (2.8)(1)* (6)′ and (2.8)*(1) (9) we see $e_{2^i-2}\otimes \bar{e}_{2^i-1}h_1$ and $e_{2^i-2}\otimes \bar{e}_1 h_i$ are not boundaries. So $\bar{E}_\infty^{1,2^{i+1}-1}(P\wedge P)$ is indeed $\mathbb{Z}/2\oplus\mathbb{Z}/2$ on $e_{2^i-2}\otimes \bar{e}_{2^i-1}h_1$ and $e_{2^i-2}\otimes \bar{e}_1 h_i$. We have $\overline{e_{2^i-2}\otimes e_1}\,h_i\equiv\{(e_{2^i-2}\otimes e_1)\,\lambda_{2^i-1}\}\bmod \bar{F}(2^i-3)$ and $\delta\left(e_{2^i}\otimes e_{2^i-1}\right)\equiv\left(e_{2^i-1}\otimes e_{2^i-1}\right)\lambda_0+\left(e_{2^i-2}\otimes e_{2^i-1}\right)\lambda_1 \bmod \bar{F}\left(2^i-3\right)$ in $H_*(P)\otimes H_*(P)\otimes\Lambda$; the latter implies $\overline{e_{2^i-1}\otimes e_{2^i-1}}\,h_0\equiv\{(e_{2^i-2}\otimes e_{2^i-1})\,\lambda_1\}$ mod $\bar{F}\left(2^i-3\right)$. Thus $Ext_A^{1,2^{i+1}-1}(P\wedge P)=\mathbb{Z}/2\oplus\mathbb{Z}/2$ on $\overline{e_{2^i-1}\otimes e_{2^i-1}}\,h_0$ and $\overline{e_{2^i-2}\otimes e_1}\,h_i$ for $i\ge 2$. This proves (2.6)(3).

This concludes our illustration of the proofs for (2.6), (2.7)(3) and (2.8)(1). We describe in the following correspondences $\alpha\leftrightarrow\alpha'$ the infinite cycles α' in the spectral sequence $\bar{E}_r^{*,*,*}(P\wedge P)$ representing the non-zero classes α in (2.6) that are not considered in the illustration of the proofs above.

(2.6)(2)* $\overline{e_{2^{i-1}-1}\otimes e_{2^{i-1}-1}}\,h_i\leftrightarrow e_{2^{i-1}-1}\otimes \bar{e}_{2^{i-1}-1}h_i$

(2.6)(5)* $\overline{e_{2^{i-1}-1}\otimes e_{2^{i-1}-1}}\,h_ih_0^2\leftrightarrow e_{2^{i-1}-3}\otimes \bar{e}_{2^{i-1}-1}h_1^2h_i.$

(2.6)(6)* $\overline{e_{2^{i-1}-1}\otimes e_{2^{i-1}-1}}\,h_ih_0\leftrightarrow e_{2^{i-1}-2}\otimes \bar{e}_{2^{i-1}-1}h_1h_i.$

$\overline{(e_{22}\otimes e_{31})\,h_1h_3}\leftrightarrow e_{22}\otimes \bar{e}_{31}h_1h_3.$

(2.6)(7)′ $\overline{e_{2^{i-1}-1}\otimes e_{2^{i-1}-1}}\,h_0^2\leftrightarrow e_{2^{i-1}-3}\otimes \bar{e}_{2^{i-1}-1}h_1^2,$

$\overline{e_{2^{i-1}-2}\otimes e_1}\,h_0h_{i-1}\leftrightarrow e_{2^{i-1}-3}\otimes \bar{e}_1 h_1h_{i-1},$

$$\overline{(e_{45}\otimes e_{63})\,h_2h_4} \leftrightarrow e_{45}\otimes \bar{e}_{63}h_2h_4.$$

(2.6)(8)* $\overline{e_{2^{i-1}-1}\otimes e_{2^{i-1}-1}}\,h_{i-1}h_0^2 \leftrightarrow e_{2^{i-1}-3}\otimes \bar{e}_{2^{i-1}-1}h_1^2h_{i-1},$

$$\overline{e_{2^{i-2}-1}\otimes e_{2^{i-2}-1}}h_1h_{i-1}^2 \leftrightarrow e_{2^{i-2}-3}\otimes \bar{e}_{2^{i-2}-1}h_2h_{i-1}^2,$$

$$\overline{e_{2^i-2}\otimes e_1}\,h_{i-2}^2h_0 \leftrightarrow e_{2^i-3}\otimes \bar{e}_1h_1h_{i-2}^2.$$

(2.6)(9)* $\overline{e_{15}\otimes e_{15}h_0^3h_5} \leftrightarrow e_{12}\otimes \bar{e}_{15}h_1^3h_5.$

In concluding this section we prove the following lemma. Recall the transfer $P \xrightarrow{\bar{t}} S^0$. Consider $P\wedge P \xrightarrow{id\wedge\bar{t}} P\wedge S^0 = P$ and its induced map $Ext_A^{s,t}(P\wedge P) \xrightarrow{(id\wedge\bar{t})_*} Ext_A^{s+1,t+1}(P)$.

LEMMA 2.9. *(1)* $(id\wedge\bar{t})_*\,\overline{(e_{2^\ell-1}\otimes e_{2^\ell-1})} = \bar{e}_{2^\ell-1}h_\ell \in Ext_A^{1,2^{\ell+1}-1}(P)$, $\ell\geq 1$.

(2) $(id\wedge\bar{t})_*\,\overline{(e_{2^\ell-2}\otimes e_1)} = \bar{e}_{2^\ell-1}h_0 \in Ext_A^{1,2^\ell}(P)$, $\ell\geq 2$.
(3) $(id\wedge\bar{t})_*\left(\overline{(e_{22}\otimes e_{31})\,h_1h_3}\right) = \hat{D}_3 = \hat{D}_3(0) \in Ext_A^{3,64}(P)$.
(4) $(id\wedge\bar{t})_*\left(\overline{(e_{45}\otimes e_{63})\,h_2h_4}\right) = \hat{D}_3(1) \in Ext_A^{3,129}(P)$.

Proof. (1) is clear. To see (2), (3) first note that $\delta\,(e_{2^\ell}) \equiv e_{2^\ell-1}\lambda_0 + e_{2^\ell-2}\lambda_1$ mod $F\left(2^\ell-2\right)$ in $H_*(P)\otimes\Lambda$. Here we recall $F(j)\subset H_*(P)\otimes\Lambda$ is $F(j) = \sum_{k\leq j} H_k(P)\otimes\Lambda$. So

(i) $\bar{e}_{2^\ell-1}h_0 \equiv \{e_{2^\ell-2}\lambda_1\}$ mod $F(2^\ell-3)$, and $2^\ell-3$ is the smallest integer with this property, that is, $\bar{e}_{2^\ell-1}h_0 \not\equiv \{e_k\lambda_{2^\ell-1-k}\}$ mod $F(k)$ for $k<2^\ell-3$.

The second conclusion is proved in [7]. We also have the following.

(ii) From (2.3)(8), $\hat{D}_3 \equiv \{e_{22}\otimes\lambda_{31}\lambda_1\lambda_7\}$ mod $F(21)$, and, by (2.3)′(8), 21 is the smallest integer with this property.

From (12) for $\overline{(e_{22}\otimes e_{31})\,h_1h_3}$ and from the definition $\overline{e_{2^\ell-2}\otimes e_1} = \left\{e_{2^\ell-2}\otimes e_1 + \sum_{2\leq j\leq 2^\ell-3} e_j\otimes e_{2^\ell-j-1}\right\}$ we see

(iii) $\overline{e_{2^\ell-2}\otimes e_1} \equiv \{e_{2^\ell-2}\otimes e_1\}$ mod $\bar{F}\left(2^\ell-3\right)$ and
$\overline{(e_{22}\otimes e_{31})\,h_1h_3} \equiv \{(e_{22}\otimes e_{31})\,\lambda_1\lambda_7\}$ mod $\bar{F}(21)$.

Since $\bar{t}_*\,(e_k) = \lambda_k$ and $Ext_A^{1,2^\ell}(P)\cong\mathbb{Z}/2$ on $\bar{e}_{2^\ell-1}h_0$, $Ext_A^{3,64}(P)\cong\mathbb{Z}/2$ on $\hat{D}_3$ (by (2.4)) it follows from (i), (ii) and (iii) that $(id\wedge\bar{t})_*\,\overline{(e_{2^\ell-2}\otimes e_1)} = \bar{e}_{2^\ell-1}h_0$ and $(id\wedge\bar{t})_*\left(\overline{(e_{22}\otimes e_{31})\,h_1h_3}\right) = \hat{D}_3$. This proves (2) and (3). (4) follows from (3) since

$$\begin{aligned}(id\wedge\bar{t})_*\,\overline{(e_{45}\otimes e_{63}h_2h_4)} &= (id\wedge\bar{t})_*\left(Sq^0\left(\overline{e_{22}\otimes e_{31}h_1h_3}\right)\right)\\ &= Sq^0\left((id\wedge\bar{t})_*\left(\overline{e_{22}\otimes e_{31}h_1h_3}\right)\right)\\ &= Sq^0\left(\hat{D}_3 = \hat{D}_3(0)\right) = \hat{D}_3(1).\end{aligned}$$

This completes the proof of Lemma 2.9.

3. Some differentials in the Adams spectral sequence

In this section we determine the differential d_2 in the Adams spectral sequence on some of the elements considered in §2. The main results are (3.4), (3.5), (3.6) and (3.8). We will abbreviate "Adams spectral sequence" by "ASS". The points (1) through (14) that might be referred to are those in §2. This numbering of points will be continued hereafter, beginning with (15) in this section.

First we recall the following basic differentials

(3.1) $d_2(h_k) = h_{k-1}^2 h_0 \neq 0$ for $k \geq 4$ in the ASS for S^0.
(3.2) $d_2(\bar{e}_{2^k-1}) = \bar{e}_{2^{k-1}-1} h_{k-1} h_0 \neq 0$ for $k \geq 4$ in the ASS for P.

(3.1) is proved in [2]. (3.2) follows from (3.1) by applying the induced map $\bar{t}_*$ in Ext groups of the transfer $P \xrightarrow{\bar{t}} S^0$, noting that $\bar{t}_*(\bar{e}_{2^\ell-1}) = h_\ell$ and $Ext_A^{2,2^k}(P) \cong \mathbb{Z}/2$ on $\bar{e}_{2^{k-1}-1} h_{k-1} h_0$ for $k \geq 4$ by (2.4).

Assume $i \geq 5$ and consider $P_{2^i+1}^\infty = P^\infty / P^{2^i}$. It is easy to see that $Ext_A^{s,t+2^i}(P_{2^i+1}^\infty) \cong Ext_A^{s,t}(P)$ for $t - s \leq 2^i - 2$ with correspondence $\bar{e}_{2^i+2^\ell-1} \leftrightarrow \bar{e}_{2^\ell-1}$ for $1 \leq \ell \leq i-1$. Thus, via this isomorphism and by (2.4), we see $Ext_A^{0,2^i+2^{i-1}-1}(P_{2^i+1}^\infty) = \mathbb{Z}/2$ on $\bar{e}_{2^i+2^{i-1}-1}$ and $Ext_A^{2,2^i+2^{i-1}}(P_{2^i+1}^\infty) = \mathbb{Z}/2$ on $\bar{e}_{2^i+2^{i-2}-1} h_{i-2} h_0$ (since $i \geq 5$). We claim

(3.3) $d_2(\bar{e}_{2^i+2^{i-1}-1}) = \bar{e}_{2^i+2^{i-2}-1} h_{i-2} h_0$ in the ASS for $P_{2^i+1}^\infty$.

To see this consider the cofibration sequence $P^{2^i} \xrightarrow{j} P \to P_{2^i+1}^\infty \xrightarrow{q} \Sigma P^{2^i}$. It is clear that $Ext_A^{0,2^\ell-1}(P^{2^i}) \underset{\cong}{\xrightarrow{j_*}} Ext_A^{0,2^\ell-1}(P)$ for $\ell \leq i$. Let $\bar{e}_{2^\ell-1} = j_*^{-1}(\bar{e}_{2^\ell-1}) \in Ext_A^{0,2^\ell-1}(P^{2^i})$ for $\ell \leq i$. The classes $\bar{e}_{2^i-1} h_{i-2}^2 h_0$ and $\bar{e}_{2^{i-1}-1} h_{i-1}^2 h_0$ are non-zero and are equal in $Ext_A^{3,2^i+2^{i-1}}(P)$ by (2.4). But they are not equal in $Ext_A^{3,2^i+2^{i-1}}(P^{2^i})$, and this is easy to see. The induced map $Ext_A^{s,t}(P_{2^i+1}^\infty) \xrightarrow{q_*} Ext_A^{s+1,t+1}(\Sigma P^{2^i}) = Ext_A^{s+1,t}(P^{2^i})$ sends the class $\bar{e}_{2^i+2^{i-1}-1}$ to $\bar{e}_{2^i-1} h_{i-1}$. From (3.1) and (3.2) we see, in the ASS for P^{2^i}, $d_2(\bar{e}_{2^i-1} h_{i-1}) = \bar{e}_{2^i-1} h_{i-2}^2 h_0 + \bar{e}_{2^{i-1}-1} h_{i-1}^2 h_0 \neq 0$. This implies $d_2(\bar{e}_{2^i+2^{i-1}-1}) \neq 0$ in the ASS for $P_{2^i+1}^\infty$. Since $Ext_A^{2,2^i+2^{i-1}}(P_{2^i+1}^\infty) = \mathbb{Z}/2$ on $\bar{e}_{2^i+2^{i-2}-1} h_{i-2} h_0$ it follows that $d_2(\bar{e}_{2^i+2^{i-1}-1}) = \bar{e}_{2^i+2^{i-2}-1} h_{i-2} h_0$. This proves (3.3).

By (2.5), $Ext_A^{0,2^i+2^{i-1}-1}(Y) = \mathbb{Z}/2$ on $\bar{e}_{2^i+2^{i-1}-1}$ and $Ext_A^{2,2^i+2^{i-1}}(Y) = \mathbb{Z}/2 \oplus \mathbb{Z}/2$ on $\bar{e}_{2^{i-1}-1} h_i h_0$ and τ where $i \geq 5$ and τ is as described (9). Let Y^k be the k-skeleton of Y. Then $Y/Y^{2^i} \simeq P_{2^i+1}^\infty$. Let $Y \xrightarrow{q_1} P_{2^i+1}^\infty$ be the pinching map. Clearly $(q_1)_*(\bar{e}_{2^i+2^{i-1}-1}) = \bar{e}_{2^i+2^{i-1}-1}$ and, by (9), $(q_1)_*(\tau) = \bar{e}_{2^i+2^{i-2}-1} h_{i-2} h_0$. From (3.3) we thus deduce the following result.

PROPOSITION 3.4. *$d_2(\bar{e}_{2^i+2^{i-1}-1}) = \tau + \epsilon \bar{e}_{2^{i-1}-1} h_i h_0$ in the ASS for Y where $\epsilon = 0$ or 1.*

PROPOSITION 3.5. *$d_2\left(\overline{e_{2^i-2}\otimes e_1}\right)=\overline{e_{2^{i-1}-2}\otimes e_1}\,h_{i-1}h_0+$ $\epsilon'\overline{e_{2^{i-2}-1}\otimes e_{2^{i-2}-1}}\,h_1h_{i-1}$ for $i\geq 5$ in the ASS for $P\wedge P$ where $\epsilon'=0$ or 1.*

COROLLARY 3.6. *$d_2\left(\overline{e_{2^i-2}\otimes e_1 h_i}\right)=0$ for $i\geq 5$ in the ASS for $P\wedge P$.*

This follows from (3.5), (3.1) and (2.6)(4) by

$$\begin{aligned}d_2\left(\overline{e_{2^i-2}\otimes e_1}\,h_i\right)&=d_2\left(\overline{e_{2^i-2}\otimes e_1}\right)h_i+\overline{e_{2^i-2}\otimes e_1}\,d_2(h_i)\\&=\overline{e_{2^{i-1}-2}\otimes e_1}\,h_0h_{i-1}h_i+\epsilon'\overline{e_{2^{i-2}-1}\otimes e_{2^{i-2}-1}}\,h_1h_{i-1}h_i\\&\qquad+\overline{e_{2^i-2}\otimes e_1}\,h_{i-1}^2h_0\\&=0+0+0=0.\end{aligned}$$

To prove (3.5) we shall need the following which will be proved in §4.

(3.7) $$d_2\left(\overline{e_{15}\otimes e_{15}}\right)=0\quad\text{in the ASS for}\quad P\wedge P.$$

Proof of (3.5): By (2.6)(7), $Ext_A^{2,2^i}(P\wedge P)$ is generated by $\overline{e_{2^{i-1}-1}\otimes e_{2^{i-1}-1}\,h_0^2}$, $\overline{e_{2^{i-1}-2}\otimes e_1\,h_0h_{i-1}}$, $\overline{e_{2^{i-2}-1}\otimes e_{2^{i-2}-1}\,h_1h_{i-1}}$ (this is zero for $i=5$) and also $\overline{(e_{45}\otimes e_{63})\,h_2h_4}$ if $i=7$. Thus in the ASS for $P\wedge P$ we have

(*) $$\begin{aligned}d_2\left(\overline{e_{2^i-2}\otimes e_1}\right)=&\,\epsilon_1\overline{e_{2^{i-1}-1}\otimes e_{2^{i-1}-1}\,h_0^2}+\epsilon_2\overline{e_{2^{i-1}-2}\otimes e_1\,h_0h_{i-1}}\\&+\epsilon_3\overline{e_{2^{i-2}-1}\otimes e_{2^{i-2}-1}\,h_1h_{i-1}}+\epsilon_4\overline{(e_{45}\otimes e_{63})\,h_2h_4}\end{aligned}$$

where $\epsilon_j=0$ or 1, $\epsilon_4=0$ if $i\neq 7$ and $\epsilon_3=0$ if $i=5$. We want to show $\epsilon_1=\epsilon_4=0$ and $\epsilon_2=1$. Applying $(id\wedge\bar{t})_*$ to (*) we obtain, by (2.9), (3.2) and the fact $h_{i-1}h_{i-2}=0$, the equation

$$\bar{e}_{2^{i-1}-1}h_{i-1}h_0^2=(\epsilon_1+\epsilon_2)\,\bar{e}_{2^{i-1}-1}h_{i-1}h_0^2+\epsilon_4\hat{D}_3(1)\quad\text{in}\quad Ext_A^{*,*}(P).$$

By (2.4), for $i=7$, $\bar{e}_{63}h_6h_0^2$ and $\hat{D}_3(1)$ are linearly independent in $Ext_A^{3,129}(P)$ and, for $i\geq 5$, $\bar{e}_{2^{i-1}-1}h_{i-1}h_0\neq 0$ in $Ext_A^{3,2^i+1}(P)$. So $\epsilon_4=0$ and either $\epsilon_1=1$, $\epsilon_2=0$ or $\epsilon_1=0$, $\epsilon_2=1$. We claim

(**) $\epsilon_1=1,\epsilon_2=0$ cannot happen; so $\epsilon_1=0$, $\epsilon_2=1$.

To prove this we note, by (2.6)(4), $d_2\left(\overline{e_{2^i-2}\otimes e_1 h_{i-1}}\right)=d_2(0)=0$. If $\epsilon_1=1$, $\epsilon_2=0$ then from this equation, equation (*) and (3.1) we get the equation

$$0=\overline{e_{2^{i-1}-1}\otimes e_{2^{i-1}-1}}\,h_{i-1}h_0^2+\overline{e_{2^i-2}\otimes e_1}\,h_{i-2}^2h_0+\epsilon_3\overline{e_{2^{i-2}-1}\otimes e_{2^{i-2}-1}}\,h_1h_{i-1}^2$$

in $Ext_A^{3,2^i+2^{i-1}}(P\wedge P)$ ($\epsilon_4=0$, as already has been shown above). This is contradictory to (2.6)(8) if $i\geq 6$. This proves (**) for $i\geq 6$. In particular, we have

(* * *) $$d_2\left(\overline{e_{62}\otimes e_1}\right)=\overline{e_{30}\otimes e_1}\,h_0h_5+\epsilon_3\overline{e_{15}\otimes e_{15}}\,h_1h_5.$$

If, for $i = 5$, $\epsilon_1 = 1$, $\epsilon_2 = 0$ so that $d_2(\overline{e_{30} \otimes e_1}) = \overline{e_{15} \otimes e_{15}}\, h_0^2$ ($\epsilon_3 = 0$ for $i = 5$) then since $d_2 d_2(\overline{e_{62} \otimes e_1}) = 0$, $d_2(h_5 h_1) = h_4^2 h_0 h_1 = 0$, $d_2(\overline{e_{15} \otimes e_{15}}) = 0$ (by (3.7)) and $\overline{e_{30} \otimes e_1} h_4 = 0$ (by (2.6)(4)), by applying d_2 to $(***)$ we obtain $\overline{e_{15} \otimes e_{15}} h_0^3 h_5 = 0$ in $Ext_A^{4,65}(P \wedge P)$. This is contradictory to (2.6)(9). Thus (**) is true for all $i \geq 5$. This completes the proof of (3.5).

The final result in this section is the following (3.8). We recall from (2.7) that $Ext_A^{0,2^{i+1}-2}(Y \wedge Y) = \mathbb{Z}/2 \oplus \mathbb{Z}/2$ on $\overline{e_{2^{i-1}-1} \otimes e_{2^i+2^{i-1}-1}}$ and $\overline{e_{2^i+2^{i-1}-1} \otimes e_{2^{i-1}-1}}$ for $i \geq 5$.

PROPOSITION 3.8.
(1) $d_2(\overline{e_{2^{i-1}-1} \otimes e_{2^i+2^{i-1}-1}}) = \alpha_2 + \epsilon \overline{e_{2^{i-1}-1} \otimes e_{2^{i-1}-1}}\, h_i h_0$ *and*

(2) $d_2(\overline{e_{2^i+2^{i-1}-1} \otimes e_{2^{i-1}-1}}) = \alpha_1 + \epsilon \overline{e_{2^{i-1}-1} \otimes e_{2^{i-1}-1}}\, h_i h_0$

in the ASS for $Y \wedge Y$ where α_1, α_2 are as described in (13), (14) and ϵ is 0 or 1 which is precisely the ϵ in (3.4).

The remainder of section is devoted to proving (3.8).

J.F. Adams shows in [1] that there is a pairing

$$(15) \qquad Ext_A^{s,t}(Y) \otimes Ext_A^{s',t'}(Y) \to Ext_A^{s+s',t+t'}(Y \wedge Y)$$
$$\alpha \otimes \beta \to \alpha \cdot \beta$$

that connects the differential d_2 in the ASS for Y to that for $Y \wedge Y$ by

$$(3.9) \qquad d_2(\alpha \cdot \beta) = d_2(\alpha) \cdot \beta + \alpha \cdot d_2(\beta).$$

We will make a study later on the pairing "$\cdot$" for $s + s' \leq 2$ and from this prove the following result in $Ext_A^{*,*}(Y \wedge Y)$ where τ is as in (2.5).

LEMMA 3.10. *(1)* $\bar{e}_{2^i+2^{i-1}-1} \cdot \bar{e}_{2^{i-1}-1} = \overline{e_{2^i+2^{i-1}-1} \otimes e_{2^{i-1}-1}}$,

$\bar{e}_{2^{i-1}-1} \cdot \bar{e}_{2^i+2^{i-1}-1} = \overline{e_{2^{i-1}-1} \otimes e_{2^i+2^{i-1}-1}}$.

(2) $\bar{e}_{2^{i-1}-1} \cdot \tau = \alpha_2$, $\tau \cdot \bar{e}_{2^{i-1}-1} = \alpha_1$.

(3) $\bar{e}_{2^{i-1}-1} h_i h_0 \cdot \bar{e}_{2^{i-1}-1} = \bar{e}_{2^{i-1}-1} \cdot \bar{e}_{2^{i-1}-1} h_i h_0 = \overline{e_{2^{i-1}-1} \otimes e_{2^{i-1}-1}}\, h_i h_0$.

(4) $\bar{e}_{2^i+2^{i-1}-1} \cdot \bar{e}_{2^{i-2}-1} h_{i-2} h_0 = \overline{e_{2^i+2^{i-1}-1} \otimes e_{2^{i-2}-1}}\, h_{i-2} h_0 = 0$ *and*

$$\bar{e}_{2^{i-2}-1} h_{i-2} h_0 \cdot \bar{e}_{2^i+2^{i-1}-1} = \overline{e_{2^{i-2}-1} \otimes e_{2^i+2^{i-1}-1}}\, h_{i-2} h_0 = 0.$$

Here the assertions

$$\overline{e_{2^{i-2}-1} \otimes e_{2^i+2^{i-1}-1}}\, h_{i-2} h_0 = 0$$

and

$$\overline{e_{2^i+2^{i-1}-1} \otimes e_{2^{i-2}-1}}\, h_{i-2} h_0 = 0$$

follow from

and

$$\delta\left(e_{2^i+2^{i-1}+2^{i-2}-1} \otimes e_{2^{i-2}-1}\right) = \left(e_{2^i+2^{i-1}-1} \otimes e_{2^{i-2}-1}\right)\lambda_{2^{i-2}-1}$$

in $H_*(Y) \otimes H_*(Y) \otimes \Lambda$ by (7) and (11).

We deduce (3.8) from (3.4), (3.9), (3.10) as follows, noting that (3.2) is also true for Y if $4 \leq k \leq i-1$ (note $i \geq 5$) which is clear.

$$\begin{aligned}
&d_2\left(\overline{e_{2^{i-1}-1} \otimes e_{2^i+2^{i-1}-1}}\right)\\
=&d_2\left(\bar{e}_{2^{i-1}-1} \cdot \bar{e}_{2^i+2^{i-1}-1}\right)\\
=&\left[d_2\left(\bar{e}_{2^{i-1}-1}\right)\right] \cdot \bar{e}_{2^i+2^{i-1}-1} + \bar{e}_{2^{i-1}-1} \cdot \left[d_2\left(\bar{e}_{2^i+2^{i-1}-1}\right)\right]\\
=&\left(\bar{e}_{2^{i-2}-1}h_{i-2}h_0\right) \cdot \bar{e}_{2^i+2^{i-1}-1} + \bar{e}_{2^{i-1}-1} \cdot \left(\tau + \epsilon\bar{e}_{2^{i-1}-1}h_ih_0\right)\\
=&0 + \bar{e}_{2^{i-1}-1} \cdot \tau + \epsilon\bar{e}_{2^{i-1}-1} \cdot \bar{e}_{2^{i-1}-1}h_ih_0\\
=&\alpha_2 + \epsilon\overline{e_{2^{i-1}-1} \otimes e_{2^{i-1}-1}}\, h_ih_0.
\end{aligned}$$

Similarly, $d_2\left(\overline{e_{2^i+2^{i-1}-1} \otimes e_{2^{i-1}-1}}\right) = \alpha_1 + \epsilon\overline{e_{2^{i-1}-1} \otimes e_{2^{i-1}-1}}\, h_ih_0$. This proves (3.8).

We proceed to study the pairing (15). Let $K_* = H_*(Y)$. We will consider the cobar constructions $\bar{B}^*(A) \otimes K_*$ for Y and $\bar{B}^*(A) \otimes K_* \otimes K_*$ for $Y \wedge Y$ where $\bar{B}^s(A) = \left(\bar{A}^*\right)^s$ $\left(\left(\bar{A}^*\right)^o = \mathbb{Z}/2\right)$ and $\bar{A}^* = \overline{\mathbb{Z}/2\left[\xi_1, \xi_2, \cdots\right]}$. The differentials in these complexes are denoted by δ and δ'. So $H^{s,*}\left(\bar{B}^*(A) \otimes K_*, \delta\right) \cong Ext_A^{s,*}(Y)$ and $H^{s,*}\left(\bar{B}^*(A) \otimes K_* \otimes K_*, \delta'\right) \cong Ext_A^{s,*}(Y \wedge Y)$. According to Adams [1], theoretically, one can construct, for each pair (s, s') of integers with $s, s' \geq 0$, a pairing

$$\left[\left(\bar{A}^*\right)^s \otimes K_*\right] \otimes \left[\left(\bar{A}^*\right)^{s'} \otimes K_*\right] \xrightarrow{\Psi_{s,s'}} \left(\bar{A}^*\right)^{s+s'} \otimes K_* \otimes K_*$$

such that these pairings satisfy the derivation formula

$$\text{(3.9)}' \qquad \delta'\left(\Psi_{s,s'}(x, y)\right) = \Psi_{s+1,s'}\left(\delta x, y\right) + \Psi_{s,s'+1}(x, \delta y)$$

and such that $\Psi_{s,s'}$ induces the pairing in (15). For $s = s' = 0$ there is a canonical pairing

$$\left[K_* = \left(\bar{A}^*\right)^o \otimes K_*\right] \otimes \left[K_* = \left(\bar{A}^*\right)^o \otimes K_*\right] \xrightarrow{\Psi_{0,0}} \left(\bar{A}^*\right)^o \otimes K_* \otimes K_* = K_* \otimes K_*$$

which is the identity map. Thus (3.10)(1) is clear. For $s + s' > 0$ there are no canonical ways to construct these pairings in the chain level. For $(s, s') = (1,0), (0,1), (0,2), (1,1)$ and (2.0) we have constructed certain pairings $\Psi_{1,0}$, $\Psi_{0,1}$, $\Psi_{0,2}$, $\Psi_{1,1}$ and $\Psi_{2,0}$ and these are described as follows.

Let $A \otimes (K = H^*(Y)) \to K$ be the A-module structure map for K and let $K_* \xrightarrow{\delta} A^* \otimes K_*$ be its $\mathbb{Z}/2$-dual, that is, the comodule structure map for K_*. We will use ℓ, m etc., to denote arbitrary elements in K_* and x, y etc., to denote arbitrary elements in $\bar{A}^*$. For $m \in K_*$ let $\Delta(m) = 1 \otimes m + \sum_j x_j \otimes m_j$ where $|x_j| > 0$. We will write

$$(*) \qquad \bar{\Delta}(m) = \sum_j x_j \otimes m_j.$$

For example, we have $\bar{\Delta}\left(e_2\right) = \xi_1 \otimes e_1$, $\bar{\Delta}\left(e_4\right) = \xi_1 \otimes e_3 + \xi_1^2 \otimes e_2 + \xi_2 \otimes e_1$, $\bar{\Delta}\left(e_{2^k-1}\right) = 0$ $(k \neq i)$, $\bar{\Delta}\left(\tilde{e}_{2^i-1}\right) = \xi_1^{2^{i-1}} \otimes e_{2^{i-1}-1}$ and $\bar{\Delta}\left(e_{2^i+2^{i-1}-1}\right) = 0$. The pairings $\Psi_{0,1}$, $\Psi_{1,0}$, $\Psi_{0,2}$ $\Psi_{1,1}$ and $\Psi_{2,0}$ that we have constructed to satisfy (3.9)′ are given as follows.

(3.11) (1) $\Psi_{0,1}[\ell \otimes (x \otimes m)] = x \otimes \ell \otimes m$

(2) $\Psi_{1,0}[(x \otimes \ell) \otimes m] = x \otimes \ell \otimes m + \sum_j x x_j \otimes \ell \otimes m_j$,

(3) $\Psi_{0,2}[\ell \otimes (x \otimes y) \otimes m] = x \otimes y \otimes \ell \otimes m$,

(4) $\Psi_{1,1}[(x \otimes \ell) \otimes (y \otimes m)] = y \otimes x \otimes \ell \otimes m + \sum_j y \otimes x x_j \otimes \ell \otimes m_j$,

(5) $\Psi_{2,0}[(x \otimes y \otimes \ell) \otimes m] = x \otimes y \otimes \ell \otimes m + \sum_j x \otimes y x_j \otimes \ell \otimes m_j$
$+\sum_j x x_j \otimes y \otimes \ell \otimes m_j + \sum_{j,k} x x_j \otimes y x_{j,k} \otimes m_{j,k}$.

Here x_j and m_j are as in $(*)$ above and $\bar{\Delta}(m_j) = \sum_k x_{j,k} \otimes m_{j,k}$.

To prove (3.10) we need only the pairings in Ext groups induced by those in (3.11)(3) and (3.11)(5). The formula in (3.11)(5) looks complicated. But if m is primitive, that is, $\bar{\Delta}(m) = 0$, then we have the following.

(3.11)(5)′ $\quad \Psi_{2,0}[(x \otimes y \otimes \ell) \otimes m] = x \otimes y \otimes \ell \otimes m \quad$ for m primitive.

From (3.11)(3) and (3.11)(5)′ we see the pairings

$$Ext_A^{0,*}(Y) \otimes Ext_A^{2,*}(Y) \longrightarrow Ext_A^{2,*}(Y \wedge Y)$$

$$Ext_A^{2,*}(Y) \otimes Ext_A^{0,*}(Y) \longrightarrow Ext_A^{2,*}(Y \wedge Y)$$

are just as simple as one can think of, since elements in $Ext_A^{0,*}(Y)$ are represented by primitives. Lemma 3.10 follows immediately. We illustrate the proof for one case in the lemma which is $\bar{e}_{2^{i-1}-1} \cdot \tau = \alpha_2$ in (3.10)(2). From (9) it is not difficult to see that τ is represented by the cycle

$$\begin{aligned} c' = {} & \xi_1 \otimes \xi_1^{2^{i-2}} \otimes e_{2^i+2^{i-2}-1} + \xi_1 \otimes \xi_1^{2^{i-1}} \otimes \tilde{e}_{2^i-1} \\ & + \xi_1 \otimes \left(\xi_1^{2^{i-1}+2^{i-2}} + \xi_2^{2^{i-2}}\right) \otimes e_{2^{i-1}+2^{i-2}-1} \\ & + \xi_1 \otimes \xi_2^{2^{i-2}} \xi_1^{2^{i-2}} \otimes e_{2^{i-1}-1} \end{aligned}$$

in $\left(\bar{A}^*\right)^2 \otimes H_*(Y)$. Similarly, from (13) we see α_2 is represented by the cycle

$$\begin{aligned} z_2' = {} & \xi_1 \otimes \xi_1^{2^{i-2}} \otimes e_{2^{i-1}-1} \otimes e_{2^i+2^{i-2}-1} + \xi_1 \otimes \xi_1^{2^{i-1}} \otimes e_{2^{i-1}-1} \otimes \tilde{e}_{2^i-1} \\ & + \xi_1 \otimes \left(\xi_1^{2^{i-1}+2^{i-2}} + \xi_2^{2^{i-2}}\right) \otimes e_{2^{i-1}-1} \otimes e_{2^{i-1}+2^{i-2}-1} \\ & + \xi_1 \otimes \xi_2^{2^{i-2}} \xi_1^{2^{i-2}} \otimes e_{2^{i-1}-1} \otimes e_{2^{i-1}-1} \end{aligned}$$

in $\left(\bar{A}^*\right)^2 \otimes H_*(Y) \otimes H_*(Y)$. From (3.11)(3) we see $\Psi_{0,2}\left(e_{2^{i-1}-1} \otimes c'\right) = z_2'$. So $\bar{e}_{2^{i-1}-1} \cdot \tau = \alpha_2$.

4. Proof of Theorem 1.2 based on Assumption 1.4

Let $M = S^{2^i-2} \cup_{2\iota} e^{2^i-1} \xrightarrow{f} P$ be as in (1.4) with $i \geq 5$ and let Y be the cofiber of f. So we have a cofiber sequence $M \xrightarrow{f} P \xrightarrow{q} Y$. The generator of $H_k(P) = \mathbb{Z}/2$ is denoted by e_k and that of $H^k(P) = \mathbb{Z}/2$ is denoted by x_k, $1 \leq k < \infty$. These notations of generators are preserved for Y except that in dimension $2^i - 1$ we use $\tilde{e}_{2^i-1}$ and $\tilde{x}_{2^i-1}$. In §2, §3 only $H_*(P)$, $H_*(Y)$ and their generators e_k, $\tilde{e}_{2^i-1}$ are considered. The Steenrod algebra acts on $H^*(P)$, $H^*(Y)$ by the following rules which are dual to (5) and (7) respectively.

(16) In $H^*(P)$, $Sq^j x_k = \binom{k}{j} x_{k+j}$.

(17) In $H^*(Y)$,

(1) $Sq^j x_k = \binom{k}{j} x_{k+j}$ for $k \neq 2^{i-1} - 1$ or for $k = 2^{i-1} - 1$ with $j \neq 2^{i-1}$,

(2) $Sq^{2^{i-1}} x_{2^{i-1}-1} = \tilde{x}_{2^i-1}$,

(3) $Sq^\ell \tilde{x}_{2^i-1} = 0$ for $\ell > 0$.

The following is easy to see from (17).

(18) $Sq^{2^i}(x_{2^{i-1}-1} \otimes x_{2^{i-1}-1}) = \tilde{x}_{2^i-1} \otimes \tilde{x}_{2^i-1}$ in $H^*(Y \wedge Y)$.

By assumption, $f^* \neq 0$ in dimension $2^i - 1$, and it is easy to see $f^* = 0$ in dimension $2^i - 2$. So $q^*(x_k) = x_k$ for $k \neq 2^i - 1$ and $q^*(\tilde{x}_{2^i-1}) = 0$. We will need to consider $M \wedge M$. The generators for $H^{2^i-2}(M) = \mathbb{Z}/2$ and $H^{2^i-1}(M) = \mathbb{Z}/2$ of the first M in $M \wedge M$ are denoted by x and y respectively and those for the second M are denoted by u and v respectively. Thus $H^*(M \wedge M)$ is generated by $x \otimes u$ in dimension $2^{i+1} - 4$, $x \otimes v$ and $y \otimes u$ in dimension $2^{i+1} - 3$ and $y \otimes v$ in dimension $2^{i+1} - 2$.

It is easy to see that there exists a stable map $S^{2^{i+1}-3} \xrightarrow{g} M \wedge M$ with $g^* \neq 0$ in dimension $2^{i+1} - 3$. In fact, there are two choices in homotopy for such a map, and we choose any one of these two. The generator of $H^{2^{i+1}-3}\left(S^{2^{i+1}-3}\right) = \mathbb{Z}/2$ will be denoted by ι. Then $g^*(x \otimes v) = g^*(y \otimes u) = \iota$. Consider the composite

$$\bar{g} : S^{2^{i+1}-3} \xrightarrow{g} M \wedge M \xrightarrow{f \wedge f} P \wedge P.$$

$\bar{g}^* = 0$ since neither $x \otimes v$ nor $y \otimes u$ is in $im(f \wedge f)^*$. So $\bar{g}$ has $AF \geq 1$ in the ASS for $P \wedge P$ where AF stands for "*Adams filtration*". For a stable $S^k \xrightarrow{\varphi} P \wedge P$ if φ has AF s then we write $AF(\varphi) = s$.

LEMMA 4.1. $AF(\bar{g}) \geq 2$.

Consider $(P \wedge P) \cup_{\bar{g}} e^{2^{i+1}-2}$ and let $\bar{q} : (P \wedge P) \cup_{\bar{g}} e^{2^{i+1}-2} \to S^{2^{i+1}-2}$ be the collapsing map. Define $\zeta \in H^{2^{i+1}-2}\left((P \wedge P) \cup_{\bar{g}} e^{2^{i+1}-2}\right)$ to be $\zeta = \bar{q}^*(\Sigma\iota)$. $\zeta \neq 0$

since $AF(\bar{g}) \geq 1$. By (2.6)(2), $Ext_A^{1,2^{i+1}-2}(P \wedge P) = \mathbb{Z}/2$ on $(e_{2^{i-1}-1} \otimes e_{2^{i-1}-1})\, h_i$. If $AF(\bar{g}) = 1$ then $\bar{g}$ is detected by $\overline{e_{2^{i-1}-1} \otimes e_{2^{i-1}-1}\, h_i}$, or equivalently, $Sq^{2^i}\left(x_{2^{i-1}-1} \otimes x_{2^{i-1}-1}\right) = \zeta$ in $H^{2^{i+1}-2}\left((P \wedge P) \cup_{\bar{g}} e^{2^{i+1}-2}\right)$. We prove (4.1) by showing that this cannot happen. The composite

$$S^{2^{i+1}-3} \xrightarrow{g} M \wedge M \xrightarrow{id \wedge f} M \wedge P \xrightarrow{f \wedge id} P \wedge P \xrightarrow{q \wedge id} Y \wedge P$$

is null homotopic since $M \xrightarrow{f} P \xrightarrow{q} Y$ is a cofiber sequence. Note that $\bar{g} = (f \wedge id)(id \wedge f)g$. So there is a map $(P \wedge P) \cup_{\bar{g}} e^{2^{i+1}-2} \xrightarrow{\overline{q \wedge id}} Y \wedge P$ extending $q \wedge id$. Consider the composite

$$(P \wedge P) \cup_{\bar{g}} e^{2^{i+1}-2} \xrightarrow{\overline{q \wedge id}} Y \wedge P \xrightarrow{id \wedge q} Y \wedge Y.$$

$(id \wedge q)^*\left(\tilde{x}_{2^i-1} \otimes \tilde{x}_{2^i-1}\right) = \tilde{x}_{2^i-1} \otimes q^*\left(\tilde{x}_{2^i-1}\right) = 0$ since $q^*\left(\tilde{x}_{2^i-1}\right) = 0$. This implies

$$\left[(id \wedge q)\left(\overline{q \wedge id}\right)\right]^*\left(\tilde{x}_{2^i-1} \otimes \tilde{x}_{2^i-1}\right) = \overline{q \wedge id}^*\left((id \wedge q)^*\left(\tilde{x}_{2^i-1} \otimes \tilde{x}_{2^i-1}\right)\right) = 0.$$

Since $q^*\left(x_{2^{i-1}-1}\right) = x_{2^{i-1}-1}$ we see

$$\begin{aligned}
&\left[(id \wedge q)\left(\overline{q \wedge id}\right)\right]^*\left(x_{2^{i-1}-1} \otimes x_{2^{i-1}-1}\right)\\
=&q^*\left(x_{2^{i-1}-1}\right) \otimes q^*\left(x_{2^{i-1}-1}\right) = x_{2^{i-1}-1} \otimes x_{2^{i-1}-1}.
\end{aligned}$$

Then in $(P \wedge P) \cup_{\bar{g}} e^{2^{i+1}-2}$

$$\begin{aligned}
&Sq^{2^i}\left(x_{2^{i-1}-1} \otimes x_{2^{i-1}-1}\right)\\
&= Sq^{2^i}\left(\left[(id \wedge q)\left(\overline{q \wedge id}\right)\right]^*\left(x_{2^{i-1}-1} \otimes x_{2^{i-1}-1}\right)\right)\\
&= \left[(id \wedge q)\left(q \wedge id\right)\right]^*\left(Sq^{2^i}\left(x_{2^{i-1}-1} \otimes x_{2^{i-1}-1}\right)\right)\\
&= \left[(id \wedge q)\left(\overline{q \wedge id}\right)\right]^*\left(\tilde{x}_{2^i-1} \otimes \tilde{x}_{2^i-1}\right) \quad \text{(by (18))}\\
&= 0 \neq \zeta.
\end{aligned}$$

This proves Lemma 4.1.

The next step is to show $AF(\bar{g}) = 2$. More precisely we will show the following (4.2). By (2.6)(6), $Ext_A^{2,2^{i+1}-1}(P \wedge P)$ is $\mathbb{Z}/2$ on $\overline{e_{2^{i-1}-1} \otimes e_{e^{i-1}-1}}\, h_i h_0$ for $i \geq 6$ and is $\mathbb{Z}/2 \oplus \mathbb{Z}/2$ on $\overline{e_{15} \otimes e_{15} h_5 h_0 + (e_{22} \otimes e_{31})\, h_1 h_3}$ for $i = 5$.

PROPOSITION 4.2. *$\bar{g} : S^{2^{i+1}-2} \xrightarrow{g} M \wedge M \xrightarrow{f \wedge f} P \wedge P$ is detected by $\overline{e_{2^{i-1}-1} \otimes e_{2^{i-1}-1}\, h_i h_0}$ for $i \geq 5$.*

Granting this we prove Theorem 1.2 as follows. First we note the following.

(19) $M \wedge M \simeq \left[\left(S^{2^{i+1}-4} \cup_{2\iota} e^{2^{i+1}-3}\right) \vee S^{2^{i+1}-3}\right] \cup e^{2^{i+1}-2}$ which is the mapping cone of $S^{2^{i+1}-3} \xrightarrow{(\eta, 2\iota)} \left(S^{2^{i+1}-4} \cup_{2\iota} e^{2^{i+1}-3}\right) \vee S^{2^{i+1}-3}$. We may assume $S^{2^{i+1}-3} \xrightarrow{g} \left[\left(S^{2^{i+1}-4} \cup_{2\iota} e^{2^{i+1}-3}\right) \vee S^{2^{i+1}-3}\right] \cup e^{2^{i+1}-2}$

$f\wedge f|S^{2^{i+1}-4}:S^{2^{i+1}-4}\to P\wedge P$ is supposed to be detected by $\left(\overline{e_{2^{i-1}-1}\otimes e_{2^{i-1}-1}}\right)h_{i-1}^2$ (since $f|S^{2^i-2}:S^{2^i-2}\to P$ is detected by $\bar{e}_{2^{i-1}-1}h_{i-1}$) which is zero by (2.6)(4). So $AF\left(f\wedge f|S^{2^{i+1}-4}\right)\geq 3$ in the ASS for $P\wedge P$. Finally note that $(f\wedge f)^*(x_{2^i-1}\otimes x_{2^i-1})=y\otimes v$ which is the class corresponding to the top cell $e^{2^{i+1}-2}$.

By (2.6)(5), $\overline{e_{2^{i-1}-1}\otimes e_{2^{i-1}-1}}\,h_ih_0^2\neq 0$ in $Ext_A^{3,2^{i+1}}(P\wedge P)$ for $i\geq 5$. We want to show for (1.2) that $d_3\left(\overline{e_{2^i-1}\otimes e_{2^i-1}}\right)=\overline{e_{2^{i-1}-1}\otimes e_{2^{i-1}-1}}\,h_ih_0^2$ in the ASS for $P\wedge P$, $i\geq 5$. We have to show $d_2\left(\overline{e_{2^i-1}\otimes e_{2^i-1}}\right)=0$ first. By (2.6)(6) we have

$$d_2\left(\overline{e_{2^i-1}\otimes e_{2^i-1}}\right)=\epsilon_1\overline{e_{2^{i-1}-1}\otimes e_{2^{i-1}-1}}h_ih_0+\epsilon_2\left(\overline{e_{22}\otimes e_{31}}\right)h_1h_3$$

where $\epsilon_j=0$ or 1, $j=1,2$ and $\epsilon_2=0$ if $i\neq 5$. In the case $i=5$ if $\epsilon_2=1$ so that $d_2\left(\overline{e_{31}\otimes e_{31}}\right)=\epsilon_1\overline{e_{15}\otimes e_{15}}h_5h_0+\left(\overline{e_{22}\otimes e_{31}}\right)h_1h_3$ then by applying $(id\wedge\bar{t})_*$ to this equation and by (2.9)(1), (3) we deduce $d_2(\bar{e}_{31}h_5)=\hat{D}_3\neq 0$ in the ASS for P. This is impossible as a direct calculation from (3.1), (3.2) shows that $d_2(\bar{e}_{31}h_5)=0$. Thus $d_2\left(\overline{e_{2^i-1}\otimes e_{2^i-1}}\right)=\epsilon_1\overline{e_{2^{i-1}-1}\otimes e_{2^{i-1}-1}}h_ih_0$ for all $i\geq 5$. If $\epsilon_1=1$ then, from (19) above, it is easy to see that this implies $\bar{g}=f\wedge f|S^{2^{i+1}-3}$ is detected by $\overline{e_{2^{i-1}-1}\otimes e_{2^{i-1}-1}}\,h_i$ which is contradictory to (4.1). So $\epsilon_1=0$, that is, we have

(∗) $\quad d_2\left(\overline{e_{2^i-1}\otimes e_{2^i-1}}\right)=0$ in the ASS for $P\wedge P$.

We stress that (∗) is for $i\geq 5$ which we deduce from (4.1) and (19) which in turn come from Assumption 1.4 for $i\geq 5$. For $i=4$, (1.4) is indeed a fact since there is a θ_4 with $2\theta_4=0$ ([13]). Thus, for $i=4$, (19) and (4.1) are also facts and therefore (∗) is a true result which is precisely (3.7) left unproved in §3.

Returning to our proof we go back to the data in (19). By (4.2) and (19) we see the composite $S^{2^{i+1}-3}\xrightarrow{(\eta,2\iota)}\left(S^{2^{i+1}-4}\cup_{2\iota}e^{2^{i+1}-3}\right)\vee S^{2^{i+1}-3}\xrightarrow{\overline{f\wedge f}}P\wedge P$ is detected by $\overline{e_{2^{i-1}-1}\otimes e_{2^{i-1}-1}}\,h_ih_0^2$ if this class is not a boundary in the ASS for $P\wedge P$ where $\overline{f\wedge f}$ is the restriction of $M\wedge M\xrightarrow{f\wedge f}P\wedge P$ to the first three cells. But this composite is null homotopic. So $\overline{e_{2^{i-1}-1}\otimes e_{2^{i-1}-1}}h_ih_0^2$ has to be a boundary. By (2.6)(3), $Ext_A^{1,2^{i+1}-1}(P\wedge P)=\mathbb{Z}/2\otimes\mathbb{Z}/2$ on $\overline{e_{2^i-1}\otimes e_{2^i-1}}h_0$ and $\overline{e_{2^i-2}\otimes e_1}h_i$ and by (2.6)(1),
$Ext_A^{0,2^{i+1}-2}(P\wedge P)=\mathbb{Z}/2$ on $\overline{e_{2^i-1}\otimes e_{2^i-1}}$. We have just shown above that $d_2\left(\overline{e_{2^i-1}\otimes e_{2^i-1}}h_0\right)=d_2\left(\overline{e_{2^i-1}\otimes e_{2^i-1}}\right)h_0=0$. By (3.6), $d_2\left(\overline{e_{2^i-2}\otimes e_1}h_i\right)$ $=0$. Thus the remaining possibility is that $\overline{e_{2^{i-1}-1}\otimes e_{2^{i-1}-1}}h_ih_0^2$ is hit via a d_3 by $\overline{e_{2^i-1}\otimes e_{2^i-1}}$, that is, $d_3\left(\overline{e_{2^i-1}\otimes e_{2^i-1}}\right)=\overline{e_{2^{i-1}-1}\otimes e_{2^{i-1}-1}}h_ih_0^2$. This proves Theorem 1.2. Another viewpoint for concluding

$$d_3\left(\overline{e_{2^i-1}\otimes e_{2^i-1}}\right)=\overline{e_{2^{i-1}-1}\otimes e_{2^{i-1}-1}}\,h_ih_0^2$$

is to note that the map $M\wedge M\xrightarrow{f\wedge f}P\wedge P$ has
$f_*(y_*\otimes v_*)=e_{2^i-1}\otimes e_{2^i-1}$ and therefore we must have $d_3\left(\overline{e_{2^i-1}\otimes e_{2^i-1}}\right)=\overline{e_{2^{i-1}-1}\otimes e_{2^{i-1}-1}}\,h_ih_0^2$.

We proceed to prove (4.2). Let $P\wedge P\xrightarrow{j}(P\wedge P)\cup_{\bar{g}}e^{2^{i+1}-2}$ be the inclusion map. To simplify we will occasionally use X to denote $(P\wedge P)\cup_{\bar{g}}e^{2^{i+1}-2}$. It is clear that $Ext_A^{s,t}(P \wedge P) \xrightarrow[\cong]{j_*} Ext_A^{s,t}(X)$ for $t-s \le 2^{i+1}-3$. The images of elements of the first Ext group in the second Ext group in this range will be denoted by the same notations. Thus, by (2.6)(6), we have the following.

(20) $Ext_A^{2,2^{i+1}-1}(X)$ is also $\mathbb{Z}/2$ on $\overline{e_{2^{i-1}-1} \otimes e_{2^{i-1}-1}} h_i h_0$ for $i \ge 6$ and is $\mathbb{Z}/2 \oplus \mathbb{Z}/2$ on $\overline{e_{15} \otimes e_{15}} h_5 h_0$ and $\overline{(e_{22} \otimes e_{31})}\, h_1 h_3$ for $i = 5$.

The image of $\overline{e_{2^i-1} \otimes e_{2^i-1}} \in Ext_A^{0,2^{i+1}-2}(P \wedge P) = \mathbb{Z}/2$ in $Ext_A^{0,2^{i+1}-2}(X)$ is also denoted by $\overline{e_{2^i-1} \otimes e_{2^i-1}}$. Since $AF(\bar{g}) \ge 2$ (by (4.1)) the induced map

$$Ext_A^{0,2^{i+1}-2}\left(X = (P \wedge P) \cup_{\bar{g}} e^{2^{i+1}-2}\right) \xrightarrow{\bar{q}_*} Ext_A^{0,2^{i+1}-2}\left(S^{2^{i+1}-2}\right) = \mathbb{Z}/2$$

of the collapsing map $X \xrightarrow{\bar{q}} S^{2^{i+1}-2}$ is onto and maps $\overline{e_{2^i-1} \otimes e_{2^i-1}}$ to zero. Let $\gamma \in Ext_A^{0,2^{i+1}-2}(X)$ be a class such that $\bar{q}_*(\gamma) = 1$. Then

$$Ext_A^{0,2^{i+1}-2}(X) = \mathbb{Z}/2 \oplus \mathbb{Z}/2$$

on $\overline{e_{2^i-1} \otimes e_{2^i-1}}$ and γ. There are two choices for γ with $\bar{q}_*(\gamma) = 1$, and we do not have a canonical choice of such a γ. We will study this more closely later. We already have shown above that $d_2\left(\overline{e_{2^i-1} \otimes e_{2^i-1}}\right) = 0$ in the ASS for $P \wedge P$, and therefore also for X. Thus no matter what we choose for γ the following is clear.

(21) $\bar{g} : S^{2^{i+1}-3} \to P \wedge P$ is detected by $\overline{e_{2^{i-1}-1} \otimes e_{2^{i-1}-1}}\, h_i h_0$ $+\epsilon\overline{(e_{22} \otimes e_{31})}\, h_1 h_3$ if and only if $d_2(\gamma) = \overline{e_{2^{i-1}-1} \otimes e_{2^{i-1}-1}}\, h_i h_0$ $+\epsilon\overline{(e_{22} \otimes e_{31})}\, h_1 h_3$ in the ASS for $X = (P \wedge P) \cup_{\bar{g}} e^{2^{i+1}-2}$ where $\epsilon = 0$ or 1 and $\epsilon = 0$ if $i \ge 6$.

(4.2) will follow if we can show $d_2(\gamma) = \overline{e_{2^{i-1}-1} \otimes e_{2^{i-1}-1}}\, h_i h_0$.

We recall the following maps.

(22) The composite $P \wedge P \xrightarrow{j} X = (P \wedge P) \cup_{\bar{g}} e^{2^{i+1}-2} \xrightarrow{\overline{q \wedge id}} Y \wedge P \xrightarrow{id \wedge q} Y \wedge Y$ is just $q \wedge q$ where $\overline{q \wedge id}$ is an extension of

$$P \wedge P \xrightarrow{q \wedge id} Y \wedge P.$$

Let $\overline{q \wedge q} = (id \wedge q)(\overline{q \wedge id})$. So $\overline{q \wedge q} : X \to Y \wedge Y$ is an extension of $q \wedge q$.

By (2.7)(3), $Ext_A^{0,2^{i+1}-2}(Y \wedge Y) \cong \mathbb{Z}/2 \oplus \mathbb{Z}/2$ represented by the classes $\overline{e_{2^{i-1}-1} \otimes e_{2^i+2^{i-1}-1}}$ and $\overline{e_{2^i+2^{i-1}-1} \otimes e_{2^{i-1}-1}}$. Our main work will be to show the following.

(23) $Ext_A^{0,2^{i+1}-2}(X) \xrightarrow{\overline{q \wedge q}_*} Ext_A^{0,2^{i+1}-2}(Y \wedge Y)$ maps γ to

$$\overline{e_{2^{i-1}-1} \otimes e_{2^i+2^{i-1}-1}} + \overline{e_{2^i+2^{i-1}-1} \otimes e_{2^{i-1}-1}}.$$

Granting this, we deduce $d_2(\gamma) = \overline{e_{2^{i-1}-1} \otimes e_{2^{i-1}-1}}\, h_i h_0$ in X as follows. By (20) we have where $\epsilon_j = 0$ or 1 and $\epsilon_2 = 0$ if $i \geq 6$. It is clear that

$$\overline{q \wedge q}_* \left(\overline{e_{2^{i-1}-1} \otimes e_{2^{i-1}-1}}\, h_i h_0\right) = \overline{e_{2^{i-1}-1} \otimes e_{2^{i-1}-1}\, h_i h_0}$$

and

$$\overline{q \wedge q}_* \left(\overline{(e_{22} \otimes e_{31})\, h_1 h_3}\right) = \overline{(e_{22} \otimes e_{31})\, h_1 h_3}$$

where the latter $\overline{e_{2^{i-1}-1} \otimes e_{2^{i-1}-1}\, h_i h_0}$ and $\overline{(e_{22} \otimes e_{31})\, h_1 h_3}$ are as in (2.7). By (23), $\overline{q \wedge q}_*(\gamma) = \overline{e_{2^{i-1}-1} \otimes e_{2^i+2^{i-1}-1} + e_{2^i+2^{i-1}-1} \otimes e_{2^{i-1}-1}}$. By (3.8),

$$\begin{aligned}
&d_2 \left(\overline{e_{2^{i-1}-1} \otimes e_{2^i+2^{i-1}-1}} + \overline{e_{2^i+2^{i-1}-1} \otimes e_{2^{i-1}-1}}\right) \\
=&\alpha_2 + \epsilon\overline{e_{2^{i-1}-1} \otimes e_{2^{i-1}-1}\, h_i h_0} + \alpha_1 + \epsilon\overline{e_{2^{i-1}-1} \otimes e_{2^{i-1}-1}\, h_i h_0} \\
=&\alpha_2 + \alpha_1 = \overline{e_{2^{i-1}-1} \otimes e_{2^{i-1}-1}\, h_i h_0} \qquad \text{(by (2.7)(2))}
\end{aligned}$$

in the ASS for $Y \wedge Y$. Then

$$\begin{aligned}
&\overline{e_{2^{i-1}-1} \otimes e_{2^{i-1}-1}\, h_i h_0} \\
=&d_2 \left(\overline{e_{2^{i-1}-1} \otimes e_{2^i+2^{i-1}-1}} + \overline{e_{2^i+2^{i-1}-1} \otimes e_{2^{i-1}-1}}\right) \\
=&d_2 \left(\overline{q \wedge q}_*(\gamma)\right) \\
=&\overline{q \wedge q}_* \left(d_2(\gamma)\right) \\
=&\overline{q \wedge q}_* \left(\epsilon_1 \overline{e_{2^{i-1}-1} \otimes e_{2^{i-1}-1}\, h_i h_0} + \epsilon_2 \overline{(e_{22} \otimes e_{31})\, h_1 h_3}\right) \\
=&\epsilon_1 \overline{e_{2^{i-1}-1} \otimes e_{2^{i-1}-1}\, h_i h_0} + \epsilon_2 \overline{(e_{22} \otimes e_{31})\, h_1 h_3}.
\end{aligned}$$

This is an equation in $Ext_A^{2,2^{i+1}-1}(Y \wedge Y)$. By (2.7)(1), $\overline{e_{2^{i-1}-1} \otimes e_{2^{i-1}-1}\, h_i h_0}$ and $\overline{(e_{22} \otimes e_{31})\, h_1 h_3}$ are linearly independent. So $\epsilon_1 = 1$ and $\epsilon_2 = 0$. Thus $d_2(\gamma) = \overline{e_{2^{i-1}-1} \otimes e_{2^{i-1}-1}\, h_i h_0}$. From (21) we see this implies $\bar{g} : S^{2^{i+1}-3} \to P \wedge P$ is detected by $\overline{e_{2^{i-1}-1} \otimes e_{2^{i-1}-1}\, h_i h_0}$. This proves (4.2).

It remains to prove (23). Recall that for any left A-module N, $Ext_A^{0,t}(N)$ is the $\mathbb{Z}/2$-dual of $Tor_{0,t}^A(\mathbb{Z}/2, N) = (N/\bar{A}N)^t$. We will describe, by this duality, in (23)′ that follows, the $\mathbb{Z}/2$-dual statement of (23) and prove it in this dual content.

For $a \in N$ its image in $N/\bar{A}N$ will be denoted by a'. Since $Ext_A^{0,2^{i+1}-2}(P \wedge P) \cong \mathbb{Z}/2$ on $\overline{e_{2^{i-1}-1} \otimes e_{2^{i-1}-1}}$ we see

(24) $\left(H^*(P \wedge P)/\bar{A}H^*(P \wedge P)\right)^{2^{i+1}-2}$ is $\mathbb{Z}/2$ generated by $(x_{2^i-1} \otimes x_{2^i-1})'$.

This can also be seen directly from (16) via Cartan formula. The following is also not difficult to see. Note that $\{x_\nu \otimes x_{2^{i+1}-2-\nu} \mid 1 \leq \nu \leq 2^{i+1}-3\}$ is a $\mathbb{Z}/2$-base for $H^{2^{i+1}-2}(P \wedge P)$.

(25) $$\left(\sum_{\nu \neq 2^i-1} \epsilon(\nu) x_\nu \otimes x_{2^{i+1}-2-\nu}\right)' = 0 \text{ in } H^*(P \wedge P)/\bar{A}H^*(P \wedge P)$$
where $\epsilon(\nu) = 0$ or 1.

We have a short exact sequence

(26) $$0 \longrightarrow H^{2^{i+1}-2}\left(S^{2^{i+1}-2}\right) \xrightarrow{\bar{q}^*} H^{2^{i+1}-2}\left(X = (P \wedge P) \cup_{\bar{g}} e^{2^{i+1}-2}\right) \xrightarrow{j^*} H^{2^{i+1}-2}(P \wedge P) \longrightarrow 0.$$

Recall $\bar{q}^*(\Sigma\iota) \in H^{2^{i+1}-2}(X)$ is denoted by ζ. Choose an element

$$\overline{x_{2^i-1} \otimes x_{2^i-1}} \in H^{2^{i+1}-2}(X)$$

such that $j^*\left(\overline{x_{2^i-1} \otimes x_{2^i-1}}\right) = x_{2^i-1} \otimes x_{2^i-1}$. There is no canonical choice of such an $\overline{x_{2^i-1} \otimes x_{2^i-1}}$. Since $Ext_A^{0,2^{i+1}-2}(X) \cong \mathbb{Z}/2 \oplus \mathbb{Z}/2$ on $\overline{e_{2^{i-1}-1} \otimes e_{2^{i-1}-1}}$ and γ, its $\mathbb{Z}/2$-dual $\left(H^*(X)/\bar{A}H^*(X)\right)^{2^{i+1}-2}$ is also $\mathbb{Z}/2 \oplus \mathbb{Z}/2$ and from (24), (26) we see $\overline{x_{2^i-1} \otimes x_{2^i-1}}'$ and ζ' are generators. We remark that, as there is no canonical choice of $\overline{x_{2^i-1} \otimes x_{2^i-1}}$, and hence $\overline{x_{2^i-1} \otimes x_{2^i-1}}'$, there is no canonical choice of γ. $Ext_A^{0,2^{i+1}-2}(Y \wedge Y)$ is $\mathbb{Z}/2 \oplus \mathbb{Z}/2$ on $\overline{e_{2^{i-1}-1} \otimes e_{2^i+2^{i-1}-1}}$ and $\overline{e_{2^i+2^{i-1}-1} \otimes e_{2^{i-1}-1}}$. Its $\mathbb{Z}/2$-dual $\left(H^*(Y \wedge Y)/\bar{A}H^*(Y \wedge Y)\right)^{2^{i+1}-2}$ is therefore also $\mathbb{Z}/2 \oplus \mathbb{Z}/2$, and $(x_{2^{i-1}-1} \otimes x_{2^i+2^{i-1}-1})'$, $(x_{2^i+2^{i-1}-1} \otimes x_{2^{i-1}-1})'$ are the generators. This can also be seen from (17). Recall $\overline{q \wedge q}$ is the composite

$$X = (P \wedge P) \cup_{\bar{g}} e^{2^{i+1}-2} \xrightarrow{\overline{q \wedge id}} Y \wedge P \xrightarrow{id \wedge q} Y \wedge Y$$

as in (22). Consider the induced maps $\overline{q \wedge q}^* : H^*(Y \wedge Y) \longrightarrow H^*(X)$ and $\overline{q \wedge q}' : H^*(Y \wedge Y)/\bar{A}H^*(Y \wedge Y) \longrightarrow H^*(X)/\bar{A}H^*(X)$. $\overline{q \wedge q}'$ in dimension $2^{i+1}-2$ is dual to $\overline{q \wedge q}_*$ in (23). Let

$$\overline{q \wedge q}'\left((x_{2^{i-1}-1} \otimes x_{2^i+2^{i-1}-1})'\right) = \delta_1 \overline{x_{2^i-1} \otimes x_{2^i-1}}' + \epsilon_1 \zeta'$$
$$\overline{q \wedge q}'\left((x_{2^i+2^{i-1}-1} \otimes x_{2^{i-1}-1})'\right) = \delta_2 \overline{x_{2^i-1} \otimes x_{2^i-1}}' + \epsilon_2 \zeta'$$

where δ_j, ϵ_j are 0 or 1. We have $j^*\overline{q \wedge q}^*\left(x_{2^{i-1}-1} \otimes x_{2^i+2^{i-1}-1}\right) = (q \wedge q)^*\left(x_{2^{i-1}-1} \otimes x_{2^i+2^{i-1}-1}\right) = q^*\left(x_{2^{i-1}-1}\right) \otimes q^*\left(x_{2^i+2^{i-1}-1}\right) = x_{2^{i-1}-1} \otimes x_{2^i+2^{i-1}-1}$, and similarly, $j^*\overline{q \wedge q}^*\left(x_{2^i+2^{i-1}-1} \otimes x_{2^{i-1}-1}\right) = x_{2^i+2^{i-1}-1} \otimes x_{2^{i-1}-1}$. From (25) we see these imply $\delta_1 = \delta_2 = 0$. So

$$\overline{q \wedge q}'\left((x_{2^{i-1}-1} \otimes x_{2^i+2^{i-1}-1})'\right) = \epsilon_1 \zeta',$$
$$\overline{q \wedge q}'\left((x_{2^i+2^{i-1}-1} \otimes x_{2^{i-1}-1})'\right) = \epsilon_2 \zeta'.$$

The claim (23) is equivalent to the following which is its $\mathbb{Z}/2$-dual statement.

(23)′ $\epsilon_1 = \epsilon_2 = 1$, that is, $\overline{q \wedge q}'\left((x_{2^{i-1}-1} \otimes x_{2^i+2^{i-1}-1})'\right) = \overline{q \wedge q}'\left((x_{2^i+2^{i-1}-1} \otimes x_{2^{i-1}-1})'\right) = \zeta'$.

This is equivalent to

(23)″ $$\overline{q \wedge q}^*\left(x_{2^{i-1}-1} \otimes x_{2^i+2^{i-1}-1}\right) \equiv \zeta \mod \bar{A}H^*(X),$$
$$\overline{q \wedge q}^*\left(x_{2^i+2^{i-1}-1} \otimes x_{2^{i-1}-1}\right) \equiv \zeta \mod \bar{A}H^*(X).$$

We will give yet another equivalent statement to this which is (23)‴ that follows.

First note, from (22), that $S^{2^{i+1}-3} \xrightarrow{\bar{g}} P \wedge P \xrightarrow{q \wedge q} Y \wedge Y$ is null homotopic. By (2.8) and (2.7) we see $E_2^{0,2^{i+1}-2}(Y \wedge Y) = 0$ in the ASS for $Y \wedge Y$. This implies that

(27) For any map $S^{2^{i+1}-2} \xrightarrow{\phi} Y \wedge Y$, $\phi^* = 0 :\ H^{2^{i+1}-2}(Y \wedge Y) \to H^{2^{i+1}-2}\left(S^{2^{i+1}-2}\right)$. Thus if $X = (P \wedge P) \cup_{\bar{g}} e^{2^{i+1}-2} \xrightarrow{(q\wedge q)_k} Y \wedge Y$, $k = 1, 2$, are any two extensions of $q \wedge q$ then $(q \wedge q)_1^* = (q \wedge q)_2^* :$ $H^{2^{i+1}-2}(Y \wedge Y) \to H^{2^{i+1}-2}(X)$.

$\overline{q \wedge q}$ in (22) is some extension of $q \wedge q$ which comes from the sequence

$$S^{2^{i+1}-3} \xrightarrow{g} M \wedge M \xrightarrow{id\wedge f} M \wedge P \xrightarrow{f\wedge id} P \wedge P \xrightarrow{q\wedge id} Y \wedge P \xrightarrow{id\wedge q} Y \wedge Y$$

with $\bar{g} = (f \wedge id)(id \wedge f)g$. If we consider the symmetrical sequence

$$S^{2^{i+1}-3} \xrightarrow{g} M \wedge M \xrightarrow{f\wedge id} P \wedge M \xrightarrow{id\wedge f} P \wedge P \xrightarrow{id\wedge q} P \wedge Y$$

which is also null homotopic then we get an extension $\overline{id \wedge q}$:
$X = (P \wedge P) \cup_{\bar{g}} e^{2^{i+1}-2} \to P \wedge Y$ of $P \wedge P \xrightarrow{id\wedge q} P \wedge Y$ and the composite

(28) $\overline{\overline{q \wedge q}} : X \xrightarrow{\overline{id\wedge q}} P \wedge Y \xrightarrow{q\wedge id} Y \wedge Y$

is another extension of $P \wedge P \xrightarrow{q\wedge q} Y \wedge Y$. From (27) we see

$$\overline{q \wedge q}^* = \overline{\overline{q \wedge q}}^* : H^*(Y \wedge Y) \to H^*(X = (P \wedge P) \cup_{\bar{g}} e^{2^{i+1}-2}).$$

Thus (23)″ is equivalent to

(23)‴ $\overline{q \wedge q}^* (x_{2^{i-1}-1} \otimes x_{2^i+2^{i-1}-1}) \equiv \zeta \mod \bar{A}H^*(X)$,
$\overline{\overline{q \wedge q}}^* (x_{2^i+2^{i-1}-1} \otimes x_{2^{i-1}-1}) \equiv \zeta \mod \bar{A}H^*(X)$.

To prove (23)‴ we need the following.

(29) (1) $\overline{q \wedge id}^* (\tilde{x}_{2^i-1} \otimes x_{2^i-1}) = \zeta$
(2) $\overline{id \wedge q}^* (x_{2^i-1} \otimes \tilde{x}_{2^i-1}) = \zeta$

where $\overline{q \wedge id}$ and $\overline{id \wedge q}$ are as in (22), (28).

Let $P \xrightarrow{q} Y \xrightarrow{q_1} \Sigma M$ be the cofiber sequence defined from the cofiber sequence $M \xrightarrow{f} P \xrightarrow{q} Y$. Consider the sequence

$$S^{2^{i+1}-3} \xrightarrow{g} M \wedge M \xrightarrow{id\wedge f} M \wedge P \xrightarrow{f\wedge id} P \wedge P \xrightarrow{q\wedge id} Y \wedge P \xrightarrow{q_1\wedge id} \Sigma M \wedge P$$

Let $g_1 = (id \wedge f)g$; so $\bar{g} = (f \wedge id)g_1$. This sequence defines a map of cofiber sequences

$$\begin{array}{ccccccc}
S^{2^{i+1}-3} & \xrightarrow{\bar{g}} & P \wedge P & \xrightarrow{j} & X = (P \wedge P) \cup_{\bar{g}} e^{2^{i+1}-2} & \xrightarrow{\bar{q}} & S^{2^{i+1}-2} \\
\downarrow g_1 & & \| & & \downarrow \overline{q \wedge id} & & \downarrow \Sigma g_1 \\
M \wedge P & \xrightarrow{f\wedge id} & P \wedge P & \xrightarrow{q\wedge id} & Y \wedge P & \xrightarrow{q_1\wedge id} & \Sigma M \wedge P.
\end{array}$$

Recall the generators in $H^{2^i-2}(M) = \mathbb{Z}/2$ and $H^{2^i-1}(M) = \mathbb{Z}/2$ of the first M in $M \wedge M$ are denoted by x and y and those of the second M are denoted by u

and v. We have $g_1^*(x \otimes x_{2^i-1}) = g^*\left((id \wedge f)^*(x \otimes x_{2^i-1})\right) = g^*(x \otimes v) = \iota$. So $\Sigma g_1^*(\Sigma x \otimes x_{2^i-1}) = \Sigma\iota$ which implies

$$\begin{aligned}[(\Sigma g_1)\bar{q}]^*(\Sigma x \otimes x_{2^i-1}) &= \bar{q}^*(\Sigma g_1^*(\Sigma x \otimes x_{2^i-1})) \\ &= \bar{q}^*(\Sigma\iota) = \zeta.\end{aligned}$$

We have $q_1^*(\Sigma x) = \tilde{x}_{2^i-1}$. So $(q_1 \wedge id)^*(\Sigma x \otimes x_{2^i-1}) = \tilde{x}_{2^i-1} \otimes x_{2^i-1}$. Then

$$\begin{aligned}\overline{q \wedge id}^*(\tilde{x}_{2^i-1} \otimes x_{2^i-1}) &= \overline{q \wedge id}^*\left((q_1 \wedge id)^*(\Sigma x \otimes x_{2^i-1})\right) \\ &= \left[(q_1 \wedge id)\left(\overline{q \wedge id}\right)\right]^*(\Sigma x \otimes x_{2^i-1}) \\ &= [(\Sigma g_1)\bar{q}]^*(\Sigma x \otimes x_{2^i-1}) = \zeta.\end{aligned}$$

This proves (29)(1). (29)(2) is similarly proved by considering the symmetrical sequence

$$S^{2^{i+1}-3} \xrightarrow{g} M \wedge M \xrightarrow{f \wedge id} P \wedge M \xrightarrow{id \wedge f} P \wedge P \xrightarrow{id \wedge q} P \wedge Y \xrightarrow{id \wedge q_1} P \wedge \Sigma M$$

which defines a map of cofiber sequences

$$\begin{array}{ccccccc} S^{2^{i+1}-3} & \xrightarrow{\bar{g}} & P \wedge P & \longrightarrow & X & \xrightarrow{\bar{q}} & S^{2^{i+1}-2} \\ \downarrow g_2 & & \| & & \downarrow \overline{id \wedge q} & & \downarrow \Sigma g_2 \\ P \wedge M & \xrightarrow{id \wedge f} & P \wedge P & \xrightarrow{id \wedge q} & P \wedge Y & \xrightarrow{id \wedge q_1} & P \wedge \Sigma M \end{array}$$

where $g_2 = (f \wedge id)g$. This proves (29).

We deduce (23)''' from (29) as follows. In $H^*(Y \wedge P)$ we have

$$(**) \qquad Sq^{2^{i-1}}(x_{2^{i-1}-1} \otimes x_{2^i-1}) = \tilde{x}_{2^i-1} \otimes x_{2^i-1} + x_{2^{i-1}-1} \otimes x_{2^i+2^{i-1}-1} + \left(w = \sum_{0<\nu<2^{i-1}} x_{2^{i-1}+\nu-1} \otimes x_{2^i+2^{i-1}-\nu-1}\right).$$

w, when considered in $H^*(P \wedge P)$, lies in $\bar{A}H^*(P \wedge P)$ by (25). So

$(***)$ $\quad \sum_\ell a_\ell x_{j(\ell)} \otimes x_{k(\ell)} = w$ in $H^*(P \wedge P)$ for some a_ℓ, $j(\ell)$, $k(\ell)$ with $a_\ell \in \bar{A}$.

Since the factor $x_{2^{i-1}+\nu-1}$ in w has $2^{i-1}+\nu-1 < 2^i-1$ we may assume $j(\ell) < 2^i-1$. So $x_{j(\ell)} \in H^*(Y)$. If $j(\ell) \neq 2^{i-1}-1$ for all ℓ then $(***)$ also holds in $H^*(Y \wedge P)$. If $j(\ell) = 2^{i-1}-1$ for some ℓ and if $a_\ell \neq Sq^{2^{i-1}} + b_\ell$ with b_ℓ decomposable then $(***)$ still holds in $H^*(Y \wedge P)$. If, for some ℓ, $a_\ell = Sq^{2^{i-1}}$ and $x_{j(\ell)} = x_{2^{i-1}-1}$ then $x_{2^{i-1}-1} \otimes x_{2^i+2^{i-1}-1}$ must be a term in $\sum_\ell a_\ell x_{j(\ell)} \otimes x_{k(\ell)}$ and so $\sum_\ell a_\ell x_{j(\ell)} \otimes x_{k(\ell)} \neq w$. These are easy to see from (16) and (17). We conclude that w in $(**)$ lies in $\bar{A}H^*(Y \wedge P)$. So

$$\tilde{x}_{2^i-1} \otimes x_{2^i-1} \equiv x_{2^{i-1}-1} \otimes x_{2^i+2^{i-1}-1} \mod \bar{A}H^*(Y \wedge P).$$

By symmetry we see

Then

$$\begin{aligned}
&\overline{q\wedge q}^{*}\left(x_{2^{i-1}-1}\otimes x_{2^i+2^{i-1}-1}\right)\\
=&\overline{q\wedge id}^{*}\left(id\wedge q\right)^{*}\left(x_{2^{i-1}-1}\otimes x_{2^i+2^{i-1}-1}\right)\\
=&\overline{q\wedge id}^{*}\left(x_{2^{i-1}-1}\otimes x_{2^i+2^{i-1}-1}\right)\\
\equiv&\overline{q\wedge id}^{*}\left(\tilde{x}_{2^i-1}\otimes x_{2^i-1}\right) \mod \bar{A}H^*(X)\\
\equiv&\zeta \mod \bar{A}H^*(X) \qquad \text{(by (29)(1))}
\end{aligned}$$

and

$$\begin{aligned}
&\overline{\overline{q\wedge q}}^{*}\left(x_{2^i+2^{i-1}-1}\otimes x_{2^{i-1}-1}\right)\\
=&\overline{id\wedge q}^{*}\left(q\wedge id\right)^{*}\left(x_{2^i+2^{i-1}-1}\otimes x_{2^{i-1}-1}\right)\\
=&\overline{id\wedge q}^{*}\left(x_{2^i+2^{i-1}-1}\otimes x_{2^{i-1}-1}\right)\\
\equiv&\overline{id\wedge q}^{*}\left(x_{2^i-1}\otimes \tilde{x}_{2^i-1}\right) \mod \bar{A}H^*(X)\\
\equiv&\zeta \mod \bar{A}H^*(X) \qquad \text{(by (29)(2))}
\end{aligned}$$

This proves $(23)'''$ and therefore (23).

5. Completion of the proof of Theorem 1.2

Let $B(k)$ denote the 2-primary k^{th} Brown-Gitler spectrum ([6]). Recall that $H^*(B(k)) \cong A/A\left\{\chi\left(Sq^j\right) \mid j > k\right\}$ where $\chi : A \to A$ is the canonical anti-automorphism. Then $\left\{\chi\left(Sq^I\right) \mid I = (i_1, \cdots, i_s) \text{ admissible}, i_1 \leq k\right\}$ is a $\mathbb{Z}/2$-base for $H^*(B(k))$ ([6, (1,3)]). In particular we have the following.

(30) Let $B = \Sigma^{7\cdot 2^{i-3}} B\left(2^{i-4}\right)$ where $i \geq 5$. Then

(1) $$H^\ell(B) = \begin{cases} 0 & \text{for } \ell < 7\cdot 2^{i-3} \text{ or } \ell > 2^i - 1 \\ \mathbb{Z}/2 & \text{for } \ell = 2^i - 2 \text{ and } \ell = 2^i - 1. \end{cases}$$

(2) $Sq^1 : H^{2^i-2}(B) = \mathbb{Z}/2 \longrightarrow H^{2^i-1}(B) = \mathbb{Z}/2$ is non-zero.

The following is proved in [14] where B is as in (30).

(5.1) There is a stable map $B \xrightarrow{\phi} S^0$ such that

$$Sq^{2^i} : H^0\left(C_\phi\right) = \mathbb{Z}/2 \to H^{2^i}\left(C_\phi\right) = \mathbb{Z}/2 \quad \text{is non-zero.}$$

Let Q denote the infinite loop space functor $\Omega^\infty\Sigma^\infty$. Kahn and Priddy prove in [8] that there is a 2-primary decomposition $QP^\infty \simeq QS^0 \times Z$ for some space Z. By dimensional reasons, the stable map ϕ in (5.1) can be realized as a space map $B \xrightarrow{\phi'} QS^0$. These implies the following.

(5.2) There is a stable map $B \xrightarrow{\bar{\phi}} P = P^\infty$ such that $\bar{\phi}^* \neq 0$ in dimension $2^i - 1$.

We make a convention. Given a stable map $\psi : U \to V$. We will simply say "ψ has AF k" to mean "ψ has AF k in the ASS for the stable groups $[U, V]_*$".

B has a CW-structure such that if B^{2^i-3} is its $(2^i - 3)$-skeleton then there is a cofiber sequence

$$B^{2^i-3} \longrightarrow B \xrightarrow{p} M = S^{2^i-2} \cup_{2\iota} e^{2^i-1} \xrightarrow{\delta} \Sigma B^{2^i-3}$$

with $AF(\delta) = 1$. This follows from (30). Note that B^{2^i-3} begins its cells in dimension $7 \cdot 2^{i-3}$. Since $i \geq 5$, from (2.4) we see $Ext_A^{s,t}(P) = 0$ for $7 \cdot 2^{i-3} \leq t - s \leq 2^i - 3$ with $s \leq 3$ except for $i = 6$, $s = 3$ and $t - s = 61$ in which case $Ext_A^{3,64}(P) \cong \mathbb{Z}/2$ on $\hat{D}_3$. It is not difficult to show (by (3.3) in [14]) that, for $i = 6$, $\bar{\phi}|B^{61} : B^{61} \to P$ cannot be detected on the top cell by $\hat{D}_3$. Thus, for each $i \geq 5$, $\bar{\phi}|B^{2^i-3} : B^{2^i-3} \to P$ has a lifting $\bar{\phi}_4 : B^{2^i-3} \to X_4$ in any minimal Adams resolution $\{X_s, q_s'\}$ for P where $q_s' : X_s \to X_{s-1}$. We fix such a resolution, and let $q = q_3 : X_3 \to P$ be the projection map. Then we have a map $M \xrightarrow{\bar{f}} P \cup_q CX_3$ such that there is a commutative diagram

$$(31) \qquad \begin{array}{ccc} B & \xrightarrow{\bar{\phi}} & P \\ \downarrow p & & \downarrow j \\ M & \xrightarrow{\bar{f}} & P \cup_q CX_3 \end{array}$$

and such that

(32) the composite $M \xrightarrow{\bar{f}} P \cup_q CX_3 \xrightarrow{\bar{q}_3} \Sigma X_3$ is the composite

$$M \xrightarrow{\delta} \Sigma B^{2^i-3} \xrightarrow{\Sigma\phi_4} \Sigma X_4 \xrightarrow{\Sigma q_3'} \Sigma X_3$$

where j is the inclusion map and $\bar{q}_3$ is the collapsing map. Since $AF(\delta) = 1$, $AF(\Sigma\phi_4) \geq 0$ and $AF(\Sigma q_s') = 1$ it follows that $AF(\bar{q}_3\bar{f}) \geq 2$. It is clear that $H^*(P \cup_q CX_3) \cong H^*(P) \oplus H^*(\Sigma X_3)$. From (5.1) and (31) we see $\bar{f}^*$, when restricted to $H^*(P)$, behaves as f^* for the assumed $M \xrightarrow{f} P$ in (1.4). The proof of Theorem 1.2 based on the assumption such an f exists as given in §4 can be carried over to $\bar{f}$ by some modifications. In the remainder of this section we describe these modifications.

For a spectrum W and an integer $r \geq 2$, $E_r^{*,*}(W)$ will denote the E_r term in the ASS for W.

We will write $\bar{P}$ to denote $P \cup_q CX_3$. Since $AF(q) = 3$ it follows that we have the following.

(33) (1) $H^*(\bar{P}) \cong H^*(P) \oplus H^*(\Sigma X_3)$,
(2) $E_r^{s,t}(\bar{P}) \cong E_r^{s,t}(P) \oplus E_r^{s,t}(\Sigma X_3)$ for $r = 2$ and $r = 3$.

It is also clear that $\bar{P} \wedge \bar{P} = (P \wedge P) \cup_{\tilde{q}} CZ$ for some spectrum Z and some map $\tilde{q} : Z \to P \wedge P$ with $AF(\tilde{q}) \geq 3$. So we also have

(2) $E_r^{s,t}(\bar{P}\wedge\bar{P})\cong E_r^{s,t}(P\wedge P)\oplus E_r^{s,t}(\Sigma Z)$ for $r=2$ and $r=3$.

Let $\bar{Y}$ be the cofiber of $M\xrightarrow{\bar{f}}\bar{P}=P\cup_q CX_3$ so that there is a cofibration $M\xrightarrow{\bar{f}}\bar{P}\xrightarrow{\bar{q}}\bar{Y}$. We still use Y to denote the cofiber of the map f in (1.4). As we do not assume (1.4) anymore it is meaningless to consider Y. All we need here are the A-modules $H^*(Y)$, $H^*(Y\wedge Y)$, $H^*(Y\wedge U)\cong H^*(Y)\otimes H^*(U)$, $H^*(V\wedge Y)\cong H^*(V)\otimes H^*(Y)$ which are existent for any spectra U and V. Then, since $AF(\bar{q}_3\bar{f})\geq 2$ where $\bar{q}_3$ is as in (32), it is not difficult to see that there are the following decompositions.

(35) (1) $H^*(\bar{Y})\cong H^*(Y)\oplus H^*(\Sigma X_3)$,
(2) $E_2^{s,t}(\bar{Y})\cong Ext_A^{s,t}(H^*(Y))\oplus Ext_A^{s,t}(\Sigma X_3)$,
(3) $H^*(\bar{Y}\wedge\bar{Y})\cong H^*(Y\wedge Y)\oplus H^*(Y\wedge\Sigma X_3)\oplus H^*(\Sigma X_3\wedge Y)$ $\oplus H^*(\Sigma X_3\wedge\Sigma X_3)$,
(4) $E_2^{s,t}(\bar{Y}\wedge\bar{Y})\cong Ext_A^{s,t}(H^*(Y\wedge Y))\oplus Ext_A^{s,t}(H^*(Y\wedge\Sigma X_3))$ $\oplus Ext_A^{s,t}(H^*(\Sigma X_3\wedge Y))\oplus Ext_A^{s,t}(\Sigma X_3\wedge\Sigma X_3)$,
(5) $H^*(\bar{Y}\wedge\bar{P})\cong H^*(Y\wedge P)\oplus H^*(Y\wedge\Sigma X_3)\oplus H^*(\Sigma X_3\wedge P)$ $\oplus H^*(\Sigma X_3\wedge\Sigma X_3)$,
(6) $E_2^{s,t}(\bar{Y}\wedge\bar{P})\cong Ext_A^{s,t}(H^*(Y\wedge P))\oplus Ext_A^{s,t}(H^*(\Sigma X_3\wedge P))$ $\oplus Ext_A^{s,t}(H^*(Y\wedge\Sigma X_3))\oplus Ext_A^{s,t}(\Sigma X_3\wedge\Sigma X_3)$,
(7) $H^*(\bar{P}\wedge\bar{Y})\cong H^*(P\wedge Y)\oplus H^*(P\wedge\Sigma X_3)\oplus H^*(\Sigma X_3\wedge Y)$ $\oplus H^*(\Sigma X_3\wedge\Sigma X_3)$,
(8) $E_2^{s,t}(\bar{P}\wedge\bar{Y})\cong Ext_A^{s,t}(H^*(P\wedge Y))\oplus Ext_A^{s,t}(P\wedge\Sigma X_3)$ $\oplus Ext_A^{s,t}(H^*(\Sigma X_3\wedge Y))\oplus Ext_A^{s,t}(\Sigma X_3\wedge\Sigma X_3)$.

We will be concerned only with the first direct summands in all the decompositions in (33) through (35). If $N=K_1\oplus\sum_{j\geq 2}K_j$, $N'=K_1'\oplus\sum_{j\geq 2}K_j'$ are two such decompositions with K_1, K_1' the first direct summands and if $N\xrightarrow{\psi}N'$ is an A-map or $Ext_A^{*,*}$-map according to the case then we say ψ preserves the first summands if $\psi(K_1)\subset K_1'$, that is, $\psi(x,0)=(y,0)$. The following (36) and (37) are also easy to see.

(36) The induced maps in $H^*(\ )$ or $Ext_A^{*,*}(\ )$ of the maps $\bar{P}\xrightarrow{\bar{q}}\bar{Y}$, $\bar{P}\wedge\bar{P}\xrightarrow{id\wedge\bar{q}}\bar{P}\wedge\bar{Y}$, $\bar{P}\wedge\bar{P}\xrightarrow{\bar{q}\wedge id}\bar{Y}\wedge\bar{P}$, $\bar{P}\wedge\bar{P}\xrightarrow{\bar{q}\wedge\bar{q}}\bar{Y}\wedge\bar{Y}$, $\bar{P}\wedge\bar{Y}\xrightarrow{\bar{q}\wedge id}\bar{Y}\wedge\bar{Y}$, $\bar{Y}\wedge\bar{P}\xrightarrow{id\wedge\bar{q}}\bar{Y}\wedge\bar{Y}$ all preserve the first summands.

(37) All the results about $Ext_A^{**}(\ )$ in §2 and all the results in §3 for $P, Y, P\wedge P, Y\wedge P, Y\wedge Y$ are also true for $\bar{P},\bar{Y},\bar{P}\wedge\bar{P},\bar{Y}\wedge\bar{P}$ and $\bar{Y}\wedge\bar{Y}$ if we restrict to the first direct summands.

Let $S^{2^{i+1}-3}\xrightarrow{g}M\wedge M$ be as in the beginning of §4 and consider the composite

$$\bar{\bar{g}}:S^{2^{i+1}-3}\xrightarrow{g}M\wedge M\xrightarrow{\bar{f}\wedge\bar{f}}\bar{P}\wedge\bar{P}.$$

parallel to the composite

$$\bar{g}:S^{2^{i+1}-3}\xrightarrow{g}M\wedge M\xrightarrow{f\wedge f}P\wedge P$$

in §4. The main results in §4 for proving Theorem 1.2 are (4.1) and (4.2) about $\bar{g}$. For $\bar{\bar{g}}$ there are analogous results (4.1)′ and (4.2)′. (4.1)′ is also stated as

(4.1)′ $AF\left(\bar{\bar{g}}\right) \geq 2$.

This follows from

(i) the proof of (4.1) in §4 with P, Y there replaced by $\bar{P}$, $\bar{Y}$ here and by using (36), (37)

and

(ii) the fact that the composite $S^{2^{i+1}-3} \xrightarrow{\bar{\bar{g}}} \bar{P} \wedge \bar{P} = (P \cup P) \cup_{\bar{q}} CZ \to \Sigma Z$ has $AF \geq 2$ which is easy to see.

(4.2)′ should be stated as follows.

(4.2)′ $\bar{\bar{g}} : S^{2^{i+1}-3} \to \bar{P} \wedge \bar{P}$ is detected by

$$\begin{aligned}&(\overline{e_{2^{i-1}-1} \otimes e_{2^{i-1}-1}} h_i h_0, \alpha) \in E_2^{2,2^{i+1}-1}\left(\bar{P} \wedge \bar{P}\right)\\ &= E_2^{2,2^{i+1}-1}(P \wedge P) \oplus E_2^{2,2^{i+1}-1}(\Sigma Z)\end{aligned}$$

for some $\alpha \in E^{2,2^{i+1}-1}(\Sigma Z)$.

The proof of (4.2)′ is modified as follows. Let $\bar{X} = \left(\bar{P} \wedge \bar{P}\right) \cup_{\bar{\bar{g}}} e^{2^{i+1}-2}$. Then we also have the following decompositions where $(P \wedge P) \cup_{\bar{g}} e^{2^{i+1}-2}$ is as in §4.

(38) (1) $H^*\left(\bar{X}\right) = H^*\left(\left(P \wedge P\right) \cup_{\bar{g}} e^{2^{i+1}-2}\right) \oplus H^*\left(\Sigma Z\right)$,

(2) $E_2^{s,t}\left(\bar{X}\right) = Ext_A^{s,t}\left(H^*\left(\left(P \wedge P\right) \cup_{\bar{g}} e^{2^{i+1}-2}\right)\right) \oplus Ext_A^{s,t}\left(\Sigma Z\right)$.

We also have extensions $\bar{X} \xrightarrow{\overline{\bar{q}\wedge id}} \bar{Y} \wedge \bar{P}$, $\bar{X} \xrightarrow{\overline{id\wedge\bar{q}}} \bar{P} \wedge \bar{Y}$ and the composites

$$\bar{X} \xrightarrow{\overline{\bar{q}\wedge id}} \bar{Y} \wedge \bar{P} \xrightarrow{id\wedge\bar{q}} \bar{Y} \wedge \bar{Y}, \quad \bar{X} \xrightarrow{\overline{id\wedge\bar{q}}} \bar{P} \wedge \bar{Y} \xrightarrow{\bar{q}\wedge id} \bar{Y} \wedge \bar{Y}$$

analogous to those in (22), (28) in §4, and the following.

(39) The induced maps in $H^*(\ \ )$ of $\overline{\bar{q} \wedge id}$ and $\overline{id \wedge \bar{q}}$ also preserve the first summands.

From (36), (37), (38) and (39) we see the proof of (4.2) in §4 also works for (4.2)′ if we always concentrate on the first summands.

In §4 the conclusion $d_3\left(\overline{e_{2^i-1} \otimes e_{2^i-1}}\right) = e_{2^{i-1}-1} \otimes e_{2^{i-1}-1} h_i h_0^2$ in (1.2) was deduced from (4.2) and point (19). (19) still holds if P (resp. f) is replaced by $\bar{P}$ (resp $\bar{f}$). In particular, we note that the restriction map of

$$M \wedge M \simeq \left[\left(S^{2^{i+1}-4} \cup_{2\iota} e^{2^{i+1}-3}\right) \vee S^{2^{i+1}-3}\right] \cup e^{2^{i+1}-2} \xrightarrow{\bar{f}\wedge\bar{f}} \bar{P} \wedge \bar{P}$$

to $S^{2^{i+1}-3}$ can be assumed to be $\bar{\bar{g}}$. Since

(i) $\bar{\bar{g}} = \bar{f} \wedge \bar{f}|\ S^{2^{i+1}-3} : S^{2^{i+1}-3} \to \bar{P} \wedge \bar{P}$ is detected by $(\overline{e_{2^{i-1}-1} \otimes e_{2^{i-1}-1}} h_i h_0, \alpha)$ (by (4.2)′),

(ii) the attaching map for the top cell $e^{2^{i+1}-2}$ in $M \wedge M$ is $S^{2^{i+1}-3} \xrightarrow{(\eta,2\iota)}$

(iii) $Ext_A^{1,2^{i+1}-1}(P \wedge P) \cong \mathbb{Z}/2 \oplus \mathbb{Z}/2$ on $\overline{e_{2^i-1} \otimes e_{2^i-1}}h_0$, $\overline{e_{2^i-2} \otimes e_1}h_i$ and $d_2\left(\overline{e_{2^i-1} \otimes e_{2^i-1}}h_0\right) = d_2\left(\overline{e_{2^i-2} \otimes e_1}h_i\right) = 0$ in the ASS for $P \wedge P$ (proved in §4), and therefore also for $\bar{P} \wedge \bar{P}$,

(iv) $E_3^{s,t}\left(\bar{P} \wedge \bar{P}\right) \cong E_3^{s,t}(P \wedge P) \oplus E_3^{s,t}(\Sigma Z)$ (by (3.4)(2))

(v) $AF\left(\bar{f} \wedge \bar{f}| \, S^{2^{i+1}-4}\right) \geq 3$,

(vi) $\left(\bar{f} \wedge \bar{f}\right)_*(y_* \otimes v_*) = e_{2^i-1} \otimes e_{2^i-1}$ where $y_* \otimes v_*$ is the top class of $H_*(M \wedge M)$ corresponding to the cell $e^{2^{i+1}-2}$.

It follows that we must have the differential

$$d_3\left(\left(\overline{e_{2^i-1} \otimes e_{2^i-1}}, 0\right)\right) = \left(\overline{e_{2^{i-1}-1} \otimes e_{2^{i-1}-1}}h_i h_0^2, \alpha h_0\right)$$

in the ASS for $\bar{P} \wedge \bar{P}$. This implies $d_3\left(\overline{e_{2^i-1} \otimes e_{2^i-1}}\right) = \overline{e_{2^{i-1}-1} \otimes e_{2^{i-1}-1}}h_i h_0^2$ in the ASS for $P \wedge P$. This proves Theorem 1.2.

6. Proof of Theorem 1.3

We begin by recalling that $E_2^{s,t}(P) = Ext_A^{s,t}(P)$ for $t - s \leq 30$ and $E_r^{s,t}(P)$ for $t - s \leq 30$, $3 \leq r \leq \infty$ are completely determined in [12]. A clear table of these can be found in [19]. In particular, we recall the following.

(6.1) (1) $E_2^{s,s+28}(P)$ is $\mathbb{Z}/2$ on $P^1\hat{g}h_0^j$ for $s = 7 + j$, $0 \leq j \leq 2$ and is zero for other s,

(2) $E_2^{s,s+29}(P)$ is $\mathbb{Z}/2$ on $\hat{k}h_0^j$ for $s = 6 + j$, $0 \leq j \leq 2$ and is zero for other s,

(3) $E_2^{5,35}(P) \cong \mathbb{Z}/2$ generated by $\hat{r}$,

(4) $d_2\left(\hat{k}h_0^j\right) = P^1\hat{g}h_0^{j+1}$, $j = 0, 1$, and $d_3(\hat{r}) = \hat{k}h_0^2$ in the ASS for P.

Let R be the fiber of the transfer $P \xrightarrow{\bar{t}} S^0$. So we have cofiber sequences

$$S^{-1} \xrightarrow{j} R \xrightarrow{q} P \xrightarrow{\bar{t}} S^0$$

$$\Sigma^{-1}P = P \wedge S^{-1} \xrightarrow{id \wedge j} P \wedge R \xrightarrow{id \wedge q} P \wedge P \xrightarrow{id \wedge \bar{t}} P \wedge S^0 = P$$

and the following long exact sequences in Ext groups:

$$\text{(40)} \quad \cdots \longrightarrow Ext_A^{s,t}\left(S^{-1}\right) \xrightarrow{j_*} Ext_A^{s,t}(R) \xrightarrow{q_*} Ext_A^{s,t}(P) \xrightarrow{\bar{t}_*} Ext_A^{s+1,t+1}\left(S^0\right) \longrightarrow \cdots.$$

$$\text{(41)} \quad \cdots \longrightarrow Ext_A^{s,t}\left(\Sigma^{-1}P\right) = Ext_A^{s,t+1}(P) \xrightarrow{(id \wedge j)_*} Ext_A^{s,t}(P \wedge R) \xrightarrow{(id \wedge q)_*} Ext_A^{s,t}(P \wedge P) \xrightarrow{(id \wedge \bar{t})_*} Ext_A^{s+1,t+1}(P) \longrightarrow \cdots.$$

$\bar{t}_*$ is onto for $t - s \geq 1$ ([9]). Thus from (40) and $Ext_A^{s,t}(P)$ for $t - s \leq 30$ ([14], [19]) we can determine $Ext_A^{s,t}(R)$ for $-1 \leq t - s \leq 29$ which are described as follows, where $Ext_A^{s,t}(R)$ for $t - s \geq 1$ is identified with $(ker\ \bar{t}_*)^{s,t} \subset Ext_A^{s,t}(P)$.

(6.2) (1) $Ext_A^{s,s-1}(R) \cong \mathbb{Z}/2$ on $e_{-1}h_0^s$, $s \geq 0$ where $\bar{e}_{-1} = j_*(1) \in Ext_A^{0,-1}(R)$.

For (2) through (7) below we only consider $0 \le t - s \le 29$.

(2) $Ext_A^{1,t}(R)$ is $\mathbb{Z}/2$ for $t = 5, 9, 11, 17, 19, 23$ on $\bar{e}_1 h_2$, $\bar{e}_1 h_3 + \bar{e}_7 h_1$, $\bar{e}_3 h_3$, $\bar{e}_1 h_4 + \bar{e}_{15} h_1$, $\bar{e}_3 h_4 + \bar{e}_{15} h_2$, $\bar{e}_7 h_4$ and is zero for other t.

(3) $Ext_A^{2,t}(R)$ is $\mathbb{Z}/2$ (resp. $\mathbb{Z}/2 \oplus \mathbb{Z}/2$) for $t = 9, 11, 12, 17, 20, 21, 23, 24, 25$ (resp. for $t = 19$) on $\bar{e}_1 h_2^2$, $\bar{e}_1 h_3 h_1 + \bar{e}_7 h_1^2$, $\bar{e}_3 h_3 h_0$, $\bar{e}_1 h_3^2$, $\bar{e}_3 h_4 h_0 + \bar{e}_{15} h_2 h_0$, $\bar{e}_1 h_4 h_2$, $\bar{e}_3 h_4 h_2 + \bar{e}_{15} h_2^2$, $\bar{e}_7 h_4 h_0$, $\bar{e}_7 h_4 h_1$ (resp. on $\bar{e}_3 h_3^2$ and $\bar{e}_1 h_4 h_1^2 + \bar{e}_{15} h_1^2$) and is zero for other t.

(4) $Ext_A^{3,t}(R)$ is $\mathbb{Z}/2$ (resp. $\mathbb{Z}/2 \oplus \mathbb{Z}/2$) for $t = 13, 19, 21, 23, 24, 26, 27$ (resp. for $t = 25$) on $e_1 h_2^3$, α_{16}, $\bar{e}_3 h_4 h_1^2 + \bar{e}_{15} h_4 h_1^3$, $\bar{e}_1 c_1$, α_{21}, $\hat{c}_0 h_4 + \bar{e}_{15} c_0$, $\bar{e}_7 h_4 h_1^2$ (resp. $\bar{e}_1 h_4 h_2^2$ and $\bar{e}_7 h_4 h_0^2$) and is zero for other t.

(5) $Ext_A^{4,t}(R)$ is $\mathbb{Z}/2$ for $t = 23, 25, 26, 27, 28$ on $\bar{e}_1 f_0$, $\bar{e}_3 f_0 + \bar{e}_1 g_1$, $\bar{e}_7 h_4 h_0^3$, $\bar{e}_1 c_1 h_2$, $\alpha_{21} h_2 = \hat{c}_0 h_4 h_1 + \bar{e}_{15} c_0 h_1$ and is zero for other t.

(6) $Ext_A^{5,t}(R)$ is $\mathbb{Z}/2$ for $t = 27, 29$ on $\bar{e}_1 f_0 h_2 = \bar{e}_1 g_1 h_1$, $\bar{e}_1 g_1 h_2$ and is zero for other t.

(7) $Ext_A^{s,t}(R) = 0$ for all $s \ge 6$ and t.

PROPOSITION 6.3. (1) *$Ext_A^{s,t}(P \wedge P) \xrightarrow{(id\wedge\bar{t})_*} Ext_A^{s+1,t+1}(P)$ is iso for $s \ge 5$, $t - s = 29$, and these groups are as described in (6.1).*

(2) *The remaining $Ext_A^{s,s+29}(P \wedge P)$ are given by: $Ext_A^{s,s+29}(P \wedge P) = 0$ for $s = 0$ and $s = 3$, $Ext_A^{1,30}(P \wedge P) = \mathbb{Z}/2$ on $\overline{e_7 \otimes e_7} h_4$, $Ext_A^{2,31}(P \wedge P) = \mathbb{Z}/2$ on $\overline{e_7 \otimes e_7} h_4 h_0$.*

(3) *$Ext_A^{s,30+s}(P \wedge P) \xrightarrow{(id\wedge\bar{t})_*} Ext_A^{s+1,31+s}(P)$ is onto for $s \ge 4$.*

(4) *$Ext_A^{s,s+29}(P \wedge R) \xrightarrow{(id\wedge q)_*} Ext_A^{s,s+29}(P \wedge P)$ is an isomorphism for $0 \le s \le 3$ and $Ext_A^{s,s+29}(P \wedge R) = 0$ for $s \ge 3$.*

We deduce Theorem 1.3 from (6.1) and (6.3) as follows. We want to show $d_r(\overline{e_{15} \otimes e_{15}}) = 0$ for all $r \ge 2$ in the ASS for $P \wedge P$. By (3.7), $d_2(\overline{e_{15} \otimes e_{15}}) = 0$. By (6.3)(2), $Ext_A^{3,32}(P \wedge P) = 0$. So $d_3(\overline{e_{15} \otimes e_{15}}) = 0$. By (6.3)(1), (3) and (6.1) we see $E_4^{s,s+29}(P \wedge P) = 0$ for $s \ge 4$. So $d_r(\overline{e_{15} \otimes e_{15}}) = 0$ for $r \ge 4$ also. This proves Theorem 1.3.

To prove (6.3) it suffices to show the following which is a part of (6.3).

(6.4) (1) $Ext_A^{s,s+28}(P \wedge R) = 0$ for $s \ge 6$.

(2) $Ext_A^{s,s+29}(P \wedge R) = 0$ for $s = 0$ and $s \ge 3$, $Ext_A^{1,30}(P \wedge R) \cong \mathbb{Z}/2$ on $\overline{e_7 \otimes e_7} h_4$ and $Ext_A^{2,31}(P \wedge R) \cong \mathbb{Z}/2$ on $\overline{e_7 \otimes e_7} h_4 h_0$.

It is easy to see that the rest of the assertions in (6.3) follow from (6.4), the knowledge of $Ext_A^{s,t}(P)$ for $t - s \le 30$ and the long exact sequence (41).

We will use the spectral sequence $\bar{E}_r^{*,*,*}(P \wedge R)$ introduced in §2 to prove (6.4). By (6.2)(7), $Ext_A^{s,t}(R) = 0$ for $s \ge 6$, $t - s \le 29$. This implies that $\bar{E}_1^{s,s+28}(P \wedge R) = \sum_j \bar{E}_1^{j,s,s+28}(P \wedge R) = \sum_j Ext_A^{s,28-j+s}(R) = 0$ for $s \ge 6$. So $Ext_A^{s,s+28}(P \wedge R) = 0$ for $s \ge 6$. This proves (6.4)(1). From (6.2) we see the set of elements listed on the left side of each arrow below is a $\mathbb{Z}/2$-base for $\bar{E}_1^{s,s+29}(P \wedge R) = \sum \bar{E}_1^{j,s,s+29}(P \wedge R)$ all $s \ge$

(6.4)* (1) $e_{30} \otimes \bar{e}_{-1} h_0^i \to e_{29} \otimes \bar{e}_{-1} h_0^{i+1}$, $i \geq 0$,
(2) $e_{25} \otimes \bar{e}_1 h_2 \to e_{21} \otimes \bar{e}_1 h_2^2$,
(3) $e_{21} \otimes (\bar{e}_1 h_3 + \bar{e}_7 h_1) \to e_{19} \otimes (\bar{e}_1 h_3 h_1 + \bar{e}_7 h_1^2)$,
(4) $e_{19} \otimes \bar{e}_3 h_3 \to e_{11} \otimes \bar{e}_3 h_3^2$,
(5) $e_{13} \otimes (\bar{e}_1 h_4 + \bar{e}_{15} h_1) \to e_{11} \otimes (\bar{e}_1 h_4 h_1 + \bar{e}_{15} h_1^2)$
(6) $e_{11} \otimes (\bar{e}_3 h_4 + \bar{e}_{15} h_2) \to e_7 (\bar{e}_3 h_4 h_2 + \bar{e}_{15} h_2^2)$,
(7) $e_7 \otimes \bar{e}_7 h_4 \to 0$,
(8) $e_{22} \otimes \bar{e}_1 h_2^2 \leftarrow e_{26} \otimes \bar{e}_1 h_2$
(9) $e_{20} \otimes (\bar{e}_1 h_1 h_3 + \bar{e}_7 h_1^2) \to e_{18} \otimes \bar{e}_1 h_1^2 h_3$,
(10) $e_{19} \otimes \bar{e}_3 h_3 h_0 \leftarrow e_{20} \otimes \bar{e}_3 h_3$,
(11) $e_{14} \otimes \bar{e}_1 h_3^2 \leftarrow e_{22} \otimes (\bar{e}_1 h_3 + \bar{e}_7 h_1)$,
(12) $e_{12} \otimes (\bar{e}_1 h_4 h_1 + \bar{e}_{15} h_1^2) \to e_{10} \otimes (\bar{e}_1 h_4 h_1^2 + \bar{e}_{15} h_1^3)$,
(13) $e_{12} \otimes \bar{e}_3 h_3^2 \to e_7 \otimes \alpha_{21}$,
(14) $e_{11} \otimes (\bar{e}_3 h_4 h_0 + \bar{e}_{15} h_2 h_0) \leftarrow e_{12} \otimes (\bar{e}_3 h_4 + \bar{e}_{15} h_2)$,
(15) $e_{10} \otimes \bar{e}_1 h_4 h_2 \to e_6 \otimes \bar{e}_1 h_4 h_2^2$,
(16) $e_8 \otimes (\bar{e}_3 h_4 h_2 + \bar{e}_{15} h_2^2) \to e_4 \otimes \bar{e}_3 h_4 h_2^2$,
(17) $e_7 \otimes \bar{e}_7 h_4 h_0 \leftarrow e_8 \otimes \bar{e}_7 h_4$,
(18) $e_6 \otimes \bar{e}_7 h_4 h_1 \to 0$,
(19) $e_{19} \otimes \bar{e}_3 h_3 h_0^2 \leftarrow e_{20} \otimes \bar{e}_3 h_3 h_0$,
(20) $e_{13} \otimes \alpha_{16} \to e_7 \otimes (\bar{e}_3 f_0 + \bar{e}_1 g_1)$,
(21) $e_{11} \otimes (\bar{e}_3 h_4 h_0^2 + \bar{e}_{15} h_2 h_0^2) \leftarrow e_{12} \otimes (\bar{e}_3 h_4 h_0 + \bar{e}_{15} h_2 h_0)$,
(22) $e_9 \otimes \bar{e}_1 c_1 \to e_5 \otimes \bar{e}_1 c_1 h_2$,
(23) $e_8 \otimes \alpha_{21} \to e_4 \otimes \alpha_{21} h_2$,
(24) $e_7 \otimes \bar{e}_1 h_4 h_2^2 \leftarrow e_{11} \otimes \bar{e}_1 h_4 h_2$,
(25) $e_7 \otimes \bar{e}_7 h_4 h_0^2 \leftarrow e_8 \otimes \bar{e}_7 h_4 h_0$,
(26) $e_6 \otimes (\hat{c}_0 h_4 + \bar{e}_{15} c_0) \leftarrow e_{21} \otimes (\bar{e}_1 h_1 h_3 + \bar{e}_7 h_1^2)$,
(27) $e_5 \otimes \bar{e}_3 h_4 h_2^2 \leftarrow e_9 \otimes (\bar{e}_3 h_4 h_2 + \bar{e}_7 h_3^2)$,
(28) $e_{10} \otimes \bar{e}_1 f_0 \to e_4 \otimes \bar{e}_1 h_2 g_1$,
(29) $e_8 \otimes (\bar{e}_1 g_1 + \bar{e}_3 f_0) \to e_6 \otimes \bar{e}_1 g_1 h_1$,
(30) $e_7 \otimes \bar{e}_7 h_4 h_0^3 \leftarrow e_8 \otimes \bar{e}_7 h_4 h_0^2$,
(31) $e_6 \otimes \bar{e}_1 c_1 h_2 \leftarrow e_{10} \otimes \bar{e}_1 c_1$,
(32) $e_5 \otimes \alpha_{21} h_2 \leftarrow e_9 \otimes \alpha_{21}$,
(33) $e_7 \otimes \bar{e}_1 g_1 h_1 \leftarrow e_9 \otimes \bar{e}_1 g_1$,
(34) $e_5 \otimes \bar{e}_1 g_1 h_2 \leftarrow e_{11} \otimes \bar{e}_1 f_0$.

From these differentials we see only $e_7 \otimes \bar{e}_7 h_4$ and $e_6 \otimes \bar{e}_7 h_4 h_1$ persist to $\bar{E}_\infty^{*,*,*}$. This proves (6.4)(2) since $\overline{e_7 \otimes e_7} h_4$ is represented by $e_7 \otimes \bar{e}_7 h_4$ and $\overline{e_7 \otimes e_7} h_4 h_0$ is represented by $e_6 \otimes \bar{e}_7 h_4 h_1$ (see (2.6)(2)*, (6)*).

References

1. J.F. Adams, *On the structure and applications of the Steenrod algebra*, Comm. Math. Helv. **32** (1958), 180-214.
2. J.F. Adams, *On the non-existence of elements of Hopf invariant one*, Ann. of Math. **72** (1960), 20-104.
3. J.F. Adams, *On the groups $J(X) - I$*, Topology **2** (1963), 181-195.

4. M.G. Barratt, J.D.S. Jones and M.E. Mahowald, *Relations amongst Toda brackets and the Kervaire invariant in dimension 62*, J. London Math. Soc. **30** (1984), 533-550.
5. A.K. Bousfield and E.B. Curtis, *A spectral sequence for the homotopy of nice spaces*, J. Trans. AMS. **151** (1970), 457-479.
6. E.H. Brown, Jr. and S. Gitler, *A spectrum whose cohomology is a certain cyclic module over the Steenrod algebra*, Topology **12** (1973), 283-295.
7. R.L. Cohen, W.H. Lin, and M.E. Mahowald, *The Adams spectral sequence of the real projective spaces*, Pacific J. Math. **134** (1988), 27-55.
8. D.S. Kahn and S.B. Priddy, *Applications of the transfer to the stable homotopy theory*, Bull. Amer. Math. Soc. **78** (1972), 981-991.
9. W.H. Lin, *Algebraic Kahn-Priddy theorem*, Pacific J. Math **96** (1981), 435-455.
10. W.H. Lin, *Some differentials in Adams spectral sequence for spheres, To appear in Trans. AMS.*.
11. A. Liulevicius, *The factorization of cyclic reduced powers by secondary cohomology operations*, Mem. AMS. **42** (1962).
12. M.E. Mahowald, *The metastable homotopy of S^n*, Memo. AMS. **72** (1967).
13. M.E. Mahowald and M.C. Tangora, *Some differential, in the Adams spectral sequence*, Topology **6** (1967), 349-369.
14. M.E. Mahowald, *A new infinite family in ${}_2\pi_*^s$*, Topology **16** (1977), 249-256.
15. J.P. May, *The cohomology of restricted Lie algebras and Hopf algebras with applications to the Steenrod algebra* (1964), Ph.D. Thesis, Princeton Univ.
16. N.Minami, *The Kervaire invariant one element and the double transfer*, Topology **34** (1995), 481-488.
17. M.C. Tangora, *On the cohomology of Steenrod algebra*, Math. Z. **116** (1970), 18-64.
18. J.S.P. Wang, *On the cohomology of the mod 2 Steenrod algebra and the nonexistence of element of Hopf invariant one III*, J. Math. **11** (1967), 480-490.
19. G.W. Whitehead, *Recent advances in homotopy theory*, Regional Conference Series in Mathematics **5** (1970).

Department of Mathematics
National Tsing Hua University
Hsinchu, Taiwan 30043

Department of Mathematics
Northwestern University
Evanston, IL 60208-0001
U.S.A.

Contemporary Mathematics
Volume **220**, 1998

Applications of the Minami-Webb Formula to Splitting Classifying Spaces

John Martino and Stewart Priddy

We obtain a simplification of the general theory of stable splittings of classifying spaces by showing that the summands occurring in the Minami-Webb formula are determined by an eigenvalue of a certain matrix. Applications include new splittings and a rederivation of several known results as well as groups whose cohomologies are isomorphic as rings and as modules over the Steenrod algebra but whose cohomologies are not isomorphic as algebras over the Steenrod algebra.

1. Introduction.

Let G be a finite group. It is now well understood that the indecomposable stable summands of the classifying space BG, localized at a prime p, correspond to certain simple modules for $\mathbf{F}_p Out(Q)$ where Q ranges over subgroups of a Sylow p-subgroup of G. As one would expect the usefulness of this theory [**BF, MP1, MP2**] depends on the complexity of G. The purpose of this note is to provide some new techniques which simplify the task of decomposing BG. One by-product is a strikingly simple example showing how the stable and unstable homotopy types of classifying spaces can subtly differ: groups G, H are given such that (1) $H^*(BG)$ and $H^*(BH)$ are isomorphic as algebras, (2) BG and BH are stably equivalent and so $H^*(BG)$ and $H^*(BH)$ are isomorphic as modules over the Steenrod algebra, (3) the cohomologies are not isomorphic as algebras over the Steenrod algebra and so BG and BH have different unstable homotopy types, completed at p. As another application we give a complete stable decomposition of the classifying space of the Borel subgroup of $SL_3(\mathbf{F}_5)$ at $p = 5$. Various known decompositions are rederived as well.

Our approach uses the Minami-Webb formula. If G acts on a geometry Δ (e.g., a simplicial complex) and the geometry is locally contractible (i.e., for every p-subgroup P of G, Δ^P is contractible), then the geometry can be used to study the cohomology of G. For example, see the work of D. Quillen [**Q**] where he uses

---

1991 *Mathematics Subject Classification.* Primary 55R35; Secondary 20J06, 55P42.

the poset $\mathcal{A}_p(G)$ of non-trivial p-subgroups. By using smaller geometries homotopy equivalent to $\mathcal{A}_p(G)$ one is led to reduction formulas for group cohomology. One such found by P. Webb [**Wb**] is

$$H^*(BG;\mathbf{F}_p) = \sum_{H\in\mathcal{C}_p(G)} \frac{f(H)}{[G:H]} H^*(BH;\mathbf{F}_p),$$

where the sum is in a Grothendieck group completion with addition the direct sum,

$$\mathcal{C}_p(G) = \{H \in G \mid H/O_p(H) \text{ is a cyclic } p'\text{-group}\}$$

(the elements of $\mathcal{C}_p(G)$ are referred to as cyclic mod-p groups), and f is a Möbius function defined by $\sum_{K\geq H} f(K) = 1$ for all $H \in \mathcal{C}_p(G)$. The Möbius function f only takes on non-zero values on maximal cyclic mod-p subgroups and intersections thereof. In [**Mi**] N. Minami showed that the sum actually occurs in the Grothendieck completion of p-local topological spaces with addition the wedge product,

$$BG_{(p)} = \bigvee_{H\in\mathcal{C}_p(G)} \frac{f(H)}{[G:H]} BH_{(p)}.$$

We will refer to this formula as the Minami-Webb formula. Suitably interpreted the Minami-Webb formula is a unique expression of $BG_{(p)}$ (see [**MP2**]). If the sums are taken over conjugacy classes in $\mathcal{C}_p(G)$ then the index $[G : H]$ is replaced by $[N_G(H) : H]$, but the Möbius function f is still defined using the poset of all subgroups in $\mathcal{C}_p(G)$.

In this paper we give some theorems and propositions which facilitate the use of the Minami-Webb formula. To do this we show how the general theory simplifies when applied to the terms of the formula. In fact an eigenvalue of a certain matrix may be used to compute the stable homotopy type of $BH_{(p)}$ where H is a cyclic mod-p group (Theorem 3.1). We use the Minami-Webb formula to derive some interesting facts about the 2-localization of $B\Sigma_6$. In the process we give the example mentioned above of two finite groups whose cohomologies are isomorphic as rings and as modules over the Steenrod algebra, but whose cohomologies are not isomorphic as algebras over the Steenrod algebra. We also give a topological proof of a result on Swan groups originally proved using group theoretic methods.

The paper is organized into the following sections. In section 2 we collect a few preliminaries. In section 3 we show how eigentheory can be used to analyze cyclic mod-p groups. In section 4 we use the Minami-Webb formula and eigentheory to analyze $B\Sigma_6$ localized at 2. Section 5 contains two examples illustrating how the eigentheory and the Minami-Webb formula may be used to compute stable summands. Section 6 is devoted to using stable splitting theory to show the metacyclic group $M_3(p)$ is a Swan group.

In the following sections all groups are finite and all spaces and spectra are localized at a prime p.

2. Preliminaries.

For the convenience of the reader we recall some background material on the general theory of stable splittings.

We refer the reader to [**MP2**] for a discussion of the correspondence between the original summands of BP and the simple modules of $\mathbf{F}_pOut(P)$. An original summand (a.k.a. dominant summand) of BP is a summand that does not occur as the summand of the classifying space of any proper subgroup of P. Recall $\{BP, BP\}$ is the ring of stable self-maps of BP and $J(P)$ is the ideal of all stable maps that factor through the classifying spaces of groups of order less than $|P|$. The quotient

$$\{BP, BP\}/J(P) \otimes \mathbf{F}_p \approx \mathbf{F}_pOut(P),$$

where $Out(P)$ is the outer automorphism group of P.

A related concept to an original summand is that of a linked summand. A summand X is *linked* in BQ if, for the corresponding idempotent $e \in \{BQ, BQ\}$, $e \notin \overline{J}_\tau(Q)$, where $\overline{J}_\tau(Q) \subseteq \{BQ, BQ\}$ is the ideal generated by maps of the form $BQ \xrightarrow{tr} BR \xrightarrow{Bh} BQ$ where $R < Q$. For $\tilde{e} \in \mathbf{F}_pOut(Q)$, if e is a summand of $\tilde{e}$ then we say that X is linked to the corresponding original summand in BQ. As with original summands each linked summand is in one-to-one correspondence with a simple module of the ring $\mathcal{R}(Q) = (\{BQ, BQ\} \otimes \mathbf{F}_p)/\overline{J}_\tau(Q)$. Note $\mathbf{F}_pOut(Q) \subseteq \mathcal{R}(Q)$ and there exists an ideal $J \subseteq \mathcal{R}(Q)$ such that $\mathbf{F}_pOut(Q) \approx \mathcal{R}(Q)/J$. Therefore, the simple modules of $\mathbf{F}_pOut(Q)$ are simple modules of $\mathcal{R}(Q)$ where J acts as zero.

The situation is particularly nice when P is an abelian p-group. This case was extensively analyzed by J. Harris and N. Kuhn [**HK**]. The transfer map plays no role in splitting BP. All the summands either originate in or are linked in BP. Furthermore, a simple $\mathcal{R}(P)$ module restricts to a simple $\mathbf{F}_pAut(P)$ module. Thus for a finite group G with an abelian Sylow p-subgroup, the multiplicity of a summand Y linked to an original summand X is exactly the same as the multiplicity of X [**MP2**]. For example, $L(2)$ is linked to $L(3)$ in $B(\mathbf{Z}/2)^3$, thus if a group G with Sylow 2-subgroup isomorphic to $(\mathbf{Z}/2)^3$ has two copies of $L(3)$ in BG then there are exactly two copies of $L(2)$ is BG. Since the transfer plays no role, the normalizer of a Sylow p-subgroup determines the classifying space BG. This gives a topological explanation of the fact, first observed by R. Swan [**Sw**], that the normalizer controls fusion in G.

We recall that a p-group P is called a *Swan group* if $H^*(BG) \approx H^*(N_G(P))$ for every finite group G with P as Sylow p-subgroup. Hence, abelian groups are Swan groups. The reader should refer to [**MP3, P**] for more explanation of Swan groups.

In general, the p-local structure of a group may be used to determine its classifying space.

THEOREM 2.1 [**MP2**]. *Let G be a finite group and let X an indecomposable summand of BP, P a Sylow p-subgroup of G. Then the number of summands stably homotopy equivalent to X in BG is* *where i ranges over the conjugacy classes Q_i of p-subgroups in which X is linked in BQ_i, M_i is the corresponding simple module in $\mathcal{R}(Q_i)$, k is the splitting field for M_i (independent of Q_i), and $\overline{W}_i = \sum w, w \in N_G(Q_i)/Q_i$*

REMARK 1. If Y is linked in BR and originates in BQ, $Q \leq R$, then Q is a retract of R [**MP2**].

REMARK 2. The element $\overline{W}_i \in \mathbf{F}_p Out(Q_i)$ equals $BQ_i \xrightarrow{Bincl} BG \xrightarrow{tr} BQ_i$ mod $J(Q_i)$.

We now describe the Minami-Webb formula [**Mi**] in more detail: let $\mathcal{C}_p(G)$ be the set of all cyclic mod-p subgroups of G. Then

$$BG = \bigvee_{(H)} \frac{f(H)}{[N_G(H) : H]} BH$$

stably, where H runs over the conjugacy classes of $\mathcal{C}_p(G)$ and $f : \mathcal{C}_p(G) \to \mathbf{Z}$ is the Möbius function given by

$$\sum_{J \leq K} f(K) = 1$$

for $J, K \in \mathcal{C}_p(G)$. This expression is unique [**MP2**]. The following lemma exhibits certain restrictions on f.

PROPOSITION 2.2. *Let $Q \leq G$ be p-subgroup and set $S = \sum \frac{f(H)}{[N_G(H):H]}$ where H runs over the conjugacy classes of $\mathcal{C}_p(G)$ with Sylow p-subgroup isomorphic to Q. Then*

1. *$S = 1$ if $Q \approx P$.*
2. *$S = 0$ if Q is not a proper retract of a subgroup conjugate to P.*

PROOF. The principal original summand $PS(Q)$ of BQ is a summand of multiplicity one in BK for every group K with Sylow p-subgroup isomorphic to Q. Thus if $Q \approx P$, (1) follows immediately. Under the hypothesis in (2) that Q is not a subretract the principal summand $PS(Q)$ is not a summand of any classifying space with Sylow p-subgroup of order greater than $|Q|$ [**MP2; Cor. 4.4**]. Hence $S = 0$ also follows immediately. □

REMARK. J. Thévenaz has shown [**Th**] the hypothesis that Q is not a subretract is unnecessary.

3. Eigenvalue One.

In this section we describe a simple way to analyze the classifying space of a cyclic mod-p group.

THEOREM 3.1. *Let $H = P \rtimes C$ be a cyclic mod-p group, X an original summand of BP, and M the corresponding simple $\mathbf{F}_p Out(P)$ module. If $x \in C$, a generator of C, acts as the matrix $\mathbf{A}$ on the module M, then*

1. *X is a summand of BH if and only if 1 is an eigenvalue of $\mathbf{A}$ and*
2. *the multiplicity of X in BH is the dimension of the eigenspace for 1 over $End(M)$.*

PROOF. Without loss of generality we may assume that C acts faithfully on P (i.e., $C \leq Aut(P)$). Let $n = |C|$. Since $\mathbf{A}^n = \mathbf{I}$, the eigenvalues of $\mathbf{A}$ are nth roots of unity in $\overline{\mathbf{F}}_p$, the algebraic closure of $\mathbf{F}_p$. The matrix $\mathbf{A}$ is diagonalizable, since if there are 1s off the diagonal in the Jordan canonical form of $\mathbf{A}$ then $\mathbf{A}^n$ has $n\lambda^{n-1}$ off the diagonal for some nth root of unity λ and $n \not\equiv 0 \bmod p$. Hence the element $\overline{C} = 1 + x + x^2 + \cdots + x^{n-1}$ acts as

$$\mathbf{I} + \mathbf{A} + \mathbf{A}^2 + \cdots + \mathbf{A}^{n-1} = n \begin{bmatrix} \mathbf{I}_d & 0 \\ 0 & 0 \end{bmatrix}$$

where d is the dimension of the eigenspace for 1. This follows from the fact that if λ is a nontrivial nth root of unity then

$$1 + \lambda + \lambda^2 + \cdots + \lambda^{n-1} = 0.$$

Therefore, $\dim \overline{C}M = d$ completing the proof. □

4. On the Cohomology and Stable Type of $B\Sigma_6$.

In this section we use the Minami-Webb formula to study $B\Sigma_6$. All spaces and spectra are localized at the prime $p = 2$.

Let $std_n : \Sigma_n \to O(n)$ be the standard orthogonal representation of the symmetric group Σ_n. We recall that $H^*(B\Sigma_4; \mathbf{F}_2) \approx \mathbf{F}_2[a, b, c]/ac$ where

$$a = w_3(std_4), b = w_2(std_4), c = w_1(std_4).$$

The 2-local stable type is given by

$$B\Sigma_4 \cong BPSL_2(F) \vee L(2) \vee B\mathbf{Z}/2$$

where $F = \mathbf{F}_q$ is a finite field with q odd such that $4 = \nu_2(q^2 - 1)$. In [**MtP**] this result is only stated for $q = 7$ but the result holds since this condition guarantees that the dihedral group D_8 is a Sylow 2-subgroup of $PSL_2(F)$.

PROPOSITION 4.1. *Let G be a finite group with Minami-Webb formula*

If P is a finite p-group, then $G \times P$ has Minami-Webb formula

$$B(G \times P) \cong \bigvee \alpha_i B(H_i \times P).$$

PROOF. The Möbius function f in the Minami-Webb formula is only non-zero on maximal and intersections of maximal cyclic mod-p subgroup of $G \times P$. Clearly, every maximal cyclic mod-p subgroup is of the form $H \times P$. Hence, every intersection of maximal cyclic mod-p subgroups of that form. Therefore, G and $G \times P$ have the same lattice of maximal cyclic mod-p subgroups and intersections thereof. The corresponding values of the Möbius functions and indices for G and $G \times P$ are thus the same. □

THEOREM 4.2. *Stably localized at 2,*
1. $B\Sigma_6 \cong B(PSL_2(F) \times \mathbf{Z}/2)$.
2. $B\Sigma_6 \cong B(D_8 \times \mathbf{Z}/2) + B(A_4 \times \mathbf{Z}/2) - B(\mathbf{Z}/2)^3$.
3. *A complete stable decomposition is given by*

$$B\Sigma_6 \cong PS(D_8 \times \mathbf{Z}/2) \vee BPSL_2(F) \vee 2L(3) \vee 2L(2) \vee B(\mathbf{Z}/2)$$

where $PS(D_8 \times \mathbf{Z}/2)$ is the principal original summand of $B(D_8 \times \mathbf{Z}/2)$.

THEOREM 4.3.
1. *The cohomology of $B\Sigma_6$ is*

$$H^*(B\Sigma_6; \mathbf{F}_2) \approx \mathbf{F}_2[w_1, w_2, w_3, u_3]/({u_3}^2 + u_3(w_3 + w_2 w_1))$$

where $w_i = w_i(std_6)$ and $i^*(u_3) = a$ under the restriction

$$i^* : H^*(B\Sigma_6; \mathbf{F}_2) \to H^*(B(\Sigma_4 \times \Sigma_2); \mathbf{F}_2).$$

2. *As algebras over $\mathbf{F}_2$ (but not as algebras over the Steenrod algebra $\mathcal{A}$)*

$$H^*(B\Sigma_6; \mathbf{F}_2) \approx H^*(B(PSL_2(F) \times \mathbf{Z}/2); \mathbf{F}_2).$$

REMARK. Σ_6 and $PSL_2(F) \times \mathbf{Z}/2$ are examples of groups whose cohomologies are isomorphic as algebras over $\mathbf{F}_2$, but not as algebras over $\mathcal{A}$. However, their $\mathcal{A}$ module structures with higher Bocksteins agree since their classifying spaces are equivalent at 2. The cohomology ring of Σ_6 was computed by A. Adem, J. Maginnis, and J. Milgram [**AMM**]. We give another derivation suitable for our purposes.

PROOF OF THEOREM 4.2. Part (2) of Theorem 4.2 is just the Minami-Webb formula for Σ_6 [**Wb; 7.5**].

For part (1) we apply Proposition 5.1 to

$$BPSL_2(F) = BD_8 + BA_4 - B(\mathbf{Z}/2)^2,$$

and observe that $B\Sigma_6$ and $B(PSL_2(F) \times \mathbf{Z}/2)$ have identical Minami-Webb formulas.

For part (3), complete stable splittings for $B(\mathbf{Z}/2)^3$ and $B(D_8 \times \mathbf{Z}/2)$ are found in [**Mt**] and [**MP1**] respectively.

$$\begin{aligned} B(\mathbf{Z}/2)^3 \cong\ & B[(\mathbf{Z}/2)^3 \rtimes (\mathbf{Z}/7 \rtimes \mathbf{Z}/3)] \vee 3X_{31} \vee 3B\mathbf{Z}/2 \vee \\ & 3X_{32} \vee 3BA_4 \vee 8L(3) \vee 8L(2). \\ B(D_8 \times \mathbf{Z}/2) \cong\ & PS(D_8 \times \mathbf{Z}/2) \vee 2X_{31} \vee 3B\mathbf{Z}/2 \vee 2X_{32} \vee 2BA_4 \vee BPSL_2(F) \\ & \vee 8L(3) \vee 8L(2). \end{aligned}$$

For the remaining group $A_4 \times \mathbf{Z}/2$ in the Minami-Webb formula in part (2), a stable decomposition can be obtained from [**CK; Th. 7.1**]. Alternatively we derive this splitting from Theorem 3.1 using the fact that $A_4 \times \mathbf{Z}/2 = [(\mathbf{Z}/2)^2 \rtimes \mathbf{Z}/3] \times \mathbf{Z}/2$, is a cyclic mod-2 group. A Sylow 2-subgroup of $A_4 \times \mathbf{Z}/2$ is isomorphic to $(\mathbf{Z}/2)^3$, and up to isomorphism there are four simple modules of $\mathbf{F}_2 GL_3(\mathbf{F}_2)$. These are the trivial module 1, the three-dimensional standard module V, its dual $V^{\#}$, and the eight-dimensional Steinberg module St. The corresponding original summands are $B[(\mathbf{Z}/2)^3 \rtimes (\mathbf{Z}/7 \rtimes \mathbf{Z}/3)]$, X_{32}, X_{31}, and $L(3)$ respectively.

Let $x \in A_4$ represent a generator of $\mathbf{Z}/3$. The element x acts as the matrix

$$\begin{bmatrix} 0 & 1 & 0 \\ 1 & 1 & 0 \\ 0 & 0 & 1 \end{bmatrix},$$

which has a one-dimensional eigenspace for the value 1. Hence X_{32} and by duality X_{31} have multiplicity one in $B(A_4 \times \mathbf{Z}/2)$. The Steinberg module St is the quotient of $H^4(B(\mathbf{Z}/2)^3; \mathbf{F}_2)$ by image of the Steenrod operations. Here $H^*(B(\mathbf{Z}/2)^3; \mathbf{F}_2) = \mathbf{F}_2[u, v, w]$ for one dimensional classes u, v, w. The action of x on generators is given by $u \mapsto v$, $v \mapsto u + v$, $u + v \mapsto u$. Thus an easy calculation shows that on St, x acts as the matrix

$$\mathbf{A} = \begin{bmatrix} 1 & 1 & 0 & 0 & 0 & 0 & 0 & 0 \\ 1 & 0 & 0 & 0 & 0 & 0 & 0 & 0 \\ 0 & 0 & 1 & 0 & 1 & 0 & 1 & 1 \\ 0 & 0 & 0 & 1 & 0 & 1 & 0 & 0 \\ 0 & 0 & 1 & 0 & 0 & 0 & 0 & 0 \\ 0 & 0 & 0 & 1 & 0 & 0 & 0 & 0 \\ 0 & 0 & 1 & 0 & 0 & 0 & 0 & 1 \\ 0 & 0 & 1 & 0 & 0 & 0 & 1 & 0 \end{bmatrix}.$$

Hence,

$$\mathbf{A} - \mathbf{I} = \begin{bmatrix} 0 & 1 & 0 & 0 & 0 & 0 & 0 & 0 \\ 1 & 1 & 0 & 0 & 0 & 0 & 0 & 0 \\ 0 & 0 & 0 & 0 & 1 & 0 & 1 & 1 \\ 0 & 0 & 0 & 0 & 0 & 1 & 0 & 0 \\ 0 & 0 & 1 & 0 & 1 & 0 & 0 & 0 \\ 0 & 0 & 0 & 1 & 0 & 1 & 0 & 0 \\ 0 & 0 & 1 & 0 & 0 & 0 & 1 & 1 \\ 0 & 0 & 1 & 0 & 0 & 0 & 1 & 1 \end{bmatrix}.$$

The first six columns of $\mathbf{A} - \mathbf{I}$ are clearly linearly independent, while the last two columns are each the sum of the third and fifth columns. So the eigenspace for 1 is of the non-original summands are accounted for by linkage. The summand $B\mathbf{Z}/2$ is linked to X_{31}, BA_4 is linked to X_{32}, and $L(2)$ is linked to $L(3)$. Hence,

$$B(A_4 \times \mathbf{Z}/2) \cong B[(\mathbf{Z}/2)^3 \rtimes (\mathbf{Z}/7 \rtimes \mathbf{Z}/3)] \vee X_{31} \vee B\mathbf{Z}/2 \vee X_{32} \vee BA_4 \vee 2L(3) \vee 2L(2).$$

Part (3) now follows from the Minami-Webb formula in part (2). □

PROOF OF THEOREM 4.3.

(1) We begin by showing the existence of a unique element u_3 which restricts in the required manner. Since $[\Sigma_6 : \Sigma_4 \times \Sigma_2]$ is odd, the restriction i^* is injective. Also $H^*(B\Sigma_4; \mathbf{F}_2)$ is detected by the elementary abelian 2-subgroups $\Delta_2 = \langle (13)(24), (12)(34) \rangle$ and $T = \langle (12), (34) \rangle$. More explicitly, if $i_\Delta : \Delta_2 \to \Sigma_4$ and $i_T : T \to \Sigma_4$ are the inclusions then it is well known that

$$i_\Delta{}^*(a) = x^2y + xy^2, i_\Delta{}^*(b) = x^2 + xy + y^2, i_\Delta{}^*(c) = 0$$

and

$$i_T{}^*(a) = 0, i_T{}^*(b) = xy, i_T{}^*(c) = x + y.$$

These facts follow since i_Δ is the regular representation while $std_4 \circ i_T$ is conjugate to the inclusion $O(1) \times O(1) \to O(2) \to O(4)$. Now we define $u_3 = tr^*(x^2y)$ where $tr^* : H^*(BE; \mathbf{F}_2) \to H^*(B\Sigma_6; \mathbf{F}_2)$ is the transfer and $E = \Delta_2 \times \Sigma_2$. Since $W = N_{\Sigma_6}(E)/E = GL_2(\mathbf{F}_2)$ we have

$$(i_\Delta \times \Sigma_2)^* tr^* = \overline{W}$$

where $\overline{W} = \Sigma c_w, w \in W$. Similarly,

$$(i_T \times \Sigma_2)^* tr^* = 0$$

([**MtP; Prop. 1.4**]). By an easy computation $\overline{W}x^2y = x^2y + xy^2$ and $a \in H^3(B(\Sigma_4 \times \Sigma_2); \mathbf{F}_2)$ is the only element which restricts to $x^2y + xy^2$ in $H^3(BE; \mathbf{F}_2)$ and to 0 in $H^3(B(T \times \Sigma_2); \mathbf{F}_2)$. Thus $i^*(u_3) = a$.

By the Whitney formula for Stiefel-Whitney classes we have

$$i^*(w_3) = a + bz, i^*(w_2) = b + cz, i^*(w_1) = c + z.$$

Computing we find $i^*(u_3{}^2 + u_3(w_3 + w_2 \cdot w_1)) = 0$ which gives the desired relation. Furthermore, it is well known from Nakaoka's computations or more explicitly by Webb that the Poincaré series of $H^*(B\Sigma_6; \mathbf{F}_2)$ is $(1 + t^3)/(1 - t)(1 - t^2)(1 - t^3)$. Thus this is the only relation. For use below we note $Sq^2(w_3) = w_3w_2 + u_3w_1{}^2$.

(2) We recall from [**MtP; Th. 2.6**] the cohomology algebra

$$H^*(BPSL_2(F); \mathbf{F}_2) \approx \mathbf{F}_2[v_2, v_3, \overline{v}_3]/(\overline{v}_3{}^2 + v_3\overline{v}_3)$$

with Steenrod operations

$$Sq^1(v_2) = v_3, Sq^2(v_3) = v_2v_3, Sq^1(\overline{v}_3) = 0, Sq^2(\overline{v}_3) = v_2\overline{v}_3.$$

An isomorphism of algebras $f : H^*(B(PSL_2(F) \times \mathbf{Z}/2); \mathbf{F}_2) \to H^*(B\Sigma_6; \mathbf{F}_2)$ may be specified by

$$f(\overline{v}_3) = u_3, f(v_3) = w_3 + w_2 \cdot w_1, f(v_2) = w_2, f(z) = w_1.$$

It is easily checked that f preserves the required relation.

Any other isomorphism g must satisfy $g(z) = w_1$. Thus by the addition of z^2 we may assume the generator v_2 is chosen so that $g(v_2) = w_2$ and $Sq^1(v_2) = v_3$ as before. If g preserves Steenrod operations then $g(v_3) = w_3 + w_2w_1$ by the Wu formula. However $g(v_2v_3) = g(Sq^2(v_3)) = Sq^2g(v_3) = Sq^2(w_3 + w_2w_1) = w_3w_2+u_3{w_1}^2+{w_2}^2w_1+w_3{w_1}^2+w_2{w_1}^3$ while $g(v_2v_3) = g(v_2)g(v_3) = w_2(w_3+w_2w_1)$. Thus an algebra isomorphism can not preserve Steenrod operations.

5. Further Examples.

In this section we give two examples showing how Theorem 3.1 and the Minami-Webb formula may be used in conjunction. All spaces and spectra are localized at the prime p.

Let $p = 5$ and consider the Borel subgroup B_3 of $SL_3(\mathbf{F}_5)$ consisting of upper triangular matrices over $\mathbf{F}_5$. Then $B_3 = U_3 \rtimes D$ where U_3 is the subgroup of upper triangular matrices with ones down the diagonal and $D = \mathbf{Z}/4 \times \mathbf{Z}/4$ is the group of matrices generated by

$$h_1 = \begin{bmatrix} 2 & 0 & 0 \\ 0 & 3 & 0 \\ 0 & 0 & 1 \end{bmatrix}, h_2 = \begin{bmatrix} 1 & 0 & 0 \\ 0 & 2 & 0 \\ 0 & 0 & 3 \end{bmatrix}$$

Inspection shows there are six maximal mod 5 cyclic subgroups $H_1, ..., H_6$ each having the form $U_3 \rtimes \mathbf{Z}/4$ where the $\mathbf{Z}/4$ is generated by $h_1, h_2, h_1h_2, {h_1}^2h_2, h_1{h_2}^2, h_1{h_2}^3$ respectively. These intersect in three subgroups K_1, K_2, K_3 of the form $U_3 \rtimes \mathbf{Z}/2$ where $\mathbf{Z}/2$ is generated by ${h_1}^2, {h_2}^2, {h_1}^2{h_2}^2$ respectively and lastly U_3 itself. The resulting Minami-Webb formula is

$$BB_3 = \sum_{i=1}^{6} \frac{1}{4} BH_i - \sum_{j=1}^{3} \frac{1}{8} BK_j - \frac{1}{8} BU_3 \tag{1}$$

From this one can obtain a complete decomposition using Theorem 3.1. We restrict ourselves to sketching the derivation of the multiplicity of the original summands, i.e., those originating in U_3. These summands, denoted by $X_{q,k}$, $0 \leq q \leq 4, 0 \leq k \leq 3$, are described in [**DP**]. They correspond to the simple $\mathbf{F}_5Out(U_3)$ modules. These modules in turn correspond to simple $\mathbf{F}_5GL_2(\mathbf{F}_5)$ modules $M_{q,k}$ under the map induced by projection $\pi : U_3 \longrightarrow U_3/Z$ where $Z \approx \mathbf{Z}/3$ is the center ([**DP; Cor. 1.5**]). If V is the standard module for $GL_2(\mathbf{F}_5)$ then $M_{q,k} = S^q(V) \otimes (det_2)^k$ where $S^q(V)$ is the q-fold symmetric product and det_2 is the determinant representation. Then

$$\pi(h_1) = \begin{bmatrix} -1 & 0 \\ 0 & 3 \end{bmatrix}, \pi(h_2) = \begin{bmatrix} 3 & 0 \\ 0 & -1 \end{bmatrix}$$

Applying Theorem 3.1 and (1) one obtains the following multiplicities for the $X_{q,k}$

Summand	Multiplicity
$X_{0,k}$	1
$X_{1,k}$	0
$X_{2,k}$	0
$X_{3,k}$	0
$X_{4,k}$	2

As a second example let $p = 2$ and define $V = (\mathbf{Z}/2)^3$ and $G_{168} = (\mathbf{Z}/2)^3 \rtimes (\mathbf{Z}/7 \rtimes \mathbf{Z}/3)$, where $\mathbf{Z}/7$ acts as $\mathbf{F}_8^*$ under the identification $V \approx \mathbf{F}_8$ and $\mathbf{Z}/3$ is generated by the Frobenius automorphism $x \mapsto x^2$. By the work of S. Mitchell [**Mt**] we have the stable splitting

$$B(\mathbf{Z}/2)^3 \cong BG_{168} \vee 3X_{31} \vee 3B\mathbf{Z}/2 \vee 3X_{32} \vee 3BA_4 \vee 8L(3) \vee 8L(2).$$

into indecomposable summands.

Applying the Minami-Webb formula we get

$$BG_{168} = B(V \rtimes \mathbf{Z}/3) + \frac{1}{3}B(V \rtimes \mathbf{Z}/7) - \frac{1}{3}BV.$$

Decompositions of $B(V \rtimes \mathbf{Z}/3)$ and $B(V \rtimes \mathbf{Z}/7)$ were given by Carlisle and Kuhn ([**CK; Th. 7.1**]). We shall rederive these results using Theorem 3.1. By the primitive normal basis theorem (see [**Da**]) it is possible to choose an element $\omega \in \mathbf{F}_8$ so that ω generates $\mathbf{F}_8^*$ and $\{\omega, \omega^2, \omega^4\}$ forms a basis for $\mathbf{F}_8$ over $\mathbf{F}_2$. Furthermore, ω, which generates $\mathbf{Z}/7$, acts on the basis $\{1, \omega, \omega^2\}$ as the matrix

$$\mathbf{A} = \begin{bmatrix} 0 & 0 & 1 \\ 1 & 0 & 1 \\ 0 & 1 & 0 \end{bmatrix}$$

since $\omega^3 + \omega + 1 = 0$. The Frobenius automorphism, which generates $\mathbf{Z}/3$, acts on the basis $\{\omega, \omega^2, \omega^4\}$ as the matrix

$$\mathbf{B} = \begin{bmatrix} 0 & 0 & 1 \\ 1 & 0 & 0 \\ 0 & 1 & 0 \end{bmatrix}.$$

The value 1 is not an eigenvalue of $\mathbf{A}$, but it is an eigenvalue of $\mathbf{B}$ with a one-dimensional eigenspace. Hence, by duality and linkage we have

$$B(V \rtimes \mathbf{Z}/7) \cong BG_{168} \vee aL(3) \vee aL(2).$$
$$B(V \rtimes \mathbf{Z}/3) \cong BG_{168} \vee X_{31} \vee B\mathbf{Z}/2 \vee X_{32} \vee BA_4 \vee bL(3) \vee bL(2).$$

Theorem 3.1 can now be used on the Steinberg module St to determine the integer scalars a and b. But suppose that the Steinberg module is not readily available. We know that BG_{168} is indecomposable so the Minami-Webb formula tells us

$$b + \frac{1}{3}a - \frac{8}{3} = 0.$$

The only non-negative integer solutions to the equation are

$$\begin{array}{ll} a=8 & b=0 \\ a=5 & b=1 \\ a=2 & b=2. \end{array}$$

Let U_3 be the subgroup of upper triangular matrices in $GL_3(\mathbf{F}_2)$ with ones on the diagonal, and for $H \leq GL_3(\mathbf{F}_2)$ by $\overline{H}$ we mean the formal sum of elements in H. The Steinberg idempotent is $e = \overline{U}_3 \cdot \overline{\Sigma}_3$. Let $R = \mathbf{F}_2 GL_3(\mathbf{F}_2)$ then as a right R module $St = eR$. Let B be the cyclic subgroup of Σ_3 generated by $\mathbf{B}$ and let C be generated by an involution so that $\Sigma_3 = BC$. Then e and $e \cdot \overline{U}_3 \cdot \overline{B}$ are both elements of $St\overline{B}$. Since $e \cdot \overline{C} = 0$ and $e \cdot \overline{U}_3 \cdot \overline{B} \cdot \overline{C} = e^2 = e$, the two elements are not equal. Thus by Theorem 2.1, $b = dim(St\overline{B})$ is at least 2. Hence $a = 2$ and $b = 2$. Therefore, we have complete decompositions of $B(V \rtimes \mathbf{Z}/7)$ and $B(V \rtimes \mathbf{Z}/3)$.

6. $M_3(p)$ Is a Swan Group.

For a p-group P, a transfer summand X is a summand of BP where the idempotent $BP \longrightarrow X \longrightarrow BP$ involves a transfer map (i.e., the idempotent is in $\overline{J}_\tau(P)$). Transfer summands are summands that are linked in proper subgroups of P. In [**MP3**] we showed that for a p-group P if the classifying space has no transfer summands, then P is a Swan group.

Let $M_3(p)$ denote the metacyclic group of order p^3 and exponent p^2.

$$M_3(p) = \langle x, y | x^{p^2} = y^p = 1, yxy^{-1} = x^{p+1} \rangle.$$

For p odd, G. Glauberman has shown $M_3(p)$ is a Swan group [**MP3; Remark 3.6**]. His argument is group theoretic and is rather *ad hoc.* Our purpose is to derive this fact from a general result in the theory of stable splittings of classifying spaces. The group $M_3(p)$ is an example of a p-group with transfer summands, which is nevertheless a Swan group.

Let P be a fixed p-group and write $BP \cong X \vee Y$ where Y is the wedge sum of all transfer summands of BP. Thus X is the wedge sum of all summands linked in BP.

PROPOSITION 6.1. *If Y is a summand of BG for every group G with P as a Sylow p-subgroup then P is a Swan group.*

PROOF. Since BG is a summand of $BN_G(P)$ it suffices to show they have the same stable type. By Theorem 2.1, the normalizer $N_G(P)$ determines whether an indecomposable summand of X is in either BG or $BN_G(P)$. They only differ in transfer summands. However we are assuming Y is a summand of both BG and

Turning now to the proof that $P = M_3(p)$ is a Swan group for p odd, we first recall from [**D**]

$$BP \cong \bigvee_{i=0}^{p-2} X_i \vee \bigvee_{j=0}^{p-2} L(2,j) \vee \bigvee_{k=0}^{p-2} L(1,1,k)$$

where the X_i are original summands of BP. Each $L(2,j)$ is a transfer summand of $BM_3(p)$. By arguments of [**DMP**], $L(2,j)$ is a summand of BG. The fate of the summands $L(1,1,k)$ are controlled by linkage either to an X_i or an $L(2,j)$. So the transfer summands consist of the $L(2,j)$s and the $L(1,1,k)$s linked to them. Therefore by [**MP1; Prop. 4.9**] the sum of all the transfer summands is a summand of the classifying space of every group with a Sylow p-subgroup isomorphic to $M_3(p)$. Hence $M_3(p)$ is a Swan group by Proposition 6.1.

PROPOSITION 6.2. *For a finite group G with P a Sylow p-subgroup, $BG \cong BN_G(P)$ if and only if $f(K) = 0$ for all $K \in \mathcal{C}_p(G)$ with a Sylow p-subgroup not isomorphic to P.*

PROOF. It is immediate that the maximal cyclic mod-p subgroups of $N_G(P)$ have P as Sylow p-subgroup. Thus if $BG \cong BN_G(P)$ then it follows by the uniqueness of the Minami-Webb formula that each $f(K) = 0$ as desired. On the other hand, if each $f(K) = 0$ then all the maximal cyclic mod-p subgroups contain P, and G and $N_G(P)$ have the same lattice of maximal cyclic mod-p subgroups and intersections thereof. □

References

[AMM] A. Adem, J. Maginnis, and J. Milgram, *Symmetric invariants and the cohomology of groups*, Mathematische Annalen **287** (1990), 391–411.

[BF] D. Benson and M. Feshbach, *Stable splittings of classifying spaces of finite groups*, Topology **31** (1992), 157–176.

[CK] D. Carlisle and N. Kuhn, *Smash products of summands of $B(\mathbf{Z}/p)^n{}_+$*, Contemporary Mathematics **96** (1989), 87–102.

[Da] H. Davenport, *Bases for finite fields*, Journal of the London Mathematical Society **43** (1968), 21–39.

[D] J. Dietz, *Stable splittings of classifying spaces of metacyclic p-groups, p odd*, Journal of Pure and Applied Algebra **90** (1993), 115–136.

[DMP] J. Dietz, J. Martino, and S. Priddy, *Cohomology of groups with metacyclic Sylow p-subgroups*, Proceedings of the American Mathematical Society **124** (1996), 2261–2266.

[DP] J. Dietz and S. Priddy, *The stable homotopy type of rank two p-groups*, Homotopy and Its Applications, Contemporary Mathematics **188** (1995), 85–95.

[HK] J. Harris and N. Kuhn, *Stable decompositions of classifying spaces of finite abelian p-groups*, Mathematical Proceedings of the Cambridge Philosophical Society **103** (1988), 427–449.

[MP1] J. Martino and S. Priddy, *The complete stable splitting for the classifying space of a finite group*, Topology **31** (1992), 143–156.

[MP2] ———, *Stable homotopy classification of $BG_p^\wedge$*, Topology **34** (1995), 633–649.

[MP3] ———, *On the cohomology and homotopy of Swan groups*, Mathematische Zeitschrift **225** (1997), 277–288.

[Mi] N. Minami, *Hecke algebras and cohomotopical Mackey functors*, to appear in the Transactions of the American Mathematical Society.

[Mt] S. Mitchell, *Splitting $B(\mathbf{Z}/p)^n$ and BT^n via modular representation theory*, Mathematische Zeitschrift **189** (1985), 1–9.

[MtP] ——— and S. Priddy, *Symmetric product spectra and splittings of classifying spaces*, American Journal of Mathematics **116** (1984), 219–232.

[P] S. Priddy, *Applications of stable classifying space theory to group cohomology*, to appear in the Proceedings of Symposia in Pure Mathematics.

[Q] D. Quillen, *Homotopy properties of the poset of non-trivial p-subgroups of a group*, Advances in Mathematics **28** (1978), 101–128.

[Sw] R. Swan, *The p-period of a finite group*, Illinois Journal of Mathematics **4** (1960), 341–346.

[Th] J. Thévenaz, private communication.

[Wb] P. Webb, *A local method in group cohomology*, Commentarii Mathematici Helvetici **62** (1987), 137–167.

Department of Mathematics and Statistics, Western Michigan University, Kalamazoo, MI 49008
E-mail address: martino@math-stat.wmich.edu

Department of Mathematics, Northwestern University, Evanston, IL 60208
E-mail address: priddy@math.nwu.edu

Contemporary Mathematics
Volume **220**, 1998

Brave new worlds in stable homotopy theory

J. P. MAY

I will give a broad survey of the general area of structured ring spectra in modern constructions of the stable homotopy category. I will give some background and history, but my main focus will be a description of work in progress of Mike Mandell, Stefan Schwede, Brooke Shipley, and myself [**25, 26**]. By "modern" I mean that there must be a smash product that gives a point-set level symmetric monoidal structure on a closed model category. The associated homotopy category that is obtained by inverting the weak equivalences must be equivalent to the classical stable homotopy category, or at least to a reasonably large full subcategory.

We define a strict ring spectrum or strict commutative ring spectrum to be a monoid or commutative monoid in any such symmetric monoidal category of spectra. In a really good category, the categories of monoids and of commutative monoids should themselves be model categories. All model categories in sight should be tensored and cotensored over either the category of spaces or the category of simplicial sets. I shall work throughout with spaces for definiteness. I shall focus on strict commutative ring spectra since their study is much deeper and more interesting than the study of merely associative strict ring spectra.

I will start with a little history. Spectra were first introduced by Lima [**22**], a student of Spanier at Chicago, in 1958. There was quite a bit of work by many people, including Adams and Puppe, trying to construct a good stable homotopy category before the problem was solved by Mike Boardman in his 1964 PhD thesis at Warwick [**4**], which is still unpublished. A popularized, and in some respects bowdlerized, version of Boardman's category was explained and written up by Adams in lectures given at Chicago in 1971 [**1**]. A parallel development of a category of simplicial spectra was begun by Kan in 1963 [**17**] although the crucial point, the definition of smash products, was only addressed

1991 *Mathematics Subject Classification.* 55P42.
The author was supported in part by NSF Grant #DMS-9423300.

later, in work of Kan and Whitehead [**18**] that was completed by Burghelea and Deleanu [**7**].

In the work that I have described so far, and in many later alternative constructions, the smash product only became associative, commutative, and unital after passage to the stable homotopy category. For that reason, ring and module spectra have traditionally been flabby homotopy theoretic objects, very useful for many applications, but hopelessly inadequate for others. In such a context, the cofiber of a map of module spectra is not in general again a module spectrum, hence one cannot do serious homotopy theory in categories of modules over ring spectra. There are many other such defects.

I began looking for a way of constructing the stable homotopy category that would have good point-set level properties almost 30 years ago. In particular, I advertised the idea that a spectrum should be a sequence of spaces E_i together with homeomorphisms $E_i \cong \Omega E_{i+1}$ at a conference in 1968 [**27**]. In 1972, Frank Quinn, Nigel Ray and I introduced E_∞-ring spectra [**28**], and we showed that Thom spectra gave naturally occurring examples.

The work with Quinn and Ray led me to develop a multiplicative elaboration of infinite loop space theory [**28, 30**] that allowed passage back and forth between E_∞-ring spaces and E_∞-ring spectra. This allowed the construction of E_∞-ring spectra from bipermutative categories, such as categories of modules over commutative rings with their direct sums and tensor products. This gave a plentiful supply of E_∞-ring spectra, such as the Eilenberg-Mac Lane spectra of commutative rings, the algebraic K-theory spectra of commutative rings and the connective topological K-theory spectra.

The passage back and forth between E_∞-ring spaces and E_∞-ring spectra was vital to calculations made in the 1970's, for example the calculation of characteristic classes for topological bundles and the calculation of the cobordism groups of topological manifolds. One point is that the operad actions in terms of which these structures were defined gave rise to computable homology operations. Another is that the point set level structures allowed the construction and descriptive analysis of new infinite loop spaces with computable invariants. For example, work of Sullivan [**39**] and many others led to a proof that, at an odd prime p, $BSTop$ is equivalent as an infinite loop space to the classifying space of kO-oriented spherical fibrations. Work of Quillen [**34**] and others led to a discrete model description of kO as an algebraic K-theory type spectrum associated to finite fields. These ideas allowed me to reduce the calculation of characteristic classes for topological bundles at odd primes to calculations in the cohomology of finite groups [**8, 29**].

Paul Goerss and Mike Hopkins [**13**] have recently developed another, obstruction theoretic, way of constructing E_∞-ring spectra. Here again, the operad actions that go into the description of E_∞-ring spectra are critical to both the theory and the calculations that make the constructions work.

During the work with Quinn and Ray, it became apparent that spectra should be defined in a coordinate-free fashion, with spaces EV indexed on finite dimensional inner product spaces V contained in a given universe U isomorphic to $\mathbb{R}^\infty$. There should be homeomorphisms $EV \cong \Omega^{W-V}EW$ whenever $V \subset W$, where $W - V$ is the orthogonal complement of V in W.

Thinking equivariantly, tom Dieck [**10**] had realized the advantages of such indexing even earlier, the V's being representations contained in a G-universe U for some compact Lie group G. Lewis and I [**21**] worked out details of the coordinate-free approach to the stable homotopy category in the early 1980's. Our focus was on the equivariant stable homotopy category, and our work gave the first and, so far, only treatment of the equivariant foundations of stable homotopy theory. A more readable account appears in "Equivariant homotopy and cohomology theory" [**31**].

Building on both the work with Quinn and Ray and the work with Lewis, a complete solution to the problem of constructing a symmetric monoidal category of spectra was finally obtained by Elmendorf, Kriz, Mandell, and myself in "Rings, modules, and algebras in stable homotopy theory" [**12**]; I will refer to that book as EKMM. The theory works in full equivariant generality, and it ties in well with all of the previous work in the area. We gave a variety of applications, and more has been done since the book was written. We call our category the category of S-modules, denoted $\mathscr{M}_S$, where S is the sphere spectrum. These are coordinate free spectra with additional structure. The additional structure allows the construction of a commutative, associative, and unital smash product.

A commutative S-algebra R is a commutative monoid in $\mathscr{M}_S$: it has a commutative, associative, and unital product $R \wedge_S R \longrightarrow R$ with unit $S \longrightarrow R$. R-modules are defined similarly in terms of maps $R \wedge_S M \longrightarrow M$. The category $\mathscr{M}_R$ of R-modules has all of the good properties of $\mathscr{M}_S$. The derived homotopy category $\mathscr{D}_R$ is obtained by inverting the weak equivalences, which are the maps of R-modules that are weak equivalences of spectra. For example, the derived category $\mathscr{D}_{Hk}$ associated to the Eilenberg-Mac Lane S-algebra Hk is equivalent to the classical derived category of chain complexes of k-modules, and the equivalence converts smash products and function Hk-modules to tensor product and hom functors. For another example, the derived categories of modules over real and complex, periodic and connective, K-theory have been analyzed algebraically by Jerome Wolbert [**41**].

Working in the derived category of MU-modules, we gave quick and easy constructions as MU-modules of spectra $(MU/I)[Y^{-1}]$, including Brown-Peterson, Johnson-Wilson, and Morava K-theory spectra. At least at odd primes, these are all easily shown to be commutative MU-ring spectra in the homotopical sense and therefore commutative ring spectra in the classical sense. Neil Strickland [**38**] has extended the results of EKMM to the 2-local case, proving for example that BP (at 2) is a commutative $MU_{(2)}$-ring spectrum. Similar arguments will apply starting with BP once Kriz's sketch proof [**19**] that BP is an E_∞ ring spectrum is completed. Maria Basterra [**3**] has worked out a rigorous construction of André Quillen cohomology for commutative S-algebras, which is the essential tool. In some combination of authorship, I expect Basterra, Kriz, Hunter, and McClure to complete this work.

A deep fact is that the category of commutative R-algebras is itself a model category, with good control on the homotopical behavior of the underlying R-modules. For any R-module E, the Bousfield localization A_E of an R-algebra A is an R-algebra. For example, $KU = ku_{KU}$ and $KO = ko_{KO}$, giving a quick proof that KU is a ku-algebra and therefore an S-algebra, and similarly for KO.

When $R = Hk$, Mike Mandell [**24**] has proven that the derived homotopy category of commutative Hk-algebras is equivalent to the algebraic derived category of E_∞-k-algebras. He has used this together with an unpublished theorem of Dwyer and Hopkins to prove that the singular cochain functor with coefficients in the algebraic closure of the field with p elements gives a contravariant equivalence from the homotopy category of p-nilpotent spaces of finite type to a full subcategory of the category of E_∞-algebras over $\bar{F}_p$. This algebraization of p-adic homotopy theory may well be the best possible p-adic analogue of Sullivan's rational homotopy theory.

There are many other applications arising from this brave new world of stable homotopy theory, but that is not the subject of this report. Rather I want to talk about brave new worlds, in the plural, and how to travel between them. Several years ago, Jeff Smith initiated a program for an alternative development of the modern foundations of stable homotopy theory based on "symmetric spectra". The full development of that theory lags a few years behind EKMM. Mark Hovey, Brooke Shipley, and Smith [**16**] are nearly done with a write-up of this version of the stable homotopy category, and they know in general how to develop the homotopy theories of algebras and of commutative algebras in their context, but it may be quite some time before that work is written up. Another alternative development, this one based on Γ-spaces, has been studied by Lydakis and Schwede [**23, 35**], who build on work of Segal [**36**], Anderson [**2**], and Bousfield and Friedlander [**6**].

In part because EKMM got there first, it has skimmed off the main applications. In fact, there have not yet been many new applications of any alternative approach, and there have not yet been any calculational applications. Moreover, calculations that are based on operad actions and space level E_∞-ring structures do not naturally live in the alternative brave new worlds. They do fit naturally into the EKMM setting, since that is an outgrowth of the earlier work which is based on a convenient conceptual repackaging of the relevant operad actions.

Nevertheless, given how central stable homotopy theory is to algebraic topology, it is of real interest to have alternative approaches that involve more elementary and less elaborate structure than is built into the S-modules of EKMM. However, it would be willfully perverse to develop such approaches in isolation. The different approaches should give mutually supportive brave new worlds in a single universe, so that one can work in whatever context seems most natural to a particular set of given data. For example, the alternative approaches seem particularly relevant to topological Hochschild homology, which can be developed in any approach, and to topological cyclic homology, whose existing construction naturally involves three different brave new worlds, as I shall briefly indicate later.

Thus what I shall talk about is work in progress of Mike Mandell, Stefan Schwede, Brooke Shipley, and myself [**25, 26**] that shows how to travel between the various brave new worlds of stable homotopy theory. One important feature of our work is its emphasis on the coordinate-free analogue of Smith's coordinatized brave new world. The idea is obvious enough from a comparison of Smith's approach with the concepts introduced by Quinn, Ray, and myself, but it has not previously been taken seriously. This approach may well lead to a pleasant alternative construction of equivariant stable homotopy categories for compact Lie groups. Examples of the kind of equivariant object I have in mind play key roles in the construction of topological cyclic homology of Lars Hesselholt and Ib Madsen [**15**] and in the proof by John Greenlees and myself [**14**] of an analogue of the Atiyah-Segal completion theorem that applies to all modules over MU.

I will begin by describing a method of constructing symmetric monoidal functor categories wholesale. The basic construction has been well-known to category theorists since the late 1960's (see Day [**9**]), but Jeff Smith is the first to have seen the relevance of the construction to stable homotopy theory. I will specialize it to describe the bottom four categories in the following "Main Diagram":

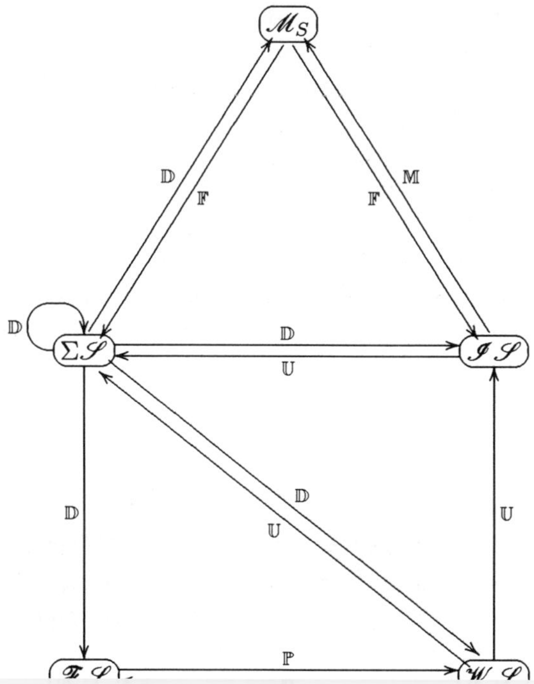

Here $\Sigma\mathscr{S}$ is Smith's category of symmetric spectra, or Σ-spectra; $\mathscr{I}\mathscr{S}$ is the category of orthogonal spectra, or $\mathscr{I}$-spectra, which is the coordinate-free analogue of Smith's category; $\mathscr{F}\mathscr{S}$ is the category of Γ-spaces, or $\mathscr{F}$-spectra, whose symmetric monoidal structure was introduced by Lydakis and whose ring objects were studied by Schwede.

I have written $\mathbb{D}$ for all functors defined on $\mathscr{I}\mathscr{S}$ since they are all defined in the same way. We write $\mathbb{D}_\Sigma$, $\mathbb{D}_{\mathscr{I}}$, $\mathbb{D}_{\mathscr{W}}$, and $\mathbb{D}_{\mathscr{M}_S}$ when we wish the notation to record the target category. Similarly, the two functors $\mathbb{F}$ are defined in the same way, and we write $\mathbb{F}_\Sigma$ and $\mathbb{F}_{\mathscr{I}}$ when we wish the notation to record the target category.

Roughly speaking, the conclusion of our work is that all approaches are essentially equivalent, so that results proven in one of these categories can be translated into conclusions applicable in another. That is, all composite endofunctors are weakly equivalent to identity functors under appropriate technical hypotheses.

The same conclusions apply to the diagram that is obtained by restricting to monoids in each of these symmetric monoidal categories. That is not quite true of the diagram that is obtained by restricting to commutative monoids. However, there is a notion of an E_∞-monoid in each of these categories, and the same conclusions do apply to the diagram that is obtained by restricting to E_∞-monoids. In the category of S-modules and symmetric spectra, at least, E_∞-monoids are equivalent to commutative monoids.

The general construction of symmetric monoidal functor categories is not limited to applications in stable homotopy theory. For example, as pointed out by Voevodsky, the construction I'm about to give has potential applications to algebraic geometry, where it may shed light on the recent importation of language from homotopy theory [**32, 40**]. In fact, this may well be the area in which the idea finds its most significant applications.

I will restrict attention to functors that take values in the category of based spaces, but everything can be adapted without difficulty to functors that take values in the category of based simplicial sets. In fact, the basic constructions apply verbatim to functors that take values in any symmetric monoidal category that is tensored and cotensored over the category of based spaces or the category of based simplicial sets.

1. Symmetric monoidal categories

We need to fix some language. A monoidal category $\mathscr{D}$ is a category together with a product $\Box = \Box_{\mathscr{D}} : \mathscr{D} \times \mathscr{D} \longrightarrow \mathscr{D}$ and a unit object $u = u_{\mathscr{D}}$ such that $\Box$ is associative and unital up to coherent natural isomorphism; $\mathscr{D}$ is symmetric monoidal if $\Box$ is also commutative up to coherent natural isomorphism.

A functor $F : \mathscr{A} \longrightarrow \mathscr{B}$ between monoidal categories is *lax monoidal* if there is a map $\lambda : u_{\mathscr{B}} \longrightarrow F(u_{\mathscr{A}})$ and a natural map

$$\phi : F(A) \,\Box_{\mathscr{B}}\, F(B) \longrightarrow F(A \,\Box_{\mathscr{A}}\, B)$$

such that all coherence diagrams relating the associativity and unit isomorphisms of $\mathscr{A}$ and $\mathscr{B}$ to the maps λ and ϕ commute. If $\mathscr{A}$ and $\mathscr{B}$ are symmetric monoidal, then F is *lax symmetric monoidal* if all coherence diagrams relating the associativity, unit, and commutativity isomorphisms of $\mathscr{A}$ and $\mathscr{B}$ commute. The functor F is *strict monoidal* or *symmetric monoidal* if λ and ϕ are isomorphisms.

LEMMA 1.1. *If $F : \mathscr{A} \longrightarrow \mathscr{B}$ is lax monoidal and M is a monoid in $\mathscr{A}$, then $F(M)$ is a monoid in $\mathscr{B}$. If $F : \mathscr{A} \longrightarrow \mathscr{B}$ is lax symmetric monoidal and M is a commutative monoid in $\mathscr{A}$, then $F(M)$ is a commutative monoid in $\mathscr{B}$.*

In the Main Diagram, all categories are symmetric monoidal, all functors except the $\mathbb{D}$ are lax symmetric monoidal, and the $\mathbb{D}$ are lax monoidal. Thus all functors preserve monoids (alias strict ring spectra) and most preserve commutative monoids. The functors $\mathbb{D}$ preserve E_∞-monoids, and this gets around the problem that they do not preserve commutative monoids.

2. $\mathscr{D}$-spaces and $\mathscr{D}$-spectra

Let $\mathscr{D}$ be a symmetric monoidal topological category with unit object u and continuous product $\Box$. We assume that $\mathscr{D}$ has an initial base object; if not, we implicitly adjoin one. The category $\mathscr{T}$ of based spaces is symmetric monoidal under the smash product with unit object S^0 and base object a one-point space.

DEFINITION 2.1. A $\mathscr{D}$-space is a based continuous functor $T : \mathscr{D} \longrightarrow \mathscr{T}$. Let $\mathscr{D}\mathscr{T}$ denote the category of $\mathscr{D}$-spaces. For $\mathscr{D}$-spaces T and T', define the "external" smash product $T \,\bar{\wedge}\, T'$ by

$$T \,\bar{\wedge}\, T' = \wedge \circ (T \times T') : \mathscr{D} \times \mathscr{D} \longrightarrow \mathscr{T};$$

thus, for objects d and e of $\mathscr{D}$, $(T \,\bar{\wedge}\, T')(d,e) = T(d) \wedge T'(e)$.

Now suppose given a based symmetric monoidal functor $S_{\mathscr{D}} : \mathscr{D} \longrightarrow \mathscr{T}$. We write $S = S_{\mathscr{D}}$ when $\mathscr{D}$ is understood. The theory actually works when $S_{\mathscr{D}}$ is *lax* symmetric monoidal, but it helps fix ideas to think in terms of the much lesser generality of a *strict* symmetric monoidal $S_{\mathscr{D}}$. The letter S is meant to suggest spheres, since our focus will be on a particular choice of S that is related to the sphere spectrum. However, each $\mathscr{D}$ can have many different choices of S.

DEFINITION 2.2. A $\mathscr{D}$-spectrum is a $\mathscr{D}$-space $T : \mathscr{D} \longrightarrow \mathscr{T}$ together with a continuous natural transformation $\sigma : T \,\bar{\wedge}\, S \longrightarrow T \circ \Box$ such that the apppropriate transitivity diagram commutes. Let $\mathscr{D}\mathscr{S}$ denote the category of $\mathscr{D}$-spectra.

We describe the categories $\mathscr{D}$ and functors $S_{\mathscr{D}}$ that we are interested in. We take S^n to be the one-point compactification of $\mathbb{R}^n$. Similarly, for a finite dimensional real inner product space V, we take S^V to be the one-point compactification of V. Note that the one-point compactification of $\{0\}$ is S^0.

EXAMPLE 2.3. Let Σ be the (unbased) category of finite sets $\underline{n} = \{1, \dots, n\}$, $n \geq 0$, and their permutations; thus there are no maps $\underline{m} \longrightarrow \underline{n}$ for $m \neq n$, and the set of maps $\underline{n} \to \underline{n}$ is the symmetric group Σ_n. The symmetric monoidal structure is given by concatenation of sets and block sum of permutations, with $\underline{0}$ as unit. The functor S sends $\underline{n}$ to S^n.

EXAMPLE 2.4. Let $\mathscr{I}$ be the (unbased) category of finite dimensional real inner product spaces and linear isometric isomorphisms; there are no maps $V \longrightarrow W$ unless $\dim V = \dim W = n$ for some $n \geq 0$, when the space of morphisms $V \longrightarrow W$ is homeomorphic to the orthogonal group $O(n)$. The symmetric monoidal structure is given by direct sums, with $\{0\}$ as unit. The functor S sends V to S^V. Define a functor $\iota : \Sigma \longrightarrow \mathscr{I}$ by sending $\underline{n}$ to $\mathbb{R}^n$ and using the standard inclusions $\Sigma_n \longrightarrow O(n)$.

EXAMPLE 2.5. We can take $\mathscr{D} = \mathscr{T}$. More generally, we can take $\mathscr{D}$ to be any based symmetric monoidal subcategory of $\mathscr{T}$ that contains S^0. The functor $S_{\mathscr{D}}$ is then the inclusion $\mathscr{D} \longrightarrow \mathscr{T}$. In particular, we let $\mathscr{W}$ be the full subcategory of $\mathscr{T}$ of based spaces homeomorphic to finite CW complexes. It turns out that $\mathscr{W}$-spaces and $\mathscr{W}$-spectra coincide.

EXAMPLE 2.6. Let $\mathscr{F}$ be the category of finite based sets $n^+ = \{0, 1, \dots, n\}$ and based maps, where 0 is the basepoint. This is the opposite of Segal's category Γ. This category is based with base object 0^+. Take $\Box$ to be the smash product of finite based sets; to be precise, we order the non-zero elements of $m^+ \wedge n^+$ lexicographically. The unit object is 1^+. The functor S sends n^+ to n^+ regarded as a discrete based space. Again, it turns out that $\mathscr{F}$-spaces and $\mathscr{F}$-spectra coincide.

3. Functors with smash product

Retaining our assumptions on $\mathscr{D}$ and $S = S_{\mathscr{D}}$, we next define $\mathscr{D}$-FSP's. The term "functor with smash product" was introduced by Bökstedt, although examples had appeared earlier. We shall see shortly that $\mathscr{D}$-FSP's give the right notion of a strict ring $\mathscr{D}$-spectrum, or $S_{\mathscr{D}}$-algebra. In fact, we shall give the category of $\mathscr{D}$-spectra a symmetric monoidal structure whose monoids are precisely the $\mathscr{D}$-FSP's.

DEFINITION 3.1. A $\mathscr{D}$-FSP is a $\mathscr{D}$-space T together with a unit map $\eta : S \longrightarrow T$ of $\mathscr{D}$-spaces and a continuous natural product map $\mu : T \bar{\wedge} T \longrightarrow T \circ \Box$ such that the appropriate unity, associativity, and centrality of unit diagrams commute. A $\mathscr{D}$-FSP is commutative if the appropriate commutativity diagram commutes.

A $\mathscr{D}$-FSP is a $\mathscr{D}$-spectrum with additional structure.

LEMMA 3.2. *A $\mathscr{D}$-FSP has an underlying $\mathscr{D}$-spectrum with structure map*

$$\sigma = \mu \circ (\mathrm{id} \bar{\wedge} \eta) : T \bar{\wedge} S \longrightarrow T \circ \Box.$$

Remark 3.3. Up to nomenclature, $\mathscr{D}$-FSP's were first introduced as follows.

(i) A $\mathscr{W}$-FSP is an FSP as introduced by Bökstedt.

(ii) A commutative $\mathscr{I}$-FSP is an $\mathscr{I}_*$-prefunctor as defined by May, Quinn, and Ray. This was the earliest definition of this sort.

(iii) A Σ-FSP is a symmetric ring spectrum as defined by Smith. Hesselholt and Madsen call these "FSP's defined on spheres".

(iv) An $\mathscr{F}$-FSP is a Gamma-ring, as defined by Lydakis and Schwede.

These different kinds of FSP's arise naturally in different applications. $\mathscr{W}$-FSP's are the least general of these examples since restriction along $S_{\mathscr{D}} : \mathscr{D} \longrightarrow \mathscr{W} \subset \mathscr{T}$ shows that a $\mathscr{W}$-FSP is a $\mathscr{D}$-FSP for the other $\mathscr{D}$. Similarly, $\mathscr{I}$-FSP's appear to be less general than Σ-FSP's since restriction along $\iota : \Sigma \longrightarrow \mathscr{I}$ shows that an $\mathscr{I}$-FSP is a Σ-FSP. The forgetful functor from $\mathscr{W}$-spaces to $\mathscr{F}$-spaces has a left adjoint "prolongation functor" $\mathbb{P}$ given explicitly by the coends

$$\mathbb{P}T(X) = \int^{n^+ \in \mathscr{F}} T(n^+) \wedge X^n.$$

The prolongation of an $\mathscr{F}$-FSP is a $\mathscr{W}$-FSP, and it restricts to give a Σ-FSP. Of course, these passages from one kind of FSP to another are functorial.

4. Symmetric monoidal categories of $\mathscr{D}$-spectra

Now assume that our given symmetric monoidal category $\mathscr{D}$ has a small skeleton $sk\mathscr{D}$. This rules out $\mathscr{T}$ but includes all the really interesting examples. Write $S = S_{\mathscr{D}}$.

THEOREM 4.1. *The category $\mathscr{D}\mathscr{S}$ of $\mathscr{D}$-spectra has a smash product $\wedge_S$ under which it is symmetric monoidal with unit S. The categories of $\mathscr{D}$-FSP's and commutative $\mathscr{D}$-FSP's are isomorphic to the categories of monoids in $\mathscr{D}\mathscr{S}$ and of commutative monoids in $\mathscr{D}\mathscr{S}$.*

It is natural to introduce alternative names for $\mathscr{D}$-FSP's.

DEFINITION 4.2. A (commutative) $S_{\mathscr{D}}$-algebra is a (commutative) monoid in $\mathscr{D}\mathscr{S}$.

The theorem asserts that these are the same as (commutative) $\mathscr{D}$-FSP's. Thus

$$\mathscr{D}\text{-FSP's} = \text{monoids in } \mathscr{D}\mathscr{S} = S_{\mathscr{D}}\text{-algebras}.$$

The difference in the notions is that $S_{\mathscr{D}}$-algebras are defined in terms of the "internal" smash product $\wedge_S$, whereas $\mathscr{D}$-FSP's are defined equivalently in terms of the more elementary external smash product $\bar{\wedge}$.

Some of the functors between categories of $\mathscr{D}$-FSP's that I described are induced by symmetric monoidal functors between their underlying categories of $\mathscr{D}$-spectra.

PROPOSITION 4.3. *The inclusion $\iota : \Sigma \longrightarrow \mathscr{I}$ induces a forgetful functor $\mathbb{U} : \mathscr{I}\mathscr{S} \longrightarrow \Sigma\mathscr{S}$, and $\mathbb{U}$ is lax symmetric monoidal.*

PROPOSITION 4.4. *Prolongation induces a functor $\mathbb{P} : \mathscr{F}\mathscr{S} \longrightarrow \mathscr{W}\mathscr{S}$, and $\mathbb{P}$ is strict symmetric monoidal.*

To prove the theorem, we first give the category $\mathscr{D}\mathscr{T}$ of $\mathscr{D}$-spaces a smash product $\wedge$ under which it is symmetric monoidal. We do this by internalizing the external smash product $T \bar{\wedge} T'$ by taking its left Kan extension along $\square$. The left Kan extension is characterized by the universal property

$$\mathscr{D}\mathscr{T}(T \wedge T', T'') \cong (\mathscr{D} \times \mathscr{D})\mathscr{T}(T \bar{\wedge} T', T'' \circ \square),$$

and it admits two equivalent explicit descriptions.

DEFINITION 4.5. Define the smash product $T \wedge T'$ of $\mathscr{D}$-spaces T and T' by letting $(T \wedge T')(d)$ be the space

$$(T \wedge T')(d) = \mathrm{colim}_{\phi : e \square f \to d} T(e) \wedge T'(f).$$

The colimit is taken over the category $\square/d$ of maps $\phi : e \square f \to d$. Equivalently, when $\mathscr{D}$ is small, $(T \wedge T')(d)$ is the coend

$$\int^{(e,f) \in \mathscr{D} \times \mathscr{D}} \mathscr{D}(e \square f, d) \wedge (T(e) \wedge T'(f)).$$

PROPOSITION 4.6. *Let $\underline{S}^0 : \mathscr{D} \longrightarrow \mathscr{T}$ be the functor that sends d to the (based) space $\mathscr{D}(u, d)$. Then $\mathscr{D}\mathscr{T}$ is symmetric monoidal under $\wedge$ with unit object $\underline{S}^0$.*

The proof is formal. The isomorphisms $S(d) \wedge S(e) \cong S(d \square e)$ specify a natural isomorphism of functors $S \bar{\wedge} S \cong S \circ \square$. By the universal property, this induces a map of $\mathscr{D}$-spaces $S \wedge S \longrightarrow S$.

LEMMA 4.7. *The $\mathscr{D}$-space $S_{\mathscr{D}}$ is a commutative monoid in the symmetric monoidal category $\mathscr{D}\mathscr{T}$.*

There is an evident notion of a (right) module T over a monoid S in any symmetric monoidal category. The universal property implies the following conceptual description of $\mathscr{D}$-spectra.

PROPOSITION 4.8. *The category of $\mathscr{D}$-spectra is isomorphic to the category of $S_{\mathscr{D}}$-modules.*

DEFINITION 4.9. For $\mathscr{D}$-spectra T and T', thought of as right and left S-modules respectively, define $T \wedge_S T'$ to be the coequalizer in the category of $\mathscr{D}$-spaces (constructed spacewise) displayed in the diagram

$$T \wedge S \wedge T' \underset{\mathrm{id} \wedge \mu'}{\overset{\mu \wedge \mathrm{id}}{\rightrightarrows}} T \wedge T' \longrightarrow T \wedge_S T',$$

where μ and μ' are the given actions of S on T and T'. Then $T \wedge_S T'$ inherits a $\mathscr{D}$-spectrum structure from the $\mathscr{D}$-spectrum structure on T or, equivalently, T'.

Our theorem describing $\mathscr{D}$-FSP's follows formally from the definition and the universal property of left Kan extensions. We return to our examples.

EXAMPLE 4.10. Consider $\mathscr{D} = \Sigma$. The functor $\underline{S}^0$ sends $\underline{0}$ to the space S^0 and sends $\underline{n}$ to a point for $n > 0$. For Σ-spaces T and T',

$$(T \wedge T')(\underline{n}) \cong \bigvee_{p=0}^{n} \Sigma_{n+} \wedge_{\Sigma_p \times \Sigma_{n-p}} T(\underline{p}) \wedge T(\underline{n-p})$$

as a Σ_n-space.

EXAMPLE 4.11. Consider $\mathscr{D} = \mathscr{I}$. The functor $\underline{S}^0$ sends $\{0\}$ to S^0 and sends V to a point if $\dim V > 0$. For an inner product space V of dimension n, choose a subspace V_p of dimension p for each $p \leq n$ and let $V - V_p$ be the orthogonal complement of V_p in V. Then, for $\mathscr{I}$-spaces T and T',

$$(T \wedge T')(V) \cong \bigvee_{p=0}^{n} O(V)_+ \wedge_{O(V_p) \times O(V-V_p)} T(V_p) \wedge T'(V - V_p)$$

as an $O(V)$-space.

EXAMPLE 4.12. Consider $\mathscr{D} = \mathscr{F}$. Here $S = S_{\mathscr{F}}$ is the unit $\underline{S}^0$ of $\wedge$, and an $\mathscr{F}$-space has a unique structure of $S_{\mathscr{F}}$-module, namely that given by the unit isomorphism $T \wedge S \cong T$. Therefore $\mathscr{F}$-spectra coincide with $\mathscr{F}$-spaces and $\wedge_S$ coincides with $\wedge$. The same phenomenon occurs in the case $\mathscr{D} = \mathscr{W}$.

The homotopy theory of $\mathscr{F}$-spectra and $\mathscr{F}$-algebras have been worked out by Bousfield and Friedlander, Lydakis, and Schwede. As I said before, the homotopy theory of Σ-spectra and S_Σ-algebras is being worked out by Hovey, Smith, and Shipley. The homotopy theory of $\mathscr{I}$-spectra and $S_{\mathscr{I}}$-algebras has not yet been studied. In fact, the desirability of such a study has only become fully apparent with our work. Note that all definitions above apply with $\mathscr{I}$ replaced by $\mathscr{I}_G$, the category of finite dimensional orthogonal representations of a compact Lie group G and all linear isometric isometries; this is a topological G-category with G acting by conjugation on morphism spaces.

5. The definition of E_∞-$\mathscr{D}$-FSP's

Our comparison results involve the notion of an E_∞-$\mathscr{D}$-FSP, which is a $\mathscr{D}$-FSP that is commutative only up to coherence conditions expressed by the action of a certain E_∞-operad $\mathscr{E}$.

Let $\mathscr{C}$ be an operad of spaces such that $\mathscr{C}(0) = *$. There is a notion of a $\mathscr{C}$-algebra in $\mathscr{D}\mathscr{S}$: an action of $\mathscr{C}$ on T is given by Σ_n-equivariant maps of $\mathscr{D}$-spectra

$$\theta_n : \mathscr{C}(n)_+ \wedge T^{(n)} \longrightarrow T, \tag{5.1}$$

where $T^{(n)}$ is the nth $\wedge_S$-power and where the smash product of a space and a $\mathscr{D}$-spectrum is defined in the evident spacewise fashion, $(X \wedge T)(d) = X \wedge T(d)$. The standard associativity, unity, and equivariance diagrams in the definition of an operad action are required to commute. We interpret $T^{(0)}$ to be $S = S_{\mathscr{D}}$, of the definition, but we can translate along the universal property of left Kan extensions to give an "external" form analogous to our original definition of a $\mathscr{D}$-FSP.

There is a canonical E_∞-operad $\mathscr{E}$ whose nth space is $B\tilde{\Sigma}_n$, where $\tilde{G}$ is the translation category of a group G. This operad contains the operad $\mathscr{M}$ that defines monoids in any symmetric monoidal category and augments to the operad $\mathscr{N}$ that defines commutative monoids.

DEFINITION 5.2. An E_∞-$\mathscr{D}$-FSP is an $\mathscr{E}$-algebra in $\mathscr{D}\mathscr{S}$.

We have a precisely analogous definition in the context of the S-modules of EKMM; I will say a little bit more about what these objects are later, but it is convenient to define E_∞-S-algebras now.

DEFINITION 5.3. An E_∞-S-algebra R is an S-module together with an action of the operad $\mathscr{E}$ given by maps $\theta : \mathscr{E}(n)_+ \wedge R^{(n)} \longrightarrow R$ such that the standard associativity, unity, and equivariance diagrams commute. Here $R^{(n)}$ is the nth $\wedge_S$-power.

In both contexts, it is reasonable to think of E_∞-objects as E_∞-monoids in the given symmetric monoidal category. E_∞-S-algebras must not be confused with E_∞-ring spectra, as originally defined in [**28**]: an E_∞-ring structure on a spectrum E is specified in terms of maps

$$\theta : \mathscr{L}(n) \ltimes E^{(n)} \longrightarrow E,$$

where $E^{(n)}$ here is the nth external smash product of spectra and $\ltimes$ is the twisted half-smash product. However, the homotopy categories of commutative S-algebras, E_∞-ring spectra, and E_∞-S-algebras are all equivalent, so that only the experts need to worry about the distinction.

6. The functors $\mathbb{D}$ and $\mathbb{E}$

All of our categories are now in place, and we can begin to compare them. We first describe functors and natural transformations that compare our categories of $\mathscr{D}$-spectra on a formal level, and we then describe how these comparison maps behave homotopically. Thus we now concentrate on the bottom square of the Main Diagram. We will come back to S-modules later.

However, our first construction applies to both $\mathscr{D}$-spectra and S-modules. It shows that the category of Σ-spectra plays a universal role: we can construct an object $\mathbb{D}(T)$ in any of our other categories from a Σ-spectrum T.

THEOREM 6.1. *Let $\mathscr{C}$ be any symmetric monoidal model category of spectra that is tensored and cotensored over based spaces and has a given sphere object S. There is a lax monoidal functor*

$$\mathbb{D} = \mathbb{D}_{\mathscr{C}} : \Sigma\mathscr{S} \longrightarrow \mathscr{C}.$$

Since $\mathbb{D}$ is lax monoidal, it carries Σ-FSP's to S-algebras in $\mathscr{C}$; $\mathbb{D}$ is not lax symmetric monoidal, but it carries E_∞-Σ-FSP's to E_∞-S-algebras in $\mathscr{C}$.

The hypotheses on $\mathscr{C}$ in the theorem just mean that it is the kind of modern category of spectra that we have been considering. The construction of $\mathbb{D}$ is based on ideas of Bökstedt [**5**].

DEFINITION 6.2. Let I be the category of finite sets $\underline{n} = \{1, \dots, n\}$, $n \geq 0$, and injective functions.

DEFINITION 6.3. Let T be a Σ-spectrum. Define

$$\mathbb{D}(T) = \operatorname{hocolim}_I \Omega^n(T(\underline{n}) \wedge S).$$

Here $\wedge$ is the "tensor" or smash product with spaces in $\mathscr{C}$, Ω^n is the "cotensor" or function object $F(S^n, -)$ in $\mathscr{C}$, and the homotopy colimit is taken in $\mathscr{C}$.

The proof that $\mathbb{D}$ is lax monoidal uses the concatenation functor $\sqcup : I \times I \longrightarrow I$ and formal properties of homotopy colimits, and similarly for the preservation of E_∞-structures. However, even when T is a commutative Σ-FSP, $\mathbb{D}(T)$ is still only an E_∞- algebra in the target category.

Now we give a construction that allows us to compare composites of $\mathbb{D}$ and forgetful functors $\mathbb{U}$ with identity functors. This only works when Σ is embedded in $\mathscr{D}$, so it doesn't apply to $\mathscr{F}$.

THEOREM 6.4. *For any $\mathscr{D}$ with an embedding $\Sigma \longrightarrow \mathscr{D}$, there is a lax monoidal functor*

$$\mathbb{E} = \mathbb{E}_{\mathscr{D}} : \mathscr{D}\mathscr{S} \longrightarrow \mathscr{D}\mathscr{S}$$

such that $\mathbb{U}\mathbb{E}_{\mathscr{D}} = \mathbb{E}_\Sigma \mathbb{U}$. Since $\mathbb{E}$ is lax monoidal, it carries $\mathscr{D}$-FSP's to $\mathscr{D}$-FSP's; $\mathbb{E}$ is not lax symmetric monoidal, but it carries E_∞-$\mathscr{D}$-FSP's to E_∞-$\mathscr{D}$-FSP's. Moreover, there are monoid and E_∞-monoid preserving natural maps

$$T \xrightarrow{\alpha} \mathbb{E}(T) \xleftarrow{\beta} \mathbb{D}_{\mathscr{D}}\mathbb{U}(T)$$

for $T \in \mathscr{D}\mathscr{S}$.

For a $\mathscr{D}$-spectrum T,

$$\mathbb{E}(T) = \operatorname{hocolim}_I \Omega^n(\Lambda_n T),$$

where $\Lambda_n T(d) = T(\underline{n} \,\Box\, d)$. The maps α are given by the inclusions of zeroth terms in homotopy colimits, and the maps β are induced by the maps

$$\Omega^n \sigma : \Omega^n(T(\underline{n}) \wedge S(d)) \longrightarrow \Omega^n(T(\underline{n} \,\Box\, d)).$$

7. Prespectra, spectra, and homotopy groups

We need some standard language to describe the homotopical behavior of α and β. A coordinatized prespectrum is a sequence of based spaces T_n and based maps $\sigma : \Sigma T_n \longrightarrow T_{n+1}$. A Σ-spectrum determines a coordinatized prespectrum by neglect of structure. A coordinate-free prespectrum consists of based spaces TV for finite dimensional sub-inner product spaces of a uni-

$\sigma : \Sigma^{W-V} TV \longrightarrow T(W)$ for $V \subset W$. An $\mathscr{I}$-spectrum determines a coordinate-free prespectrum indexed on any U by neglect of structure. A coordinate-free prespectrum T determines a coordinatized prespectrum with $T_n = T(\mathbb{R}^n)$.

A prespectrum of either sort is a spectrum if the adjoints

$$T_n \longrightarrow \Omega T_{n+1} \quad \text{or} \quad TV \longrightarrow \Omega^{W-V} TW$$

of its structure maps are homeomorphisms. The unmodified term "spectrum" will only be used in this sense. Let $\mathscr{P}$ and $\mathscr{S}$ be the categories of coordinate-free prespectra and spectra indexed on a fixed universe U. The suspension spectrum $\Sigma^\infty X$ of a based space X has Vth space $Q\Sigma^V X$, where $QX = \cup \Omega^V \Sigma^V X$. We have an adjunction

$$\mathscr{S}(\Sigma^\infty X, E) \cong \mathscr{T}(X, \Omega^\infty E), \tag{7.1}$$

where Ω^∞ is the zeroth space functor. The forgetful functor $\mathscr{S} \longrightarrow \mathscr{P}$ has a left adjoint spectrification functor $L : \mathscr{P} \longrightarrow \mathscr{S}$.

The prespectrum determined by a non-trivial $\mathscr{D}$-spectrum is never a spectrum. Thus $\mathscr{D}$-spectra are never spectra in the sense just defined.

DEFINITION 7.2. The homotopy groups of a coordinatized prespectrum $T = \{T_n, \sigma_n\}$ are defined by

$$\pi_q(T) = \operatorname{colim} \pi_{n+q}(T_n).$$

A map $f : T \longrightarrow T'$ is said to be a π_*-isomorphism if it induces an isomorphism on homotopy groups. More strongly, a map $f : T \longrightarrow T'$ of prespectra is a spacewise weak equivalence if each $f : T_n \longrightarrow T'_n$ is a weak homotopy equivalence. A prespectrum T is said to be

(i) an inclusion prespectrum if each $T_n \longrightarrow \Omega T_{n+1}$ is an inclusion.
(ii) convergent if $T_n \longrightarrow \Omega T_{n+1}$ is an $(n + \lambda(n))$-equivalence for some non-decreasing sequence $\{\lambda(n)\}$ with infinite limit.
(iii) an Ω-spectrum if each $T_n \longrightarrow \Omega T_{n+1}$ is a weak homotopy equivalence.
(iv) an almost Ω-spectrum if each $T_n \longrightarrow \Omega T_{n+1}$ for $n > 0$ is a weak homotopy equivalence.

These notions have evident reformulations for coordinate-free prespectra, which are equivalent to the notions for the underlying coordinatized prespectrum. We apply these definitions to Σ-spectra or $\mathscr{I}$-spectra by applying them to their underlying prespectra.

The importance of inclusion prespectra is given by the following fact.

LEMMA 7.3. *For an inclusion prespectrum T, the natural map $\iota : T \longrightarrow LT$ is a π_*-isomorphism.*

8. Homotopical comparisons between Σ, $\mathscr{I}$, and $\mathscr{F}$-spectra

THEOREM 8.1. *If T is convergent, then α is a π_*-isomorphism and β is a spacewise weak equivalence and therefore a π_*-isomorphism.*

By the derived category of any of our categories of $\mathscr{D}$-spectra or $\mathscr{D}$-FSP's, we understand the category that is obtained by adjoining formal inverses to the "weak equivalences". Provided that we restrict attention to Ω-$\mathscr{D}$-spectra, a weak equivalence just means a spacewise weak equivalence when $\mathscr{D}$ is Σ or $\mathscr{I}$. I'll come back to what a weak equivalence means more generally later.

COROLLARY 8.2. *The functors*

$$\mathbb{D}_{\mathscr{I}} : \Sigma\mathscr{S} \longrightarrow \mathscr{I}\mathscr{S} \quad \text{and} \quad \mathbb{U} : \mathscr{I}\mathscr{S} \longrightarrow \Sigma\mathscr{S}$$

induce equivalences between the respective derived categories of Ω-spectra, Ω-FSP's, and Ω-E_∞FSP's.

PROOF. The required natural spacewise weak equivalences are obtained by restriction from the diagrams

$$\mathrm{Id}_{\mathscr{I}\mathscr{S}} \xrightarrow{\alpha} \mathbb{E}_{\mathscr{I}} \xleftarrow{\beta} \mathbb{D}_{\mathscr{I}}\mathbb{U}$$

and

$$\mathrm{Id}_{\Sigma\mathscr{S}} \xrightarrow{\alpha} \mathbb{E}_{\Sigma} \xleftarrow{\beta} \mathbb{D}_{\Sigma} = \mathbb{U}\mathbb{D}_{\mathscr{I}}. \quad \square$$

We conclude that the more richly structured $\mathscr{I}$-FSP's are really no less general than the Σ-FSP's.

The methods I have been discussing apply equally well to give a comparison between the categories of Σ-spectra and of $\mathscr{W}$-spectra. We build $\mathscr{F}$-spectra into the picture by comparing them to $\mathscr{W}$-spectra via homotopical analysis of the prolongation functor. Working simplicially, this analysis has been carried out by Anderson [**2**], Bousfield and Friedlander [**6**], and Lydakis [**23**]. Their conclusions carry over to the topological setting. The prespectra associated to $\mathscr{F}$-spectra are always connective, which gives an intrinsic limitation to this approach.

9. The category of S-modules

Let $S = \Sigma^\infty S^0$ be the sphere spectrum. There is a symmetric monoidal category $\mathscr{M} = \mathscr{M}_S$ of S-modules, where an S-module is a spectrum with certain additional structure. We have used the term "$S_{\mathscr{D}}$-module" in the diagram spectrum context: from now on, S-modules mean only the kinds of spectra defined in EKMM (=[**12**]). The smash product in $\mathscr{M}$ will be denoted $\wedge_S$. I won't repeat the definitions here. They are based on the idea that prespectra and spectra indexed on inner product spaces should have functoriality in linear isometries built into their structure. Since S-modules and $\mathscr{I}$-spectra are both defined in terms of constructions involving naturality on spaces of linear isometries, it is natural to think of the category of $\mathscr{I}$-spectra as intermediate between the categories of

10. The functor $\mathbb{F}$ from S-modules to $\mathscr{D}$-spectra

Returning to the Main Diagram, we describe its functors $\mathbb{F}$. The sphere spectrum S is not q-cofibrant in $\mathscr{M}$. This causes the appearance of almost Ω-$\mathscr{D}$-spectra in the next result.

THEOREM 10.1. *There are lax symmetric monoidal functors $\mathbb{F}_{\mathscr{D}} : \mathscr{M} \longrightarrow \mathscr{D}\mathscr{S}$, $\mathscr{D} = \Sigma$ and $\mathscr{D} = \mathscr{I}$, such that $\mathbb{F}_{\Sigma} = \mathbb{U} \circ \mathbb{F}_{\mathscr{I}}$. Moreover, these functors take values in the respective subcategories of almost Ω-Σ-spectra and almost Ω-$\mathscr{I}$-spectra.*

Before defining the functors, we point out a canonical misconception.

SCHOLIUM 10.2. For an S-module M and any $\mathscr{D}$, one has a $\mathscr{D}$-space that sends d to $\Omega^\infty(M \wedge_S \Sigma^\infty S(d))$. It is a standard mistake to imagine that this gives a $\mathscr{D}$-FSP when M is an S-algebra. To obtain a product, one would need that $(\Sigma^\infty, \Omega^\infty)$ is an adjoint pair relating $\mathscr{T}$ to $\mathscr{M}$, which is not the case. In fact, as observed by Lewis **[20]**, this cannot be the case for a category as nicely structured as $\mathscr{M}$. This is closely related to the fact that S is not q-cofibrant: if it were, we could deduce that its zeroth space QS^0 is a commutative monoid, which is clearly false.

The construction that we shall give is a variant of this mistaken idea. Let S_S be the canonical q-cofibrant approximation of S and let $S_S^{(n)}$ be its n-fold $\wedge_S$-power, with $S_S^{(0)} = S$.

DEFINITION 10.3. Let M be an S-module. Define

$$\mathbb{F}_{\Sigma}(M)(\underline{n}) = \mathscr{M}(S_S^{(n)}, M \wedge S^n).$$

Elements of Σ_n act by conjugation on the specified space of S-maps.

The analogue for $\mathscr{I}$ needs a construction. We may view $n \longmapsto S_S^{(n)}$ as giving a contravariant strict symmetric monoidal functor $\Sigma \longrightarrow \mathscr{M}$ over the constant functor at S, with each map $S_S^{(n)} \longrightarrow S$ a weak equivalence.

LEMMA 10.4. *This functor extends to a continuous contravariant strict symmetric monoidal functor $\mathscr{I} \longrightarrow \mathscr{M}$ over the constant functor at S such that each of its maps $S_S^{(V)} \longrightarrow S$ is a weak equivalence.*

DEFINITION 10.5. Let M be an S-module. Define

$$\mathbb{F}_{\mathscr{I}}(M)(V) = \mathscr{M}(S_S^{(V)}, M \wedge S^V).$$

Maps of $\mathscr{I}$ act by conjugation on the specified spaces of S-maps.

In both cases, we obtain the $S_{\mathscr{D}}$-module structure required of a $\mathscr{D}$-spectrum by use of the tensor adjunction for the category $\mathscr{M}$ together with smash products of maps. The proof that $\mathbb{F}_{\mathscr{D}}$ is lax symmetric monoidal is formal. It is a basic fact from EKMM that $\mathscr{M}(S_S, N)$ is naturally weakly equivalent to the zeroth space $\Omega^\infty N$, and this leads to the proof that the $\mathbb{F}_{\mathscr{D}}(M)$ are almost Ω-$\mathscr{D}$-spectra.

11. The functor $\mathbb{M}$ and the composite $\mathbb{M}\mathbb{D}$

I shall say just a little about the proofs of the following results, which finally complete the Main Diagram by describing the functor $\mathbb{M}$.

THEOREM 11.1. *There is a lax symmetric monoidal functor* $\mathbb{M} : \mathscr{I}\mathscr{S} \longrightarrow \mathscr{M}$. *If* T *is an inclusion* $\mathscr{I}$*-spectrum, then the homotopy groups of* $\mathbb{M}(T)$ *are naturally isomorphic to the homotopy groups of* T.

THEOREM 11.2. *There is a monoidal natural isomorphism* $\iota : \mathbb{M}\mathbb{D}_{\mathscr{I}} \longrightarrow \mathbb{D}_{\mathscr{M}}$. *When applied to an* E_∞*-*Σ*-FSP,* ι *is an isomorphism of* E_∞*-*S*-algebras.*

The functor $\mathbb{M}$ is the composite of three functors:

$$\mathscr{I}\mathscr{S} \xrightarrow{P} \mathscr{P}[\mathbb{L}] \xrightarrow{L} \mathscr{S}[\mathbb{L}] \xrightarrow{S\wedge_{\mathscr{L}}(-)} \mathscr{M}.$$

The functors P and L are restrictions of the evident functors

$$\mathscr{I}\mathscr{S} \xrightarrow{P} \mathscr{P} \xrightarrow{L} \mathscr{S}.$$

Here P takes an $\mathscr{I}$-spectrum to the underlying coordinate-free prespectrum indexed on our given universe $U \cong \mathbb{R}^\infty$, and L is the spectrification functor.

The categories $\mathscr{P}[\mathbb{L}]$ and $\mathscr{S}[\mathbb{L}]$ are certain subcategories of the categories of prespectra and spectra. Their objects, called $\mathbb{L}$-prespectra or $\mathbb{L}$-spectra, have extra structure defined in terms of functoriality on linear isometries. The functor P takes values in $\mathscr{P}[\mathbb{L}]$, and the functor L preserves the additional structure. The details were already quite explicit in the early work of Quinn, Ray, and myself, where it was shown that the spectrum associated to a commutative $\mathscr{I}$-FSP is an E_∞-ring spectrum. The last functor, $S \wedge_{\mathscr{L}} (-)$, is defined and studied in EKMM; it converts $\mathbb{L}$-spectra to weakly equivalent S-modules.

The isomorphism $\mathbb{M}\mathbb{D}_{\mathscr{I}} \cong \mathbb{D}_{\mathscr{M}}$ is very illuminating. I promised to say something about what weak equivalences of symmetric spectra are. The definition in terms of localization theory that is used in the formal development of the theory of Σ-spectra [**16**] is technical and makes it difficult to identify concretely which maps are weak equivalences. However, Brooke Shipley [**37**] has given the following curious looking description.

THEOREM 11.3. *A map* f *of* Σ*-spectra is a weak equivalence if and only if* $\mathbb{D}_\Sigma f$ *is a* π_**-isomorphism.*

Obviously $\mathbb{D}_\Sigma f$ is a π_*-isomorphism if and only if $\mathbb{D}_{\mathscr{I}} f$ is a π_*-isomorphism. We now see that this is true if and only if $\mathbb{D}_{\mathscr{M}} f$ is a π_*-isomorphism. In $\mathscr{M}$, the weak equivalences are the π_*-isomorphisms.

COROLLARY 11.4. *A map* f *of* Σ*-spectra is a weak equivalence if and only if*

Thus the definition of a weak equivalence of Σ-spectra is that dictated by consistency with classical stable homotopy theory: the associated map of S-modules, or equivalently spectra, must induce an isomorphism of homotopy groups. This also strongly suggests that the right definition of a weak equivalence of $\mathscr{I}$-spectra is just a π_*-isomorphism. This in turn suggests that the homotopy theory of $\mathscr{I}$-spectra will work somewhat differently, and perhaps more simply, than the homotopy theory of Σ-spectra.

I have now described all of the functors in the diagram, but that is only the beginning of our work. The essential point is the comparison of such composites as $\mathbb{FM} : \mathscr{I} \longrightarrow \mathscr{I}$ and $\mathbb{MF} : \mathscr{M}_S \longrightarrow \mathscr{M}_S$ with identity functors. I will say nothing about that here.

12. Topological cyclic homology

Let T be a monoid in any of our symmetric monoidal categories. We can mimic the standard complex for the calculation of Hochschild homology and define $THH(T)$ to be the geometric realization of the evident simplicial object with p-simplices the $(p+1)$st smash power $T^{(p+1)}$. Since this simplicial object is cyclic, there is an action of the circle group G on the object $THH(T)$ so constructed. The construction of topological cyclic homology requires a more sophisticated circle action. Although details are work in progress, it seems that the ideas here may lead to significant clarification of the construction.

Certainly Hesselholt and Madsen [**15**] begin with a Σ-FSP T and construct from it an "orthogonal G-spectrum" ($\mathscr{I}_G$-spectrum). It seems to me that their construction is equivalent to applying $\mathbb{D}_{\mathscr{I}_G}$ to the naive $\mathscr{I}_G$-spectrum $THH(T)$ constructed in the category of Σ-spectra. This effectively adds in the more sophisticated circle action they require. They then apply the G-equivariant version of the functor $\mathbb{M}$ to construct the G-spectrum version of $THH(T)$ from which $TC(T)$ is constructed. The proof that $THH(T)$ is cyclotomic is really an argument on the $\mathscr{I}_G$-spectrum level. Passage to Lewis-May G-spectra or, equivalently, EKMM S_G-modules, then allows the use of the established foundations of equivariant stable homotopy theory. It would be illuminating to have fully developed foundations already on the $\mathscr{I}_G$-spectrum level.

13. Recent developments

The preceding was prepared around the time of the conference [March, 1997], and a brief update may be appropriate [November 30, 1997]: this is a snapshot of work in progress. All four categories in the square displayed in the Main Diagram are model categories, and the weak equivalences of $\mathscr{I}$-spectra are indeed just the π_*-isomorphisms. The four forgetful functors $\mathbb{U}$ all have strict symmetric monoidal left adjoint prolongation functors $\mathbb{P}$, and the categories of symmetric and orthogonal spectra, for example, are model adjoint equivalent in the sense of Quillen [**33**, 4.7]. That doesn't render the sketch above obsolete since the arguments with $\mathbb{D}$ and $\mathbb{E}$ give the non-formal part of the proof. The proofs of all of our equivalences of homotopy categories, including those involving S-modules, give weak equivalences between their hammock localizations of Dwyer and Kan [**11**]. This means that our equivalences of brave new worlds give equivalences of their "homotopy theories", not just of their homotopy categories. This remains true for monoids and, where the theory works, commutative monoids. In particular, remarkably, the hammock localizations of the categories of commutative orthogonal ring spectra and of commutative S-algebras are weakly equivalent.

Bibliography

1. J.F. Adams. Lectures on generalised cohomology. Springer Lecture Notes in Mathematics Vol. 99, 1969, 1-138.
2. D. W. Anderson. Chain functos and homology theories. Springer Lecture Notes in Mathematics Vol. 249. 1971, 1-12.
3. M. Basterra. André-Quillen cohomology of commutative S-algebras. Preprint. 1997.
4. J.M. Boardman. Stable homotopy theory. PhD Thesis, Warwick University, 1964.
5. M. Bökstedt. Topological Hochschild homology. Preprints, 1985 and later.
6. A. K. Bousfield and E. M. Friedlander. Homotopy theory of Γ-spaces, spectra, and bisimplicial sets. Springer Lecture Notes in Mathematics Vol. 658. 1978, 80-130.
7. D. Burghelea and A. Deleanu. The homotopy category of spectra. I, II, III. Illinois J. Math. 11(1967), 454-473p; Math. Ann. 178(1968), 131-144; Math. Z. 108(1969), 154-170.
8. F. R. Cohen, T. J. Lada, and J. P May. The Homology of Iterated Loop Spaces. Lecture Notes in Mathematics Vol. 533. Springer-Verlag 1976.
9. B. Day. On closed categories of functors. Reports of the Midwest Category Seminar IV, Lecture Notes in Mathematics Vol. 137. Springer-Verlag, 1970, pp 1-38.
10. T. tom Dieck. Bordism of G-manifolds and integrality theorems. Topology 9(1970), 345-358.
11. W. G. Dwyer and D. M. Kan. Calculating simplicial localizations. J. Pure and Applied Algebra 18(1980), 17-35.
12. A. D. Elmendorf, I. Kriz, M. A. Mandell, and J. P. May. Rings, modules, and algebras in stable homotopy theory. Surveys and Monographs in Mathematics Vol. 47. 1997. American Mathematical Society.
13. P. Goerss and M.J. Hopkins. In preparation.
14. J. P. C. Greenlees and J. P. May. Localization and completion theorems for MU-module spectra. Annals of Math. To appear.
15. L. Hesselholt and I. Madsen. On the K-theory of finite algebras over Witt vectors of perfect fields. Topology 36(1997), 29-101.
16. M. Hovey, B. Shipley, and J. Smith. Symmetric spectra. In preparation.
17. D.M. Kan. Semisimplicial spectra. Illinois J. Math. 7(1963), 463-478.
18. D.M. Kan and G.W. Whitehead. Orientability and Poincaré duality in general homology theories. Topology 3(1965), 231-270.
19. I. Kriz. Towers of E_∞-ring spectra with an application to BP. Preprint. 1995
20. L. G. Lewis, Jr. Is there a convenient category of spsectra? J. Pure and Applied Algebra 73(1991), 233-246.
21. L. G. Lewis, Jr., J. P. May, and M. Steinberger (with contributions by J. E. McClure). Equivariant stable homotopy theory. Springer Lecture Notes in Mathematics Vol. 1213. 1986.
22. E.I. Lima. The Spanier-Whitehead duality in new homotopy categories. Summa Brasil Math. 4(1959), 91-148.
23. M. Lydakis. Smash-products and Gamma-spaces. Preprint. Bielefeld. 1996.
24. M. Mandell. E_∞ algebras and p-adic homotopy theory. PhD Thesis. The University of

25. M. A. Mandell, J. P. May, S. Schwede and B. Shipley. Diagram spectra and generalized FSP's. In preparation.
26. M. A. Mandell, J. P. May, S. Schwede and B. Shipley. Diagram spectra and S-modules. In preparation.
27. J.P. May. Categories of spectra and infinite loop spaces. Lecture Notes in Mathematics Vol. 99, Springer-Verlag 1969, 448-479.
28. J. P. May (with contributions by F. Quinn, N. Ray, and J. Tornehave). E_∞-ring spaces and E_∞-ring spectra. Springer Lecture Notes in Mathematics Vol. 577. 1977.
29. J.P. May. Infinite loop space theory. Bull. Amer. Math. Soc. 83(1977), 456-494.
30. J.P. May. Multiplicative infinite loop space theory. J. Pure and Applied Algebra 26(1983), 1–69.
31. J. P. May, et al. Equivariant homotopy and cohomology theory. CBMS Regional Conference Series in Mathematics, Number 91. American Mathematical Society. 1996.
32. F. Morel and V. Voevodsky. Homotopy category of schemes over a base. Preprint. 1996.
33. D. G. Quillen. Homotopical algebra. Springer Lecture Notes in Mathematics Vol. 43. 1967.
34. D. G. Quillen. On the cohomology and K-theory of the general linear groups over a finite field. Annals of Math. (2)96(1972), 552-586.
35. S. Schwede. Stable homotopical algebra and Γ-spaces. Preprint, 1997.
36. G. Segal. Categories and cohomology theories. Topology 13(1974), 293-312.
37. B. Shipley. Symmetric ring spectra and topological Hochschild homology. Preprint. 1997.
38. N. Strickland. Products on MU-modules. Preprint, 1997.
39. D. Sullivan. Geometric topology. I. Localization, periodicity, and Galois symmetry. MIT. Mimeographed notes. Circa 1972.
40. V. Voevodsky. The Milnor conjecture. Preprint. 1996.
41. J. Wolbert. Classifying modules over K-theory spectra. J. Pure and Applied Algebra. To appear.

THE UNIVERSITY OF CHICAGO, CHICAGO, IL 60637
E-mail address: may@math.uchicago.edu

Contemporary Mathematics
Volume **220**, 1998

The Stable Splitting of $B_{M_{12}}$ and Related Groups

R. James Milgram and C. Overton

Introduction

The sporadic group M_{12} is very interesting. For one thing it has the same 2-Sylow subgroup as $G_2(q)$ or ${}^3D_4(q)$ for $q \equiv 3,5 \bmod (8)$. But there is no non-trivial homomorphims from M_{12} to $G_2(q)$, q as before. None-the-less [**M**] (see also [**BW**]) shows that $H^*(B_{G_2(q)}\mathbb{F}_2) \subset H^*(M_{12};\mathbb{F}_2)$.

The way in which this embedding was achieved strongly suggested first, that $B_{G_2(q)}$ is the dominant summand of $B_{Syl_2(M_{qw})}$ and second, that the exact structure of the stable splitting of $B_{M_{12}}$ should be very interesting as well.

In this note which includes, among other things, many of the main results of [**O**], we determine the (2-local) stable splittings of $B_{M_{12}}$ as well as of many associated groups. In all the results we now state we are only talking about the structure at 2. Also, it should be mentioned that we use the notation of the Atlas, [**C**], for groups and their subgroups, so, for example the group $(\mathbb{Z}/n)^s$ is written n^s, and a semi-direct product $A \times_\alpha B$ is written $A{:}_\alpha B$ where $\alpha\colon B \longrightarrow Aut(A)$ is the twisting homomorphism.

THEOREM. *$B_{G_2(q)}$ is stably irreducible. In fact it is the principal dominant summand of $B_{Syl_2(M_{12})}$.*

We don't know when a finite group has the property that its classifying space is a single dominant summand in the classifying space of its 2-Sylow subgroup but here is a second example: the principal dominant summand of B_{2^2} is $B_{\mathcal{A}_4}$ where $\mathcal{A}_4$ is the alternating group on four letters.

THEOREM. *$B_{M_{12}}$ stably splits as the wedge sum of 3 irreducibles:*

$$B_{M_{12}} \simeq_s B_{G_2(q)} \vee B_{\mathcal{A}_4} \vee X_{E,S_1}$$

where

$$E \cong Q_8 * Q_8 \subset Syl_2(M_{12})$$

and X_{E,S_1} is a non-principal dominant summand.

1991 *Mathematics Subject Classification.* Primary 20C05, 20J06, 55P42, 55R35, 55R40.

The first author was partially supported by a grant from the N.S.F. and the Regents Profes-

REMARK. : It will be seen in [**M2**] that one of the roles that $B_{\mathcal{A}_4}$ plays in $B_{M_{12}}$ is to construct an element, $A \in \pi_{14}^s(B_{M_{12}})$ which, under the composite

$$Q(B_{M_{12}}) \hookrightarrow Q(B_{\mathcal{S}_{12}}) \hookrightarrow Q(S^0)$$

where $Q(X) = \lim_{n\mapsto\infty}(\Omega^n\Sigma^n X)$, which maps the stable homotopy of $B_{M_{12}}$ to the stable homotopy of spheres, goes to an element of Kervaire invariant one in the stable 14-stem.

This above result is part (2) of theorem 5.3 in the text. Part (a) of 5.3 describes the complete stable splitting for $H = Syl_2(M_{12})$

THEOREM. *The complete stable splitting of B_H is given as*

$$B_H \simeq_s B_{G_2(q)} \vee X_{K,S} \vee 2(X_{E,S_1} \vee X_{E,S_2}) \vee B_{L_3(2)} \vee 2B_{\mathcal{A}_4} \vee 8(L(3) \vee L(2)) \vee 3B_2^\infty$$

where $L(i)$ is the Steinberg summand for B_{2^i} and $K = 4^2{:}_{-1}2$ is a second index 2 subgroup of H.

REMARK. The notation $4^2{:}_{-1}2$ for K says that K is the semi-direct product

$$(\mathbb{Z}/4)^2 \times_{-1} \mathbb{Z}/2$$

where the non-trivial element in the $\mathbb{Z}/2$ acts on the $(\mathbb{Z}/4)^2$ to invert each element.

REMARK. Catalano in his thesis (U. of Minnesota, 1995), determines the Poincaré series for all the dominant summands of the groups of order 32 with the exception of the groups E and K. The results here complete the determination of these summands as well.

The key step in determining the splitting of any group is to determine the lattice of subgroups and the outer automorphism groups of each subgroup. The first three sections do this for the group H. Then the later sections, using the techniques of [**MP**] and [**BF**], complete the determination of the stable splittings.

We would like to thank M. Feshbach and S. Priddy for many very useful conversations during the course of this work.

1. The group $Syl_2(M_{12})$

The Sylow 2-subgroup of M_{12}, $Syl_2(M_{12}) = H$, can be described as a semidirect product $4^2{:}\,2^2$ with the 4^2 generated by c, d, the 2^2 by k, s where k acts to invert both c and d while s interchanges them.

The index 2 subgroups of H.

A list of the conjugacy classes of subgroups of H can be extracted from [**BR**]. But since we will need slightly more detail than is given there we include the following list of subgroups. To begin note that $H' = 4 \times 2 = \langle cd, c^2\rangle$ so $H/H' = 2^3$ with generators c, k, s. There are thus 7 non-trivial homomorphisms $H \longrightarrow 2$ with the 7 corresponding kernels being the index 2 subgroups of H. In the table that follows the homomorphism is listed as a triple (a, b, c) where a is the value on c, b

on k, and c on s.

(1.1)

homomorphism	kernel	name
$(1,0,0)$	$\langle cd, c^2, k, s\rangle = D_8 * D_8$	E
$(0,1,0)$	$\langle cd, c, s\rangle = 4 \wr 2$	
$(0,0,1)$	$\langle cd, c, k\rangle = 4^2{:}_{-1}\, 2$	K
$(1,1,0)$	$\langle cks, ck, c^2\rangle = 8{:}\, Aut(8)$	
$(1,0,1)$	$\langle cks, c^2, k\rangle = (8{:}_5\, 2) : 2$	W
$(0,1,1)$	$\langle cd, c, ks\rangle = 4 \wr 2$	
$(1,1,1)$	$\langle cs, c^2, ck\rangle = 8{:}\, Aut(8)$	

Here W is described as follows. Set $M_{16} = 8{:}_5\, 2$ with presentation

(1.2) $$M_{16} = \{x, y \mid x^8 = y^2 = 1,\ yxy = x^5\}$$

then $Out(M_{16}) = 2^2$ with generators

(1.3) $$\alpha = \begin{Bmatrix} x & \mapsto & xy \\ y & \mapsto & y \end{Bmatrix}, \qquad \beta = \begin{Bmatrix} x & \mapsto & x^{-1} \\ y & \mapsto & y \end{Bmatrix}$$

and W is the split extension $M_{16}{:}_\alpha\, 2$. Also, note that $D_8 * D_8 = Q_8 * Q_8$ as is well known, [**Q1**]. Finally, note that since H only has one copy of each of E, K, and W we have

COROLLARY 1.4. *The index two subgroups E, K, and W are characteristic in H. Also, $V = \langle c, d\rangle$ is characteristic, being the centralizer of H' in H.*

The remaining subgroups of H.

We list the subgroups of order 16 in H. From [**BR**] or direct calculation we find that there are six isomorphism classes.

$$4^2 = V,\ D_8 * 4,\ D_8 \times 2,\ SD_{16},\ M_{16},\ D_{16},$$

where the semi-dihedral group $SD_{16} = 8{:}_3\, 2$. It is also easy to verify the isomorphism $D_8 * 4 \cong Q_8 * 4$. All five groups of order 8 are subgroups of H as are the two groups of order 4.

In more detail we have the following description of the subgroups of the groups of order 32 contained in H.

LEMMA 1.5.

1. *The group $4 \wr 2$ has three subgroups of index two: $D_8 * 4$, 4^2, and M_{16}.*
2. *The group W has three subgroups of index two: $D_8 \times 2$ and two copies of M_{16}.*
3. *Write $Hol(8) = 8{:}\, Aut(8)$ as*

$$\{x, y, z \mid x^8 = y^2 = z^2 = (xy)^2 = 1,\ yxy = x^5, zxz = x^{-1}\},$$

then $Hol(8)$ contains two copies of D_{16}, two copies of SD_{16}, $\langle xy, z\rangle$, $\langle x, yz\rangle$, one copy of M_{16}, one copy of $D_8 \times 2$ and one copy of $D_8 * 4$, $\langle x^2, xz, x^2y\rangle$.

(The subgroups are obtained by determining L/L' for each of these groups and checking the kernels of the associated homomorphisms. Thus, $4 \wr 2$ and W both have commutator quotients 2^2 while the others have quotient 2^3. The calulations are all direct.)

The inclusion results in (1.5) can be summarized in the following table.

(1.6)

Subgroup	Contained in
V	$4 \wr 2,\ K$
$D_8 * 4$	$4 \wr 2,\ Hol(8),\ E$
$D_8 \times 2$	$Hol(8),\ W,\ E,\ K$
SD_{16}	$Hol(8)$
M_{16}	$W,\ Hol(8),\ 4 \wr 2$
D_{16}	$Hol(8)$

We also need the following observation.

LEMMA 1.7. *The intersection of each $4 \wr 2$ with E is a copy of $D_8 * 4$ as is the intersection of each $Hol(8)$ with E.*

(This is direct.)

Some useful splittings of H.

There are (at least) three non-conjugate splittings of the form $H = L{:}\,2$,

$$(1.8)\qquad \begin{aligned} H &= K{:}\,\langle s\rangle \\ &= (4 \wr 2){:}\,\langle k\rangle \\ &= E{:}\,\langle ck\rangle \end{aligned}$$

The subgroup E can also be written in two ways as a split extension of the form $Q_8{:}\,2^2$:

$$(1.9)\qquad \begin{aligned} E &= \langle cd, c^{-1}dks\rangle{:}\,\langle k, c^2\rangle \\ &= \langle c^{-1}d, cds\rangle{:}\,\langle k, c^2\rangle. \end{aligned}$$

These last two descriptions of E give H as a split extension in two ways:

$$(1.10)\qquad \begin{aligned} H &= Q_8{:}\,D_8 \subset Hol(Q_8) \\ &= \langle cd, c^{-1}dks\rangle{:}\,\langle k, c\rangle \\ &= \langle c^{-1}d, cds\rangle{:}\,\langle k, c\rangle \end{aligned}$$

which, in both cases, embed D_8 as a two-Sylow subgroup of $Aut(Q_8) = \mathcal{S}_4$.

The conjugacy classes of maximal 2-elementaries in H.

There are five conjugacy classes of maximal elementary subgroups $2^3 \subset H$ with representatives

$$
(1.11)\qquad \begin{aligned}
V_1 &= \langle c^2, d^2, k\rangle\\
V_2 &= \langle c^2, d^2, ck\rangle\\
V_3 &= \langle c^2, d^2, cdk\rangle\\
V_4 &= \langle (cd)^2, k, s\rangle\\
V_5 &= \langle (cd)^2, cdk, s\rangle
\end{aligned}
$$

where V_1 and V_3 are normal in H, while the Weyl groups of V_2, V_4 and V_5 are each 2^2. These five conjugacy classes fuse in M_{12} into three conjugacy classes

$$
(1.12)\qquad \{V_1, V_2\},\ \{V_3, V_4\},\ V_5.
$$

The first and second of these conjugacy classes have Weyl group $\mathcal{S}_4$ in M_{12}, (there are exactly two non-isomorphic actions of $\mathcal{S}_4$ on 2^3, the first – the action on V_1 – leaves a vector invariant while the second – the action on V_3 – leaves a 2^2 invariant). The Weyl group of V_5 is $\mathcal{A}_4$ in M_{12} – the action which leaves a vector invariant.

We also note that H is $Syl_2(G_2(q))$ for $q \equiv 3, 5 \bmod (8)$ and there the 2^3's fuse into two conjugacy classes

$$
(1.13)\qquad \{V_1, V_2, V_5\},\ \{V_3, V_4\}
$$

the first with Weyl group $\mathcal{S}_4$ (fixing a vector), and the second with Weyl group $L_3(2)$.

2. The groups $Out(L)$ for $L \subset H$ of index 2

The following lemma is contained in [**BF**] and provides a basic principle in determining the splitting summands for a p-group.

LEMMA 2.1. *Let P be a p-group and $L \subset P$ any index p subgroup. Then the principal dominant of L is a splitting summand of P if and only if L is a split summand of P.*

REMARK. A sharpened version of 2.1 occurs in J. Martino's thesis, [**Ma**], Theorem 2.11. He shows that for $Q \subset P$ any subgroup, then the principal dominant of Q is a summand of B_P if and only if Q is a retract of P.

Note that none of the index two subgroups of H are split in H so none of them contribute their principal dominant to H. We now show that the only ones which have non-principal dominant summands are K and E.

LEMMA 2.2. *$Hol(8)$, W, and $4 \wr 2$ all have the property that $Out(L)$ is a 2-group so each has only one dominant summand, the principal dominant.*

PROOF. For $4 \wr 2 = \langle c, d\rangle \colon \langle s\rangle$ we have $4^2 = \langle c, d\rangle$ is characteristic and

$$
\{s, c^{-1}ds, cd^{-1}s, c^2d^2s\}
$$

For $Hol(8) = \langle cs, c^2, ck\rangle$ the commutator subgroup is $4 = \langle cd\rangle$ with centralizer $M_{16} = \langle cs, c^2\rangle$ which is thus characterstic. It follows that under the map

$$Aut(Hol(8)) \longrightarrow GL_3(2) = Aut(2^3)$$

induced from factoring out by the commutator subgroup, the image of $Aut(Hol(8))$ lies in the subgroup S_4 leaving the plane $\langle cs, c^2\rangle$ invariant. But since c^2 is the only elment in this plane which lifts to an elment of order 2 in $Hol(8)$ it follows that the image is contained in D_8 and (2.2) follows for this case as well.

For $W = \langle sc, c^2, k\rangle$ we see that the commutator subgroup is $2^2 = \langle c^2, d^2\rangle$. Thus, $W/W' = 4 \times 2$ and the image of $Aut(W)$ in $Aut(W/W')$ is a 2-group. □

The group $Out(H)$.

As we have seen $V = 4^2 \subset H$ is characteristic. Consequently it is taken to itself by any automorphism of H and we have an exact sequence

$$M \xrightarrow{\triangleleft} Aut(H) \xrightarrow{\pi} GL_2(\mathbb{Z}/4)$$

where M is the subgroup of the automorphism group which acts as the identity on V. We obtain M and $Im(\pi)$ as follows.

LEMMA 2.3.
1. *The subgroup M has order 16, while the intersection of M with $Inn(H)$ has order 8. A representative for the non-trivial outer automorphism in M is $k \mapsto cdk$, $s \mapsto s$, while c, d are fixed.*
2. *The image of π is the normalizer of the subgroup*

$$N = \left\langle \begin{pmatrix} 0 & 1 \\ 1 & 0 \end{pmatrix}, \begin{pmatrix} 3 & 0 \\ 0 & 3 \end{pmatrix} \right\rangle \subset GL_2(\mathbb{Z}/4)$$

which is given as the (degree 4) extension of N by the matrices $\begin{pmatrix} 0 & 1 \\ -1 & 0 \end{pmatrix}$, $\begin{pmatrix} 1 & 2 \\ 2 & 1 \end{pmatrix}$.

PROOF. Any automorphism in M fixes c, d, and on k, s has the form $k \mapsto c^i d^j k$, $s \mapsto c^m d^n s$ because only s in the set $\{s, k, sk\}$ interchanges c, d, and only k inverts both c and d. Since the image is again 2^2 it follows that

$$\begin{aligned} ks &\mapsto c^i d^j k c^m d^n s \\ &= c^{i-m} d^{j-n} ks \\ &= c^m d^n s c^i d^j k \\ &= c^{m+j} d^{n+k} sk \end{aligned}$$

so necessary and sufficient conditions on i, j, m, n are $m + n \equiv 0 \bmod (4)$, $i - j \equiv 2m \bmod (4)$. Thus $|M| = 16$. Now, the inner automorphisms in M are clearly conjugation by the elements of $\langle c, d\rangle$, which has order 8 since the center of H is a copy of $\mathbb{Z}/2$, $\langle c^2 d^2\rangle$.

We see that the form of the inner automorphisms is

$$k \mapsto c^{2i} d^{2j} k, \qquad s \mapsto (c^2 d^2)^{\epsilon} (cd^{-1})^{i+j} s$$

so the automorphism $k \mapsto cdk$, $s \mapsto s$ is outer.

Finally, the description of $Im(\pi)$ is pretty much direct. Note that the image of k, $\begin{pmatrix} 3 & 0 \\ 0 & 3 \end{pmatrix}$ is in the center of $GL_2(\mathbb{Z}/4)$. Consequently, the centralizer of N is given as the set of invertible matrices m for which $\pi(s)m\pi(s) = m$, i.e., since $\pi(s) = \begin{pmatrix} 0 & 1 \\ 1 & 0 \end{pmatrix}$, these are all m so that $m_{11} = m_{22}$, $m_{12} = m_{21}$. This gives $\begin{pmatrix} 1 & 2 \\ 2 & 1 \end{pmatrix}$. On the other hand, the elements in the normalizer of N are given by the equation $\alpha\pi(s)\alpha^{-1} = \pm\pi(s)$. But $\begin{pmatrix} 0 & 1 \\ -1 & 0 \end{pmatrix} = \alpha$ conjugates $\pi(s)$ to $-\pi(s)$ and the result follows. □

COROLLARY 2.4. $Out(H) = 2^3$.

COROLLARY 2.5. *B_H has a single dominant summand.*

The determination of $Out(E)$.

LEMMA 2.6. *There are two and only two Q_8's contained in E.*

PROOF. In the description of $Q_8 * Q_8 = E$ in (1.10) we have

$$c^2 = s^2 = (cdk)^2 = k^2 = [c^2, cdk] = [s, cdk] = [c^2, k] = [s, k] = 1,$$
$$[c^2, s] = [cdk, k] = (cd)^2,$$

with $(cd)^2$ central. It follows that there are exactly 12 elements of order 4 in E, the six classes

$$c^2s, cd, c^2scdk, c^2sk, c^3d, cds$$

and their respective third powers. Checking the six elements above we see there are precisely two Q_8's that can be constructed using 2 at a time. □

COROLLARY 2.7. $Out(E) = Out(Q_8) \wr 2 = \mathcal{S}_3 \wr 2$.

(Since the center of Q_8 is fixed by any element in $Aut(Q_8)$, one can act independently on each Q_8 or exchange them, so the corollary is clear.)

COROLLARY 2.8. *The dominant summands for E are of the form $X_{E,\mathbb{F}_2} \vee 4X_{E,S_1} \vee 4X_{E,S_2}$ where $X_{E,\mathbb{F}_2}$ is the principal dominant.*

PROOF. We have

$$\begin{aligned} \mathbb{F}_2(\mathcal{S}_3) &= \mathbb{F}_2(3 :_{-1} 2) \\ &= \mathbb{F}_2(3) \odot_{-1} 2 \\ &= (\mathbb{F}_2 \oplus \mathbb{F}_2(\zeta_3)) \odot_{-1} 2 \\ &= \mathbb{F}_2(2) \oplus \mathbb{F}_2(\zeta_3) \odot 2 \end{aligned}$$

where the twisting in the second summand is via the Galois action on $\mathbb{F}_4 = \mathbb{F}_2(\zeta_3)$. But $\mathbb{F}_4 \odot 2$ is a simple algebra with center $\mathbb{F}_2$ and is thus isomorphic to $M_2(2)$. Consequently, we have $\mathbb{F}_2(\mathcal{S}_3) = \mathbb{F}_2(2) \oplus M_2(\mathbb{F}_2)$, and the wreath product splits as follows when we note that $M_2(2) \otimes M_2(2) = M_4(2)$:

$$\mathbb{F}_2(\mathcal{S}_3 \wr 2) = [\mathbb{F}_2(\mathcal{S}_3) \otimes \mathbb{F}_2(\mathcal{S}_3)] \odot 2$$

since the 2 must act as an inner automorphism on $M_4(\mathbb{F}_2(2))$, and it acts to exchange the two copies of $M_2(2) \otimes \mathbb{F}_2(2) = M_2(\mathbb{F}_2(2))$. The corollary follows. □

COROLLARY 2.9. *There are precisely 6 copies of $D_8 * 4 \subset H$, all contained in E. They break up into 4 conjugacy classes in H, two of which are normal. For the non-normal classes the Weyl group $\mathbb{Z}/2$ acts trivially on Q_8 and as inversion on the 4. For the normal classes the Weyl group is 2^2, generated by the automorphism above and the element acting trivially on the 4 and as the non-trivial outer automorphism of order 2 on Q_8.*

PROOF. From the analysis of the subgroup structure in (1.5), (1.7), we see that every $D_8 * 4 \subset E$. But since $D_8 * 4 = Q_8 * 4$ and there are exactly two Q_8's in E it follows that there are exactly 6 copies of $Q_8 * 4$ contained in E. Moreover, in E each of these is normal with Weyl group $\mathbb{Z}/2$ acting as described in the corollary. Finally, the action of H/E on E is as the diagonal outer automorphism on $Q_8 * Q_8$ which finishes the proof. □

COROLLARY 2.10. *There are exactly two copies of Q_8 in H, the two copies in E.*

PROOF. The groups W and K contain no copies of Q_8. The group $4 \wr 2$ contains exactly one which is contained in $D_8 * 4$ so is contained in E as well. Finally, the subgroups SD_{16} and $D_8 * 4$ in $Hol(8)$ each contain one copy of Q_8 but from the description of these subgroups in (1.5), (1.7) it is clear that this Q_8 lies in the intersection of all three and consequently lies in E. □

COROLLARY 2.11. *There are two conjugacy classes of Q_8's in H and each is non-split.*

The determination of $Out(K)$.

$K = 4^2{:}2$ with the 4^2 characteristic since it is the only 4^2 contained in K. Thus we have the commutative diagram

$$\begin{array}{ccc} Aut(K) & \xrightarrow{RES} & Aut(4^2) \\ \downarrow{\scriptstyle \pi} & & \downarrow{\scriptstyle \pi} \\ Out(K) & \xrightarrow{res} & GL_2(\mathbb{Z}/4)/\{\pm I\} \end{array}$$

since $Aut(4^2) = GL_2(\mathbb{Z}/4)$, and the action of k maps $Inn(K)$ to $\pm I$.

LEMMA 2.12.
1. *RES: $Aut(K) \longrightarrow Aut(4^2)$ is onto with kernel 4^2.*
2. *res is onto with kernel 2^2.*
3. *$Out(K)$ is given as a semi-direct product $2^5{:}\mathcal{S}_3$.*

PROOF. To show a) note that $\alpha \in GL_2(\mathbb{Z}/4)$ is in the image of $Aut(K)$ it and only if it centralizes $-I$, but $-I$ is central in $GL_2(4)$. The kernel of this projection consists of all automorphism ϕ so that $\phi(b) = b$, $\phi(c) = c$, and consequently, $\phi(k) = b^i d^j k$. Note that $\phi^2(k) = b^{2i} c^{2j} k$, $\phi^3(k) = b^{3i} c^{3j} k$ so the kernel is indeed 4^2.

To show b) note that $k \longrightarrow b^{2i} c^{2j} k$ is inner, but $\phi_b\colon k \mapsto bk$, $\phi_c\colon k \mapsto ck$, $\phi_{bc}\colon k \mapsto cdk$ are outer.

Finally, to show c) write $GL_2(4)$ as an extension $2^4 \triangleleft GL_2(4) \longrightarrow GL_2(2) = \mathcal{S}_3$ where the elements of the normal 2^4 have the form $I + 2M$, $M \in M_2(4)$. Factoring out by $\langle -I \rangle = \langle I + 2I, I \rangle$ we get a 2^3 with generators

$$\begin{array}{llll} \psi_1\colon & k \mapsto k & c \mapsto c^{-1} & b \mapsto b \\ \psi_2\colon & k \mapsto k & c \mapsto cd^2 & b \mapsto b \\ \psi_3\colon & k \mapsto k & c \mapsto c & b \mapsto bc^2 \end{array}$$

Now, a direct calculation shows that $\langle \psi_1, \psi_2, \psi_3, \phi_b, \phi_c \rangle = 2^5$ after factoring out by conjugation by c, d. □

COROLLARY 2.13. *The dominant summands for* K *have the form* $X_{K,\mathbb{F}_2} \vee 2X_{K,S}$ *where, as before* $X_{K,\mathbb{F}_2}$ *denotes the principal dominant.*

PROOF. Since the subgroup 2^5 only contributes elements in the Jacobson radical to the group ring $\mathbb{F}_2(Out(K))$ we can ignore it, and since $\mathbb{F}_2(\mathcal{S}_3) = \mathbb{F}_2(2) \oplus M_2(\mathbb{F}_2)$ the corollary follows. □

3. The groups $Out(L)$ for the groups of order 16 in H

We now determine the outer automorphism groups for the six isomorphism classes of subgroups of index 4 in H as far as we need them. Of course $Aut(4^2) = Out(4^2) = GL_2(\mathbb{Z}/4)$ is given as a semi-direct product

$$Aut(4^2) = Out(4^2) = 2^4 \colon \mathcal{S}_3.$$

LEMMA 3.1. *The four subgroups* $D_8 \times 2$, SD_{16}, M_{16} *and* D_{16} *all satisfy the condition that* $Aut(L)$ *is a finite 2-group.*

PROOF. We have already verified this for M_{16} in (1.3). In SD_{16} and D_{16} the subgroup 8 is characteristic, so the result is direct. Finally, for $D_8 \times 2$ the subgroup 4×2 is characteristic with $Aut(4 \times 2)$ a finite 2-group so the lemma follows. □

For the remaining case $D_8 * 4 = Q_8 * 4$ and there is only one Q_8 subgroup in $D_8 * 4$ so Q_8 is characteristic. Consequently there is a short exact sequence

$$2 \triangleleft Aut(D_8 * 4) \xrightarrow{\pi} Aut(Q_8) = \mathcal{S}_4 \longrightarrow 1$$

and $Out(D_8 * 4)$ is the product $2 \times \mathcal{S}_3$.

COROLLARY 3.2. *The only index* 4 *subgroups of* H *which can contribute dominant summands to* H *are* V *and* $D_8 * 4$. *The dominant summands in these two cases have the form*

$$X_{V,\mathbb{F}_2} \vee 2X_{V,S} \textit{ for } V, \textit{ and } X_{D_8*4,\mathbb{F}_2} \vee 2X_{D_8*4,S} \textit{ for } D_8 * 4.$$

4. A preliminary analysis of the stable splitting for H

We have that K, E, V, and $D_8 * 4$ all satisfy the hypothesis that they are not subretracts in H. For K, E, V, this is easy since they are normal with $H/L \subset$

LEMMA 4.1. *No dominant summand for V or $D_8 * 4$ is a summand of B_H. The dominant summands of B_H coming from K and E are $X_{K,S}$ and $2(X_{E,S_1} \vee X_{E,S_2})$ respectively.*

PROOF. We consider the elements $\bar{W}_i = \sum_{x \in N_H(L_i)/L_i} c_x$ acting on the non-trivial modules M associated to the splitting summands according to 4.4 of [**MaP**]. The action of H/E on E embeds the non-trivial element as the diagonal element of order two in the $\mathcal{S}_3 \times \mathcal{S}_3 \subset Out(E) = \mathcal{S}_3 \wr 2$. Note, from the proof of (2.8), that

$$\mathbb{F}_2(\mathcal{S}_2 \wr 2) \cong \mathbb{F}_2(D_8) \oplus M_4(\mathbb{F}_2(2)) \oplus M_4(\mathbb{F}_2(2)).$$

This implies that the quotient by the radical is

$$\mathbb{F}_2 \oplus M_4(\mathbb{F}_2) \oplus M_4(\mathbb{F}_2).$$

Consequently, a direct calculation shows that the image of this non-trivial element in the quotient is the triple

$$(1, \begin{pmatrix} J & 0 \\ 0 & J \end{pmatrix}, \begin{pmatrix} 0 & I \\ I & 0 \end{pmatrix}) \tag{4.2}$$

where $J = \begin{pmatrix} 0 & 1 \\ 1 & 0 \end{pmatrix}$. Thus, the rank of $I + \{ck\}$ is $(0, 2, 2)$ and the summands coming from E are determined. Similarly, for K the outer automorphism coming from H/K embeds in $\mathcal{S}_3$ as a non-trivial element of order 2, thus giving $(1, J)$ as its image in $\mathbb{F}_2(\mathcal{S}_3)$, so $I + \{s\}$ has rank $(0, 1)$.

Since the automorphism which is the identity on Q_8 and -1 on the 4 in $Q_8 * 4$ restricts to I in $Aut(Q_8 * 4)/J$ it follows that no dominant summand of $Q_8 * 4$ is a summand of B_E, so, since every $Q_8 * 4$ in H is contained in E the proof is complete in this case.

Finally, we must verify the result for V. Note that there is only one copy of V contained in H and it is contained in K, where the non-trivial element of K/V acts as -1 in V. But this is in the radical of $Out(V)$, and consequently the argument above verifies the result. □

Next we check the dominant summands coming from the 2^3's. From [**HK**] we have

$$B_{2^3} \simeq_s B_{J_1} \vee 3X_{3,1} \vee 3X_{3,2} \vee 8L(3) \vee Y \tag{4.3}$$

where Y consists of summands coming from 2^2, B_{J_1} is the principal dominant, $L(3)$ comes from the Steinberg representation $L_3(2) \longrightarrow L_8(2)$, $X_{3,1}$ is associated to $id: L_3(2) \longrightarrow L_3(2)$ and $X_{3,2}$ is associated to $*: L_3(2) \longrightarrow L_3(2)$, $(*(a) = a^{-t})$. The summand $L(3)$ is 10 connected. (A construction of all the Steinberg summands $L(n)$ is given in [**MP**].)

We also need the summands coming from 2^2. From [**W**] or [**HK**] we have

$$B_{2^2} \simeq_s B_{A_4} \vee 2L(2) \vee 2B_2 \tag{4.4}$$

where the Steinberg summand $L(2)$ is three connected.

From the discussion in (1.11) we note that the hypothesis of 4.6 of [**MaP**] are satisfied for each of the five conjugacy classes of 2^3's in H.

It follows that the number of $L(3)$ summands of B_H is given as

$$\sum_{V_i} 8/|N_H(V_i)| = 1+1+2+2+2 = 8. \tag{4.5}$$

Moreover, since the $L(2)$'s in B_{2^3} are linked to $L(3)$ it follows that we have at least 8 copies of $L(2)$ as splitting summands for B_H as well.

The situation for B_{J_1}, $X_{3,1}$ and $X_{3,2}$ is similar. The key thing to remark is that the hypothesis of 4.4 of [**MaP**] is satisfied for each 2^3. Consequently (2.1) implies that B_{J_1} does not occur as a summand of B_H. For $X_{3,i}$, note that each of the Weyl groups of the V_i in H contains the group 2^2 acting on 2^3 via the four matrices

$$\begin{pmatrix} 1 & 0 & 0 \\ 0 & 1 & 0 \\ 0 & 0 & 1 \end{pmatrix}, \begin{pmatrix} 1 & 0 & 1 \\ 0 & 1 & 0 \\ 0 & 0 & 1 \end{pmatrix}, \begin{pmatrix} 1 & 0 & 0 \\ 0 & 1 & 1 \\ 0 & 0 & 1 \end{pmatrix}, \begin{pmatrix} 1 & 0 & 1 \\ 0 & 1 & 1 \\ 0 & 0 & 1 \end{pmatrix} \tag{4.6}$$

the sum of which is already the zero matrix. Consequently, the five sums are identically zero, so these dominant summands do not occur in B_H.

Finally, the three splittings $H = L{:}\,2$ give at least three copies of B_2 as summands of B_H since they give independent maps of $H/H' = 2^3$. But since $H_1(3B_2) = 2^3$ as well, it follows that there are exactly three B_2's and all remaining summands are at least 1-connected.

Summarizing, we have

LEMMA 4.7.

$$B_H \simeq_s X_{H,\mathbb{F}_2} \vee 2(X_{E,S_1} \vee X_{E,S_2}) \vee X_{K,S} \vee 8(L(3) \vee L(2)) \vee 3B_2 \vee Y$$

where $X_{H,\mathbb{F}_2}$ is the principal dominant of H and Y is a wedge of dominant summands coming from D_8 and 2^2.

REMARK. From the proof of (2.8) there are three irreducible $\mathcal{S}_8 \wr 2$-modules, the first, a copy of $\mathbb{Z}_2$, corresponding to the first summand $\mathbb{F}_2(D_8)$ in $\mathbb{F}_2(\mathcal{S}_3 \wr 2)$, the second, S_1, corresponding to the copy of $M_4(\mathbb{F}_2(2))$ which arises as $(M_2(2) \otimes \mathbb{F}_2(2) \oplus \mathbb{F}_2(2) \otimes M_2(2)) \odot 2$, and the third, S_2, coming from the remaining $M_4(\mathbb{F}_2(2))$. Also, S is defined in (2.13) as the non-dominant summand corresponding to the $M_2(\mathbb{F}_2(2^5))$ summand of $\mathbb{F}_2(Out(K))$.

PROOF. The only subgroups we have not yet discussed that are excluded above are Q_8, 8 4 and 4×2. In the case of Q_8, (2.11) shows that we can again apply 4.4 of [**MaP**] to conclude that no dominant summands occur. In the cases of 8 and 4 there are no splitings so the dominant summands are B_8 and B_4 respectively. But neither one is 1-connected.

It remains to exclude 4×2. 4×2 only has the principal dominant so it remains to note that 4×2 is not split in any subgroup of H. But the subgroups of order 16 which contain a copy of 4×2 are V, $D_8 \times 2$, $D_8 * 4$, and M_{16}, and (4.7) follows.□

REMARK 4.8. From the splittings $H = Q_8{:}\,D_8$, $H = V{:}\,2^2$ it follows that the and (4.4) describes the splitting for B_{2^2}. We have seen that at least 8 copies of $L(2)$ occur in B_H. It remains to get the exact number of copies of $L(2)$, $B_{\mathcal{A}_4}$ and $B_{L_3(2)}$. To do this it seems easier to use a cohomology calculation. Indeed,

$$\begin{aligned} H^*(B_{\mathcal{A}_4}) &= \mathbb{F}_2[d_2, d_3](1, c_3), \\ & Sq^1(c_3) = d_2^2, \\ H^*(B_{L_3(2)}) &= \mathbb{F}_2[d_2, d_3](1, c_3'), \\ & \beta_4(c_3') = d_2^2, \end{aligned}$$

so $H^4(B_{L_3(2)}; \mathbb{Z}) = \mathbb{Z}/4$ while $H^4(B_{\mathcal{A}_4}; \mathbb{Z}) = \mathbb{Z}/2$ so cohomology with $\mathbb{Z}/4$ coefficients will distinguish these cases, and since $L(2)$ is three connected a dimension count in $H^4(H)$ will serve to determine the exact number of these summands.

5. The cohomology of H and the complete stable splitting

The cohomology of H is determined in [**AM**], pp. 258-261, [**FM**], by noting that there is an index two embedding $H \subset 2 \wr 2 \wr 2$ and using the Gysin sequence together with the well known structure of $H^*(2 \wr 2 \wr 2)$. The result is that $H^*(B_H)$ has 7 generators, c, $\pi^* tr(x_1)$, e in dimension one, $\pi^*\Gamma(x)$, $\pi^* tr(w_1)$ in dimension 2, $\pi^* tr(x_1 w_2)$ in dimension three and $\pi^*\Gamma(w)$ in dimension 4. They restrict to the five detecting 2^3's according to the table

(5.1)

gen.\group	V_1	V_2	V_3	V_4	V_5
c	h	h	h	τ	τ
$\pi^* tr(x_1)$	0	h	0	0	0
e	0	0	0	h	h
$\pi^*\Gamma(\bar{x})$	h^2	0	0	$\tau^2 + \tau h$	0
$\pi^* tr(w_1)$	n	n	n	0	0
$\pi^* tr(x_1 w_2)$	$\tau^2 h + \tau h^2$	$\lambda^2 h + \lambda h^2$	0	0	0
$\pi^*\Gamma(w)$	v	v	v	v	v

where $H^*(2^3) = \mathbb{F}_2[\lambda, \tau, h]$, $n = \tau^2 + \tau h$ and $v = d_4 - d_2^2$, and $d_2 = \tau^2 + \tau h + h^2$.

The Poincaré series for $H^*(H; \mathbf{F}_2)$ is equal to

$$\text{(5.2)} \qquad \frac{1}{(1-t)^3} = 1 + 3t + 6t^2 + 10t^3 + 15t^4 + \cdots,$$

from [**AM**], pg. 259.

THEOREM 5.3.

1. *The complete stable splitting of B_H is given as follows*

$$B_H \simeq_s X_{H,\mathbb{F}_2} \vee X_{K,S} \vee 2(X_{E,S_1} \vee X_{E,S_2}) \vee B_{L_3(2)} \vee 2B_{\mathcal{A}_4} \vee 8(L(3) \vee L(2)) \vee 3B_2.$$

2. *The complete stable splitting of $B_{M_{12}}$ is given as*

$$B_{M_{12}} \simeq_s X_{H,\mathbb{F}_2} \vee B_{\mathcal{A}_4} \vee X_{E,S_2}$$

where, to be definite X_{E,S_2} corresponds to the idempotent in $M_2(\mathbb{F}_2) \otimes M_2(\mathbb{F}_2) \subset (\mathbb{F}_2 \oplus M_2(\mathbb{F}_2)) \otimes (\mathbb{F}_2 \oplus M_2(\mathbb{F}_2))$.

3. $X_{H,\mathbb{F}_2} = B_{G_2(q)}$ *for* $q \equiv 3, 5 \bmod (8)$.

PROOF. To begin we see that a basis for $H^2(H)$ is given as c^2, $\pi^* tr(x_1)^2$, e^2, corresponding to the elements coming from the three copies of B_2, and ce, $\pi^*\Gamma(\bar{x})$, $\pi^*(tr(w_1))$. Squaring these elements and checking the tables we have that $(ce)^2 = Sq^1(c^2 e)$, $(\pi^*\Gamma(\bar{x}))^2 = Sq^1(c\pi^*\Gamma(\bar{x}))$. Thus there are exactly two copies of $B_{\mathcal{A}_4}$ and one copy of $B_{L_3(2)}$ in the splitting of B_H.

We have accounted for every summand except possibly further copies of $L(2)$. But $L(2)$ has Poincaré series $\frac{t^4}{(1-t)(1-t^3)}$. From [**Q**] or [**FM**] the Poincaré series for $H^*(B_{G_2(q)})$ is

$$\frac{(1+t^3)(1+t^5)(1+t^6)}{(1-t^4)(1-t^6)(1-t^7)} = 1 + t^3 + t^4 + t^5 + \cdots, \tag{5.4}$$

and we must have that the generator in dimension three and the generator in dimension four are connected by a Bochstein. Consequently, $B_{G_2(q)}$ cannot have any $L(2)$ summands and either $X_{H,\mathbb{F}_2}$, $X_{K,S}$, X_{E,S_1}, or X_{E,S_2} starts in dimension three and has one class in dimension four as well. On the other hand each B_2 contributes one element in dimension 4 as does each copy of $B_{\mathcal{A}_4}$, $B_{L_3(2)}$. But this accounts for seven generators showing that we cannot have more than eight copies of $L(2)$. The splitting of B_H is now determined.

We now prove (c). To begin, the normalizer of K in $G_2(q)$ has the form $K{:}\mathcal{S}_3$ from [**FM**] and the corresponding element $\bar{W}$ has the form

$$\left(I + \begin{pmatrix} 0 & 1 \\ 1 & 0 \end{pmatrix}\right)\left(I + \begin{pmatrix} 0 & 1 \\ 1 & 1 \end{pmatrix} + \begin{pmatrix} 1 & 1 \\ 1 & 0 \end{pmatrix}\right) = 0$$

so $X_{K,S}$ is not a summand of $B_{G_2(q)}$. Next, we claim that the normalizer of E in $G_2(q)$ has the form $E{:}(3^2{:}2)$ with the $3^2{:}2 \subset Out(E)$. Indeed, the centralizer of an involution in $G_2(q)$ is $(SL_2(q) * SL_2(q)){:}2$ from [**FW**], and $E = Syl_2(SL_2(q) * SL_2(q))$ with Weyl group 3×3, [**FM**], so the claim follows. It follows that the corresponding $\bar{W}$'s are given as

$$\begin{aligned}&\left(I + \begin{pmatrix} J & 0 \\ 0 & J \end{pmatrix}\right) \times \left(I + \begin{pmatrix} N & 0 \\ 0 & I \end{pmatrix} + \begin{pmatrix} N^2 & 0 \\ 0 & I \end{pmatrix}\right) \\ &\qquad \times \left(I + \begin{pmatrix} I & 0 \\ 0 & N \end{pmatrix} + \begin{pmatrix} I & 0 \\ 0 & N^2 \end{pmatrix}\right) = 0\end{aligned} \tag{5.5}$$

where $N = \begin{pmatrix} 0 & 1 \\ 1 & 1 \end{pmatrix}$, $J = \begin{pmatrix} 0 & 1 \\ 1 & 0 \end{pmatrix}$ for the first representation and

$$\begin{aligned}&\left(I + \begin{pmatrix} 0 & J \\ J & 0 \end{pmatrix}\right) \times \left(I + \begin{pmatrix} N & 0 \\ 0 & N \end{pmatrix} + \begin{pmatrix} N^2 & 0 \\ 0 & N^2 \end{pmatrix}\right) \\ &\qquad \times \left(I + \begin{pmatrix} 0 & I \\ I & I \end{pmatrix} + \begin{pmatrix} I & I \\ I & 0 \end{pmatrix}\right) = 0\end{aligned} \tag{5.6}$$

for the second representation. Thus the stable splitting for $B_{G_2(q)}$ does not involve $X_{K,S}$ or X_{E,S_i}. Similarly, since the cohomology of $G_2(q)$ starts in dimension three it cannot have any B_2, $B_{\mathcal{A}_4}$ or $B_{L_3(2)}$ summands. Also, since the three and 4 class

It remains to determine the stable splitting for M_{12}. From [**AMM**] or [**AM**], pp. 261-263, we have that the Poincaré series for $H^*(M_{12})$ has the form $1 + t^2 + 3t^3 + 2t^4 + \cdots$. It follows that the splitting has the form $B_{G_2(q)} \vee Y \vee Z$ where Y is either $B_{\mathcal{A}_4}$ or $B_{L_3(2)}$ and Z is a sum of dominants coming from E or K. We first determine that $Z = X_{E,S_2}$ and then use the cohomology of M_{12} to determine Y.

The normalizer of K in M_{12} is the same as its normalizer in $G_2(q)$ so the summand X_{K,S_1} is not involved in M_{12}. However, the normalizer of E in M_{12} is the involution centralizer, $Q_8{:}\mathcal{S}_4 = E{:}\mathcal{S}_3$ where the element of order three in $\mathcal{S}_3$ acts diagonally on $Q_8 * Q_8$. It follows that the $\bar{W}$ terms for E and M_{12} become

$$\left(I + \begin{pmatrix} J & 0 \\ 0 & J \end{pmatrix}\right)\left(I + \begin{pmatrix} N & 0 \\ 0 & N \end{pmatrix} + \begin{pmatrix} N^2 & 0 \\ 0 & N^2 \end{pmatrix}\right) = 0 \tag{5.7}$$

for the sum representation and

$$\left(I + \begin{pmatrix} 0 & J \\ J & 0 \end{pmatrix}\right)\left(I + \begin{pmatrix} 0 & N \\ N & N \end{pmatrix} + \begin{pmatrix} N^2 & N^2 \\ N^2 & 0 \end{pmatrix}\right) = \begin{pmatrix} 0 & 0 & 0 & 0 \\ 0 & 1 & 1 & 0 \\ 0 & 1 & 1 & 0 \\ 0 & 0 & 0 & 0 \end{pmatrix} \tag{5.8}$$

which has rank one.

6. The cohomology of M_{12} and the completion of the proof of (5.3)

Here is the explicit structure of the cohomology ring for M_{12} as discussed in [**AM**]. The result is

THEOREM 6.1 [**AMM**]. *$H^*(M_{12};\mathbb{F}_2)$ has 8 generators, α_2, β_3, c_3, m_3, d_4, n_5, d_6 and d_7, (where the subscripts denote the dimension) with restriction images given as follows:*

(6.2)

gen.\group	V_1	V_3	V_5
α	0	h^2	d_2
β	0	0	d_3
c	0	h^3	$\tau^3 + \tau^2\lambda + \lambda^3$
m	d_3	0	0
d_4	d_4	d_4	d_4
n	$d_3 d_3$	0	0
d_6	d_6	d_6	d_6
d_7	d_7	d_7	d_7

where

$$\begin{aligned} H^*(2^3) &= \mathbb{F}_2[h, \tau, \lambda], \\ d_2 &= \tau^2 + \tau\lambda + \lambda^2, \\ d_3 &= \tau^2\lambda + \lambda^2\tau, \\ d_4 &= h^4 + h^2 d_2 + h d_3 + d_2^2, \\ d_6 &= h^4 d_2 + h^2 d_2^2 + h d_2 d_3 + d_3^2, \\ d_7 &= h^4 d_3 + h^2 d_2 d_3 + h d_3^2. \end{aligned}$$

In particular, the class $\alpha^2 = Sq^1(c)$ so it follows that $Y = B_{\mathcal{A}_4}$. The proof of (5.3) is complete. □

COROLLARY 6.3. *The Poincaré series of the summand X_{E,S_2} is given as*

$$\frac{t^6(1+t+t^2+t^3+t^4+t^6-t^7)}{(1-t^7)(1-t^3)(1+t^2)(1-t)}.$$

PROOF. The Poincaré series for $H^*(M_{12})$ is given in [**AMM**] as

$$\frac{(1+t^3)^2}{(1-t^2)(1-t^3)(1-t^4)}$$

while the Poincaré series for $H^*(G_2(q))$ is given in (5.4), and (4.8) shows that the Poincaré series for $H^*(\mathcal{A}_4)$ is $\frac{1+t^3}{(1-t^2)(1-t^3)}$. But

$$P_{X_{E,S_2}}(t) = P_{M_{12}}(t) - P_{G_2(q)}(t) - P_{\mathcal{A}_4}(t) + 1$$

and (6.3) follows. □

References

[AM] A.Adem, R.J.Milgram, *Cohomology of Finite Groups*, vol. 309, Grundlehren der Math. Wiss., Springer-Verlag, Berllin, 1994.

[AMM] A.Adem, J.Maginnis, R.J. Milgram, *The geometry and cohomology of the Mathieu group M_{12}*, J. of Algebra **139** (1991), 90-133.

[BF] D.Benson, M.Feshbach, *Stable splittings of classifying spaces of finite groups*, Topology **31** (1992), 157-176.

[BR] F.Buekenhout, S.Rees, *The subgroup structure of the Mathieu group*, Math. Comp. **50** (1988), 595-605.

[C] J.H.Conway, R.T.Curtis, S.P.Norton, R.A.Parker, R.A.Wilson, *Atlas of Finite Groups*, Clarendon Press - Oxford, 1985.

[FM] P.Fong, R.J.Milgram, *On the geometry and cohomology of the simple groups $G_2(q)$ and ${}^3D_4(q)$*, Proceedings of Symposia in Pure Mathematics, vol. 63, AMS, 1998, pp. 221-244.

[FW] P.Fong, W.J.Wong, *A characterization of the finite simple groups $PSp(4,q)$, $G_2(q)$, $D_4^2(q)$, I*, Nagoya J. Math **36** (1969), 143-184.

[HK] J.Harris, N.Kuhn, *Stable decompositions of classifying spaces of finite Abelian p-groups*, Math. Proc. Camb. Phil.Soc. **103** (1988), 427-449.

[Ma] J. Martino, *Stable splittings and the Sylow 2-subgroups of $SL_3(\mathbb{F}_2)$*, Thesis, Northwestern University, 1988.

[MaP] J. Martino, S.Priddy, *The complete stable splitting for the classifying space of a finite group*, Topology **31** (1992), 143-156.

[M] R.J.Milgram, *On the geometry and cohomology of the simple groups $G_2(q)$ and ${}^3D_4(q)$: II*, Proceedings of Symposia in Pure Mathematics, vol. 63, AMS, 1998, pp. 397-418.

[M2] R.J.Milgram, *The 14 dimensional Kervaire invariant and the sporadic group M_{12}*, Asian J. Math **1** (1997), 314-329.

[MP] S.Mitchell, S.Priddy, *Some splittings derived from the Steinberg module*, Topology **22** (1983), 285-298.

[MP2] S.Mitchell, S.Priddy, *Symmetric product spectra and splittings of classifying spaces*, Amer. J. Math. **106** (1984), 219-232.

[N] G. Nishida, *Stable homotopy type of classifying spaces of finite groups*, Algebraic and Topological Theories, 1985, pp. 391-404.

[O] C. Overton, *Thesis*, Stanford University, 1996.

[Q] D.Quillen, *Cohomology of groups*, Proceedings of the International Congress of Mathe-

[W] C. Witten, *Ph. Thesis: Self-maps of classifying spaces of finite groups and classification of low-dimensional Poincaré duality spaces*, Stanford, 1978.

Department of Mathematics, Stanford University, Stanford, CA, 94305
E-mail address: mailgram@math.stanford.edu

Contemporary Mathematics
Volume **220**, 1998

On the Kervaire invariant problem

Norihiko Minami

ABSTRACT. We review some of the recent results concerning the Kervaire invariant one problem, including the author's own, from some global and philosophical viewpoints.

Among all, we highlight two global conjectures about $\pi_*^s(S^0)$ which present completely different predictions about the Kervaire invariant problem.

What is the Kervaire invariant problem?

Let M^n be a n-dimensional closed smooth manifold whose normal bundle is stably trivial, and let Ψ be its framing, i.e. an explicit trivialization of the stable normal bundle. Then, as is well known, the Pontryagin-Thom construction induces the isomorphism:

THEOREM A. *$\{[M^n, \Psi]\}/\sim \cong \pi_n^s(S^0)$, where $\sim$ is given by the framed bordism.*

A naive question here is whether we could take a simple representative $[M^n, \Psi]$ for an element in $\pi_*^s(S^0)$ or not. Then, the basic work of Kervaire-Milnor [**39**] using surgery techniques, and W. Browder [**21**] using more homotopy theoretic techniques imply the following:

THEOREM B. [**21**] *Any $\alpha \in \pi_n^s(S^0)$ is represented by a framed homotopy sphere Σ^n as long as $n \neq 2^{j+1} - 2$. Moreover, in dimension $2^{j+1} - 2$ every element is represented by a framed homotopy sphere if and only if $h_j^2 \in \mathrm{Ext}_{\mathcal{A}_*}^{2,2^{j+1}}(\mathbb{Z}/2, \mathbb{Z}/2)$ is not a permanent cycle in the Adams spectral sequence for $\pi_*^s(S^0)_{(2)}$.*

REMARK 0.1. (i) Because of the above theorem, an element $\alpha \in \pi_*^s(S^0)$ is called a Kervaire invariant one element if and only if it is detected by $h_j^2 \in \mathrm{Ext}_{\mathcal{A}_*}^{2,2^{j+1}}(\mathbb{Z}/2, \mathbb{Z}/2)$ for some j.
(ii) Kervaire-Milnor [**39**] extensively studied the group θ^n of h-cobordism classes of homotopy n-spheres, which turns out to be the set of all diffeomorphism classes of differentiable structures on the topological n sphere when $n \geq 5$, thanks to Smale's work [**88, 89**] (here the former is his famous solution of the generalized Poincare conjecture) for $n \geq 5$. The idea of their approach was to define the subgroup

1991 *Mathematics Subject Classification.* Primary 55Q10, 55Q45, 55T15; Secondary 55R12,

$bP^{n+1} \subset \theta^n$ consisting of homotopy n-spheres which are boundaries of parallelizable manifolds, and then to establish an exact sequence

$$0 \to bP^{n+1} \to \theta^n \xrightarrow{T_n} \pi_n^s(S^0)/J_*(\pi_n(SO)), \tag{0.1}$$

where $J : SO \to Q_1S^0 \simeq Q_0S^0$ is the Whitehead J map and $T_n : \theta^n \to \pi_n^s(S^0)/J_*(\pi_n(SO))$ is induced by the Pontryagin-Thom construction. Here, combining the positive solution [**77, 94, 4**] of the Adams conjecture and Browder's work [**21**], we can reduce the problem of understanding (0.1) to the computation of $\pi_*^n(S^0)/J_*(\pi_n(SO))$, up to the Kervaire invariant one elements:

$$bP^{n+1} = \begin{cases} 0 & \text{if } n \text{ is even,} \\ 0 & \text{if } n = 2^{j+1}-3 \text{ with } h_j^2 \text{ a permanent cycle,} \\ \mathbb{Z}/2 & \text{if } n = 4k+1 \text{ and } n \text{ does not satisfy the above condition,} \\ \mathbb{Z}/(a_k 2^{2k-2}(2^{2k-1}-1)\cdot n_k) & \text{if } n = 4k-1, \end{cases}$$

$$\text{Coker } T_n = \begin{cases} \mathbb{Z}/2 & \text{if } n = 2^{j+1}-2 \text{ with } h_j^2 \text{ a permanent cycle,} \\ 0 & \text{if else,} \end{cases}$$

where a_k equals 1 or 2 according as k is even or odd, n_k denotes the numerator of $(B_k/4k)$ (here B_k denotes the k-th Bernoulli number).

For more details, the reader is referred to [**39, 43, 21**].

(ii) Just like the mod 2 cohomology of a space, $\mathrm{Ext}_{\mathcal{A}_*}^{*,*}(\mathbb{Z}/2, \mathbb{Z}/2)$ is acted by the Steenrod algebra, except Sq^0 does *not* act as the identity [**46, 55, 81**]. This point turns out to be very crucial in our discussion. (See §3.)

Now, *the Kervaire invariant problem*, which we abbreviate as KIP for the rest of this paper for simplicity, asks when we have Kervaire invariant one elements in $\pi_*^s(S^0)$. At this stage, recall the adjoint isomorphism

$$\pi_*^s(S^0) \cong \lim_{n\to\infty} \pi_*(\Omega^n S^n) \cong \pi_*\left(\lim_{n\to\infty}(\Omega^n S^n)\right)$$

and the fundamental Barratt-Priddy-Quillen theorem [**13, 78**]

$$\Omega B\left(\coprod_{n=0}^{\infty} B\Sigma_n\right) \simeq Q(S^0) := \lim_{n\to\infty}(\Omega^n S^n), \tag{0.2}$$

where B is the classifying space functor of a topological monoid, Ω is the pointed loops space functor, and Σ_n is the symmetric group of n letters (c.f. Adams' book [**6**]). Then these naturally tempt us to restate KIP in terms of classifying spaces of finite groups or not. Fortunately, work of Priddy [**76**] (see also [**63**]) and Snaith-Tornehave [**91**] make it possible:

THEOREM C. *Any $\alpha \in \pi_n^s(S^0)$ is represented by a framed homotopy sphere Σ^n as long as $n \neq 2^{j+1}-2$. Moreover, in dimension $2^{j+1}-2$ every element is represented by a framed homotopy sphere if and only if the image of the Hurewicz homomorphism*

$$\pi_{2^{j+1}-2}^s(BGL_3(\mathbb{F}_2)) \to H_{2^{j+1}-2}(BGL_3(\mathbb{F}_2); \mathbb{Z}/2)$$

is trivial. Here we note that $BGL_3(\mathbb{F}_2)$ *is the dominant summand of* BD_8 *and* $B\Sigma_4$ **[72]***:*

$$\Sigma^\infty BD_8 \simeq \Sigma^\infty BGL_3(\mathbb{F}_2) \vee 2L(2) \vee 2\Sigma^\infty B\mathbb{Z}/2,$$
$$\Sigma^\infty B\Sigma_4 \simeq \Sigma^\infty BGL_3(\mathbb{F}_2) \vee L(2) \vee \Sigma^\infty B\mathbb{Z}/2.$$

A stronger version of KIP is also well-known among homotopy theorists: *The strong Kervaire invariant problem*, which we abbreviate as SKIP, asks when there exists θ_j with an order 2, where $\theta_j \in \pi^s_{2^{j+1}-2}(S^0)$ is any element with Kervaire invariant one. In the formulation of Theorem C, the existence of θ_j with order 2 is the same thing as the existence of an order 2 element in $\pi^s_{2^{j+1}-2}(BGL_3(\mathbb{F}_2))$ with a nontrivial mod 2 Hurewicz image.

Whichever the case, it is well-understood among homotopy theorists that KIP (or SKIP) is an extremely difficult problem. The current status of the problem is as follows:

THEOREM D. **[54, 11, 41]** *For* $j \leq 5$, *there exists* θ_j *with order* 2.

It should be noticed that not much progress has been achieved since **[48, 10]**, except for the work of Kochman **[41]** which proved θ_5, originally constructed in **[11]**, may be chosen to be of order 2. So, still nothing is known for $j \geq 6$.

Because of its immense difficulty, we would like to look after some global generalizations of KIP. A natural first step would be to come up with an odd primary analogue of KIP. Traditionally, the Kervaire invariant problem for an odd prime p asks when $b_j = <\underbrace{h_j, \cdots, h_j}_{p}> \in \mathrm{Ext}^{2, qp^{j+1}}_{\mathcal{A}_*}(\mathbb{Z}/p, \mathbb{Z}/p)$ is a permanent cycle, where $q = 2(p-1)$. (Here and after, we use this standard notation.) Just as in the case $p = 2$, we abbreviate it as KIP, and let $\theta_j \in \pi^s_{qp^{j+1}-2}(S^0)$ be any element which is detected by b_j if such an element exists. Similarly, SKIP asks when θ_j exists as an order p element. Then we have the following nice result of Ravenel **[80]**:

THEOREM E. **[80]** (i) *For* $p \geq 5$, θ_j *does not exist for* $j \geq 1$.
(ii) *For* $p = 3$, θ_1 *does not exist, but* θ_2 *exists.*

In fact, for an odd prime p, Toda **[96]** had already shown θ_1 does not exist, and this fact was used as a starting point of the proof in **[80]**. On the other hand, θ_0 always exists as $\theta_0 = \beta_1$.

Now the purpose of this paper is to review some of the recent results concerning KIP, including the author's own, from some global and philosophical points of views.

Looking at Theorem E, we are tempted to suspect there are just finitely many θ_j's for $p = 2,\ 3$. However, before we believe such a viewpoint, we would like to have some more similarities between the (original) mod 2 θ_j elements and the odd primary θ_j elements.

On the other hand, looking at Theorem D, we are tempted to believe θ_j exists for all j when $p = 2$.

Whichever viewpoint we choose, we would like to have some global picture about $\pi^s_*(S^0)$ which may justify our choice. The rest of the paper is organized emphasizing such global and philosophical points:

§4 Milgram's point of view.
§5 The Barratt-Mahowald hypothesis - optimistic version.
§6 The Adams filtration 2 case.

In §1 and §2, we search for similarities between the 2 primary θ_j elements and the odd primary θ_j elements. In §3 and §4, we search for some possible global pictures about $\pi_*^s(S^0)$ which may justify the viewpoint that there are just finitely many θ_j's for $p = 2$. To the contrary, in §5, we search for some possible global picture about $\pi_*^s(S^0)$ which may justify the viewpoint that θ_j exists for all j when $p = 2$. In §6, we discuss η_j elements which are known to have some relationship with θ_j elements. A secret ambition of the author is to attract attentions of group theorists to the following "dream problem" suggested by one of the two contradicting viewpoints:

QUESTION 4.2. *Can we use Ravenel's theorem* [**80**] *(and NDC) to explain the Feit-Thompson theorem* [**26**]*? More generally, can we use NDC to explain the finiteness of sporadics?*

In spite of such an (possible) excitement, we should emphasize that these two conflicting viewpoints are both extreme. Such is the case, future study of the subject might reveal some true global picture of $\pi_*^s(S^0)$ somewhere in between. However, for that to happen, KIP should be solved first.

Finally, the author would like to express his gratitude to Mark Mahowald, Jim Milgram and Doug Ravenel, for sharing their insights with him. The author also would like to express his heartfelt gratitude to the referee for this careful reading of a preliminary version of this paper.

1. The Curtis-Madsen conjecture

In this section, we review the author's recent result [**69**], where a reformulation of KIP is given for any p.

To motivate it, we first recall the well-known conjecture of Curtis and Madsen:

THE CURTIS-MADSEN CONJECTURE. *The mod 2 unstable Hurewicz map*

$$\pi_n^s(S^0) \cong \pi_n(Q_0S^0) \to H_n(Q_0S^0; \mathbb{Z}/2)$$

detects exactly the Hopf invariant one elements, the Kervaire invariant one elements, and nothing else.

At this stage, we would like to recall the Quillen-Tornehave splitting [**77, 97**], for which purpose we digress a little bit now to discrete models of QS^0 and J. Good references here are [**56, 27, 47, 16**]. For simplicity, we assume p is odd. Then the Barratt-Priddy-Quillen Theorem (0.2) may be completed to the following commutative diagram:

$$\begin{array}{ccc} \Omega B(\coprod_{n=0}^{\infty} B\Sigma_n) & \xrightarrow{\simeq} & QS^0 = \coprod_{n\in\mathbb{Z}} Q_nS^0 \\ {\scriptstyle \Omega B(\coprod_{n=0}^{\infty} Br_n)}\downarrow & & \downarrow{\scriptstyle \coprod_{n=0}^{\infty} \iota_n} \\ \Omega B(\coprod_{n=0}^{\infty} BGL_n(\mathbb{F}_q)) & \xrightarrow{\simeq} & J = \coprod_{n\in\mathbb{Z}} J_n, \end{array}$$

where q is chosen to be a prime so that q mod p^2 generates $(\mathbb{Z}/p^2)^*$, $r_n : \Sigma_n \to GL_n(\mathbb{F}_q)$ is the canonical permutation representation.

For readers' benefit, we note that Quillen [**79**] proved a homotopy equivalence

$$J_0 = BGL_\infty(\mathbb{F}_q)^+ \xrightarrow{\simeq} F\Psi^q,$$

where $(?)^+$ is the Quillen plus construction [**6**, p.84-86] [**16**, 2.10.1] [**7**, IX] of a based space $(?)$ with respect to a perfect normal subgroup of $\pi_1((?))$ (which is $GL_\infty(\mathbb{F}_q)$ in this case), the equality is an immediate consequence of the group completion theorem [**6**, 3.2.2], and $F\Psi^q$ is the homotopy fiber of $1-\psi^q : BU \to BU$ where ψ^q corresponds to the Adams operation in the complex K-theory [**3**] [**16**, p.47]. This homotopy equivalence actually shows up as the left vertical arrow in the following commutative diagram

$$\begin{array}{ccccc}
BGL_\infty(\mathbb{F}_q)^+ & \xrightarrow{i} & BGL_\infty(\overline{\mathbb{F}}_q)^+ & \xrightarrow{1-\psi^q} & BGL_\infty(\overline{\mathbb{F}}_q)^+ \\
\downarrow & & \downarrow{\scriptstyle BL} & & \downarrow{\scriptstyle BL} \\
F\Psi^q & \longrightarrow & BU & \xrightarrow{1-\psi^q} & BU,
\end{array}$$

where the upper ψ^q is induced by the Frobenius map $\overline{\mathbb{F}} \to \overline{\mathbb{F}}$; $x \mapsto x^q$, BL (which is homology isomorphism over a ring where q is invertible) is induced by the Brauer lifting, and the left vertical map (which turns out to be a homotopy equivalence) is obtained using the fact that the Brauer lifting of a finite group G induces a factorization

$$R_{\mathbb{F}_q}(G) \to R(G)^{\psi^q} \subset R(G),$$

where $R_{\mathbb{F}_q}(G)$ (resp. $R(G)$) is the Grothendieck group of finite dimensional representations of G over $\mathbb{F}_q$ (resp. $\mathbb{C}$) and $R(G)^{F\Psi_q}$ is the fixed point subring of $R(G)$ with respect to the action of the Adams operation ψ^q. For more details, the reader is referred to Quillen's original articles [**77, 78**] (see also [**16**, 2.9]).

We now go back to our business; following the standard convention, we set

$$\begin{aligned}
SG &:= Q_1S^0 \\
J_\otimes &:= J_1 \\
J_\oplus &:= J_0
\end{aligned}$$

Then the infinite loop spaces $Coker J_\otimes$ and $Coker J_\oplus$ are defined as the fiber of the fiber sequences:

$$(1.1) \qquad \begin{aligned}
Coker J_\otimes &\to SG \xrightarrow{\iota_1} J_\otimes \\
Coker J_\oplus &\to Q_0S^0 \xrightarrow{\iota_0} J_\oplus
\end{aligned}$$

Now the fundamental Quillen-Tornehave splitting theorem claims the upper fiber sequence in (1.1) splits:

$$SG \simeq Coker J_\otimes \times J_\otimes.$$

This is proved via the diagram

$$Q_0S^0 \qquad\qquad SG$$

Here $e : J_{\oplus} \to SG$ is the "exponential map" induced by

$$\Omega B\left(\coprod_{n=0}^{\infty} BGL_n(\mathbb{F}_q)\right) \xrightarrow{\Omega B(\coprod_{n=0}^{\infty} Be_n)} \Omega B\left(\coprod_{n=0}^{\infty} B\Sigma_{q^n}\right),$$

where $e_n : GL_n(\mathbb{F}_q) \to \Sigma_{q^n}$ is defined by regarding $(\mathbb{F}_q)^n$ as a finite set of cardinality q^n.

From the Quillen-Tornehave splitting, we see the lower fiber sequence in (1.1) also splits:

$$Q_0S^0 \simeq Coker J_{\oplus} \times J_{\oplus}. \tag{1.2}$$

In fact, this immediately follows from the homotopy commutative diagram

$$\begin{array}{ccc} SG = Q_1S^0 & \xrightarrow[\simeq]{*[-1]} & Q_0S^0 \\ {\scriptstyle \iota_1}\downarrow & & \downarrow{\scriptstyle \iota_0} \\ J_1 = J_{\otimes} & \xrightarrow[\simeq]{*[-1]} & J_0 = J_{\oplus}, \end{array}$$

where $*[-1]$ is the "addition" by the basepoints $[-1]$ in the "ring" spaces $Q_{-1}S^0$ and J_{-1}, and is a homotopy equivalence with $*[1]$ as its homotopy inverse. From this, we also see

$$Coker J_{\oplus} \simeq Coker J_{\otimes}. \tag{1.3}$$

The above argument also works for $p = 2$ by choosing slightly different groups to describe J. For this and more details about the above discussion, the reader is referred to [**56**]. We also note that Miller-Priddy [**61**] conjecture that (1.3) is in fact an equivalence of infinite loop spaces. The book of Fiedorowicz and Priddy [**27**] is a nice account about the relationships between the homology of classical groups over finite fields and the infinite loop spaces.

We now return to the Curtis-Madsen conjecture. Since $J_{\oplus}$ is well-understood and carries the Hopf invariant one elements, (1.2) implies the Curtis-Madsen conjecture is essentially equivalent to the conjecture that the mod 2 unstable Hurewicz map

$$\pi_n(Coker J_{\oplus}) \to H_n(Coker J_{\oplus}; \mathbb{Z}/2)$$

detects exactly the Kervaire invariant one elements and nothing else.

Since the Quillen-Tornehave splitting (1.2) is valid at any prime p, we might have a chance to reformulate KIP for any p, via the ($Coker J_{\oplus}$ version of) Curtis-Madsen conjecture.

However, no matter which prime we choose, the Curtis-Madsen conjecture appears to be an extremely difficult problem (c.f. [**98**]). To avoid this difficulty, we consider its stable analogue instead; let $cokerj_{\oplus}$ denote the spectrum corresponding to the (additive) infinite loop space $Coker J_{\oplus}$ so that $Coker J_{\oplus} = \Omega^{\infty} cokerj_{\oplus}$. Actually, there is a fiber sequence of spectra

$$cokerj_{\oplus} \to S^0 \to j_{\oplus},$$

where $j_{\oplus}$ is the spectrum corresponding to the (additive) infinite loop space $J = \coprod_{n\in\mathbb{Z}} J_n$ and the second map is the unit map. Then we have following commutative involving Hurewicz maps: diagram:

$$\begin{array}{ccccc}\pi_*(Coker J_\oplus) & \longrightarrow & H_*(Coker J_\oplus;\mathbb{Z}_{(p)}) & \longrightarrow & H_*(Coker J_\oplus;\mathbb{Z}/p)\\ \cong\downarrow & & \downarrow & & \downarrow\\ \pi_*^s(coker j_\oplus) & \longrightarrow & H_*(coker j_\oplus;\mathbb{Z}_{(p)}) & \longrightarrow & H_*(coker j_\oplus;\mathbb{Z}/p),\end{array}$$

Now we have the following:

THEOREM 1.1. [**69**] *For any prime p,, the stable integral Hurewicz map*

$$\pi_n(Coker J_\oplus)_{(p)} \cong \pi_n^s(coker j_\oplus)_{(p)} \to H_n(coker j_\oplus;\mathbb{Z}_{(p)})$$

detects exactly the Kervaire invariant one elements and nothing else.

This result offers a restatement of KIP at any prime p. However, we have to admit that the relation between Theorem 1.1 and the Curtis-Madsen conjecture is not clear, though Theorem 1.1 was clearly motivated by the Curtis-Madsen conjecture. The following problem appears to be relevant:

QUESTION 1.2. Does the unstable integral Hurewicz map

$$\pi_*(Coker J_\oplus) \to H_*(Coker J_\oplus;\mathbb{Z}_{(p)})$$

have an infinite image?

It would be quite interesting if a new infinite family of the stable homotopy groups of the sphere (other than the possible Kervaire invariant one family) is constructed to enjoy this property.

2. Some work of Selick and Hemmi

In this section, we recall some works of Selick [**85**] and Hemmi [**29**], both of whom have given reformulations of SKIP for an arbitrary prime p.

Let $W(n)$ denote the homotopy-theoretic fiber of $E^2 : S^{2n-1} \to \Omega^2 S^{2n+1}$, and let f denote the composite

$$S^{2np-3} \to W(n) \to S^{2n-1}$$

where the first map is a generator of $\pi_{2np-3}(W(n)) \cong \mathbb{Z}/p$. Also, let $P^n(p) = S^{n-1} \cup_p e^n$, the mod p Moore space. Suppose $n \neq 1, 2$, or 4 (resp. $n \neq 1$) if $p = 2$ (resp. p is odd), so that $f \neq 0$. Then Selick [**85**] showed the following:

THEOREM 2.1. [**85**] *For any prime p, the following conditions are equivalent:*

1. *$f = pg$ for some g.*
2. *There exists $h : P^{2np-2}(p) \to \Omega^2 S^{2n+1}$ such that $h_* \neq 0$ on H_{2np-2}.*
3. *$\pi_{qn-2}^s(S^0)$ has an order p element of Kervaire invariant one.* □

If $p = 2$ then $f = [\iota_{2n-1}, \iota_{2n-1}]$. Thus the Selick's result is a generalization of the well-known reformulation of the original 2 primary SKIP [**50, 10**]. For $p = 2$, Fred Cohen [**22, 23**] have some other reformulations of SKIP. Now based upon Theorem 2.1, Hemmi [**29**] has recently given another reformulation of the SKIP for any p, using Stasheff's concepts of A_n-spaces [**92**]. Here an A_2-space is a space equipped with a continuous product and the two-sided identity element (but not

Working in the category of pointed spaces and base-point preserving continuous maps, we review some more basic facts related to A_n-spaces. First, Stasheff's work in [**92**] may be roughly summarized as follows (the exposition here follows Adams [**6**, p.33-34]):

1. Define a sequence of parameter spaces K_m for $m \geq 2$ such that K_m is a $(m-2)$-cell E^{m-2} with an explicitly described subdivision of its boundary; in particularK_2 is a point, K_3 is a unit interval $I = [0,1]$ and K_4 is the disc E^2 with its boundary subdivided as a pentagon (whose five vertices correspond to five four-fold products:
$(wx)(yz),\ ((wx)y)z,\ (w(zy))z,\ w((xy)z),\ w(x(yz)))$.
2. An A_n-space is defined to be a space X provided with maps
$$M_r : K_r \times X^r \to X$$
for $2 \leq r \leq n$ satisfying suitable conditions. Forexample, an A_2-space is a space provided with a map $M_2 : X \times X \to X$ such that the composite
$$X \xrightarrow{i_j} X \times X \xrightarrow{M_2} X \qquad (j = 1,2)$$
is the identity map, where $i_1(x) = (x,*)$, $i_2(x) = (*,x)$ for any $x \in X$ ($*$ is the fixed base-point of X) (Thus, an A_2 space is nothing but an H-space); an A_3-space X provided with a map $M_2 : X \times X \to X$ and an explicit homotopy (associative homotopy) between $M_2 \circ (M_2 \times 1)$ and $M_2 \circ (1 \times M_2)$; and so on.
3. More precisely, the definition is by induction. Suppose that X is an A_{n-1}-space, so that it comes provided with maps $M_2, M_3, \cdots, M_{n-1}$. In terms of these maps Stasheff defines a map from $(\partial K_n) \times X^n$ to X. Then X is an A_n-space if it comes provided with a map
$$M_n : K_n \times X^n \to X$$
extending this map of $(\partial K_n) \times X^n$.
The maps M_n for $n \geq 4$ are called "higher homotopies."
4. A space X is an A_∞-space if it has maps M_n for all $n \geq n$ which makes it an A_n space for all n.
5. An A_∞-space is an adequate substitute (usable by homotopy theorists) for a space with a product which is strictly associative.

Having obtained the concept of an A_n-space, it is a matter of course to seek an appropriate concept of an A_n-map between two A_n-spaces. A naive guess would be to take those maps which commute with all the A_n-structures. However, such a concept, which is called an A_n-homomorphism in [**92**], turns out to be too rigid; a map homotopic to an A_n-homomorphism is not necessarily an A_n-homomorphism; there are many H-maps which are not A_2-homomorphism. In fact, let M_X and M_Y be respectively the product maps, i.e. M_2, of A_2-spaces X and Y. Then $f : X \to Y$ is called an H-map (resp. A_2-homomorphism) when the following diagram is pointed homotopy commutative (resp.strictly commutative):

$$\begin{array}{ccc} X \times X & \xrightarrow{f \times f} & Y \times Y \\ {\scriptstyle M_X}\downarrow & & {\scriptstyle M_Y}\downarrow \\ X & \xrightarrow{f} & Y. \end{array}$$

Fortunately, generalizing [**102, 92**], Iwase-Mimura [**33**] have arrived at an appropriate concept of an A_n-map. We now list some of its properties from [**33**, p.196]:

1. A map homotopic to an A_n-map is an A_n-map.
2. A composition of A_n-maps is an A_n-map.
3. An A_n-homomorphism is an A_n-map.
4. The localization map of an A_n-space is an A_n-map. The localization of an A_n-map is an A_n-map.
5. If a homotopy equivalence is an A_n-map, so is its homotopy inverse.
6. The homotopy fibre of an A_n-map admits an A_n-structure.
7. The pull-back of two A_n-maps admits an A_n-structure.
8. A map is an A_2-map, A_3-map or A_∞-map iff it is an H-map, an H-map preserving homotopy associativity or a loop map, respectively.

In spite of these nice properties, the actual definition is quite complicated.

The final concept needed here is due to Hemmi [**29**]. Suppose X is an A_{n-1}-space and $f : X \to Y$ is a map. Then X is called an *A_n-space corelative to f* if there is a map

$$M_n : K_n \times X^n \to Y$$

which enjoys certain compatible conditions. This concept enjoys the following properties:

1. X is an A_n-space if and only if X is an A_n-space corelative to the identity map of X.
2. Let X be an A_n-space. Then, for any $f : X \to Y$, X is an A_n-space corelative to f. (In fact, let $\overline{M}_n : K_n \times X^n \to X$ be a part of the A_n-structure of X. Then the composite $M_n := f \circ \overline{M}_n : K_n \times X^n \to X \to Y$ will do the job.)
3. Suppose $f : X \subset Y$ is an injection and X is an A_n-space corelative to f with $M_n : K_n \times X^n \to Y$ a part of its structure. If we can compress M_n to inside X,

$$\begin{array}{ccc} & \overset{?}{\dashrightarrow} & X \\ & & \downarrow{\scriptstyle f} \\ K_n \times X^n & \xrightarrow[M_n]{} & Y \end{array}$$

then X is an A_n-space extending its underlying A_{n-1}-space structure.

REMARK 2.2. (i) An A_2-space is nothing but an H-space, and thus the classical well-known result may be restated that S^n is an A_2-space if and only if there is a Hopf invariant one element in $\pi_n^s(S^0)_{(2)}$. Of course, the famous theorem of Adams [**1**] claims this happens only if $n = 1, 3, 7$. We notice that the same conclusion holds even if we replace S^n with its localization at 2, $(S^n)_{(2)}$.
(ii) However, the situation drastically changes if we consider the localization $(S^n)_{(p)}$ at an odd prime p. In fact, the localized odd sphere $(S^{2k-1})_{(p)}$ is not only an H-space, but actually an A_{p-1} space [**2, 92**]. Furthermore, such an A_{p-1}-structure is induced from $\Omega^2 S^{2k-1}$, which is of course an A_∞-space, via the double suspension map $E^2 : (S^{2k-1}) \to (\Omega^2 S^{2k+1})$. The essential idea here is to notice the fact

A_{p-1}-structure maps $\{M_i : K_i \times (S^{2k-1})^i_{(p)} \to (S^{2k-1})_{(p)}\}_{i \le p-1}$ on $(S^{2k-1})_{(p)}$ so that the following diagram is homotopy commutative:

$$\begin{array}{ccc} K_i \times (S^{2k-1})^i_{(p)} & \xrightarrow{1_{K_i} \times (E^2)^i} & K_i \times (\Omega^2 S^{2k+1})^i_{(p)} \\ M_i \big\downarrow & & \mu_i \big\downarrow \\ (S^{2k-1})_{(p)} & \xrightarrow{E^2} & (\Omega^2 S^{2k+1})_{(p)}. \end{array}$$

With respect to this A_{p-1} structure on $(S^{2k-1})_{(p)}$, $E^2 : (S^{2k-1})_{(p)} \to (\Omega^2 S^{2k+1})_{(p)}$ becomes an A_{p-1}-map.

(iii) For p odd, it is also known $(S^{2k-1})_{(p)}$ becomes an A_p-space if and only if $k \mid (p-1)$ [**93, 100**]. (Actually, this condition turns out to be the same as an A_∞-space, i.e. a loop space, in this particular case.) Thus, unlike the case $p = 2$ (see (i)), this condition is not the same as the existence of the mod p Hopf invariant one element in $\pi^s_{qk-1}(S^0)_{(p)}$ where $q = 2(p-1)$, In fact, it is well-known [**46, 86**] that a mod p Hopf invariant one element exists in $\pi^s_{qk-1}(S^0)_{(p)}$ only if $k = 1$.

(iv) In spite of the problem mentioned in (iii), Hemmi [**29**] has proved the following two conditions are the same for any prime p:

- There exists a mod p Hopf invariant one element in $\pi^s_{qn-1}(S^0)$,
- The p-localized sphere $(S^{2n-1})_{(p)}$ possesses an A_p-space structure so that $E^2 : (S^{2n-1})_{(p)} \to (\Omega^2 S^{2n+1})_{(p)}$ is an A_{p-1}-map.

Here we use the convention that any map is an A_1-map. Thus the claim is the well-known fact mentioned in (i) when $p = 2$.

Let $\iota : S^{2n-1} \to P^{2n}(p)$ be the inclusion. Then, along the line of Remark 2.2 (iv), Hemmi [**29**] has given another reformulation of the SKIP for any p :

THEOREM 2.3. [**29**] *For any prime p, the following conditions are equivalent:*

1. *There exists an A_p-form on $(S^{2n-1})_{(p)}$ corelative to $\iota : (S^{2n-1})_{(p)} \to (P^{2n}(p))_{(p)}$ so that $\Sigma^2 : (S^{2n-1})_{(p)} \to (\Omega^2 S^{2n+1})_{(p)}$ is an A_{p-1} map.*
2. *$\pi^s_{qn-2}(S^0)$ has an order p element of a p-primary Kervaire invariant one.*

□

Hemmi's results, Remark 2.2 (iv) and Theorem 2.3, suggest some *p-localized* (or *p-adic*) objects might be a "right" geometric object to describe the p-local stable homotopy groups of the sphere. However, even if that is possible, it is still unclear whether such a geometric view point could lead to a "geometric interpretation" of the p-primary Kervaire invariant or not. In fact, unlike SKIP, such a restatement using the A_n-theoretical concepts has not been presented for KIP. Thus the first question we would like to propose in this direction is the following:

QUESTION 2.4. Is there any p-local (or p-completed) geometrical interpretation of KIP? If so, can we use the A_n-theoretical concepts to describe it?

3. The new doomsday conjecture

In the previous two sections, we have observed some similarities between the (original) 2 primary θ_j elements and the odd primary θ_j elements. Thus, looking at Theorem E, we are tempted to suspect that there might be just finitely many θ_j's for $p = 2, 3$. However, any such a guess would not be so appealing unless it is a consequence of a reasonable conjecture about a global picture of $\pi_*^s(S^0)$. In this section, we recall such a conjecture about the behavior of the Adams spectral sequence, which has been proposed by the author [**67, 68**].

We begin by approximating the Adams tower (from which the classical Adams spectral sequence is obtained) by some tractable objects. For this purpose, we use the classifying spaces of the (rank s) torus T^s and the (rank s) elementary abelian p-group $(\mathbb{Z}/p)^s$. Consider the following commutative diagram of fiber bundles:

$$\begin{array}{ccccc}
\mathbb{Z}/p & \longrightarrow & T & \longrightarrow & T/(\mathbb{Z}/p) \\
\| & & \downarrow & & \downarrow \\
\mathbb{Z}/p & \longrightarrow & ET & \xrightarrow{p_1} & ET/(\mathbb{Z}/p) = BZ/p \\
\downarrow & & \downarrow{\scriptstyle p} & & \downarrow{\scriptstyle p_2} \\
* & \longrightarrow & ET/T & = & BT.
\end{array}$$

Using the Becker-Gottlieb transfer [**15**][**6**, Chap.4], we find

$$p : * \simeq ET \xrightarrow{p_1} B\mathbb{Z}/p \xrightarrow{p_2} BT$$

induces the composite of the transfers

$$t_{\mathbb{C}} : \Sigma BT_+ \xrightarrow{t_2} B\mathbb{Z}/p_+ \xrightarrow{t_1} S^0$$

in the homotopy category of CW-spectra, where an interpretation of t_1 is given in the paragraph just before Theorem 3.2. Here and for the rest of this section, we will write X to stand for $\Sigma^\infty X$ for simplicity.

Define $\overline{H\mathbb{Z}/p}$ by the following cofiber sequence

$$\overline{H\mathbb{Z}/p} \xrightarrow{j} S^0 \xrightarrow{\eta} H\mathbb{Z}/p,$$

where η is the unit map. Then t_1 factorizes uniquely as

$$t_1 : B\mathbb{Z}/p_+ \xrightarrow{t} \overline{H\mathbb{Z}/p} \xrightarrow{j} S^0.$$

From these data, we have the following commutative diagram of towers over S^0:

$$\begin{array}{ccccc}
\vdots & & \vdots & & \vdots \\
\downarrow{\scriptstyle 1\wedge t_{\mathbb{C}}} & & \downarrow{\scriptstyle 1\wedge t_1} & & \downarrow{\scriptstyle 1\wedge j} \\
\Sigma^2 BT_+^2 & \xrightarrow{(t_2)^{\wedge 2}} & B(\mathbb{Z}/p)_+^2 & \xrightarrow{t^{\wedge 2}} & \overline{H\mathbb{Z}/p}^{\wedge 2} \\
\downarrow{\scriptstyle 1\wedge t_{\mathbb{C}}} & & \downarrow{\scriptstyle 1\wedge t_1} & & \downarrow{\scriptstyle 1\wedge j} \\
\Sigma BT_+ & \xrightarrow{t_2} & B(\mathbb{Z}/p)_+ & \xrightarrow{t} & \overline{H\mathbb{Z}/p} \\
\downarrow{\scriptstyle t_{\mathbb{C}}} & & \downarrow{\scriptstyle t_1} & & \downarrow{\scriptstyle j} \\
S^0 & = & S^0 & = & S^0.
\end{array}$$

tower by the complex iterated transfer tower and the iterated transfer tower. To describe the situation concisely, we set

$$PH_*(X;\mathbb{Z}/p) := \mathrm{Ext}^{0,*}_{\mathcal{A}_*}(\mathbb{Z}/p, H_*(X;\mathbb{Z}/p)) \quad (\subseteq H_*(X;\mathbb{Z}/p))$$

for any spectrum X. Since the mod p Hurewicz image of X is contained in $PH_*(X;\mathbb{Z}/p)$, the above commutative diagram of towers induces the following commutative diagram:

(3.1)

$$\begin{array}{ccccc}
 & & & & \pi^s_*(S^0) \\
 & & & & \uparrow{\scriptstyle (j^{\wedge s})_*} \\
\pi^s_*(\Sigma^s BT^s_+) & \xrightarrow{(t_2^{\wedge s})_*} & \pi^s_*(B(\mathbb{Z}/p)^s_+) & \xrightarrow{(t^{\wedge s})_*} & \pi^s_*(\overline{H\mathbb{Z}/p}^{\wedge s}) \\
\downarrow & & \downarrow & & \downarrow \\
PH_*(\Sigma^s BT^s_+;\mathbb{Z}/p) & \xrightarrow{(t_2^{\wedge s})_*} & PH_*(B(\mathbb{Z}/p)^s_+;\mathbb{Z}/p) & \xrightarrow{(t^{\wedge s})_*} & \mathrm{Ext}^{s,*+s}_{\mathcal{A}_*}(\mathbb{Z}/p,\mathbb{Z}/p).
\end{array}$$

Our hope here is that horizontal maps are closer to the identities, so that the left and middle objects would reflect some global properties of the right objects.

As a first step, let us consider the complex iterated transfer tower. For this, there is a recent result of Hubbuck [**30**]:

THEOREM 3.1. [**30**] *For any positive integer* s,

$$\pi^s_*(\Sigma^s BT^s_+) \to PH_*(\Sigma^s BT^s_+;\mathbb{Z}/2)$$

has a finite image.

Hubbuck's proof [**30**] heavily depends upon the work of Baker-Clarke-Ray-Schwartz [**14**], which was proved using the K-theory. Hoping that the complex iterated transfer tower would give a nice approximation for the Adams tower, we arrive at the following obvious conjecture suggested by Theorem 3.1:

THE DOOMSDAY CONJECTURE. *For each prime* p *and a positive integer* s, *only finitely many elements in* $\mathrm{Ext}^{s,*}_{\mathcal{A}_*}(\mathbb{Z}/p,\mathbb{Z}/p)$ *are permanent cycles.*

In fact, this conjecture was originally presented around the mid to late 60's; it was made by Joel Cohen, named by Michael Barratt, and appears in [**57**, p.199, Conjecture 73]. J. Cohen was motivated by the celebrated Hopf invariant one theorem of Adams [**1**] and its odd-primary analogue [**46, 86**], which verifies the doomsday conjecture for $s = 1$. However, the doomsday conjecture was doomed when Mahowald [**49**] showed h_1h_j is a permanent cycle for any j (Note: This is the famous Mahowald η_j element; see §6.), a counter-example to the case $p = 2$, $s = 2$, and soon Ralph Cohen [**24**] found a similar odd-primary counter-example for $s = 3$, by showing h_0b_j is a permanent cycle for any j. (By the way, there was another meaning for "doomsday" in addition to the obvious one which compares each filtration degree s to a "world" and the stem degree $t - s$ to "time". For this, see §5.)

In search of a better conjecture, we now turn our attention to the iterated transfer tower. For this, we note that the adjoint of the s-th iterated transfer

$$(t_1)^{\wedge s} : B(\mathbb{Z}/p)^s_+ = (BZ/p_+)^{\wedge s} \to (S^0)^{\wedge s} = S^0$$

is no other than the composite

$$B(\mathbb{Z}/p)^s \xrightarrow{Breg} B\Sigma_{p^s} \xrightarrow{DL_{p^s}} Q_{p^s}S^0,$$

where $reg : (\mathbb{Z}/p)^s \to \Sigma_{p^s}$ is the regular representation and DL_{p^s} is the $B\Sigma_{p^s}$-component of the map $\coprod_{n\in\mathbb{N}} B\Sigma_n \to \coprod_{n\in\mathbb{Z}} Q_n S^0 = Q(S^0)$ which was the input of the Barratt-Priddy-Quillen theorem (0.2). Since the mod p homology of a symmetric group is generated by the images from various elementary abelian p subgroups, it is tempting to suspect the same thing holds for the stable homotopy of symmetric groups. Although such a hope is too much to be true, we wish the difference between the homology and the stable homotopy to be very small. In this direction, we have proven the following:

THEOREM 3.2. [**68**] *The stable Hurewicz map*

$$\pi^s_{2n-s}(B(\mathbb{Z}/p)^s_+) \to PH_{2n-s}(B(\mathbb{Z}/p)^s_+; \mathbb{Z}/p)$$

has a trivial image, when $\nu_p(n)$ (:= the largest integer $\nu_p(n)$ s.t. $p^{\nu_p(n)} \mid n$) becomes sufficiently large for fixed p and s. More explicitly, this happens if

$$\begin{cases} 2^{\nu_2(n)+2-2s} > \nu_2(n) + 2 + \left[\frac{5[\frac{s-1}{2}]}{2}\right] - \alpha_2\left(\left[\frac{s-1}{2}\right]\right) & \text{if } p = 2 \\ p^{\nu_p(n)+2-2s} > (p-1)\nu_p(n) + 1 + p\left[\frac{s-1}{2}\right] - \alpha_p\left(\left[\frac{s-1}{2}\right]\right) & \text{if } p \text{ is odd} \end{cases}$$

The basic idea of the proof is to show there is no BP_*BP-primitive element with a non-trivial Thom reduction image under the assumption. However, unlike Theorem 3.1, the proof is not K-theoretic. In fact, in terms of the chromatic terminology [**82**], we must make use of the v_s-periodic information to prove the statement for $B(\mathbb{Z}/p)^s$:

A SKETCH OF THE PROOF OF THEOREM 3.2. The techniques we used are

1. The BP-Adams operations ψ^{p+1} [**73, 9**].
2. The determination of $BP_*(B(\mathbb{Z}/p)^s)$ by [**35, 36, 83, 71**].
3. Kameko's reduction [**38**].
4. A "dual" Landweber exact functor theorem, due to Johnson-Landweber-Yoshimura [**34**].

As 3) and 4) and less well-known, let us present some ideas about them:

First, Kameko's reduction basically tells us when

$$Sq^0 : PH_n(B(\mathbb{Z}/2)^s; \mathbb{Z}/2) \to PH_{2n+s}(B(\mathbb{Z}/2)^s; \mathbb{Z}/2)$$
$$x_{i_1} \otimes x_{i_2} \otimes \cdots x_{i_s} \mapsto x_{2i_1+1} \otimes x_{2i_2+1} \otimes \cdots x_{2i_s+1}$$

is well-defined and an isomorphism. It should be pointed out that Kameko [**38**] used Wood's solution [**101**] of the Peterson's conjecture [**74**]. Even though Kameko worked only for $p = 2$, we can still obtain a similar (but slightly different) odd primary analogue. (As the notation suggests, this Sq^0 is seen to commute that of the Ext, using a technique in [**8**].)

Second, Johnson-Landweber-Yoshimura [**34**] roughly say that the chromatic comodules M_n of Miller-Ravenel-Wilson [**62**] are injective in the category, whose objects consist of BP_*BP-comodules and morphisms simply consist of BP_* module maps. In fact, this technical assumption was needed to apply the Landweber's

(known only for $s = 1$). Instead, we approach as follows:
(i) Using 2) 3), show any element in $BP_*(B(\mathbb{Z}/p)^s_+)$, which could map to $(\mathcal{P}^0)^{n(s)} \operatorname{Ext}^{s,t}_{\mathcal{A}_*}(\mathbb{Z}/p, \mathbb{Z}/p)$, has a gigantic order.
(ii) Using 1) 2) 4), show any element in $BP_*(\wedge^s B\mathbb{Z}/p)$, which are in the relevant dimensions (as in (i)) and annihilated by $\psi^{p+1} - 1$ (this is a necessary condition to be contained in $PBP_*(\wedge^s B\mathbb{Z}/p)$), has a relatively small order.

The details of the proof can be found in [**68**]. □

For $s = 1$, together with the Kahn-Priddy theorem [**37**], Theorem 3.2 reproves the Hopf invariant one theorem [**1, 46, 86**].

Now, let us specialize to the caes $p = s = 2$, which is the case of the 2-primary KIP (see Theorem B). Suppose $\theta_j \in \pi^s_{2^{j+1}-2}(S^0)$ exists, then Theorem C implies there exists certain $\tilde{\theta}_j \in \pi^s_{2^{j+1}-2}(BGL_3(\mathbb{F}_2))$ with a nontrivial mod 2 homology. Let $i : (\mathbb{Z}/2)^2 \to GL_3(\mathbb{F}_2)$ be an inclusion of a subgroup. Since the mod 2 cohomology of $B\mathcal{A}_4$ (= the principal indecomposable summand of $B(\mathbb{Z}/2)^2$) and $BGL_3(\mathbb{F}_2)$ are rather similar, there were some hope that $\tilde{\theta}_j$ can be always lifted to $B(\mathbb{Z}/2)^2$:

$$\tag{3.2} \begin{array}{ccc} & & \Sigma^\infty B(\mathbb{Z}/2)^2_+ \\ & \overset{?}{\nearrow} & \big\downarrow {\scriptstyle \Sigma^\infty Bi} \\ S^{2^{j+1}-2} & \xrightarrow[\tilde{\theta}_j]{} & \Sigma^\infty BGL_3(\mathbb{F}_2)_+ \end{array}$$

We note that such a lift ? may be interpreted as a double transfer lift of an appropriate (possibly different from θ_j) Kervaire invariant one element, since ? is always detected by $x_{2^j-1} \otimes x_{2^j-1} \in H_{2^{j+1}-2}(B\mathbb{Z}/2 \times B\mathbb{Z}/2; \mathbb{Z}/2)$, where $x_n \in H_n(B\mathbb{Z}/2; \mathbb{Z}/2)$ is the generator. Furthermore, since $B(\mathbb{Z}/2)^2$ is easy to handle, a traditional approach to KPI [**48**] was to construct an appropriate element of the stable homotopy of $B(\mathbb{Z}/2)^2$.

In spite of such a hope, Theorem 3.2 specializes to the following:

THEOREM 3.3. [**65**] *If the Kervaire invariant one element $\theta_j \in \pi^s_{2^{j+1}-2}(S^0)$ exists and factors through the double transfer $B(\mathbb{Z}/2)^2_+ \to S^0$, then $j \leq 4$.*

COROLLARY 3.4. [**65, 64**] *The Kervaire invariant one element θ_j is not represented by a framed hypersurface for $j \leq 4$.*

A recent work of Lin and Mahowald [**45**] claims the conclusion of Theorem 3.3 is best possible, i.e. θ_4 does factorize the double transfer.

We now return to the problem of finding a new conjecture. Recall that the Steenrod algebra acts on $\operatorname{Ext}^{*,*}_{\mathcal{A}_*}(\mathbb{Z}/p, \mathbb{Z}/p)$ just like the usual action on the mod p cohomology of a space, *except* $Sq^0 \neq 1$ ($\mathcal{P}^0 \neq 1$ for p odd):

$$\begin{cases} Sq^0 : \operatorname{Ext}^{s,t}_{\mathcal{A}_*}(\mathbb{Z}/2, \mathbb{Z}/2) \to \operatorname{Ext}^{s,2t}_{\mathcal{A}_*}(\mathbb{Z}/2, \mathbb{Z}/2) & \text{if } p = 2 \\ \mathcal{P}^0 : \operatorname{Ext}^{s,t}_{\mathcal{A}_*}(\mathbb{Z}/p, \mathbb{Z}/p) \to \operatorname{Ext}^{s,pt}_{\mathcal{A}_*}(\mathbb{Z}/p, \mathbb{Z}/p) & \text{if } p \text{ is odd} \end{cases}$$

As was pointed out by May [**55**], this is described as

$$[t_1|t_2|\cdots|t_s] \mapsto [t_1^p|t_2^p|\cdots|t_s^p]$$

in the cobar complex. In particular, we have $\mathcal{P}^0\big|_{\operatorname{Ext}^{s,2t+1}_{\mathcal{A}_*}(\mathbb{Z}/p,\mathbb{Z}/p)} = 0$ (strictly speaking, this should be regarded as a definition).

Looking at the way how Sq^0 and $\mathcal{P}^0$ change the total degree t in conjunction with Theorem 3.2, it is now a matter of course to speculate the following:

THE NEW DOOMSDAY CONJECTURE. [**67, 68**] *For each s, there exists some integer $n(s)$ such that no nontrivial element in the image of*

$$(\mathcal{P}^0)^{n(s)}\left(\mathrm{Ext}^{s,*}_{\mathcal{A}_*}(\mathbb{Z}/p,\mathbb{Z}/p)\right) \subseteq \left(\mathrm{Ext}^{s,p^{n(s)}*}_{\mathcal{A}_*}(\mathbb{Z}/p,\mathbb{Z}/p)\right)$$

is a nontrivial permanent cycle.

For the reset of this paper, we abbreviate "new doomsday" (resp. the new doomsday conjecture) by ND (resp. NDC) for simplicity.

Notice that Mahowald's η_j-family is *not* a counter-example to NDC. In fact, η_j is detected by h_1h_j in the Adams E_2-term, and $Sq^0(h_ah_b) = h_{a+1}h_{b+1}$.

However, we should still explain why we used Sq^0 and $\mathcal{P}^0$ to formulate NDC; just looking at Theorem 3.2, we could have formulated it purely in terms of the filtration degree s and the total degree t of the Adams E_2-term.

Well, this is because we would like to compare Sq^0 and $\mathcal{P}^0$ with the Mahowald's root invariant R [**50, 53**]. Suppose $p = 2$ for simplicity. Then the Mahowald's root invariant is of the form $R : \pi^s_n(S^0)_{(2)} \to \pi^s_{2n+?}(S^0)_{(2)}$, where ? depends upon each element in $\pi^s_n(S^0)_{(2)}$ and the image is usually not uniquely defined; it is usually a coset.

Now the following iteration of R and Sq^0 is very suggestive:

$$\begin{array}{ll} 2 \overset{R}{\mapsto} \eta \overset{R}{\mapsto} \nu \overset{R}{\mapsto} \sigma \overset{R}{\mapsto} \theta_3 = \sigma^2 \overset{R}{\mapsto} \theta_4 \overset{R}{\mapsto} \theta_5 ? & \\ h_0 \overset{Sq^0}{\mapsto} h_1 \overset{Sq^0}{\mapsto} h_2 \overset{Sq^0}{\mapsto} h_3 \qquad h_3^2 \overset{Sq^0}{\mapsto} h_4^2 \overset{Sq^0}{\mapsto} h_5^2, & \end{array}$$

where any element of $\pi^s_*(S^0)$ in the above line is represented in the Adams spectral sequence E_2-term by the corresponding lower line element, and $a \overset{R}{\mapsto} b$ indicates that the root invariant of a contains b. As we can see, the actions of Sq^0 and R are very similar. Sq^0 acts almost as if it is an induced map of R in the Adams spectral sequence, except for $h_3^2 \neq Sq^0(h_3)$ when $\theta_3 \in R(\sigma)$, an Adams filtration jump by 1. Of course, $Sq^0(h_3) = h_4$, which is not a permanent cycle by the Hopf invariant one differential. Observing that the Hopf invariant one differentials are typical examples of the ND phenomena, we may speculate that the ND phenomena caused this Adams filtration jump of the Mahowald root invariant at σ. Likewise, we may speculate that some iterations of the root invariant applied to θ_4 to have Adams filtration strictly higher than 2, for NDC suggests that the higher Kervaire invariant elements θ_j's do not exist. In this way, we are led to speculate the following:

R.I. DOOMSDAY CONJECTURE. [**67**] *For each s, there exists some integer $m(s)$ such that, for any element of the stable homotopy groups of the sphere $f \in \pi_*(S^0)$ with Adams filtration s, its $m(s)$-fold iterated root invariant $R^{m(s)}(f)$ has Adams filtration strictly higher than s.*

Philosophically, (R.I.) NDC claims that geometry (homotopy, $\pi^s_*(S^0)$ and R) and algebra (homological algebra, $\mathrm{Ext}^{*,*}_{\mathcal{A}_*}(\mathbb{Z}/p,\mathbb{Z}/p)$ and $\mathcal{P}^0$) are hopelessly different. Consequently, there ought to be lots of lots of differentials (= deviations of geometry from algebra) in the Adams spectral sequence for $\pi^s_*(S^0)$.

< 14. (By the way, the first discrepancy is caused by the first Hopf invariant one differential $d_2(h_4) = h_0 h_3^2 \neq 0$, which is also responsible for the discrepancy: $R(\sigma) = \sigma^2$; $Sq^0(h_3) = h_4 \neq h_3$, which was mentioned above.) As a consequence, initially there was a real hope that the deviation between this purely algebraic calculation and the actual homotopy groups would be quite small. From this point of view one aspect of this "New Doomsday Conjecture" is the implication that the differences between the two objects become greater and greater as dimensions increase. Incidently, these remarks are supported by recent calculations of R. Bruner and M. Tangora where they show that the existence of the Sq^i's in Ext implies a vast number of differentials in the range from $60 < t - s < 110$. Furthermore, W. H. Lin has recently announced that only finitely many h_i^3's are permanent cycles. Although we would like to regard it as a supporting conjecture of NDC, it might not be so. This is because such a phenomenon was already conjectured in mid 60's by Mahowald, who thought all the 2 primary θ_j's to exist at that time.

Now, it is quite natural to ask how much we may regard Theorem 3.2 as a supporting evidence of NDC. Essentially, the problem is how much the iterated transfer tower approximates the Adams tower. Historically, there have been many efforts to clarify this situation [**84, 87, 17, 65**], highlighted by the celebrated Kahn-Priddy theorem [**37**]. However, Singer [**87**] showed they do not give a faithful approximation, i.e. $Ph_1 \in \mathrm{Ext}_{\mathcal{A}_*}^{5,14}(\mathbb{Z}/2, \mathbb{Z}/2)$ does not come from the 5-fold iterated transfer. So, the problem is whether such a discrepancy is of finite nature or not. Now there is a possible way to get around Singer's counter-example. To see this, we recall that Singer [**87**] considered the so-called Singer homomorphism

$$\oplus_s PH_*(B(\mathbb{Z}/p)^s; \mathbb{Z}/p)/GL_s(\mathbb{Z}/p) \to \oplus_s \mathrm{Ext}_{\mathcal{A}_*}^{s,*+s}(\mathbb{Z}/p, \mathbb{Z}/p),$$

which is induced by $(t^{\wedge s})_*$ in (3.1). Singer discovered this is not surjective when $p = 2, * = 9, s = 5$ (but conjectured this is injective). However, from the ND philosophy, especially by the way how we used Kameko's reduction in the proof of Theorem 3.2, we might have to take into account the $\mathcal{P}^0$-actions here. In this way, we are led to propose

QUESTION 3.5. Is the localized Singer homomorphism

$$\oplus_s (\mathcal{P}^0)^{-1} PH_*(B(\mathbb{Z}/p)^s; \mathbb{Z}/p)/GL_s(\mathbb{Z}/p) \to \oplus_s (\mathcal{P}^0)^{-1} \mathrm{Ext}_{\mathcal{A}_*}^{s,*+s}(\mathbb{Z}/p, \mathbb{Z}/p)$$

an isomorphism?

If this question is answered affirmatively, then our Theorem 3.2 might be regarded as a supporting evidence of the new doomsday conjecture, in spite of Singer's example [**87**].

On the other hand, if we try to disprove NDC, the obvious candidate is KIP. In view of Theorem 3.3, Theorem C and the lifting diagram 3.2, the problem is how much $B\mathcal{A}_4$ and $BGL_3(\mathbb{F}_2)$ are similar. As was shown by Lin and Mahowald [**45**], θ_4 does factorize $B\mathcal{A}_4$, which might suggest these two are not so different. Theorem 3.3 actually claims there is no BP_*BP primitive element for $B\mathcal{A}_4$, which corresponds to a Kervaire invariant one element for $j \geq 5$. However, we have recently shown that there *does exit* an order 2 BP_*BP-primitive element, which carries the Kervaire invariant one, in $BP_{2^{j+1}-2}BD_8$ (and consequently in $BP_{2^{j+1}-2}BGL_3(\mathbb{F}_2)$) for any j [**70**]. Thus, from the viewpoint of the BP-theory, $B\mathcal{A}_4$ and $BGL_3(\mathbb{F}_2)$ are not so similar as much as we want.

By the way, such a limitation of the BP_*-theory in disproving the existence of the 2 primary θ_j's was already observed in another context: In view of [**58**], it is very natural to expect the differential

$$(3.3) \qquad d_3\left(\beta_{2^i/2^i}\right) = \alpha_1 \beta^2_{2^{i-1}/2^{i-1}} + \textit{other non-essential terms}$$

in the Adams-Novikov spectral sequence of the sphere (which is constructed using the BP_*-theory, rather than the mod p homology theory which was used to construct the traditional Adams spectral sequence), where $\beta_{2^i/2^i}$ is a certain element in the Adams-Novikov E_2-term which essentially corresponds to the Kervaire invariant one element. Following Ravenel's success for the odd primary case [**80**], we are tempted to disprove the existence of the 2 primary Kervaire invariant one element by showing the right hand side of (3.3) is nonzero. However, in their famous paper [**62**], Miller-Ravenel-Wilson showed this is not possible by showing $\beta^2_{2^{i-1}/2^{i-1}} = 0$.

These results may be interpreted as showing that BP_*-theory cannot detect the possible non-existence of an element of (the 2-primary) Kervaire invariant one in any kind of direct way, so one would probably have to go to a much deeper theory to have a chance of proving non-existence.

Meanwhile, Milgram [**59**] has recently proposed a philosophy which might let one suspect there are just finitely many θ_j's. We will briefly review Milgram's philosophy in the next section.

REMARK 3.6. Interestingly, Theorem 3.1 [**30**] and Theorem 3.2 [**68**] have been obtained completely independently without knowing each other.

4. Milgram's point of view

In this section, we briefly review Milgram's philosophy [**59**], especially his suspicion about the Kervaire invariant one elements. His philosophy is based upon the O'Nan-Scott theorem (c.f. [**44**]):

THEOREM 4.1. *A maximal subgroup of Σ_m has one of the following six forms:*
(i) $\Sigma_n \times \Sigma_m \subset \Sigma_{n+m}$
(ii) $\Sigma_m \wr \Sigma_k \subset \Sigma_{mk}$
(iii) $\Sigma_m \wr \Sigma_n \subset \Sigma_{m^n}$
(iv) $Aff_n(p) \subset \Sigma_{p^n}$
(v) $G \subset \Sigma_n$ *where* $T \triangleleft G \to L$ *with* T *a non-abelian simple group but not* $\mathcal{A}_n$, $L \subset Aut(T)$, *and the associated action is primitive.*
(vi) $(G \wr \Sigma_n) \cdot Out(G) \subset \Sigma_{|G|^{n-1}}$ *where* G *is a simple group, not* $\mathcal{A}_n$.

Noticing that (i)-(iv) correspond to well-known operations in $Q(S^0)$, Milgram asked the roles of (v) and (vi), especially that of sporadic finite simple groups, in homotopy theory. Among all, Milgram [**60**] proved θ_3 can be constructed from M_{12}. This led Milgram to propose

MILGRAM'S POINT OF VIEW. *Maximal subgroups of finite symmetric groups play essential roles in the homotopy theory of $Q(S^0)$.*

Since, in view of Theorem B and Theorem C, θ_j's play exceptionally important

MILGRAM'S POINT OF VIEW ON KIP. *One would hope that within this context one could find, somehow, a minimal subgroup responsible for a given homotopy class. Then, it is likely that some of the sporadics would be the groups responsible for classes like the θ_i's, and consequently, the fact that there are only 26 sporadics leads to a suspicion that there are only a finite number of θ_i's.*

For more details abot Milgram's point of view, the reader is urged to consult Milgram's own account [**59**] and [**7**, Chapter VIII].

It should be pointed out that Milgram's current interest in KIP is purely 2 primary, because no geometric interpretation of KIP is known when p is odd. However, if Question 2.4 is answered affirmatively, then Milgram's point of view might provide some conceptual explanation of KIP even when p is odd. For this, let us recall that Ravenel's theorem [**80**] claims there is no θ_j for $j > 0$ for $p \geq 5$ and suggests the existence of θ_j for $p = 3$ is more difficult than the case $p = 2$ (see Theorem E). Then, if sporadic finite simple groups are also responsible for θ_j's even when p is odd, Ravenel's theorem might be a consequence of the well-known fact that p-Sylow subgroups of these sporadics are much simpler when p is odd. (Of course, this fact makes the odd primary cohomology of those sporadics easier to calculate, as was exploited in [**95**].) Of course, all of these result from our hopes to apply group theory to prove results in homotopy theory. But, if we change the direction, we immediately realize the following "dream problem" is waiting for us along the way:

QUESTION 4.2. *Can we use Ravenel's theorem* [**80**] *(and NDC) to explain the Feit-Thompson theorem* [**26**]*? More generally, can we use NDC to explain the finiteness of sporadics?*

Note that the first part is a reminiscence of Atiyah's problem which asks a proof of the Feit-Thompson theorem using topology.

5. The Barratt-Mahowald hypothesis - optimistic version

In the mid 60's, Barratt and Mahowald had come up with various conjectures about the global picture of $\pi_*^s(S^0)_{(2)}$. Their most optimistic version claims $\pi_*^s(S^0)_{(2)}$ is essentially understood by using the Steenrod algebra and KO-theory. Putting in another way, this hypothesis predicts all the non-trivial differentials (= deviation of geometry from algebra) of the Adams spectral sequence for $\pi_*^s(S^0)$ would be essentially obtained by KO-theoretical analysis. So, philosophically, they predicted the deviation of geometry (homotopy, $\pi_*^s(S^0)$ and R) from algebra (homological algebra, $\mathrm{Ext}_{\mathcal{A}_*}^{*,*}(\mathbb{Z}/p, \mathbb{Z}/p)$ and $\mathcal{P}^0$) is a sort of minimum. They realized, for this optimistic hypothesis to happen, all the θ_j's should exist as a special case, and wondered whether KIP is the essential obstacle for this optimistic hypothesis or not.

While they remained neutral about this hypothesis, it was no secret that this optimistic hypothesis was their sentimental favorite one (see [**48**]). This is because they hoped the global picture of $\pi_*^s(S^0)_{(2)}$ to be as simple as possible (just like any one of us!); they know, if θ_j does not exist (which would violate their optimistic hypothesis), then the world of $\pi_*^s(S^0)_{(2)}$ will be too chaotic...almost the end of the world. Such is the case, when J. Cohen proposed his conjecture, it was a sort of natural for Barratt to christen it as the "doomsday conjecture" (see Conjecture 73 in p.199 of [**57**]).

Nowadays, people might say KO-theory can tell us only the v_1-periodic phenomena; we must study all the v_n-periodic phenomena according to the chromatic philosophy (c.f. [**82**]). However, it is too naive to jump to such a conclusion.

To illustrate the situation, we recall the n-th Morava stabilizer algebra S_n, whose computability madc thc chromatic approach so successful (see [**62**] for instance). Many things are known about S_n, including the Krull dimension, which is 1 or 0:

$$\mathbb{Z}/p \subset S_n \iff p-1 \mid n,$$

and so, for fixed p, there are infinitely many n such that $p-1 \mid n$. Let n be such and write $n = j(p-1)$. It would be quite tempting to search for global properties of $\pi_*^s(S^0)_{(p)}$, using mod p cyclic subgroups $\mathbb{Z}/p \subset S_{j(p-1)}$ for $j = 1, 2, \cdots$. However, except for the first one $\mathbb{Z}/p \subset S_{p-1}$, it might be difficult to get useful global informations out of these infinitely many mod p cyclic subgroups.

To see this, we first recall the restriction map

$$f^* : \mathrm{Ext}^*_{BP_*BP} \to H^*(\mathbb{Z}/p; \mathbb{F}_{p^n}), \ (n = j(p-1))$$

considered in [**80**][**81**, p.249]. For $j = 1$, f^* played an important role in Ravenel's proof [**80**] of the nonexistence of the odd primary Kervaire invariant one elements. For any j, f^* is induced by the composite Hopf algebroid map

$$\begin{aligned} BP_*BP &= BP_*[t_1, t_2, \cdots] = \mathbb{Z}_{(p)}[v_1, v_2, \cdots][t_1, t_2, \cdots] \\ &\xrightarrow{f} \Sigma(n) = K(n)[t_1, t_2, \cdots]/(t_i^{p^n} - v_n^{p^i-1} t_i) \\ &\quad = \mathbb{Z}/p[v_n, v_n^{-1}][t_1, t_2, \cdots]/(t_i^{p^n} - v_n^{p^i-1} t_i) \\ &\xrightarrow{t} S(n) := \Sigma(n) \otimes_{K(n)_*} \mathbb{F}_{p^n} = \mathbb{F}_{p^n}[e_i; i \geq 1]/(e_i^{p^n} - e_i) \\ &\xrightarrow{\rho} \mathcal{C} = \mathbb{F}_{p^n}[t]/(t^p - t), \end{aligned}$$

where $\mathcal{C}$ is the linear dual of the group ring $\mathbb{F}_{p^n}[\mathbb{Z}/p]$, f and t are characterized by

$$f(v_i) = \begin{cases} v_n & \text{if } i = n, \\ 0 & \text{if } i \neq n, \end{cases}$$
$$t(v_n) = 1,$$
$$t(t_i) = e_i,$$

and ρ is induced by the inclusion map $\mathbb{Z}/p \subset S_n$ $(n = j(p-1))$. (For the precise definitions and notations, the reader is referred to the standard references of the BP-theory [**62, 81, 82**].) Now the point is that ρ annihilates $e_1, \cdots e_{j-1}$, and e_j is the first element which is not annihilated [**82**, Lemma 8.5.2]. Let c be the Hopf Algebroid conjugate map, and $C(H)$ be the cobar complex of a Hopf algebroid H [**81**, A1.2.11]. Then the above discussion implies $f^* : \mathrm{Ext}^*_{BP_*BP} \to H^*(\mathbb{Z}/p; \mathbb{F}_{p^n})$, $(n = j(p-1))$ is induced by a Hopf algebroid map

$$f : BP_*BP = BP_*[c(t_1), c(t_2), \cdots] \to \mathcal{C} = \mathbb{F}_{p^n}[t]/(t^p - t),$$

such that

$$\begin{cases} 0 & \text{if } 1 \leq i \leq j-1, \end{cases}$$

Next, we recall the Thom reduction [**62, 81**]

$$\Psi : \mathrm{Ext}^*_{BP_*BP} \to \mathrm{Ext}^*_{\mathcal{A}_*}.$$

(This map plays a key role in the proof of Theorem 6.1.) Ψ is induced by the Hopf algebroid map

$$g : BP_*BP = \mathbb{Z}_{(p)}[v_1, v_2, \cdots][c(t_1), c(t_2), \cdots]$$
$$\to \mathcal{A}_* = \begin{cases} \mathbb{Z}/2[\xi_1, \xi_2, \cdots] & \text{if } p = 2, \\ \mathbb{Z}/p[\xi_1, \xi_2, \cdots] \otimes E(\tau_0, \tau_1, \cdots) & \text{if } p \text{ is odd,} \end{cases}$$

such that

$$g(v_i) = 0$$
$$g(c(t_i)) = \begin{cases} \xi_i^2 & \text{if } p = 2, \\ \xi & \text{if } p \text{ is odd,} \end{cases}$$

as the (co) homology of $C(g) : C(BP_*BP) \to C(\mathcal{A}_*)$.

To understand $\mathrm{Ext}^*_{\mathcal{A}_*}$ better, we use Priddy's [**75**, §7] canonical chain homotopy equivalence of differential algebras between $C(\mathcal{A}_*)$ and (the anti-algebra of) the lambda algebra Λ [**18, 19**], where Λ is a differential graded algebra generated by λ_i $(i = 0, 1, \cdots)$ (resp. λ_i $(i = 1, 2, \cdots)$ and μ_i $(i = 0, 1, \cdots)$) for $p = 2$ (resp. p odd), subject to certain "Adem relations." Writing $C(\mathcal{A}_*)$ as $T(\bar{\mathcal{A}}_*)$, the tensor algebra of the augmentation ideal $\bar{\mathcal{A}}_*$ of $\mathcal{A}_*$, Priddy's map

$$h : C(\mathcal{A}_*) = T(\bar{\mathcal{A}}_*) \to \Lambda$$

is characterized on a monomial x by

$$h(x) = \begin{cases} \lambda_{k-1} & \text{if } x = \xi_1^k \ (k > 0), \\ -\mu_{k-1} & \text{if } p \text{ is odd and } x = \tau_0 \xi_1^k \ (k \geq 0), \\ 0 & \text{if otherwise.} \end{cases}$$

Now we see the Thom reduction

$$\Psi : \mathrm{Ext}^*_{BP_*BP} \to \mathrm{Ext}^*_{\mathcal{A}_*}$$

is nothing but the (co) homology of the composite

$$h \circ C(g) : C(BP_*BP) \to \Lambda,$$

which annihilates any monomial (with respect to the generators $c(t_i)$, $i \geq 1$) involving any $c(t_i)$ $(i \geq 2)$.

The above discussion about Ψ suggests, among the generators $c(t_i)$ $(i > 0)$, only $c(t_1)$ is essential, as far as those elements in $\pi_*^s(S^0)$ which are detected in the same filtration both in the Adams-Novikov spectral sequence (whose E_2-term is $\mathrm{Ext}^*_{BP_*BP}$) and the Adams spectral sequence (whose E_2-term is $\mathrm{Ext}^*_{\mathcal{A}_*}$) concerns. (Many interesting elements in $\pi_*^s(S^0)$, including the Hopf invariant one elements and the Kervaire invariant one elements, have the same filtration.) However, we have already seen the mod p cyclic group $\mathbb{Z}/p \subset S_n$ $(n = j(p-1))$ is useless to detect an information concerning $c(t_1)$ through $C(f)$ except for $j = 1$. Therefore, if we change the view point and if we still believe elementary abelian p subgroups would tell us any global information, then we might hope the mod p cyclic subgroup $\mathbb{Z}/p \subset S_{p-1}$ would recover useful global pictures about $\pi_*^s(S^0)_{(p)}$. Especially, when

$p = 2$, $\mathbb{Z}/2 \subset S_1$ is generated by the complex conjugate and its homotopy fixed point of K is KO. Thus, such a hope specializes to the optimistic version of the Barratt-Mahowald picture of $\pi_*^s(S^0)_{(2)}$.

Recently, Hopkins and Miller [**32**] have constructed a p-local spectrum EO_{p-1}, as a homotopy fixed point spectrum of a certain spectrum using this subgroup $\mathbb{Z}/p \subset S_{p-1}$ (and some Galois group). In particular, EO_{p-1} is essentially KO when $p = 2$ (with the $\mathbb{Z}/2$ generated by the complex conjugate), and, for an odd prime p, EO_{p-1} carries the information of the Ravenel's odd primary Kervaire invariant one theorem [**80**] (see Theorem E in the introduction). Since EO_{p-1} generalize KO, we may now generalize the Barratt-Mahowald's hypothesis to all primes, so as not to contradict Ravenel's odd primary Kervaire invariant theorem:

THE GENERALIZED BARRATT-MAHOWALD HYPOTHESIS. *For each p, EO_{p-1} would tell us essentially everything about the differentials of the Adams spectral sequence.*

Just like NDC, there is no counter-example known for this hypothesis. For $p = 2$, it is very impressive that even Barratt-Mahowald's most optimistic version has never been challenged for thirty years.

6. The Adams filtration 2 case

We first recall the well-known result for the Adams 2-line elements.

THEOREM 6.1. [**54, 62**] *Up to finitely many exceptions, only θ_j's and η_j's are possible infinite cycles among Adams spectral sequence filtration 2 elements, where these elements are represented in the Adams E_2 term as follows:*

p	2	*odd*
θ_j	h_j^2	$b_j = \underbrace{\langle h_j, \cdots, h_j \rangle}_{p}$
η_j	$h_1 h_j$	$h_0 h_j$

*Here h_i is represented in the cobar complex by $[\xi_1^{p^n}]$ and b_j is the p-fold Massey product of h_j (see [**81**, A1.4] for a nice account of Massey product).*

PROOF. This is most effectively shown by calculating the Thom reduction $\Psi : \mathrm{Ext}^*_{BP_*BP} \to \mathrm{Ext}^*_{A_*}$, which was reviewed in the last section. See [**81**, 5.4.6] for details. □

There is a well-known relationship between these families:

$$\text{There exists } \theta_j \text{ such that } p\theta_j = 0 \Longrightarrow \eta_{j+1} \text{ also exists,}$$

from which, Mahowald [**49**, p.250, second paragraph] hoped the existence of his 2 primary η_j's would be a strong evidence for the existence of the 2 primary θ_j's.

Now the current status of these infinite families is as follows:

p	2	3	5
θ_j	???	??	No (Ravenel [**80**])
η_j	Yes (Mahowald [**49**])	??	?

QUESTION 6.2. How much can we regard the existence of η_{j+1} as a supporting evidence for the existence of θ_j? In other words, how much can we regard the nonexistence of θ_j as a supporting evidence for the nonexistence of η_{j+1}?

There is a preliminary result [**66**] in this direction. But we will not review it here, because it is rather technical. We simply note that the result in [**66**] suggests a possible use of Hopkins-Miller EO_{p-1} for this problem.

References

1. J. F. Adams, *On the non-existence of elements of Hopf invariant one*, Ann. of Math. **72** (1960), 20–104.
2. J. F. Adams, *The sphere, considered as an H-space mod p*, Quart. J. Math. Oxford **12** (1961), 52–60.
3. J. F. Adams, *Vector fields on spheres*, Ann. of Math. **75** (1962), 603–632.
4. J. F. Adams, *On the groups J(X) IV*, Topology **5** (1966), 21–71.
5. J. F. Adams, *Stable homotopy and generalized homology*, University of Chicago Press, Chicago, (1974).
6. J. F. Adams, *Infinite Loop Spaces*, Annnals of Mathematics Studies, **90** (1978).
7. A. Adem and R. J. Milgram, *Cohomology of Finite Groups*, Springer-Verlag, Berlin Heidelberg (1994).
8. D. W. Anderson and D. M. Davis, *A vanishing theorem in homological algebra*, Comm. Math. Helv. **48** (1973), 318–327.
9. S. Araki, *Multiplicative operations in BP cohomology*, Osaka J. Math.**12** (1975), 343–356.
10. M. Barratt, J. Jones and M. Mahowald, *The Kervaire invariant problem*, Contemp. Math. **19** (1983), 9–22.
11. M. Barratt, J. Jones and M. Mahowald, *Relations amongst Toda brackets and the Kervaire invariant in dimension 62*, J. London Math. Soc. **30** (1985), 533–550.
12. M. G. Barratt, J. D. S. Jones, and M. E. Mahowald, *The Kervaire invariant and the Hopf invariant*, Springer Lecture Notes in Mathematics **1286** (1987), 135–173.
13. M. G. Barratt and S. Priddy, *On the homology of non-connected monoids and their associated groups*, Comm. Math. Helv. **47** (1972), 1–14.
14. A. Baker, F. Clarke, N. Ray, and L. Schwartz, *On the Kummer congruences and the stable homotopy of BU*, Trans. Amer. Math. Soc. **316** (1989), 385–432.
15. J. C. Becker and D. H. Gottlieb, *The transfer map and fiber bundles*, Topology **14** (1975), 1–12.
16. D. J. Benson, *Representations and cohomology II*, Cambridge studies in advanced mathematics **31**, Cambridge University Press (1991).
17. J. M. Boardman, *Modular representations on the homology of powers of real projective space*, Algebraic Topology; Oaxtepec 1991 (ed M. C. Tangora), American Mathematical Society, Contemp. Math. **146** (1993).
18. A. K. Bousfield, E. B. Curtis, D. M. Kan, D. G. Quillen, D. L. Rector and J. W. Schlesinger, *The mod p lower central series and the Adams spectral sequence*, Topology **5** (1966), 331–342.
19. A. K. Bousfield and D. M. Kan, *The homotopy spectral sequence of a space with coefficients in a ring*, Topology **11** (1972), 79–106.
20. E. H. Brown, *Cohomology theories*, Annals of Math. **75** (1962), 467–484.
21. W. Browder, *The Kervaire invariant of a framed manifold and its generalization*, Annals of Math. **90** (1969), 157–186.
22. F. Cohen, *A course in some aspects of classical homotopy theory*, Springer Lecture Notes in Mathematics **1286** (1987) 1–92.
23. F. Cohen, *On the Whitehead square, Cayley-Dickson algebras, and rational functions*, Bol. Soc. Mat. Mexicana **37** (1992), 55–62.
24. R. Cohen, *Odd Primary Infinite Families in Stable Homotopy Theory*, Memoirs Amer. Math. Soc. **242** (1981).
25. A. D. Elmendorf, I. Kriz, M. A. Mandell, and J. P. May, *Rings, modules, and algebras in stable homotopy theory*, Mathematical Surveys and Monographs **47**, American Mathematical Society (1997).

26. W. Feit and J. G. Thompson, *Solvability of groups of odd order*, Pacific J. Math. **13** (1963), 775–1029.
27. Z. Fiedorowicz and S. Priddy, *Homology of classical groups over finite fields and their associated infinite loop spaces*, Springer Lecture Notes in Mathematics **674** (1978).
28. M. J. Greenberg and J. R. Harper, *Algebraic Topology: A First Course.* Benjamin/Cummins, Reading, Massachusetts (1981),
29. Y. Hemmi, *Higher homotopy associativity and the mod p Kervaire invariant*, preprint.
30. J. R. Hubbuck, *On numerical forms*, preprint.
31. M. J. Hopkins and M. E. Mahowald.
32. M. J. Hopkins and H. R. Miller, in preparation.
33. N. Iwase and M. Mimura, *Higher Homotopy Associativity*, Springer Lecture Notes in Mathematics **1370** (1989), 193–220.
34. D. C. Johnson, P. S. Landweber, and Z. Yoshimura, *Injective BP_*BP-comodules and localizations of Brown-Peterson Homology*, Illinois J. Math. **25** (1981), 599–610.
35. D. C. Johnson and W. S. Wilson, *The Brown-Peterson Homology of Elementary p-groups*, Amer. J. Math. **107** (1985), 427–453.
36. D. C. Johnson, W. S. Wilson, and D.-Y. Yan, *The Brown-Peterson Homology of Elementary p-groups, II*, Topology Appl. **59** (1994), 117–136.
37. D. S. Kahn and S. B. Priddy, *Applications of the transfer to stable homotopy theory*, Bull. Amer. Math. Soc. **78** (1972), 981–987.
38. M. Kameko, *Products of projective spaces as Steenrod modules*, Ph. D. Thesis, Johns Hopkins University (1990).
39. M. A. Kervaire and J. W. Milnor, *Groups of homotopy spheres-I*, Ann. of Math. **77** (1963), 504–537.
40. K. Knapp, *Im(J)-theory and the Kervaire invariant*, to appear in Math. Z.
41. S. O. Kochman, *Stable homotopy groups of spheres*, Springer Lecture Notes in Mathematics **1423** (1990).
42. P. S. Landweber, *Homological properties of comodules over $MU_*(MU)$ and $BP_*(BP)$*, Amer. J. Math. **98** (1976), 591–610.
43. J. P. Levine, *Lectures on groups of homotopy spheres*, Springer Lecture Notes in Mathematics **1126** (1985), 62–95.
44. M. Liebeck, C. E. Praeger and J. A. Saxl, *A classification of the maximal subgroups of the finite alternating and symmetric groups*, J. Algebra **111** (1987), 365–383.
45. W. H. Lin and M. E. Mahowald, *The Adams spectral sequence for Minami's theorem*, prreprint.
46. A. L. Liulevicius, *The factorization of cyclic reduced power by secondary cohomology operations*, Mem. Amer. Math. Soc. **42** (1962).
47. Ib Madsen and R. J. Milgram, *The classifying spaces for surgery and cobordism of manifolds*, Annals of Mathematical Studies **92** Princeton University Press (1979).
48. M. E. Mahowald, *Some remarks on the Kervaire invariant problem from the Homotopy point of view*, Proc. Symp. Pure Math. XXII, Amer. Math. Soc. (1971), 165–169.
49. M. E. Mahowald, *A new infinite family in* ${}_2\pi_*^s$, Topology **16** (1977), 249–256.
50. M. E. Mahowald, *The metastable homotopy of S^n*, Memoir Amer. Math. Soc. **72** (1967).
51. M. E. Mahowald, *The Image of J in the EHP sequence*, Ann. of Math. **116** (1982), 65–112.
52. M. E. Mahowald, *Some remarks on a Theorem of Minami*, preprint.
53. M. E. Mahowald and D. C. Ravenel, *The root invariant in Homotopy Theory*, Topology **32** (1993), 865–898.
54. M. E. Mahowald and M. C. Tangora, *Some differentials in the Adams spectral sequence* , Topology **6** (1967), 349–369
55. J. P. May, *A general algebraic approach to Steenrod operations*, Springer Lecture Notes in Mathematics **168** (1970), 153–231.
56. J. P. May, *E_∞ Ring Spaces and E_∞ Ring Spectra*, with contributions by F. Quinn, N. Ray, and J. Tornehave, Springer Lecture Notes in Mathematics **577** (1977).
57. R. J. Milgram, *Problems presented to the 1970 AMS summer colloquium in algebraic topology*, Proc. Symp. Pure Math. XXII, Amer. Math. Soc. (1971) 187–201.

60. R. J. Milgram, *The 14-dimensional Kervaire invariant and the sporadic group M_{12}*, preprint.
61. H. R. Miller and S. B. Priddy, *On G and the stable Adams conjecture*, Springer Lecture Notes in Mathematics **658** (1978), 331–348.
62. H. R. Miller, D. C. Ravenel and W. S. Wilson, *Periodic phenomena in the Adams-Novikov spectral sequence,* Annals of Math. **106** (1997), 469–516.
63. N. Minami, *Multiplicative homology operations and transfer,* Memoirs. Amer. Soc. **457** (1991).
64. N. Minami, *On the double transfer*, Contemp. Math. **146** (1993), p.339-347.
65. N. Minami, *The Kervaire invariant one element and the double transfer*, Topology **34** (1995), 481–488.
66. N. Minami, *On the odd primary Adams 2-line elements*, preprint.
67. N. Minami, *The Adams spectral sequence and the triple transfer*, Amer. J. Math. **117** (1995), 965–985.
68. N. Minami, *The iterated transfer analogue of the new doomsday conjecture*, to appear in Trans. Amer. Math. Soc.
69. N. Minami, *On the Hurewicz image of the Cokernel J spectrum*, to appear in Fund. Math.
70. N. Minami, *On some BP_*BP-primitive elements related to the Kervaire invariant problem*, to appear in the Fields's Institute Homotopy Theory Proceedings.
71. S. A. Mitchell, *A proof of the Conner-Floyd conjecture* , Amer. J. Math.**106** (1984), 889–891.
72. S. Mitchell and S. Priddy, *Symmetric product spectra and splittings of classifying spaces*, Amer. J. Math. **106** (1984), 219–232.
73. S. P. Novikov, *The methods of algebraic topology from the viewpoint of cobordism theories (Russian)*, Izv. Akad. Nauk SSSR ser. Mat. **31** (1967), 855–951 (English transl. Math. USSR-Izv. (1967), 827–913).
74. F. P. Peterson, *$\mathcal{A}$-generators for certain polynomial algebras*, Math. Proc. Cambridge Phil. Soc. **105** (1989), 311–312.
75. S. B. Priddy, *Koszul resolutions*, Trans. Amer. Math. Soc. **152** (1970), 39–60.
76. S. B. Priddy, *Homotopy splittings of G and G/O*, Comm. Math. Helv. **53** (1978), 470–484.
77. D. Quillen, *The Adams conjecture*, Topology **10** (1971), 67–80.
78. D. Quillen, *On the completion of a simplicial monoid*, preprint.
79. D. Quillen, *On the cohomology and K-theory of the general linear groups over a finite field*, Ann. of Math. **96** (1972), 552–586.
80. D. C. Ravenel, *The nonexistence of odd primary Arf invariant elements in stable homotopy*, Math. Proc. Cambridge Philos. Soc. **83** (1978), 429–443.
81. D. C. Ravenel, *Complex Cobordism and Stable Homotopy Groups of Spheres*, Academic Press, Orlando, Florida, (1986).
82. D. C. Ravenel, *Nilpotence and Periodicity in Stable Homotopy Theory*, Annals of Math. Series, **128**, Princeton University Press (1992).
83. D. C. Ravenel and W. S. Wilson, *The Morava K-theories of Eilenberg-MacLane spaces and the Conner-Floyd conjecture*, Amer. J. Math. **102** (1980), 691–748.
84. F. W. Roush, *Transfer in generalized cohomology theories*, Ph. D. thesis, Princeton Univ. (1971).
85. P. Selick, *A reformulation of the Arf invariant one mod p problem and applications to atomic spaces*, Pacific J. Math. **108** (1983), 431–450.
86. N. Shimada and T. Yamanoshita, *On the triviality of the mod p Hopf invariant*, Japan J. Math. **31** (1961) 1–24.
87. W. M. Singer, *The transfer in homological algebra*, Math. Z. **202** (1989), 493–523.
88. S. Smale, *Generalized Poincaré conjecture in dimensions greater than four*, Ann. of Math. **74** (1961), 391–406.
89. S. Smale, *On the structure of manifolds*, Amer. J. Math. **84** (1962), 387–399.
90. J. Smith, *Specters of symmetry.*
91. V. Snaith and J. Tornehave, *On $\pi_*^s(BO)$ and the Arf invariant of framed manifolds*, Contemp. Math. **12** (1982), 299–313.
92. J. D. Stasheff, *Homotopy associativity of H-spaces, I,II*, Trans. Amer. Math. Soc. **108** (1963), 272–312.
93. D. Sullivan, *Geometric topology I. Localization, periodicity and Galois symmetry*, MIT preprint (1971).
94. D. Sullivan, *Genetics of homotopy theory and the Adams conjecture*, Annals of Math. **100** (1974), 1–79.

95. M. Tezuka and N. Yagita, *On odd prime components of cohomologies of sporadic simple groups and the rings of universal stable elements*, J. Algebra **183** (1996), 483–513.
96. H. Toda, *An important relation in the homotopy groups of spheres*, Proc. Japan Acad. **43** (1967), 893–942.
97. J. Tornehave, *The splitting of spherical fibration theory at odd primes*, preprint.
98. R. J. Wellington, *The unstable Adams spectral sequence for iterated loop spaces*, Memoirs. Amer. Soc. **258** (1982).
99. W. S. Wilson, *Brown–Peterson Homology, an Introduction and Sampler*, Regional Conference Series in Math. **48**, American Mathematical Society, Providence, Rhode Island, 1980.
100. C. Wilkerson, *Sphere which are not loop spaces mod p*, Proc. Amer. Math. Soc. **39** (1973), 616–618.
101. R. M. W. Wood, *Steenrod squares of polynomials and the Peterson conjecture*, Math. Proc. Cambridge Phil. Soc. **105** (1989), 307–309.
102. A. Zabrodsky, *Homotopy associativity and finite CW complexes*, Topology **9** (1970), 121–128.

Department of Mathematics, The University of Alabama, Box 870350 Tuscaloosa AL 35487-0350 U.S.A

E-mail address: norihiko@gp.as.ua.edu

Contemporary Mathematics
Volume **220**, 1998

Schur Q-functions and a Kontsevich-Witten genus

Jack Morava

ABSTRACT. The Virasoro operations in Witten's theory of two-dimensional topological gravity have a homotopy-theoretic interpretation as endomorphisms of an **ordinary** cohomology theory with coefficients in a localization of I. Schur's ring Δ of Q-functions. The resulting theory has many of the features of a vertex operator algebra.

Introduction

Smooth complex curves of genus $g > 1$ form a stack $\mathcal{M}_g$ with a compactification defined by adjoining the divisor of stable singular curves (with double points, but only finitely many automorphisms). Ideas from string theory have led to great progress in understanding the topology of these moduli objects in the large genus limit. This note is concerned with an algebra introduced by Schur in connection with the classification of projective representations of symmetric groups, and its relevance to these spaces. This algebra appears in disguise in classical work on the Riemann moduli space and in more recent work of Witten and Kontsevich on the intersection theory of its Deligne-Mumford compactification $\overline{\mathcal{M}}_g$, but its significance has become clear only in retrospect, as integral cohomology emerges from the fog of $\mathbb{Q}$.

The central construction of topological gravity is a partition function which can be defined geometrically by a family

$$\tau_g : \overline{\mathcal{M}}_g \to \mathbf{MU}\hat{\otimes}\mathbb{Q}[v]$$

of maps to the complex cobordism spectrum tensored with the rationals, and the main result of this paper is the construction of a morphism

$$\mathrm{kw} : \mathbf{MU} \to \mathbf{H}(\Delta[q_1^{-1}])$$

of ring-spectra, which (when composed with τ) sends the fundamental class of the moduli space of stable curves to a highest-weight vector for a naturally defined Virasoro action on $\Delta_{\mathbb{Q}}$. This Kontsevich-Witten genus can be constructed in purely algebraic terms from the theory of formal groups, but its natural context is the topology of moduli spaces. The first section below is a summary of some background from algebraic geometry, emphasizing homotopy theory, while the Hopf algebra of

1991 *Mathematics Subject Classification.* Primary 14H10, Secondary 55N35, 81R10.

The author was supported in part by the NSF.

Q-functions is the topic of the second. This whole subject is in many ways still quite mysterious, and a final section argues that some features are most natural in an equivariant context.

I would like to thank M. Ando, A. Baker, F. Cohen, C. Y. Dong, C. Faber, T. Józefiak, M. Karoubi, E. Looijenga, G. Mason, S. Morita, and U. Tillmann for conversations and correspondence about the material in this paper.

§1 Background from geometry

1.1. Teichmüller theory describes the Riemann moduli space $\mathcal{M}_g$ as the quotient of a complex $3(g-1)$-dimensional cell by an action of the mapping class group

$$\Gamma_g = \pi_0 \mathrm{Diff}^+(\Sigma_g)$$

of isotopy classes of orientation-preserving diffeomorphisms of a closed surface Σ_g; this action is properly discontinuous and almost free, in the sense that its isotropy groups are finite, so the resulting quotient is an orbifold. The components of the group of diffeomorphisms are contractible, and the map

$$B\mathrm{Diff}^+(\Sigma_g) = B\Gamma_g \to \mathcal{M}_g$$

from the homotopy-theoretic quotient of the Teichmüller action to the geometric quotient is a rational homology isomorphism.

There is no such essentially topological description of the compactification $\overline{\mathcal{M}}_g$, but there are very interesting (proper) forgetful maps

$$\Phi^n_g : \overline{\mathcal{M}}^n_g \to \overline{\mathcal{M}}_g$$

defined on the moduli stack of stable curves marked with n ordered smooth points. [A marked curve is stable, as above, if it has only finitely many automorphisms; the loss of the marking may render an irreducible component of the curve unstable, and the forgetful map is understood to contract such a component to a point.] According to the conventions of Quillen, such a proper complex-oriented map between homology manifolds defines an element

$$[\Phi^n_g] \in MU^{-2n}_{\mathbb{Q}}(\overline{\mathcal{M}}_g)$$

of the ring (tensored with $\mathbb{Q}$) of complex cobordism classes of maps to the moduli space. These cobordism classes can be identified in terms of ordinary cohomology by the ring isomorphism

$$MU^*_{\mathbb{Q}}(X) \to H^*(X, \mathbb{Q}[t_k | k \geq 1])$$

which sends $[\Phi : V \to X]$ to the characteristic number polynomial

$$\sum_I \Phi_* m_I(T^*_\Phi) t^I ,$$

where T^*_Φ is the formal cotangent bundle along Φ and Φ_* is the Gysin or transfer map in cohomology; $I = i_1, \ldots$ is a multi-index, $t^I = \prod t_k^{i_k}$, and m_I denotes the characteristic class associated to the monomial symmetric function indexed by the partition of $|I| = \sum k i_k$ defined by I. [The t_k correspond to the complete symmetric functions h_k but we keep the standard notation.] This differs from the usual [24 §6.2] definition, which is expressed in terms of the normal bundle rather than the cotangent bundle, but the two constructions are related by a straightforward change of basis for the characteristic classes, and the choice above leads most directly to the formulas of Witten.

1.2. The action of a diffeomorphism of the surface Σ_g on its homology defines an integral symplectic representation of Γ_g, and thus a map

$$B\mathrm{Diff}^+(\Sigma_g) \to BSp(2g, \mathbb{Z}) \ .$$

The group $Sp(2g, \mathbb{Z})$ acts with finite isotropy on the contractible symmetric space $Sp(2g, \mathbb{R})/\mathbb{U}(g)$, with quotient the space $\mathcal{A}_g$ of g-dimensional principally polarized abelian varieties, so the homomorphism induced by this map on rational homology agrees with that defined by the construction which assigns to a Riemann surface its Jacobian. By stability theorems of Harer and Ivanov, these maps are suitably compatible for increasing g, and we can think of them as taking values in the classifying space $BSp(\mathbb{Z})$ of the infinite integral symplectic group. The composition of the obvious maps

$$BSp(\mathbb{Z}) \to BSp(\mathbb{R}) = B\mathbb{U} \to B(\mathbb{U}/\mathbb{O}) = Sp(\mathbb{H})/\mathbb{U}$$

with a final Bott isomorphism is a rational homology isomorphism which (away from two [15 §3.15]) splits a copy of $Sp(\mathbb{H})/\mathbb{U}$ off the group completion $BSp(\mathbb{Z})^+$, cf. [3,25].

1.3.1. One of the main contentions of this paper is that the theory of topological gravity defines maps to the complex cobordism spectrum which are natural analogues, for the compactifications $\overline{\mathcal{M}}_g$, of the classical Abel-Jacobi map described above. The physicists' definitions are motivated by ideas from statistical mechanics: using the language of cobordism, let

$$\tau_g = \sum_{g \geq 0} [\Phi_g^n] \frac{v^n}{n!} \in MU^0_{\mathbb{Q}[v]}(\overline{\mathcal{M}}_g) \ ,$$

with v a bookkeeping variable of (cohomological) degree two. The element τ_g is the class of the space of configurations defined by an indefinite number of distinct but unordered smooth points on a stable curve of genus g; it is tempting to think of this as the space of states of the 'Mumford gas' of free particles on a Riemann surface, with existence as their only attribute. The homotopy class representing this tau-function induces a homomorphism

$$\tau_{g*} : H_*(\overline{\mathcal{M}}_g, \mathbb{Q}) \to H_*(\mathbf{MU}, \mathbb{Q}[v]) = \mathbb{Q}[t_k | k \geq 1][[v]] \ ;$$

the image of the fundamental class $[\overline{\mathcal{M}}_g]$ of the moduli space under this homomorphism is essentially Witten's free energy F_g. [It is convenient to interpret homology with $\mathbb{Q}[v]$ coefficients to be v-adically completed; for a finite complex this implies no change at all.]

1.3.2. To be more precise about this, we will need the slightly more sensitive characteristic number homomorphism which assigns to $[\Phi] \in MU^{-2n}(X)$ the class

$$\mathrm{cl}\ [\Phi] = \sum \Phi_* m_I(T^*_\Phi) t_0^{n-l(I)} t^I \ ,$$

where $l(I) = \sum_{k>0} i_k$ is the length (or number of parts) of the partition I. The ring $\mathbb{Q}[t_k | k \geq 0][t_0^{-1}]$ of cohomology coefficients has now been enlarged to include an invertible polynomial generator of degree zero; this corresponds [cf. §1.5 below] to a natural extension of the Landweber-Novikov where d_i is a finite sequence of nonnegative numbers with $\sum d_i = |I|$ in which k appears i_k times; it is convenient to think of this sequence as having exactly n terms, with zero appearing $i_0 = n - l(I)$ times. The stable cotangent bundle T^*_Φ along Φ^n_g is the sum of the cotangent line bundles L_i of the modular curve at its marked points, and the characteristic class $m_I(T^*_\Phi)$ is obtained by substituting Euler classes $\mathbf{e}(L_i)$ for the formal variables x_i. Witten's characteristic number [26 §2.4]

$$\langle\tau^I\rangle = (\prod_{1\le i\le n} \mathbf{e}(L_i)^{d_i})[\overline{\mathcal{M}}^n_g]$$

is invariant under permutations of the marked points, so

$$m_I(T^*_\Phi)[\overline{\mathcal{M}}^n_g] = \frac{n!}{I!}\langle\tau^I\rangle\ ,$$

where $I! = \prod_{k\ge 0} i_k!$. This rational number is the image of

$$\Phi^n_{g*} m_I(T^*_\Phi) \in H^*(\overline{\mathcal{M}}_g, \mathbb{Q})$$

under the Gysin homomorphism defined by the map from $\overline{\mathcal{M}}_g$ to a point; the latter is just evaluation on the fundamental class of $\overline{\mathcal{M}}_g$, and the free energy [26 §2.16] can thus be recognized as the sum over g and n of terms of the form

$$\mathrm{cl}\ ([\Phi^n_g]/n!)[\overline{\mathcal{M}}_g] = \sum \langle\tau^I\rangle \frac{t^I}{I!}\ .$$

The classical Jacobian is similarly a configuration space of divisors, and the sum

$$j_g = \sum_{n\ge 0} [\mathrm{SP}^n C_g] v^n \in MU^0_{\mathbb{Q}[v]}(\mathcal{M}_g)$$

of symmetric powers of the modular family of curves is the pullback of the product of the universal torus bundle in $MU^{-2g}(BSp(2g,\mathbb{Z}))$ by the coupling constant

$$v^g \sum_{n\ge 0} [\mathbb{C}P(n)] v^n\ .$$

It is somewhat surprising that no expression seems to be known for the class of this universal bundle.

1.4. The construction τ can be extended to stable curves which are not necessarily connected. From the point of view of homotopy theory the space of unordered configurations of points in X is

$$Q(X) = \coprod_{n\ge 0} E\Sigma_n \times_{\Sigma_n} X^n$$

but over the rationals this has the homotopy type of the infinite symmetric product $\mathrm{SP}^\infty(X)$. We will therefore take

$$Q(\overline{\mathcal{M}}) = Q(\coprod_{g\ge 0} \overline{\mathcal{M}}_g)$$

as a model for the space of stable curves, connected or not; its rational homology is the symmetric tensor algebra on $\oplus_{g\ge 0} H_*(\overline{\mathcal{M}}_g, \mathbb{Q})$, with $\overline{\mathcal{M}}_0$ and $\overline{\mathcal{M}}_1$ interpreted as one-point spaces. It is natural to think of the fundamental homology class of the orbifold $SP^n\overline{\mathcal{M}}_g$ as the quotient in this tensor algebra of $[\overline{\mathcal{M}}_g]^n$ by $n!$, so

$$[Q(\overline{\mathcal{M}})] = \exp(\sum_{g\ge 0} [\overline{\mathcal{M}}_g] v^{3(g-1)})$$

defines a kind of fundamental class for $Q(\overline{\mathcal{M}})$ with coefficients in $\mathbb{Q}[v]$. Because no marked curve is stable when $3(g-1)+n < 0$, surfaces of small genus play a slightly anomalous role in these formulas; for example, it is useful to define $[\overline{\mathcal{M}}_1] = -\frac{1}{12}$.

The hard Lefschetz theorem (or the theory of mixed Hodge structures) defines an action of sl_2 on the rational homology of a projective orbifold, with its fundamental class as a highest weight vector, so the homology of $Q(\overline{\mathcal{M}})$ inherits an action of $sl_2[v, v^{-1}]$ in which the class $[Q(\overline{\mathcal{M}})]$ has conformal weight zero [13 §2.6]. The construction of τ_g extends multiplicatively to define an element

$$\tau \in MU^0_{\mathbb{Q}[v]}(Q(\overline{\mathcal{M}}))$$

which sends this fundamental class to the partition function

$$\tau_W = \tau_*[Q(\overline{\mathcal{M}})] = \exp(\sum_{g\geq 0} F_g) \in \mathbb{Q}[t_k | k \geq 1][[t_0, v]] .$$

These results from physics suggest that the (immensely complicated) moduli space of curves has quite interesting homotopy-theoretic approximations, but (unlike the somewhat similar situation in algebraic K-theory) we do not yet understand these stabilizations in terms of universal mapping properties. That τ takes the fundamental class of the moduli space to a highest-weight vector for the natural endomorphisms of $\Delta_{\mathbb{Q}}$ is the central result of Kontsevich-Witten theory, but it is not yet a characterization.

1.5. The parameters t_k have an intrinsic interpretation as polynomial generators for the extended Landweber-Novikov Hopf algebra

$$S = \mathbb{Z}[t_k |\ k \geq 0][t_0^{-1}]$$

of cooperations in complex cobordism. The universal stable cohomology operation is a ring homomorphism

$$MU^*(X) \to MU^* \otimes S$$

and the characteristic number homomorphism of §1.1 can be defined as the composition of this map with the Thom map from cobordism to ordinary cohomology. The universal operation sends the Euler class $\mathbf{e}$ of a complex line bundle to

$$t(\mathbf{e}) = \sum_{k\geq 0} t_k \mathbf{e}^{k+1} ;$$

it follows that the algebra S represents the group of formal origin-preserving diffeomorphisms of the line, and that the characteristic number homomorphism induces the classifying map for the universal formal group law

$$X +_S Y = t(t^{-1}(X) + t^{-1}(Y))$$

of additive type. The Lie algebra of formal vector fields on the line is closely

§2 Background from algebra

2.1. The simplest definition of the algebra Δ of Q-functions is as the quotient of the polynomial algebra $\mathbb{Z}[q_k|k \geq 1]$ by the ideal generated by the coefficients of the relation

$$q(T^{\frac{1}{2}})q(-T^{\frac{1}{2}}) = 1 ,$$

where

$$q(T^{\frac{1}{2}}) = \sum_{k\geq 0} q_k T^{\frac{1}{2}k}$$

is a generating function with $q_0 = 1$ [9 §7]. This algebra has a natural grading, which does not fit very comfortably with the conventions of algebraic topology; we will assign the formal variable $T^{\frac{1}{2}}$ cohomological degree one, but we will not assume that elements of odd degree anticommute. The diagonal homomorphism

$$q_i \mapsto \sum_{i=j+k} q_j \otimes q_k$$

makes Δ into a bicommutative Hopf algebra over $\mathbb{Z}$, and the relation

$$(-1)^i q_i^2 = 2q_{2i} + 2 \sum_{i-1\geq k\geq 1} (-1)^{k-1} q_k q_{2i-k}$$

implies that the square of a generator can be expressed as a sum of monomials in which no q_k appears with exponent greater than one. It follows that Δ is a free module over the integers, with a basis of square-free monomials; similarly, $\Delta[\frac{1}{2}]$ is a polynomial algebra generated by the elements q_{2k+1}, while the reduction of Δ modulo two is an exterior algebra. Being torsion-free, Δ embeds in $\Delta_{\mathbb{Q}} = \Delta \otimes \mathbb{Q}$, and its defining relations imply that the power series

$$\log q(T^{\frac{1}{2}}) = \sum_{i\geq 0} \frac{2x_k}{2k+1} T^{k+\frac{1}{2}}$$

contains only odd powers of $T^{\frac{1}{2}}$. Newton's identity

$$(2k+1)q_{2k+1} = 2(x_0 q_{2k} + x_1 q_{2k-2} + \cdots + x_k)$$

shows that the classes $2x_k$ are integral, e.g. $2x_0 = q_1$ and $2x_1 = 3q_3 - q_1 q_2$.

2.2. The generators q_i of Δ can be interpreted as specializations at $t = -1$ of Hall-Littlewood symmetric functions $q_k(\Lambda; t)$ of the eigenvalues of a positive-definite self-adjoint matrix Λ, defined by

$$q_{\Lambda,t}(T^{\frac{1}{2}}) = \sum q_k(\Lambda; t) T^{\frac{1}{2}k} = \det \frac{1 - \Lambda^{-1} T^{\frac{1}{2}} t}{1 - \Lambda^{-1} T^{\frac{1}{2}}} ;$$

the primitive element x_k thus becomes the power sum $tr\ \Lambda^{-2k-1}$. The canonical symmetric bilinear $\mathbb{Z}[t]$-valued form on the algebra of Hall-Littlewood functions defines a positive-definite inner product on $\Delta_{\mathbb{Q}}$ regarded as a quotient of that algebra [18 III §8.12]. The product and coproduct of Δ are dual with respect to this bilinear form; in fact $\Delta[\frac{1}{2}]$ is a positive self-adjoint Hopf algebra in the sense of Zelevinsky [18 I §5 ex 25]. The classical Q-functions are the orthogonalization of the basis of square-free monomials with respect to this inner product.

2.3. The inner product on $\Delta_{\mathbb{Q}}$ defines a skew bilinear form on the complexification of its space of primitives, which allows us to interpret $\Delta_{\mathbb{Q}}$ as a standard representation of a Heisenberg algebra, or alternately [21, appendix] of the group of antiperiodic loops on the circle. The latter group has two connected components, and in some ways [9, appendix 8] it is natural to think of Δ as $\mathbb{Z}\times\mathbb{Z}/2\mathbb{Z}$-graded algebra, with q_k in homological degree $(k-1,1)$; but in other contexts the $\mathbb{Z}/2\mathbb{Z}$-grading appears as an action of the Galois group of $\mathbb{Q}(\sqrt{2})$ over $\mathbb{Q}$.

If, using the notation of [27 §1.7], we write

$$2^{\frac{1}{2}}\alpha_{-k-\frac{1}{2}} = -x_k$$

when $k > 0$ then the operators

$$L_0 = \sum_{k>0} \alpha_{-k}\alpha_k + \frac{1}{16} \ , \ L_n = \frac{1}{2}\sum \alpha_k\alpha_{n-k} \ , \ n \neq 0$$

define a twisted [5 §9.4] charge one representation of the Virasoro algebra on $\Delta_{\mathbb{Q}}$. There appear to be deep general connections between self-dual Hopf algebras and vertex operator algebras [1,9].

2.4. It is striking that these Q-functions have been important in algebraic topology for more than a generation: in Cartan's 1960 Seminar [2 §17] the integral homology of $Sp(\mathbb{H})/\mathbb{U} = \Omega^5\mathbb{O}$ appears as the quotient Δ^+ of a polynomial algebra on generators q_k^+ of degree $2k$, modulo the ideal generated by the coefficients of the relation

$$q^+(T)q^+(-T) = 1 \ ,$$

where

$$q^+(T) = \sum_{k\geq 0} q_k T^k \ ;$$

the algebras Δ and Δ^+ thus differ only by a doubling of the grading. The cohomology of the corresponding finite-dimensional homogeneous spaces has been studied more recently along similar lines [11].

In light of the results of Karoubi cited in §1.2, an algebra of Q-functions forms a substantial part of the stable integral homology of the space of Abelian varieties; the primitives x_k^+ are dual to Mumford's classes λ_{2k+1}. The stable rational cohomology of the Riemann moduli space contains a polynomial algebra on the Mumford classes κ_k of degree $2k$, and the dual homology Hopf algebra contains the polynomial algebra [19] on dual primitives $\hat{\kappa}_k$. The map of §1.2 kills the even classes $\hat{\kappa}_{2k}$, while

$$\lambda_{2k-1} \mapsto (-1)^k \frac{B_k}{2k}\kappa_{2k-1} \ ,$$

where B_k is the kth Bernoulli number [23 §6.2]; the even Mumford classes are detected by the constructions of [22]. Stabilization implies that

$$\hat{\kappa}_k \in H_{2k}(\mathcal{M}_g, \mathbb{Q})$$

for sufficiently large g, and (since these classes vanish on decomposables) it follows from the characteristic number formula of §1.1 that

$$T_{\mathbb{Q}*}(\hat{\kappa}_k) = vt_{k+1} \ ;$$

2.5. The importance of the theory of symmetric functions in the work of Kontsevich and Witten was discovered by Di Francesco, Itzykson, and Zuber [4 §3.2], but it was Józefiak [12] who saw the connection with Q-functions. In our formalism, their map

$$t_k \mapsto -(2k-1)!!\ tr\ \Lambda^{-2k-1} = -(2k-1)!!x_k\ ,$$

[where the 'odd' factorial $(2k+1)!!$ is the product of the odd integers less than or equal to $2k+1$, with $(-1)!! = 1$ by convention] defines a homomorphism from the extended Landweber-Novikov algebra to $\Delta[\frac{1}{2}]$. The image of the partition function τ_W under this map satisfies a large family of differential equations [17] which can be summarized impressionistically by the assertion that

$$\tilde{\tau}_W(x_i) = \tau_W(x_i + \delta_{i,1})$$

is an sl_2-invariant highest weight vector for the natural Virasoro action on (a completion of) $\Delta_{\mathbb{Q}}$. The formal series τ_W is divergent at $x_i = \delta_{i,1}$, but this claim can be reformulated precisely as

$$\tilde{L}_n \tau_W = 0\ ,\ n \geq -1\ ,$$

where $\tilde{L}_n$ is the linear shift of L_n defined by the map which sends x_1 to $x_1 - 1$, cf. [14]. This defines a charge one vertex operator algebra embedded in a completion of $\Delta_{\mathbb{Q}}$, cf. [5].

There is reason [17 §7, 18 III §7 ex 7] to expect that these results generalize to Hall-Littlewood functions at other roots of unity.

§3 The Kontsevich-Witten genus

3.1. Quillen's theorem establishes a bijection between one-dimensional formal group laws over a commutative ring A, and homomorphisms from the complex cobordism ring to A, so we can define a formal group law

$$X +_Q Y = \mathrm{kw}^{-1}(\mathrm{kw}(X) + \mathrm{kw}(Y))$$

over the localization $\Delta[q_1^{-1}]$, and hence a Q-function valued genus of complex manifolds, by specifying its exponential series to be

$$\mathrm{kw}^{-1}(T) = \sum_{k \geq 0} (2k-1)!! x_0^{-1} x_k T^{k+1}\ .$$

The homomorphism

$$\mathrm{kw} : MU \to S \to \Delta[q_1^{-1}]$$

classifying this group law factors through the map classifying the universal additive law of §1.5 and therefore defines a group law of additive type.

3.2. This Kontsevich-Witten genus is defined by a nonstandard orientation on an ordinary cohomology theory represented by the generalized Eilenberg-Mac Lane space associated to $\Delta[q_1^{-1}]$; its vertex algebra structure is the source of Witten's Virasoro operations.

Some properties of this genus are unfamiliar to the point of pathology: not only is its exponential series integral, for example, but modulo an odd prime it is polynomial as well. The point of this section is that the orientation defining this genus can be interpreted as the formal completion of a coordinate, in the sense of [7 §1.8], on a $\mathbb{T}$-equivariant cohomology theory taking values in sheaves of modules over a certain abelian group object in the category of ringed spaces. This global object is semi-classical, and some of the strangeness of the Kontsevich-Witten genus is a property not of the group object itself, but of a rather inconvenient coordinate on it.

It seems simplest to present this construction in two parts: the first step will define a local version over a basic group object $\mathcal{Q}$. The genus itself will then be defined by a family of group objects parametrized by a Grassmannian of positive-definite self-adjoint matrices.

3.3.1. The basic idea comes from the classical theory of functions of one complex variable: if g is an entire function, of exponential type in every right half-plane, then under certain circumstances there is an asymptotic relation

$$\sum_{n\in\mathbb{Z}} g(n)z^n \sim 0 \; ;$$

such equations may look more familiar written in the form

$$\sum_{n\geq 0} g(n)z^n \sim -\sum_{n\geq 1} g(-n)z^{-n} \; .$$

If $G(z)$ is the left-hand sum, $\check{G}(z^{-1})$ will denote the sum on the right. When g is rational [e.g. $g = 1$] such relations are familiar, but a less trivial example is defined by

$$g(w) = \Gamma(1+\alpha w)^{-1}$$

with $0 < \alpha < 2$. The resulting (Mittag-Leffler) function

$$\exp_\alpha(z) = \sum_{n\geq 0} \frac{z^n}{\Gamma(1+\alpha n)}$$

thus has the asymptotic expansion

$$\exp_\alpha(z) \sim -\sum_{n\geq 1} \frac{z^{-n}}{\Gamma(1-\alpha n)}$$

for z outside a sector of angular width $\alpha\pi$ centered on the positive real axis, cf. [8 §11.3.23]. This function is especially interesting when α is a rational number between zero and one, but we will be concerned only with $\exp_{\frac{1}{2}}(z)$ which, up to normalization, is the Laplace transform of Gaussian measure. Using the duplication formula for the gamma function, its asymptotic expansion takes the form

$$\exp_{\frac{1}{2}}(z) \sim -\pi^{-\frac{1}{2}} z^{-1} \sum_{n\geq 0} (2n-1)!!(-2z^2)^{-n}$$

for z outside a sector of width $\frac{1}{2}\pi$ centered on the positive real axis; in particular the expansion is valid along the entire imaginary axis. The odd function

$$\sin_{\frac{1}{2}}(z) = -\frac{i}{2}[\exp_{\frac{1}{2}}(iz) - \exp_{\frac{1}{2}}(-iz)]$$

therefore satisfies

$$\sin_{\frac{1}{2}}(x) \sim \pi^{-\frac{1}{2}} x^{-1} \sum (2n-1)!!(2x^2)^{-n}$$

3.3.2. Now consider the behavior on the real line of the entire function

$$\epsilon(z) = (\frac{\pi}{2z})^{\frac{1}{2}} \sin_{\frac{1}{2}}((\frac{z}{2})^{\frac{1}{2}})$$

defined by the power series

$$\frac{\pi^{\frac{1}{2}}}{2} \sum_{n \geq 0} \frac{(-\frac{1}{2}z)^n}{\Gamma(n + \frac{3}{2})} = \sum_{n \geq 0} \frac{(-z)^n}{(2n+1)!!} .$$

The argument above implies that

$$\epsilon(x) \sim \check{\epsilon}(x^{-1}) = \sum_{n \geq 0} (2n-1)!! x^{-n-1}$$

for x large and positive; it follows that ϵ is monotone decreasing for positive x. On the other hand it is clear from its power series expansion that ϵ is monotone decreasing on the negative real axis, so

$$\epsilon : (\mathbb{R}, 0) \to (\mathbb{R}_+, 1)$$

is a bijection. The open interval

$$\mathcal{Q} = (0, \infty)^+ - \{1\}$$

can now be made an abelian group by interpreting ϵ to be the exponential of a group law defined in a neighborhood of infinity on the projective line: if x and y are large (i. e. nonzero) real numbers, then

$$x +_\infty y = \frac{xy}{x+y}$$

will again be a large real number. It is easy to see that the resulting composition on $P^1(\mathbb{R}) - \{0\}$ is associative, with ∞ as identity element; the inverse of x is just $-x$. We can therefore construct a formal group law $+_{\infty\epsilon}$ over $\mathbb{R}$ by requiring that

$$\epsilon(x) +_{\infty\epsilon} \epsilon(y) \sim \epsilon(x +_\infty y)$$

for x and y large and positive. This group law as the completion at the identity of the analytic composition

$$X +_\epsilon Y = \epsilon(\epsilon^{-1}(X) +_\infty \epsilon^{-1}(Y))$$

with translation-invariant one-form $d\epsilon^{-1}(T)^{-1}$; its exponential is the series $\check{\epsilon}$. The involution

$$[-1]_\epsilon(T) = \epsilon(-\epsilon^{-1}(T))$$

interchanges $(0,1)$ and $(1,\infty)$, identifying $\mathcal{Q}$ with the group completion of the semigroup defined on $(0,1)$ by $+_\epsilon$. This group object is smooth, but not analytic; its exponential map has trivial domain of convergence.

3.3.3. The usual sheaf of smooth real-valued functions defines the structure of a ringed space on $\mathcal{Q}$, but its law of addition is also compatible with the sheaf of real-analytic functions formally completed at the origin. The composition

$$x \mapsto \Psi(x) = \exp(-\epsilon^{-1}(x)^{-1}) : \mathcal{Q} \to \mathbb{R}_+^\times$$

is a homomorphism to the real multiplicative group; pulling the sheaf over $\mathbb{G}_m(\mathbb{R})$ defined by $\mathbb{T}$-equivariant K-theory back along this map yields a $\mathbb{T}$-equivariant cohomology theory taking values in the category of sheaves of modules over $\mathcal{Q}$. This is a kind of ordinary equivariant cohomology theory with coefficients in $\mathcal{Q}$; its Chern-Dold character sends the standard one-dimensional representation of the circle to Ψ, regarded as a section of the structure sheaf. The Hirzebruch genus

$$\frac{x}{\epsilon(x^{-1})} = \pi^{-\frac{1}{2}} \frac{(2x)^{\frac{1}{2}}}{\sin_{\frac{1}{2}}((2x)^{-\frac{1}{2}})}$$

of the associated complex orientation is the reciprocal of a power series in x^{-1} with trivial constant term. It defines a function on $P^1(\mathbb{R})$ analytic aside from a jump discontinuity at 0.

3.4. By replacing the function ϵ with

$$\epsilon_\Lambda(x) = (\frac{\pi}{2x})^{\frac{1}{2}}\ tr\ \sin_{\frac{1}{2}}((\frac{x}{2})^{\frac{1}{2}}\Lambda)$$

we obtain a $\mathbb{T}$-equivariant theory taking values in sheaves over a family of abelian group objects $\mathcal{Q}_\Lambda$ parametrized by equivalence classes of self-adjoint positive-definite matrices Λ. It is remarkable that if Λ is an $np \times np$ matrix and t is a pth root of unity, then it follows from the definition of §2.2 that

$$q_{\Lambda,t}(T^{\frac{1}{2}}) = q_{\Lambda^{-1},t^{-1}}(T^{-\frac{1}{2}})\ ;$$

in particular, if Λ is an endomorphism of an even-dimensional vector space, and $\check{q}(T^{\frac{1}{2}})$ denotes the generating function for the Hall-Littlewood functions at -1 associated to the matrix Λ^{-1}, then

$$\check{q}(T^{-\frac{1}{2}}) = q(T^{\frac{1}{2}})\ .$$

It seems reasonable to expect that this $\mathbb{T}$-equivariant theory is a real version of a theory over the quotient of some completion of $\Delta \otimes \check{\Delta}$ by the coefficients of the relation

$$\check{q}(T^{-\frac{1}{2}})q(-T^{\frac{1}{2}}) = 1\ .$$

References

1. A. Baker, Vertex algebras in algebraic topology, Friends of the Monster Conference, to appear; available at www.maths.gla.ac.uk
2. H. Cartan, J. C. Moore et al, Seminaire Cartan 1959-60
3. F. R. Cohen, On mapping-class groups from a homotopy-theoretic point of view, in Algebraic topology: new trends, Barcelona 1994, ed. C. Broto, C. Casacuberta, G. Mislin; Progress in Math. no. 136, Birkhäuser
4. P. Di Francesco, C. Itzykson, J.-B. Zuber, Polynomial averages in the Kontsevich model, CMP 151 (1993) 193-219
5. I. Frenkel, Y. Zhu, Vertex operator algebras associated to representations of affine and Virasoro algebras, Duke Math. J. 66 (1992) 123-168
6. —, J. Lepowsky, A. Meurman, Vertex operator algebras and the Monster, Academic Press Pure and Applied Mathematics no. 134 (1988)
7. V. Ginzburg, M. Kapranov, E. Vasserot, Elliptic algebras and equivariant elliptic cohomology I, preprint, Northwestern (1995)
8. E. Hille, Analytic function theory, Blaisdell (1965)
9. P. N. Hoffman, J. F. Humphreys, Projective representations of the symmetric group, Oxford Mathematical Monographs (1992)
10. Jing Naihuan, Vertex operators and Hall-Littlewood symmetric functions, Advances in Mathematics 87 (1991) 226-248
11. T. Józefiak, Schur Q-functions and cohomology of isotropic Grassmannians, Math. Proc. Camb. Phil. Soc. 109 (1991) 471-478
12. —, Symmetric functions in the Kontsevich-Witten intersection theory of the moduli space of curves, Lett. in Math. Phys. 33 (1995) 347-351
13. V. G. Kac, Vertex algebras for beginners, AMS University Lecture Series 10 (1996)
14. —, A.S. Schwartz, Geometric interpretation of the partition function of $2D$ gravity, Phys. Lett. 257 (1991) 329-334
15. M. Karoubi, Théorie de Quillen et homologie du groupe orthogonal, Annals of Math. 112 (1980) 207-257
16. T. Katsura, Y. Shimizu, K. Ueno, Complex cobordism ring and conformal field theory over $\mathbb{Z}$, Math. Ann. 291 (1991) 551-571
17. E. Looijenga, Intersection theory on Deligne-Mumford compactifications [after Witten and Kontsevich], Sem. Bourbaki no. 768, Asterisque 216 (1993)
18. I. MacDonald, Symmetric functions and Hall polynomials, 2nd ed., Oxford Mathematical Monographs (1995)
19. E. Miller, The homology of the mapping-class group, J. Differential Geometry 28 (1986) 1-14
20. J. Morava, Topological gravity and algebraic topology, Proc. of the Adams symposium, LMS Lecture Notes 176 (1992)
21. —, Quantum generalized cohomology, in Operads: proceedings of renaissance conferences, ed. J.L. Loday, J.D. Stasheff, A.A. Voronov, Contemporary Math. 202 (1997) 407-419
22. S. Morita, A linear representation of the mapping class group of orientable surfaces and characteristic classes of surface bundles, University of Tokyo preprint UTMS 96-17
23. D. Mumford, Towards an enumerative geometry of the moduli space of curves, in Arithmetic and Geometry, ed. M. Artin, J. Tate; Progress in Mathematics no. 36, Birkhäuser (1983)
24. D. Quillen, Elementary proofs of some results of cobordism theory using Steenrod operations, Adv. in Math 7 (1971) 29-56
25. U. Tillmann, On the homotopy of the stable mapping-class group, Inv. Math. 130 (1997) 257-276
26. E. Witten, Two-dimensional gravity and intersection theory on moduli space, Surveys in Differential Geometry 1 (1991) 243
27. —, On the Kontsevich model, and other models of two-dimensional gravity, preprint, IASSNS-HEP-91/24

Department of Mathematics, Johns Hopkins University, Baltimore, Maryland 21218
E-mail address: jack@math.jhu.edu

Contemporary Mathematics
Volume **220**, 1998

STABLE SPLITTINGS OF CLASSIFYING SPACES OF AMALGAMS OF FINITE GROUPS

DANIEL JUAN-PINEDA

ABSTRACT. We prove that the classifying spaces of certain amalgams of finite groups stably split as wedge of spectra coming from finite subgroups as classifying spaces of finite groups do. We apply this result to the Bianchi group Γ_{-1}.

1. INTRODUCTION

Let G be a finite group, due to extensive work in the last decade, see for example [1] for a summary, the stable splitting of the classifying space BG of G is now well understood: stably, there is a wedge decomposition

$$BG \cong X_1 \vee X_2 \vee \cdots \vee X_r,$$

where the X_i are p_i-local spectra for the different prime divisors p_i of the order of G. In this work we prove that the techniques of finite groups apply to some infinite groups which are amalgams of groups.

We define a *twin group*, let G be a finite group as before, and let P be a Sylow subgroup of G. Let $\Gamma = G *_P G$ be the amalgam of G with itself along the subgroup P. We call such a Γ a *twin group*. The hypotheses on G imply that the spectrum BP stably splits as $BG \vee Z$.

Main Theorem. *Let $\Gamma = G *_P G$ be a twin group. Assume that stably $BP \cong BG \vee Z$ for a p-local spectrum Z. Then there is a p-local stable splitting*

$$B\Gamma \cong BG \vee \Sigma Z.$$

As an application we have the Bianchi group Γ_{-1}: let $\mathcal{O}_{-1}$ denote the ring of integers in $\mathbb{Q}(i)$, the rationals with the square root of -1 adjoined. The Bianchi group Γ_{-1} is defined as $PSL_2(\mathcal{O}_{-1})$. This group is not a twin group, nor an amalgam of finite groups, nonetheless its classifying space may be studied with these techniques:

1991 *Mathematics Subject Classification.* Primary: 55R35; Secondary: 55R40.
Key words and phrases. splittings, classifying spaces, spectra, amalgams.

Theorem 1. *Let Γ be the Bianchi group Γ_{-1}. Then $B\Gamma$ stably splits 3-locally as*

$$B\Gamma \cong B\mathbf{S}_3 \vee \Sigma^2 Y,$$

where $\mathbf{S}_3$ is the symmetric group on three letters, and Y is the stable summand in the decomposition of $B\mathbb{Z}/3 \cong B\mathbf{S}_3 \vee \Sigma Y$. On the other hand the 2-local splitting is

$$B\Gamma \cong 2(B\mathbf{A}_4 \vee \overline{S}_p^4 \mathbb{S}^0 \vee P^\infty),$$

where $\mathbf{A}_4$ is the alternating group on four letters, P^∞, the real infinite projective space and $\overline{S}_p^4\mathbb{S}^0$ is the quotient $S_p^4\mathbb{S}^0/S_p^2\mathbb{S}^0$ (see section 4).

We may say that Γ_{-1} behaves like a finite group as $SL_3(\mathbb{Z})$ does, see [11].

We take a few lines to thank Professor F. Cohen for suggesting the splitting of these amalgams, S. Priddy and E. Berkove for their valuable comments.

2. Preliminaries

In this paper we will fix the following conventions. All (co)-homology groups are with $\mathbb{Z}/p$ coefficients for p a fixed prime, all spectra are implicitly localized at p.

2.1. **Amalgams and their (co)-homology.** We first review some facts about amalgams of groups and their (co)-homology, the main reference for this part is [4]. Let G_1, G_2, C be groups and let $\alpha_i : C \to G_i$ be monomorphisms, $i = 1, 2$. Then, the amalgamated product of G_1 and G_2 along C (written $G_1 *_C G_2$) is the pushout diagram:

$$\begin{array}{ccc} C & \xrightarrow{\alpha_1} & G_1 \\ {\scriptstyle\alpha_2}\downarrow & & \downarrow \\ G_2 & \longrightarrow & \Gamma \end{array}$$

This is the group theoretic analog of "pasting two spaces along a common subspace". In fact the above situation has its topological realization with $K(\pi, 1)$-spaces:

Theorem 2 (Whitehead). *Any amalgamation diagram (2.1) can be realized by a pushout diagram*

$$\begin{array}{ccc} Y & \xrightarrow{i_1} & X_1 \\ {\scriptstyle i_2}\downarrow & & \downarrow \\ X_2 & \longrightarrow & X \end{array}$$

of $K(\pi,1)$-spaces, namely: $Y = K(C,1), X_i = K(G_i,1)$ and $Y = X_1 \cap X_2$, $X = X_1 \cup X_2$, and both i_1, i_2 are inclusions.

As a corollary we have that given $\Gamma = G_1 *_C G_2$, where $C \hookrightarrow G_1$ and $C \hookrightarrow G_2$, there is Mayer-Vietoris sequence associated to this amalgamation:

Corollary 3. *(Mayer-Vietoris Seq.) Let Γ be an amalgam of groups as above, then we have a long exact cohomology sequence*

$$\cdots H^n(C) \xrightarrow{\delta} H^{n+1}(\Gamma) \xrightarrow{\beta} H^{n+1}(G_1) \oplus H^{n+1}(G_2) \xrightarrow{\alpha} H^{n+1}(C) \to \cdots .$$

We may also say a few more about the maps, β is the restriction, whereas $\alpha = res_C^{G_1} - res_C^{G_2}$. We also have the corresponding homology exact sequence.

We also have that the map δ is multiplicative

Lemma 4. *In the long exact sequence (3), the map δ respects cup products in $H^*(\Gamma)$. Let $i : C \to \Gamma$ be the inclusion map. Then for $u \in H^p(\Gamma)$ and $v \in H^q(C)$, we have that*

$$\delta(i^*(u) \cup v) = u \cup \delta(v), \quad \textit{and}$$
$$\delta(i^*(v) \cup u) = \delta(v) \cup u.$$

2.2. **Stable Splittings of Finite Groups.** Let G be a finite group of order $|G| = p_1^{n_1} p_2^{n_2} \dots p_r^{n_r}$, with the p_i are different primes. Let BG be its classifying space. Then there is a stable splitting of p_i local spectra:

$$BG \cong BG_{p_1} \vee BG_{p_2} \vee \cdots \vee BG_{p_r},$$

where the BG_{p_i}'s are the different p_i-localizations [5] of BG. Furthermore, transfer arguments give that BG_{p_i} is a stable summand of BP_i, where P_i is a Sylow p_i-subgroup of G, see [10] for more details.

We specialize to the following situation. As before, let $G := G_1 = G_2$, be a *finite* group of order $|G| = p_1^{n_1} p_2^{n_2} \dots p_r^{n_r}$ and $P \subseteq G$ be a p-Sylow subgroup, for $p = p_1$. Assume further that G has a stable splitting

$$BG \cong BG_{p_1} \vee BG_{p_2} \vee \cdots \vee BG_{p_r},$$

by the above BG_p is a stable wedge summand of BP.

Moreover, the above splittings are equivalent to finding idempotents in the ring of stable homotopy classes of maps $\{BG, BG\}$. The correspondence is as follows: for such an idempotent $e \in \{BG, BG\}$, we get the corresponding wedge summand X by taking the telescope $X = eBG = \{Tel(BG \xrightarrow{e} BG \xrightarrow{e} \cdots)\}$, see [9] for more details.

Definition 5. *Let G, P be finite groups as above. Let $\Gamma = G *_C G$ be*

The above was motivated by the following observation. Let G_1, G_2 be finite groups with BG_1, BG_2 their corresponding classifying spaces. Then $B(G_1 * G_2) \cong BG_1 \vee BG_2$, i.e. the classifying space of a free product is the one-point union of the corresponding classifying spaces. Hence, if both BG_1 and BG_2 have stable splittings, so does $G_1 * G_2$. We analyse a slight variation of this. The general amalgam situation is still unclear at this point as the inclusions of C into G_1 and G_2 are quite arbitrary.

Let now $K \subseteq Aut(H)$ be a subgroup,where H is a finite group. It is known that idempotents in $\mathbb{Z}/p[K]$ can be lifted to idempotents in $\mathbb{Z}_p[K]$ and the latter acts on our spectra as they are assumed completed. These idempotents give splittings of BH:

Proposition 6. *(see [6]) Let $\pi_1, \cdots \pi_s$ be orthogonal idempotents in $\mathbb{Z}/p[K]$ with $\sum \pi_i = 1$. Then we have a stable splitting $BK \cong \vee_{i=1}^s X_i$, where $H^*(X_i) = (H^*(BH))\pi_i$.*

3. Splitting Amalgams

Let $\Gamma = G *_P G$ be a twin group. The identity $id : G \to G$ defines a group homomorphism $\Gamma \stackrel{\varphi}{\to} G$ which is a left inverse for $i_1 : G \to \Gamma$ (sending G to one of its copies in Γ). Passing to classifying spaces, and to the stable category we obtain:

Proposition 7. *Let $\Gamma = G *_P G$ be a twin group. Then BG is stable summand of $B\Gamma$.*

In other words, let $e \in \{B\Gamma, B\Gamma\}$ be the idempotent corresponding to the summand BG in the splitting of $B\Gamma$. From the above proposition e is given by the composition $B\Gamma \stackrel{B\varphi}{\to} BG \stackrel{Bi_1}{\to} B\Gamma$. We just need to identify the wedge summand corresponding to $1 - e$.

Assume that BP has a stable splitting

$$BP \cong X_1 \vee \cdots \vee X_r,$$

where the X_i are p-local spectra, assume further that

$$BG \cong X_1 \vee \cdots \vee X_s,$$

where $s \leq r$ is a p-local splitting for BG. Hence, p-locally we have that

$$BP \cong BG \vee Z,$$

where $Z = X_{s+1} \vee \cdots \vee X_r$. On the other hand, such a decomposition naturally splits $H_*(G)$, and the inclusion $i_2 : P \hookrightarrow G$, by transfer arguments, gives a split surjection $H_*(P) \to H_*(G)$. Moreover, the cofiber sequence

$$BP \stackrel{Bi_2}{\to} BG \to C_{Bi_2},$$

where C_{Bi_2} is the mapping cone of $BP \overset{Bi_2}{\to} BG$, gives that $C_{Bi_2} \cong \Sigma Z$. Furthermore, the Mayer-Vietoris sequence associated to Γ gives that the image of $\delta : H_n(\Gamma) \to H_{n-1}(P)$ is precisely the cohomology of ΣZ.

We summarize this in

Main Theorem. *Let* $\Gamma = G *_P G$ *be a twin group. Assume that stably* $BC \cong BG \vee Z$ *for a p-local spectrum* Z. *Then there is a p-local stable splitting*

$$B\Gamma \cong BG \vee \Sigma Z.$$

Proof. This now follows from the above discussion and Mayer-Vietoris sequence, the image of $1-e$ in $H_*(B\Gamma)$ is precisely $H_*(C_{Bi_2}) = H_*(\Sigma Z)$. □

4. Example

The Bianchi group Γ_{-1}. see [3] for more details and [2] for its *mod*-2-cohomology. Let $\mathcal{O}_{-1}$ denote the ring of integers in $\mathbb{Q}(i)$, the rationals with the square root of -1 adjoined. Then the Bianchi group is defined as $\Gamma_{-1} = PSL_2(\mathcal{O}_{-1})$. This group also admits a description as an amalgam of groups. We will write $\mathbf{S}_3, \mathbf{A}_4, \mathbb{Z}/2, \mathbb{Z}/3, \mathbf{D}_2$ for the symmetric group on three letters, the alternating group on four letters, the cyclic group of order 2, the cyclic group of order 3, and the four Klein group $\mathbb{Z}/2\times\mathbb{Z}/2$, respectively. We now define $\Gamma_1 = \mathbf{S}_3 *_{\mathbb{Z}/3} \mathbf{A}_4$, $\Gamma_2 = \mathbf{S}_3 *_{\mathbb{Z}/2} \mathbf{D}_2$, and $M = \mathbb{Z}/2 * \mathbb{Z}/3$. Then it is known [7] that $\Gamma_{-1} \cong \Gamma_1 *_M \Gamma_2$, i.e. although Γ_{-1} is not an amalgam of finite groups, it is an amalgam of groups which are amalgams of *finite* groups over groups of this kind.

The stable splittings of the finite groups of the above list are known see for example [9] and [8]. They are as follows. At the prime $p = 3$ we have that the ones we need are $\mathbf{S}_3$ and $\mathbb{Z}/3$. It is known that 3-locally we have a stable splitting (obtained after one suspension) $B\mathbb{Z}/3 \cong B\mathbf{S}_3 \vee \Sigma Y$ where $B\mathbf{S}_3$ and ΣY correspond to suitable idempotents π_1, π_2 in the group ring $\mathbb{Z}/3[\mathbb{Z}/2]$.

The case $p = 2$ (see [10] for details). $B\mathbf{A}_4$ is a dominant summand of $B\mathbf{D}_2$. The other summands are as follows: Let $S_p^n\mathbb{S}^0$ be the *nth*-symmetric product spectrum of the 0-sphere $\mathbb{S}^0$. We now let $\overline{S}_p^4\mathbb{S}^0$ be the quotient spectrum $S_p^4\mathbb{S}^0/S_p^2\mathbb{S}^0$ and denote by P^∞ the infinite dimensional real projective space. Finally, we recall from [10]

Lemma 8. *(Mitchell-Priddy* [10]*) The* 2*-local stable splitting for* $\mathbf{D}_2$ *is given as:*

We wil apply the above splittings to Γ_{-1}. We will work out at the prime $p = 3$ in detail. The Mayer-Vietoris sequence (3) associated to Γ_1 looks like

$$\cdots H^{q-1}(\mathbb{Z}/3) \xrightarrow{\delta} H^q(\Gamma_1) \xrightarrow{\beta} H^q(\mathbf{S}_3) \oplus H^q(\mathbf{A}_4) \xrightarrow{\alpha} H^q(\mathbb{Z}/3) \to \cdots.$$

From the Leray-Serre spectral sequence associated to the extension

$$1 \to \mathbf{D}_2 \to \mathbf{A}_4 \to \mathbb{Z}/3 \to 1,$$

with $\mathbb{Z}/3$-coefficients, we have that $res^{\mathbf{A}_4}_{\mathbb{Z}/3}$ is an isomorphism. We also have that $res^{\mathbf{S}_3}_{\mathbb{Z}/3}$ injects into the $\mathbb{Z}/2$-invariants. Hence $\delta = 0$ and α is surjective in the above sequence. Hence $res^{\Gamma_1}_{\mathbf{S}_3} : H^*(\Gamma_1) \to H^*(\mathbf{S}_3)$ is an isomorphism.

A similar argument shows that $res^{\Gamma_2}_{\mathbf{S}_3} : H^*(\Gamma_2) \to H^*(\mathbf{S}_3)$ is also an isomorphism. We now take $\Gamma = \mathbf{S}_3 *_{\mathbb{Z}/3} \mathbf{S}_3$, observe that this is a twin group. The inclusions $i_s : \mathbf{S}_3 \to \Gamma_s$, $s = 1, 2$ induce a group homomorphism $\phi : \Gamma \to \Gamma_{-1}$. By naturality of the Mayer-Vietoris sequence, we have a commutative diagram

$$\begin{array}{ccccccc} ..H^{q-1}(\mathbb{Z}/3) & \to & H^q(\Gamma) & \to & H^q(\mathbf{S}_3) \oplus H^q(\mathbf{S}_3) & \to & H^q(\mathbb{Z}/3).. \\ \uparrow & & \uparrow \phi^* & & \uparrow & & \uparrow \\ ..H^{q-1}(M) & \to & H^q(\Gamma_{-1}) & \to & H^q(\Gamma_1) \oplus H^q(\Gamma_2) & \to & H^q(M)... \end{array}$$

The vertical maps are restrictions except ϕ^*. We appeal to the 5-lemma and summarize:

Corollary 9. *Let Γ be the Bianchi group Γ_{-1}. Then $B\Gamma$ stably splits 3-locally as*

$$B\Gamma \cong B\mathbf{S}_3 \vee \Sigma^2 Y,$$

where $\mathbf{S}_3$ is the symmetric group on three letters, and Y is the stable summand in the decomposition of $B\mathbb{Z}/3 \cong B\mathbf{S}_3 \vee \Sigma Y$.

We turn now to the prime $p = 2$. Let Γ_1 and Γ_2 be as above, the following computations are performed in detail in [2]. This time the Mayer -Vietoris sequence for Γ_1 breaks up into ring isomorphisms

$$H^q(\Gamma_2) \cong H^q(\mathbf{S}_3) \oplus H^q(\mathbf{A}_4).$$

The computations for Γ_2 are similar, they give a ring isomorphism

$$H^*(\Gamma_2) \cong H^*(\mathbf{D}_2).$$

We now take the Mayer-Vietoris sequence for Γ_{-1}. This time the sequence breaks up into short exact sequences

$$0 \to H^q(\Gamma_{-1}) \xrightarrow{\beta} H^q(\Gamma_1) \oplus H^q(\Gamma_2) \xrightarrow{\alpha} H^q(\mathbb{Z}/2) \to 0.$$

And this last is an isomorphism over $\mathcal{A}_2$ (the Steenrod algebra), thus giving

$$H^*(\Gamma_{-1}) \cong H^*(\mathbf{D}_2) \oplus H^*(\mathbf{A}_4).$$

As the isomorphisms are induced by maps we have

Corollary 10. *The 2-local splitting of the Bianchi group Γ_{-1} is*

$$B\Gamma \cong B\mathbf{A}_4 \vee B\mathbf{D}_2.$$

References

[1] Benson D. Stably splitting BG, *Bull. of the Amer. Math. Soc.* vol. 33, number 2, april (1996) 189-198.

[2] Berkove E. *Ph. D. Thesis* University of Wisconsin-Madison, 1996.

[3] Berkove E. and Juan-Pineda D. On the K-theory of Bianchi groups, *Bol. Soc. Mat. Mexicana* **3** vol. 2 (1996) 15-29.

[4] K. Brown, *Cohomology of Groups*, Springer-Verlag Graduate Texts in Mathematics 87, 1982.

[5] Bousfield, A. K. and Kan D. M.. Homotopy Limits Completions and Localizations, *Springer-Verlag* LNM 304 1972.

[6] Cohen F. Splitting certain suspensions via self-maps, *Ill. J. of Math.* **20** (1976), 336-347.

[7] B. Fine, *Algebraic Teory of Bianchi Groups*, *M. Dekker*, New York 1989.

[8] Harris J. and Kuhn N. Stable decompositions of classifying spaces of finite abelian p-groups. *Math. Proc. Camb. Phil. Soc.* **103** (1988), 427-449.

[9] Martino J. and Priddy S. The complete stable splitting for the classifying space of a finite group, *Topology* Vol. 31 No. 1 (1992) 143-156.

[10] Mitchell S. A. and Priddy S. B. Symmetric product of spectra and splittings of classifying spaces. *Amer. Jour. of Math.* **106** (1984), 219-232.

[11] Minami N. On the classifying spaces of $SL_3(\mathbb{Z})$, $ST_3(\mathbb{Z})$ and finite Chevalley groups, *Contemporary Math.* **158** (1994) 175-185.

[12] Spanier E. *Algebraic Topology* Springer, New York (1966).

[13] Whitehead G. *Elements of Homotopy Theory*, Springer, New York (1978).

Universidad Nacional Autónoma de México, Instituto de Matemáticas-Unidad Morelia, Nicolás Romero 150, col. centro, Morelia, Michoacán, MEXICO 58000

E-mail address: djuan@zeus.ccu.umich.mx

Contemporary Mathematics
Volume **220**, 1998

What we still don't know about loop spaces of spheres

Douglas C. Ravenel
University of Rochester

ABSTRACT. We describe a program for computing the Morava K-theory of certain iterated loop spaces of spheres, based on a hypothetical duality with the computation of that of the Eilenberg-Mac Lane spaces by Wilson and the author [**RW80**]. Under this duality the duals of the bar spectral sequence and facts about $K(n)_*(B\mathbf{Z}/p)$ used in [**RW80**] are the Eilenberg-Moore spectral sequence and facts about $K(n)^*(\Omega^2 S^{2m+1})$ respectively. The program depends on the existence of a new geometric structure in those loop spaces dual to the cup product. We gèt a precise answer under these hypotheses.

Iterated loop spaces of spheres have played a central role in homotopy theory for many years. They have been thoroughly studied but there are still some open questions concerning them.

We will denote $\Omega^{m+1}S^{s+1}$ by $L_{m,s}$. Its homology has long been known for many years and is given in [**CLM76**]. Its BP homology is known only for $m \leq 1$; see [**Rav93**]. Its Morava K-theory was computed for $m = 1$ by Yamaguchi [**Yam88**] and for $m = 2$ by Tamaki (unpublished). In §5 we will describe a speculative program to compute the Morava K-theory more generally in a way that is analogous to the Ravenel-Wilson computation [**RW80**] of the Morava K-theory of Eilenberg-Mac Lane spaces, which is reviewed in §4. It depends on a hypothetical new geometric structure on the stable Snaith summands of iterated loop spaces.

1. Notation

- p is an odd prime. (See §5.4 for the case $p = 2$.)
- h denotes either ordinary mod p homology H or $K(n)$.
- $h^{*,*}(X) = \mathrm{Tor}^{h^*(X)}(h^*, h^*)$, the E_2-term of the Eilenberg-Moore spectral sequence converging (in favorable circumstances) to $h^*(\Omega X)$.
- For a loop space X, $h_{*,*}(X) = \mathrm{Tor}^{h_*(X)}(h_*, h_*)$, the E_2-term of the bar spectral sequence converging to $h_*(BX)$.
- $K_m = K(\mathbf{Z}/(p), m)$.
- $L_{m,s} = \Omega^{m+1}S^{s+1}$ for $0 \leq m < s$.
- $E(x)$ is the exterior algebra on x.
- $P(x)$ is the polynomial algebra on x.
- $T(x)$ is the truncated polynomial algebra of height p on x.

1991 *Mathematics Subject Classification.* Primary 55P35; Secondary 55R20, 55R25, 55T20

2. Two old splitting theorems

2.1. The James splitting. We begin by recalling the splitting theorem of James [**Jam55**]

$$\Sigma L_{0,s} = \Sigma \Omega S^{s+1} \simeq \bigvee_{i>0} S^{si+1}. \tag{2.1}$$

This means projection onto the ith summand gives a map

$$\Sigma\Omega S^{s+1} \longrightarrow S^{si+1},$$

which is adjoint to

$$\Omega S^{s+1} \xrightarrow{\quad H_i \quad} \Omega S^{si+1}, \tag{2.2}$$

the ith James-Hopf map. James also showed that ΩS^{s+1} is homotopy equivalent to a CW complex with a single cell in each dimension divisible by s. He identified the ks-skeleton as a certain topological quotient of the k-fold Cartesian product of S^s. It is called the reduced product or the kth James construction and is denoted by $J_k S^s$, i. e.,

$$J_k S^s = S^s \cup e^{2s} \cup \ldots \cup e^{ks} \simeq (\Omega S^{s+1})^{ks}.$$

Earlier Hopf ([**Hop30**] and [**Hop35**]) studied the map of (2.2) for $i = 2$. When s is odd, its fiber is S^s, i.e., we have a fibration

$$S^{2m-1} \longrightarrow \Omega S^{2m} \longrightarrow \Omega S^{4m-1}. \tag{2.3}$$

Serre [**Ser53**] showed that this splits after localization at any odd prime p, i.e., that

$$\Omega S^{2m}_{(p)} \simeq S^{2m-1}_{(p)} \times \Omega S^{4m-1}_{(p)},$$

so

$$\pi_{2m+k}(S^{2m}_{(p)}) \cong \pi_{2m-1+k}(S^{2m-1}_{(p)}) \oplus \pi_{2m+k}(S^{4m-1}_{(p)}).$$

In other words, from the point of view of computing homotopy groups, *even dimensional spheres are uninteresting at odd primes*; one need only study odd dimensional spheres.

James also identifed the p-local fiber of (2.2) when s is even and i is a power of p. We get a fiber sequence

$$J_{p^j-1} S^{2s}_{(p)} \longrightarrow \Omega S^{2s+1}_{(p)} \xrightarrow{\quad H_{p^j} \quad} \Omega S^{2sp^j+1}_{(p)}$$

This is of particular interest for $j = 1$. We regard $J_{p-1} S^{2s}_{(p)}$ as a p-local substitute for S^{2s}, and denote it by $\widehat{S}^{2s}$. (Note that for $p = 2$, $\widehat{S}^{2s}$ is simply $S^{2s}_{(2)}$.) Thus we have a p-local fiber sequence

$$\widehat{S}^{2s} \longrightarrow \Omega S^{2s+1} \xrightarrow{\quad H_p \quad} \Omega S^{2ps+1}. \tag{2.4}$$

Toda [**Tod62**] showed that there is also a p-local fiber sequence

$$S^{2s-1} \longrightarrow \Omega\widehat{S}^{2s} \longrightarrow \Omega S^{2ps-1}; \tag{2.5}$$

for $p = 2$ this coincides with (2.3). The long exact sequences of homotopy groups associated with these two fiber sequences are called the *EHP sequence*. They give us a wonderful inductive procedure for computing the homotopy groups of spheres. Some additional references for this are [**Whi53**], [**Mah67**], [**BCK$^+$66**], [**Mah82**], [**Gra84**], [**Gra85**], [**MR87**], [**Rav86**, Chapter 1] and [**MR93**].

2.2. The Snaith splitting. In [**Sna74**] Snaith generalized (2.1) and gave a *stable* splitting of $L_{m,s}$ of the form

$$\Sigma^\infty \Omega^{m+1} S^{s+1} \simeq \bigvee_{i>0} \Sigma^{i(s-m)} D^i_{m,s}, \tag{2.6}$$

where each stable summand $D^i_{m,s}$ is a certain finite spectrum with bottom cell in dimension 0, defined in terms of configuration spaces. When an element $x \in H_*(\Omega^{m+1}S^{s+1})$ comes from the i the Snaith summand, we will say that its *Snaith degree*, denoted by $||x||$, is i.

By the James, splitting, $D^i_{0,s} = S^0$ for all i and s. One also knows that $D^1_{m,s} = S^0$ for all m and s.

One also has pairings

$$D^i_{m,s} \wedge D^j_{m,s} \longrightarrow D^{i+j}_{m,s} \tag{2.7}$$

having degree one on the bottom cell.

These summands also have the following properties after localization at a prime p.

- The stable summands of $\Omega^2 S^{2s+1}$ are

$$D^{pi+e}_{1,2s} = \begin{cases} \text{pt.} & \text{if } e \not\equiv 0, 1 \bmod (p) \\ \Sigma^{(p-2)i} B_i & \text{if } e = 0 \text{ or } 1, \end{cases} \tag{2.8}$$

  where B_i is the ith Brown-Gitler spectrum. For $p = 2$ this fact is due to Mahowald [**Mah77**] and Brown-Peterson [**BP78**]. For odd primes part of it was proved by Ralph Cohen [**Coh81**] and the rest by David Hunter and Nick Kuhn [**HK**].
- For every odd m, $D^i_{m,2s}$ is contractible unless i is congruent to 0 or 1 mod p. The bottom p-local cell of $D^{pi}_{m,2s}$ has dimension $(p-2)i$. (For even m it has dimension 0.)
- The top cell of $D^i_{m,2s}$ is in dimension $m(i - \alpha_p(i))$ where $\alpha_p(i)$ denotes the sum of the digits in the p-adic expansion of i.
- For fixed i and m, the p-local homotopy type of $D^i_{m,2s}$ depends only on the congruence of s modulo $p^{f(m)}$ for a certain arithmetic function f. In particular $f(1) = 0$ and $f(2) = 1$; see [**CCKN83**] for more information. Using this fact we can define $D^i_{m,2s}$ for all integers s. In particular we can define a ring spectrum

3. Some new spectra

3.1. A colimit of Snaith summands. The pairings (2.7) with $j = 1$ give us maps $D^i_{m,2s} \to D^{i+1}_{m,2s}$. For even m we define

$$D^\infty_{m,2s} = \underset{\overrightarrow{i}}{\text{hocolim}}\, D^i_{m,2s}. \tag{3.1}$$

For odd m this would be contractible at odd primes, so we need to modify the definition to get something interesting. Consider the composite

$$S^{p-2} \wedge D^{pi}_{m,2s} \longrightarrow D^p_{m,2s} \wedge D^{pi}_{m,2s} \longrightarrow D^{p(i+1)}_{m,2s}.$$

This has degree 1 on the bottom cell, which lies in dimension $(p-2)(i+1)$, so we define

$$D^\infty_{m,2s} = \underset{\overrightarrow{i}}{\text{hocolim}}\, \Sigma^{i(2-p)} D^{pi}_{m,2s}. \tag{3.2}$$

QUESTION 1. *What is the spectrum $D^\infty_{m,2s}$ defined by (3.1) and (3.2)?*

The James splitting (2.1) gives

$$D^\infty_{0,2s} = S^0$$

and (2.8) gives

$$D^\infty_{1,2s} = \underset{\overrightarrow{i}}{\text{hocolim}}\, B_i = H/p, \tag{3.3}$$

the mod p Eilenberg-Mac Lane spectrum. This latter fact plays a central role in the proofs of the nilpotence theorem of Devinatz-Hopkin-Smith [**DHS88**] (see also [**Rav92**, Chapter 9]) and of Nishida's theorem [**Nis73**].

The spectrum $D^\infty_{1,2s}$ can be identified with the Thom spectrum of certain p-local spherical fibration over $\Omega^2 S^3$. For $p = 2$ it is the vector bundle given by the double loop map $\Omega^2 S^3 \to BO$ extending the nontrivial map from S^1. The identification of the corresponding Thom spectrum with $H/2$ (which does *not* require knowing that $D^{2i}_{1,2s}$ is a Brown-Gitler specturm) is originally due to Mahowald [**Mah79**].

To see (3.3) computationally, note that

$$H_*(\Omega^2 S^{2s+1}) = \begin{cases} P(e_0, e_1, \dots) & \text{for } p = 2 \\ E(e_0, e_1, \dots) \otimes P(f_1, f_2 \dots) & \text{for } p \text{ odd} \end{cases}$$

where $|e_i| = 2sp^i - 1$, $|f_i| = 2sp^i - 2$ and both generators have Snaith degree p^i. The action of the Steenrod algebra is given by the following formulas, with all other operations on the generators being zero.

$$\begin{aligned} \mathrm{Sq}^1(e_{i+1}) &= e_i^2 && \text{for } p = 2 \\ \beta(e_i) &= f_i && \text{for } p \text{ odd} \\ \mathcal{P}^1(f_{i+1}) &= f_i^p. \end{aligned}$$

In the mapping the Snaith summands to $D^\infty_{1,2s}$, we have

$$\begin{aligned} e_0 &\mapsto 1, \\ f_1 &\mapsto 1, \\ f_{i+1} &\mapsto \xi_i, \\ \text{and} \quad e_{i+1} &\mapsto \tau_i \end{aligned}$$

for odd primes (where ξ_i and τ_i denote the standard generators of the dual Steenrod algebra) and

$$\begin{array}{rrcl} & e_0 & \mapsto & 1 \\ \text{and} & e_i & \mapsto & \xi_i \end{array}$$

for $p = 2$.

We can use (3.3) to get information about $D^{\infty}_{2k+1,2s}$ in the following way. We have the $2k$-fold suspension map

$$\Omega^2 S^{2s-2k+1} \longrightarrow \Omega^{2k+2} S^{2s+1},$$

which induces maps of Snaith summands

$$D^i_{1,2s-2k} \longrightarrow D^i_{2k+1,2s},$$

making the spectrum $D^{\infty}_{2k+1,2s}$ a module spectrum over $D^{\infty}_{1,2s-2k}$, i.e. a generalized mod p Eilenberg-Mac Lane spectrum.

However, *the homotopy type of* $D^{\infty}_{m,2s}$ *for even* m *is far less obvious.* $D^{\infty}_{2,0}$ is the Thom spectrum for the complex vector bundle induced by the composite

$$\Omega^3_0 S^3 = \Omega^3_0 SU(2) \longrightarrow \Omega^3_0 SU = BU.$$

Thus MU is a module spectrum over $D^{\infty}_{2,0}$, which is *not* an Eilenberg-Mac Lane spectrum. In §6 we will see that certain telescopes are module spectra over it.

For odd primes we have

$$H_*(\Omega^3 S^{2ps+1}) = P(u_k\colon k \geq 0) \otimes E(x_{i,j}\colon i > 0, j \geq 0) \otimes P(y_{i,j}\colon i > 0, j \geq 0),$$

with

$$\begin{array}{rcl} |u_k| & = & 2sp^{k+1} - 2 \\ ||u_k|| & = & p^k \\ |x_{i,j}| & = & 2sp^{i+j+1} - 2p^j - 1 \\ ||x_{i,j}|| & = & p^{i+j} \\ |y_{i,j}| & = & 2sp^{i+j+2} - 2p^{j+1} - 2 \\ ||y_{i,j}|| & = & p^{i+j+1}, \end{array}$$

where $||x||$ denotes the Snaith degree of x. For $p = 2$ we have a similar description with $y_{i,j} = x^2_{i,j}$.

For odd primes the action of the Steenrod algebra A is given by

$$\begin{array}{rcl} \beta(u_i) & = & x_{i,0} \quad \text{for } i > 0 \\ \mathcal{P}^{p^k}(u_i) & = & \begin{cases} u^p_{i-1} & \text{if } k = 0 \\ 0 & \text{otherwise} \end{cases} \\ \beta(x_{i,j}) & = & y_{i,j-1} \quad \text{for } j > 0 \\ \mathcal{P}^{p^k}(x_{i,j}) & = & \begin{cases} x_{i-1,j+1} & \text{if } k = j \\ 0 & \text{otherwise} \end{cases} \end{array}$$

For $p = 2$ we have

$$\mathrm{Sq}^{2^k}(u_i) = \begin{cases} x_{i,0} & \text{if } k = 0 \\ u_{i-1}^2 & \text{if } k = 1 \\ x_{i-k+1,k-2}^2 & \text{otherwise} \end{cases}$$

$$\mathrm{Sq}^{2^k}(x_{i,j}) = \begin{cases} x_{i,j-1}^2 & \text{if } k = 0 \text{ and } j > 0 \\ x_{i-1,j+1} & \text{if } k = j+1 \\ 0 & \text{otherwise.} \end{cases}$$

More generally, $D_{2m,2s}^{\infty}$ is the Thom spectrum for the stable complex vector bundle induced by the composite

$$\Omega_0^{2m+1} S^{2m+1} \longrightarrow Q_0 S^0 = B\Sigma_{\infty}^{+} \xrightarrow{(s-m)\rho} BU$$

where ρ denotes the standard complex representation of the infinite symmetric group Σ_∞. (For the equivalence of $Q_0 S^0 = B\Sigma_\infty^+$, see [**Ada78**].) It follows that MU is a module spectrum over each $D_{2m,2s}^{\infty}$.

3.2. A colimit of dual Snaith summands. It is convenient here to restate the p-local stable Snaith splitting [**Sna74**]

$$L_{m,2s} \simeq \bigvee_{i>0} \Sigma^{2si} C_{m,2s}^{i} \tag{3.4}$$

where $C_{m,2s}^i$ is a certain suspension of the summand $D_{m,2s}^i$ of (2.6). It has top cell in dimension $-m\alpha_p(i)$, where $\alpha_p(i)$ is the sum of the digits in the p-adic expansion of i. As before the p-local homotopy of $C_{m,2s}^i$ for fixed m and i depends only on the congruence of s modulo $p^{f(m)}$. When s is divisible by $p^{f(m)}$, we abbreviate $C_{m,2s}^i$ by C_m^i. L_m will denote the spectrum

$$\bigvee_{i>0} C_m^i.$$

Knowing its cohomology is equivalent to knowing that of the space $L_{m,2s}$ for s highly divisible in light of the Snaith splitting. We can use this isomorphism to define cup products in $h^*(L_m)$.

The Hopf map

$$L_{m,2s} \xrightarrow{H_p} L_{m,2ps},$$

induces maps

$$C_{m,2s}^{pi} \longrightarrow C_{m,2ps}^{i}$$

having degree 1 on the top cell. For $m = 1$ and $p = 2$ this is due to [**CMM78**], and the general case is due to [**Kuh**]. For s highly divisible by p this gives us

$$C_m^{pi} \longrightarrow C_m^i.$$

This leads to an inverse system of spectra, but we want to look instead at the homotopy direct limit of the Spanier-Whitehead duals. Let

$$\tilde{K}_m = \underset{\overrightarrow{DH}}{\mathrm{hocolim}}\, DC_m^{p^j}.$$

It has bottom cell in dimension m.

CONJECTURE 1. *For $p = 2$ the spectrum $\tilde{K}_m$ has the suspension spectrum of the Eilenberg-Mac Lane space K_m as a retract. For an odd prime p $\tilde{K}_m$ has a nontrivial retract of the suspension spectrum of the Eilenberg-Mac Lane space K_m as a retract.*

For $m = 1$ and $p = 2$ this can be deduced from a theorem of Carlsson [**Car83**], which is needed for Miller's proof of the Sullivan conjecture [**Mil84**]. Kuhn [**Kuh**] has recently shown that $H^*(\tilde{K}_m)$ is an unstable A-module having $H^*(K_m)$ as a summand for all m, and has identified the other summands as well.

4. The Ravenel-Wilson computation of $K(n)_*(K_m)$

Conjecture 1 implies that $K(n)_*(\tilde{K}_m)$ and $K(n)_*(K_m)$ have a common nontrivial summand. In this section we will describe the computation of the latter by Wilson and the author [**RW80**]. The approach used there is also good for computing the ordinary homology of the Eilenberg-Mac Lane spaces, and with it one can get complete information *with no prior knowledge of Steenrod operations.* For more details, see Wilson's primer [**Wil82**]. In the next section we will suggest a way in which this approach might carry over to a computation of $K(n)_*(L_{m,2s})$.

We have the bar spectral sequence with

$$E_2^{*,*} = h_{*,*}(K_m) \implies h_*(K_{m+1}).$$

It always collapses for $h_* = H_*$ but not for $h_* = K(n)_*$. Since $K_1 = B\mathbf{Z}/p$ is an S^1-bundle over $\mathbf{C}P^\infty$, $h_*(K_1)$ can be also calculated with the Gysin sequence. This result implies that in the bar spectral sequence for $m = 0$ there are both nontrivial differentials and multiplicative extensions at E_∞.

A key computational ingredient for this bar spectral sequence is the cup product map

$$K_\ell \wedge K_m \longrightarrow K_{\ell+m},$$

(induced by the cup product of the fundamental classes in $H^*(K_\ell \times K_m)$) which induces a spectral sequence pairing

$$h_*(K_\ell) \otimes h_{*,*}(K_m) \longrightarrow h_{*,*}(K_{\ell+m}) \tag{4.1}$$

known as the *circle product.* The image of $x \otimes y$ under this map is denoted by $x \circ y$.

This enables us to determine the behavior of the bar spectral sequence for each $m > 0$ once we know it for $m = 0$. In particular we find that

$$\overline{K(n)}_*(K_m) = 0 \qquad \text{iff} \quad m > n,$$

and it is concentrated in even dimensions for $m \leq n$.

In more detail, K_0 is the discrete group of order p so $h_*(K_0)$ is its group ring over h_*, which is isomorphic to $T(u)$. It follows that for each h,

$$h_{*,*}(K_0) = E(s) \otimes T(a_i : i \geq 0) \tag{4.2}$$

with $|s| = 1$ and $|a_i| = 2p^i$. The bar spectral sequence collapses for $h = H$, and for $h = K(n)$, there is a differential which gives

$$E^0h^*(K_1) = E_\infty = T(a_i : 0 \leq i < n). \tag{4.4}$$

There is a multiplicative extension

$$a_{n-1}^p = v_n a_0. \tag{4.5}$$

Now here is a useful mnenomic device not used in [**RW80**]. *We can ignore the multiplicative extension and still get the right answer in the spectral sequence.* We pretend that $E_\infty = h_*(K_1)$, and find that the resulting Tor group (using our transpotent formula $\tau(V(x)) = a_0 \circ x$) is

$$E(s \circ a_i : 0 \leq i < n) \otimes T(a_j \circ a_{i+j+1} : 0 \leq j, 0 \leq i < n).$$

We can regard this as the E_1-term of the bar spectral sequence. Our formulas give differentials

$$d_r(a_{n-1-i} \circ a_n) = -v_n s \circ a_{n-1-i}$$

(this is a d_1 for $i = n-1$) which leads to the correct value of E_∞.

For $h = H$ the bar spectral sequence collapses and we have

$$H_*(K_2) = H_{*,*}(K_1) = T(b_i : i \geq 0) \otimes T(a_j \circ a_{i+j+1} : i, j \geq 0) \tag{4.6}$$

with $|b_i| = 2p^i$. The element b_i is in the image of the map $\mathbf{C}P^\infty \to K_2$ and it is the dual (with respect to basis described below) of $\mathcal{P}^{\Delta_i} x_2$, where $x_m \in H^m(K_m)$ is the fundamental class. The element $a_j \circ a_{i+j+1}$ is dual to $Q_j Q_{i+j+1} x_2$. The basis in question is that of monomials in generators obtained by the action on x_m of suitable Milnor basis elements of the Steenrod algebra.

5. A speculative computation of $K(n)_*(L_m)$

5.1. A possible analog of the cup product.

CONJECTURE 2. *The cup product pairing*

$$K_\ell \wedge K_m \longrightarrow K_{\ell+m}$$

has an analog

$$\tilde{K}_\ell \wedge \tilde{K}_m \longrightarrow \tilde{K}_{\ell+m}$$

which also has degree one on the bottom cell and similar formal properties.

I had hoped for a similar copairing

$$C_\ell^i \wedge C_m^i \longleftarrow C_{\ell+m}^i$$

but this cannot exist as the following example illustrates. Let $p = 2$ and $i = 2$. Then

$$C_k^i = \Sigma^{-k} RP_{-k}^0$$

and the desired map up to suspension is

$$RP_{-\ell}^0 \wedge RP_{-m}^0 \longleftarrow RP_{-\ell+m}^0,$$

which cannot exist for A-module reasons. Perhaps there is a copairing of the form

$$C_\ell^i \wedge C_m^j \longleftarrow C_{\ell+m}^{i+j}.$$

The 'dual' of the bar spectral sequence is the Eilenberg-Moore spectral sequence, in which we have

$$E_2^{*,*} = h^{*,*}(L_m) \implies h^*(L_{m+1}).$$

Tamaki ([**Tam94**] and [**Tam**]) has shown that it converges for Morava K-theory. Like the bar spectral sequence, it collapses for all m for $h^* = H^*$ but not for $h^* = K(n)^*$.

If Conjecture 2 holds, one could hope for spectral sequence pairings

$$h^*(L_\ell) \otimes h^{*,*}(L_m) \longrightarrow h^{*,*}(L_{\ell+m}),$$

analogous to (4.1), which I will refer to as the *cocircle product.* Again we denote the image of $x \otimes y$ by $x \circ y$ As in the Ravenel-Wilson computation, these should enable us to use the behavior of the spectral sequence for $m = 0$ to determine its behavior for all $m > 0$. This leads to an explicit description of $K(n)^*(L_m)$ in all cases. It agrees with results of Yamaguchi for $m = 1$ and with an unpublished computation of Tamaki for $m = 2$. We will give more details in the next subsections.

5.2. An analog of the Ravenel-Wilson computation. I believe that the computation of $h^*(L_m)$ by induction on m via the Eilenberg-Moore spectral sequence, which converges by Tamaki's theorem, is parallel to that of $h_*(K_m)$ by the bar spectral sequence. To emphasize this analogy, I will use notation for elements in $h^*(L_m)$ similar to that for the corresponding the elements in $h_*(K_m)$, except that there will be an extra index in the former related to the Snaith degree.

Here is a table illustrating this. It includes a description of the Verschiebung V when it is nontrivial. Our indexing convention is the following. *All indices are assumed to be nonnegative, with any additional conditions stated explicitly.* When there are two subscripts, the Snaith degree is determined by their sum, and the Verschiebung lowers the first index.

Eilenberg-Mac Lane spaces	Loop spaces of spheres
$u \in h_0(K_0)$	$u_j \in h^0(L_0)$ $\|\|u_j\|\| = p^j$ $V(u_{j+1}) = u_j$ $H(u_j) = u_{j+1}$
$s \in h_1(K_1)$	$s_j \in h^{-1}(L_1)$ $\|\|s_j\|\| = p^j$ $H(s_j) = s_{j+1}$
$a_i \in h_{2p^i}(K_1)$ $V(a_{i+1}) = a_i$	$a_{i,j} \in h^{-2p^i}(L_1)$ $\|\|a_{i,j}\|\| = p^{i+j+1}$ $V(a_{i+1,j}) = a_{i,j}$ $H(a_{i,j}) = a_{i,j+1}$
$b_i \in h_{2p^i}(K_2)$ $b_0 = s \circ s$ $V(b_{i+1}) = b_i$	$b_{i,j} \in h^{-2p^i}(L_2)$ $b_{0,j} = s_j \circ s_j$ $\|\|b_{i,j}\|\| = p^{i+j}$ $V(b_{i+1,j}) = b_{i,j}$ $H(b_{i,j}) = b_{i,j+1}$

The element u_j is the unit for the cocircle product in Snaith degree p^j.

The following formulas should be compared with (4.2–4.5). We have

$$h^{*,*}(L_0) = E(s_j) \otimes T(a_{i,j}) \tag{5.1}$$

The Eilenberg-Moore spectral sequence collapses for $h = H$, and for $h = K(n)$, there is a differential

$$d_r(a_{n,j}) = v_n s_{n+j+1}, \tag{5.2}$$

which gives

$$E_\infty = E(s_j : j \leq n) \otimes T(a_{i,j} : i < n). \tag{5.3}$$

This is compatible with Yamaguchi's computation [**Yam88**] of $K(n)_*(L_1)$. There is a multiplicative extension (proved in [**Rav93**])

$$a_{n-1,j}^p = v_n a_{0,j+n}. \tag{5.4}$$

As before, (5.4) can be ignored when computing Tor. Using (5.3) as a substitute for $h^*(L_1)$, and assuming that a generator $V(x)$ of Snaith degree p^j has desuspension $s_j \circ V(x)$, and transpotent $a_{0,j} \circ x$, we get

$$\begin{aligned} h^{*,*}(L_1) &= T(b_{i,j} : j \leq n) \otimes \\ &\quad E(s_{i+j+1} \circ a_{i,j} : i < n) \otimes \\ &\quad T(a_{k,i+j+1} \circ a_{i+k+1,j} : i < n). \end{aligned} \tag{5.5}$$

There are differentials and extensions formally implied by (5.2) and (5.4), and we get

$$\begin{aligned} h^*(L_2) &= T(b_{i,j} : j \leq n) \otimes \\ &\quad E(s_{i+j+1} \circ a_{i,j} : i + j < n) \otimes \\ &\quad T(a_{k,i+j+1} \circ a_{i+k+1,j} : i + k + 1 < n). \end{aligned} \tag{5.6}$$

Note that $b_{i,j}$ is a permanent cycle because there is no suitable target for a nontrivial differential on it. The resulting value of $K(n)_*(\Omega^3 S^{odd})$ is equivalent to Tamaki's (unpublished).

In ordinary cohomology we have

$$\begin{aligned} H^*(L_1) &= E(s_j) \otimes T(a_{i,j}) \\ H^*(L_2) &= \Gamma(b_{0,j}) \otimes E(s_{i+j+1} \circ a_{i,j}) \\ &\qquad \otimes \Gamma(a_{0,i+j+1} \circ a_{i+1,j}) \\ &= T(b_{k,j}) \otimes E(s_{i+j+1} \circ a_{i,j}) \\ &\qquad \otimes T(a_{k,i+j+1} \circ a_{k+i+1,j}) \end{aligned}$$

In terms of Dyer-Lashof operations acting on the fundamental class $x_{-m} \in H_{-m}(L_m)$,

$$\begin{aligned}
s_j &\quad \text{is dual to} \quad Q_1^j x_{-1}, \\
a_{i,j} &\quad \text{is dual to} \quad Q_0^i \beta Q_1^j x_{-1}, \\
b_{k,j} &\quad \text{is dual to} \quad Q_0^k \beta Q_1^j x_{-2}, \\
s_{i+j+1} \circ a_{i,j} &\quad \text{is dual to} \quad Q_1^i \beta Q_1^j x_{-2} \quad \text{and} \\
a_{k,i+j+1} \circ a_{k+i+1,j} &\quad \text{is dual to} \quad Q_0^k \beta Q_1^i \beta Q_1^j x_{-2}.
\end{aligned}$$

5.3. A computational conjecture. Now I will state a conjecture about the structure of the 'Hopf Ring' $K(n)^*(L_*)$.

Here is some more notation. For a sequence of n zeroes and ones

$$I = (i_0, i_1, \ldots i_{n-1}),$$

let

$$\begin{aligned}
\lambda(I) &= \max(s : i_s > 0), \\
|I| &= i_0 + i_1 + \ldots + i_{n-1}, \\
\text{and} \quad ||I|| &= i_0 + p i_1 + \ldots + p^{n-1} i_{n-1}.
\end{aligned}$$

For $t \geq \lambda(I)$, let

$$a_{I,t} = a_{0,t}^{\circ i_0} \circ a_{1,t-1}^{\circ i_1} \circ \cdots a_{n-1,t+1-n}^{\circ i_{n-1}} \in K(n)^{-2||I||}(L_{|I|})$$

For a sequence of $n+1$ nonnegative integers

$$J = (j_0, j_1, \ldots j_n),$$

let

$$\begin{aligned}
\lambda(J) &= \max(s : j_s > 0), \\
|J| &= j_0 + j_1 + \ldots + j_n, \\
\text{and} \quad ||J|| &= j_0 + p j_1 + \ldots + p^n j_n.
\end{aligned}$$

For $t \geq \lambda(J)$, let

$$b_{t,J} = b_{t,0}^{\circ j_0} \circ b_{t-1,1}^{\circ j_1} \circ \cdots \circ b_{t-n,n}^{\circ j_n}. \in K(n)^{-2||J||}(L_{2|J|}).$$

CONJECTURE 3.

$$\bigotimes_{m \geq 0} K(n)^*(L_m) = E(s_t \circ b_{t,J} \circ a_{I,t-1} \colon t \leq n) \otimes T(b_{t,J} \circ a_{I,t-1})$$

with all possible values of I, J and t, subject to the multiplicative extension of (5.4). If I and J consist entirely of zeroes, then $t \geq 0$, and $s_t \circ b_{t,J} \circ a_{I,t-1}$ and $b_{t,J} \circ a_{I,t-1}$ are understood to be s_t and u_t respectively. For $n = \infty$ (ordinary homology) this should lead to the usual description of $H^*(L_m)$.

Now consider the Hopf map $H : L_{m,2s} \to L_{m,2ps}$. For $m \equiv 0$ this induces maps $C_m^{pi} \to C_m^i$. In cohomology we have

Since s_j and $b_{i,j}$ are trivial for $j > n$, only the generators in the limit

$$\varinjlim_i h^*(C_m^{p^i})$$

are m-fold cocircle products of $a_{i,j}$s, so the limit contains a summand isomorphic to the vector space spanned by the generators of $K(n)_*(K_m)$, as one would expect in light of Conjecture 1.

Next we note that Conjecture 3 leads to a description of $K(n)^*(L_{m,2s})$ for general s as follows. In the Snaith splitting 3.4, each summand $C^i_{m,2s}$ is up to suspension the Thom spectrum of a certain complex vector bundle over a certain configuration space. The stable fiber homotopy type of this bundle depends only on the congruence class of s module $p^{f(m)}$ for a known function f. Our divisibility condition on s meant that we were considering the cases where the bundle is fiber homotopically trivial. Since $K(n)$ is complex orientable, it follows that $K(n)^*(C^i_{m,2s})$ is Thom isomorphic to $K(n)^*(C^i_m)$, which is described in principle by Conjecture 3.

It may be possible to prove Conjecture 3 without proving the other two. The former describes the answer in the language (the cocircle product) provided by Conjecture 2. Without this language it could be translated into a description of the differentials and multiplicative extensions occuring in the Eilenberg-Moore spectral sequence. Without the cocircle product we still have a spectral sequence of Hopf algebras which must respect the Snaith splitting and the Hopf map. The conjecture says that there all the differentials are formally implied (via the Hopf algebra structure or the Hopf map) by those occuring in Snaith degree p^{n+1}. The differentials in Snaith degree p^{n+1} may be detected by the Milnor operation Q_n. The absence of any other differentials could be forced on us by all the structure at hand if we do the required bookkeeping carefully enough.

5.4. Computations for $p = 2$. In this section we will describe the computation for $p = 2$. *Again it is understood that all indices range over the nonnegative integers unless otherwise specified.* The notation here is simpler. The table of §5.2 gets replaced by

Eilenberg-Mac Lane spaces	Loop spaces of spheres
$u \in h_0(K_0)$	$u_j \in h^0(L_0)$ $\lVert u_j \rVert = 2^j$ $V(u_{j+1}) = u_j$ $H(u_j) = u_{j+1}$
$a_i \in h_{2^i}(K_1)$ $V(a_{i+1}) = a_i$	$a_{i,j} \in h^{-2^i}(L_1)$ $\lVert a_{i,j} \rVert = 2^{i+j}$ $V(a_{i+1,j}) = a_{i,j}$ $H(a_{i,j}) = a_{i,j+1}$
$b_i = a_i \circ a_i$	$b_{i,j} = a_{i,j} \circ a_{i,j}$

In the odd primary case we had $a_i \circ a_i = 0$ and $a_{i,j} \circ a_{i,j} = 0$ due to sign considerations. For $p = 2$ these products need not vanish, and it is convenient to use the indicated notation for them.

In ordinary homology we have

$$\begin{aligned} H_*(K_0) &= E(u) \\ H_*(K_1) = H_{*,*}(K_0) &= \Gamma(a_0) \\ &= E(a_i \colon i \geq 0) \end{aligned}$$

where $a_i \in H_{2^i}(K_1)$ is dual to $\mathrm{Sq}^{\Delta_i} x_1 \in H^{2^i}(K_1)$, with $x_m \in H^m(K_m)$ being the fundamental class. More generally we have

$$\begin{aligned} H_*(K_m) &= \Gamma(a_0 \circ a_{i_1} \circ a_{i_1+i_2} \circ \dots \circ a_{i_1+\dots+i_{m-1}}) \\ &= E(a_{i_0} \circ a_{i_0+i_1} \circ a_{i_0+i_1+i_2} \circ \dots \circ a_{i_0+\dots+i_{m-1}}) \end{aligned}$$

where

$$a_{i_0} \circ a_{i_0+i_1} \circ a_{i_0+i_1+i_2} \circ \dots \circ a_{i_0+\dots+i_{m-1}} \in H_{2^{i_0}+2^{i_0+i_1}+\dots+2^{i_0+\dots+i_{m-1}}}(K_m)$$

is dual to

$$\mathrm{Sq}^{\Delta_{i_0}+\Delta_{i_0+i_1}+\dots+\Delta_{i_0+\dots i_{m-1}}} x_m.$$

Similarly for loop spaces we have

$$H^*(L_m) = E(a_{i_0, i_1+\dots+i_m} \circ a_{i_0+i_1, i_2+\dots+i_m} \circ \dots \circ a_{i_0+\dots+i_{m-1}, i_m})$$

where $a_{i,j} \in H^{-2^i}(L_1)$ has Snaith degree 2^{i+j}. In terms of Dyer-Lashof operations on the fundamental class $x_{-m} \in H_{-m}(L_m)$, the element

$$a_{i_0, i_1+\dots+i_m} \circ a_{i_0+i_1, i_2+\dots+i_m} \circ \dots \circ a_{i_0+\dots+i_{m-1}, i_m}$$

is dual to

$$Q_0^{i_0} Q_1^{i_1} \dots Q_m^{i_m} x_{-m}.$$

For $h = K(n)$ we have differentials

$$\begin{aligned} d_r(a_{n+1}) &= v_n a_0 \\ \text{and} \qquad d_r(a_{n+1,j}) &= v_n a_{0,i+n+1}, \end{aligned}$$

and multiplicative extensions

$$\begin{aligned} a_n^2 &= v_n a_1 \\ \text{and} \qquad a_{n,j}^2 &= v_n a_{1,n+j}. \end{aligned}$$

Thus we have (using the same mnemonic device as before)

$$\begin{aligned} h_*(K_0) &= E(u) \\ h_{*,*}(K_0) &= E(a_i \colon i \geq 0) \\ E^0 h_*(K_1) &= E(a_{i+1} \colon 0 \leq i < n) \\ h_{*,*}(K_1) &= E(a_{i_0} \circ a_{1+i_0+i_1} \colon i_1 < n) \\ E^0 h_*(K_2) &= E(a_{1+i_0} \circ a_{2+i_0+i_1} \colon i_0 + i_1 < n-1) \\ &\vdots \\ E^0 h_*(K_m) &= E(a_{1+i_0} \circ a_{2+i_0+i_1} \circ \dots \circ a_{m+i_0+\dots+i_{m-1}} \colon \\ &\qquad i_0 + \dots + i_{m-1} < n + 1 - m) \end{aligned}$$

We can show that $a_i \circ a_i = 0$ by induction on i as follows. Since a_1 is primitive, $a_1 \circ a_1$ is primitive and there are no primitives of its dimension in $K(n)_*(K_2)$. Inductively $a_i \circ a_i$ is primitive and hence zero.

For loop spaces we have

$$\begin{aligned}
h^*(L_0) &= E(u_j) \\
h^{*,*}(L_0) &= E(a_{i,j}) \\
E^0h^*(L_1) &= E(a_{0,j}\colon j \leq n) \otimes E(a_{1+i,j}\colon i < n) \\
h^{*,*}(L_1) &= \Gamma(a_{0,j} \circ a_{0,j}\colon j \leq n) \otimes \Gamma(a_{0,1+i_1+j} \circ a_{1+i_1,j}\colon i_1 < n) \\
&= E(b_{i_0,j}\colon j \leq n) \otimes E(a_{i_0,1+i_1+j} \circ a_{1+i_0+i_1,j}\colon i < n) \\
E^0h^*(L_2) &= E(b_{i_0,j}\colon j \leq n) \otimes E(a_{0,1+i_1+j} \circ a_{1+i_1,j}\colon i_1 + j < n) \\
&\quad \otimes E(a_{1+i_0,1+i_1+j} \circ a_{2+i_0+i_1,j}\colon i_0 + i_1 < n-1) \\
&\vdots
\end{aligned}$$

Notice that the last factor of $E^0h^*(L_1)$ and of $E^0h^*(L_1)$ resemble $E^0h_*(K_1)$ and $E^0h_*(K_2)$ respectively, but the other factors in the former have no analogs in the latter.

6. Triple loop spaces and telescopes

In this section we will outline the computation of $h^*(L_2)$ without relying on the conjectures stated above. We want to do this for some additional homology theories h which we now describe. The first is $k(n)$, connective Morava K-theory.

The next is $y(n)$, which for $p = 2$ is the Thom spectrum associated with the composite map

$$\Omega J_{p^n-1}(S^2) \to \Omega^2 S^3 \to BO$$

Here $J_k(S^2)$ denotes the kth James reduced product on S^2 (which is the same as the $2k$-skeleton of ΩS^3), and the last map is the double loop map induced by the nontrivial element of $\pi_1(BO)$. For odd primes $y(n)$ is the Thom spectrum associated with a certain p-local spherical fibration induced from one over $\Omega^2 S^3$ which Thomifies to H/p.

Then we have

$$H_*(y(n)) = \begin{cases} P(\xi_1, \dots, \xi_n) & \text{for } p = 2 \\ E(\tau_0, \dots \tau_{n-1}) \otimes P(\xi_1, \dots, \xi_n) & \text{for } p \text{ odd} \end{cases}$$

as comodule algebras over the dual Steenrod algebra. There is a map $y(n) \to k(n)$ inducing a surjection in ordinary mod p homology. It is an equivalence through dimension $2p^{n+1} - 4$.

The spectrum $Y(n)$ is the telescope obtained from $y(n)$ by iterating a map

$$v_n : \Sigma^{2p^n-2} y(n) \to y(n),$$

which is obtained as follows. We have a fiber sequence of spaces

$$\Omega^3 S^{2p^n+1} \to \Omega J_{p^n-1}(S^2) \to \Omega^2 S^3$$

which Thomifies to a stable map

$$\Omega^3 S^{2p^n+1} \to y(n).$$

The bottom cell of the source gives us an element of $v_n \in \pi_{2p^n-2}(y(n))$ which (since $y(n)$ is a ring spectrum) gives us the desired self map. *This map makes $Y(n)$ a module spectrum over $D^{\infty}_{2,0}$.*

Note that $K(n)$ and $k(n)$ are module spectra over $Y(n)$ and $y(n)$ respectively, while $Y(n)_*$ and $y(n)_*$ are (unnaturally) modules over $K(n)_*$ and $k(n)_*$.

Now consider the Atiyah-Hirzebruch spectral sequence for $h^*(L_1)$, where we have

$$E_2 = H^*(L_1, h^*).$$

For each theory other than H^* the first differential is induced by the Milnor operation Q_n, and we have (modulo v_n-torsion for $h = y(n)$ and $k(n)$)

$$E_{2p^n} = E(s_j : j \leq n) \otimes T(a_{i,j} : i < n).$$

We want to show that there are no other differentials. Note that this spectral sequence has the following structure.

- All differentials must respect the Snaith splitting, i.e., they must preserve Snaith degree.
- It is a spectral sequence of Hopf algebras. If s_j or $a_{i,j}$ supports a nontrivial differential, its target must be a linear combination (over h^*) of other such generators of the same Snaith degree.
- Differentials must commute with the Hopf map H, which raises the index j by one.

Concerning the Hopf map, we remark that Tamaki's spectral sequence for the homology of a space of the form $\Omega^3\Sigma^3 X$ is functorial on X while the Hopf map is not. However Tamaki has shown **[Tam]** that his spectral sequence coincides from E_2 onward with the Eilenberg-Moore spectral sequence, which is natural with respect to all loop maps including the Hopf map H.

Now the primitives of Snaith degree p^k surviving to E_{2p^n} are

$$\begin{cases} \{s_k, a_{0,k-1}, a_{1,k-2}, \ldots a_{k-1,0}\} & \text{for } k \leq n \\ \{a_{0,k-1}, a_{1,k-2}, \ldots a_{n-1,k-n}\} & \text{for } k > n. \end{cases}$$

Thus within each Snaith degree the dimensions of these elements are within $2p^{n-1}$ of each other, so there is no room for any more differentials. Hence we have (subject to the multiplicative extensions of (5.4))

$$\begin{aligned} Y(n)^*(L_1) &= Y(n)^* \otimes_{K(n)^*} K(n)^*(L_1) \\ &= E(s_j : j \leq n) \otimes T(a_{i,j} : i < n) \\ \text{and} \quad y(n)^*(L_1) &= y(n)^* \otimes_{k(n)^*} k(n)^*(L_1) \\ & \qquad \text{modulo } v_n\text{-torsion.} \end{aligned}$$

Now we can proceed to the computation of $h^*(L_2)$. We will study Tamaki's formulation of the Eilenberg-Moore spectral sequence. In the case $h^* = Y(n)^*$, $h^*(L_1)$ is a free module over the coefficient ring, so the E_2-term is Tor as before. In the case $h^* = y(n)^*$ we can make similar computations modulo v_n-torsion. With this understanding we have

$$E_2 = \Gamma(\sigma s_j : j \leq n) \otimes E(\sigma a_{i,j} : i < n) \otimes \Gamma(\tau a_{i,j} : i < n).$$

If we can prove this for $j=0$, the Hopf map will give it to us for $j>0$. The element $\gamma_{p^{n-i-1}}(\tau a_{i,0})$ has Snaith degree p^{n+1}, and there are no differentials (in either the Atiyah-Hirzebruch spectral sequence or the Eilenberg-Moore spectral sequence) in lower Snaith degrees. In ordinary cohomology one can show that

$$Q_n\gamma_{p^{n-i-1}}(\tau a_{i,0}) = \sigma a_{n-i-1,i+j+1},$$

so a similar differential occurs in the Atiyah-Hirzebruch spectral sequence. The failure of such a differential to occur in the Eilenberg-Moore spectral sequence would lead to the wrong description of $h^*(C_2^{p^{n+1}})$, so the differentials of (6.1) must occur.

This means that $E_{2p^{n-1}}$ is a subquotient of

$$(6.2)\quad T(\gamma_{p^k}(\sigma s_j) : j\le n)\otimes E(\sigma a_{i,j} : i+j<n)\otimes T(\gamma_{p^k}(\tau a_{i,j}) : k+i<n-1).$$

In Snaith degree p^ℓ for $\ell\ge n+1$, the generators listed are

$$(6.3)\quad \begin{aligned} &\{\gamma_{p^{\ell-i-j-1}}(\tau a_{i,j})\in h^{-2p^{\ell-j}-2p^{\ell-i-j}-2p^{\ell-i-j-1}} : i<n-1, \ell-j<n\}\\ &\bigcup\{\gamma_{p^{\ell-j}}(\sigma s_j)\in h^{-2p^{\ell-j}} : j\le n\};\end{aligned}$$

The dimensions shown are the ones the elements would have if they survived. Differentials in this spectral sequence raise dimensions. Elements in the first family are in dimensions so high that they cannot support any nontrivial differentials.

Elements of the second family of (6.3) (where there is no restriction on $\ell-j$) occur in lower dimensions. Each is annihilated by some iterate of the Hopf map, so it cannot support a differential hitting an element of the first family. The only remaining possibility is a differential of the form

$$d_r(\gamma_{p^{\ell-j_1}}(\sigma s_{j_1})) = x\gamma_{p^{\ell-j_2}}(\sigma s_{j_2})$$

for some $x\in h^*$ and $j_1<j_2$. Since we have a spectral sequence of Hopf algebras, the target of the first such differential must be primitive. (This is dual to the statement that a pth power can support a differential only if its pth root supports an earlier differential.) However since $j_2\le n$ and $\ell\ge n+1$ the target above is never primitive.

It follows that no other differentials can occur so E_∞ (modulo v_n-torsion) is the Hopf algebra of (6.2).

References

[Ada78] J. F. Adams. *Infinite Loop Spaces*. Number 90 in Annals of Mathematics Studies. Princeton University Press, Princeton, 1978.

[BCK+66] A. K. Bousfield, E. B. Curtis, D. M. Kan, D. G. Quillen, D. L. Rector, and J. W. Schlesinger. The mod-p lower central series and the Adams spectral sequence. *Topology*, 5:331–342, 1966.

[BP78] E. H. Brown and F. P. Peterson. On the stable decomposition of Ω^2S^{r+2}. *Transactions of the American Mathematical Society*, 243:287–298, 1978.

[Car83] G. Carlsson. G. B. Segals's Burnside ring conjecture for $(Z/2)^k$. *Topology*, 22:83–103, 1983.

[CCKN83] F. R. Cohen, R. L. Cohen, N. J. Kuhn, and J. A. Neisendorfer. Bundles over configuration spaces. *Pacific Journal of Mathematics*, 104:47–54, 1983.

[CLM76] F. R. Cohen, T. Lada, and J. P. May. *The Homology of Iterated Loop Spaces*, volume 533 of *Lecture Notes in Mathematics*. Springer-Verlag, New York, 1976.

[CMM78] F. R. Cohen, M. E. Mahowald, and R. J. Milgram. The stable decomposition for the double loop space of a sphere. In *Algebraic and geometric topology (Proc. Sympos. Pure Math., Stanford Univ., Stanford, Calif., 1976), Part 2,*, volume XXIII of *Proc. Sympos. Pure Math.* American Mathematical Society, 1978.

[Coh81] R. L. Cohen. *Odd primary infinite families in stable homotopy theory*, volume 242 of *Memoirs of the American Mathematical Society*. American Mathematical Society, Providence, Rhode Island, 1981.

[DHS88] E. Devinatz, M. J. Hopkins, and J. H. Smith. Nilpotence and stable homotopy theory. *Annals of Mathematics*, 128:207–242, 1988.

[Gra84] B. Gray. Unstable families related to the image of J. *Math. Proc. Cambridge Phil. Soc.*, 96:95–113, 1984.

[Gra85] B. Gray. On Toda's fibrations. *Math. Proc. Cambridge Phil. Soc.*, 97:289–298, 1985.

[Hop30] H. Hopf. Über die Abbildungen der dreidimensionalen Sphäre auf die Kugelfläche. *Mathematische Annalen*, 104:639–665, 1930.

[Hop35] H. Hopf. Über die Abbildungen von Sphären auf Sphären niedrigerer Dimension. *Fundamenta Mathematica*, 25:427–440, 1935.

[HK] D. J. Hunter and N. J. Kuhn. Characterizations of spectra with $\mathcal{U}$-injective cohomology which satisfy the Brown-Gitler property. Preprint, 1997, `http://hopf.math.purdue.edu/pub/HunterD-Kuhn/hunterkuhn2.dvi`.

[Jam55] I. M. James. Reduced product spaces. *Annals of Mathematics*, 62:170–197, 1955.

[Kuh] N. J. Kuhn. New cohomological relations among loop spaces, symmetric products, and Eilenberg-Mac Lane spaces. Preprint, 1996, `ftp://hopf.math.purdue.edu/pub/Kuhn/loopspaces.dvi`.

[Kuh82] N. J. Kuhn. The geometry of the James-Hopf maps. *Pacific Journal of Mathematics*, 102:397–412, 1982.

[Mah67] M. E. Mahowald. *The Metastable Homotopy of S^n*, volume 72 of *Memoirs of the American Mathematical Society*. American Mathematical Society, Providence, Rhode Island, 1967.

[Mah77] M. E. Mahowald. A new infinite family in ${}_2\pi_*^S$. *Topology*, 16:249–256, 1977.

[Mah79] M. E. Mahowald. Ring spectra which are Thom complexes. *Duke Mathematical Journal*, 46:549–559, 1979.

[Mah82] M. E. Mahowald. The image of J in the EHP sequence. *Annals of Mathematics*, 116:65–112, 1982.

[Mil84] H. R. Miller. The Sullivan conjecture on maps from classifying spaces. *Annals of Mathematics*, 120:39–87, 1984. Correction in *Ibid.* 121(1985), 605–609.

[MR87] M. E. Mahowald and D. C. Ravenel. Toward a global understanding of the homotopy groups of spheres. In Samuel Gitler, editor, *The Lefschetz Centennial Conference: Proceedings on Algebraic Topology*, volume 58 II of *Contemporary Mathematics*, pages 57–74, Providence, Rhode Island, 1987. American Mathematical Society.

[MR93] M. E. Mahowald and D. C. Ravenel. The root invariant in homotopy theory. *Topology*, 32:865–898, 1993.

[Nis73] G. Nishida. The nilpotence of elements of the stable homotopy groups of spheres. *Journal of the Mathematical Society of Japan*, 25:707–732, 1973.

[Rav86] D. C. Ravenel. *Complex Cobordism and Stable Homotopy Groups of Spheres.* Academic Press, New York, 1986.

[Rav92] D. C. Ravenel. *Nilpotence and periodicity in stable homotopy theory*, volume 128 of *Annals of Mathematics Studies.* Princeton University Press, Princeton, 1992.

[Rav93] D. C. Ravenel. The homology and Morava K-theory of $\Omega^2 SU(n)$. *Forum Mathematicum*, 5:1–21, 1993.

[RW80] D. C. Ravenel and W. S. Wilson. The Morava K-theories of Eilenberg-Mac Lane spaces and the Conner-Floyd conjecture. *American Journal of Mathematics*, 102:691–748, 1980.

[Ser53] J.-P. Serre. Groupes d'homotopie et classes de groupes abeliens. *Annals of Mathematics*, 58:258–294, 1953.

[Sna74] V. Snaith. Stable decomposition of $\Omega^n\Sigma^n X$. *Journal of the London Mathematical*

[Tam94] D. Tamaki. A dual Rothenberg-Steenrod spectral sequence. *Topology*, 33:631–662, 1994.

[Tod62] H. Toda. *Composition Methods in Homotopy Groups of Spheres*. Number 49 in Annals of Mathematics Studies. Princeton University Press, Princeton, 1962.

[Whi53] G. W. Whitehead. On the Freudenthal theorems. *Annals of Mathematics*, 57:209–228, 1953.

[Wil82] W. S. Wilson. *Brown-Peterson homology: an introduction and sampler*. Number 48 in C. B. M. S. Regional Conference Series in Mathematics. American Mathematical Society, Providence, Rhode Island, 1982.

[Yam88] A. Yamaguchi. Morava K-theory of double loop spaces of spheres. *Mathematische Zeitschrift*, 199:511–523, 1988.

Contemporary Mathematics
Volume **220**, 1998

A topological calculus for formal power series

Nigel Ray

ABSTRACT. We propose geometric models for performing computations with formal power series over a commutative ring, including reciprocation, substitution, reversion, and Lagrange inversion. The models are based on a family of complex Bott-Samelson varieties which may be realized as manifolds of flags satisfying appropriate restrictions. We discuss the relationship of the geometric computations with multiple complex cobordism theory, focussing on the dual of the Landweber-Novikov algebra and raising delicate issues concerning the construction of explicit cobordisms. We outline extensions of the calculus to Hurwitz series, appealing to the Faà di Bruno algebra of algebraic combinatorics.

1. Introduction

The methods of formal power series have permeated algebraic topology since the work of Hirzebruch in the 1950s, and have often centered around cobordism theory. The applications became more specific following the work of Novikov and Quillen on the relationship between formal group laws and complex cobordism theory in the 1960s, and have motivated many recent developments in stable homotopy theory [**14**]. At their heart lies the interpretation of Landweber-Novikov operations in terms of automorphisms of the universal formal group law.

Our purpose here is to develop geometrical aspects of the dual of the Landweber-Novikov algebra in this context. With the help of certain Bott-Samelson varieties, interpreted as *bounded flag manifolds* in [**4**], we introduce geometrical models which codify the substitution of one formal power series in another, as well as the computation of the substitutional inverse (or *reversion*) whenever it exists. One component of the models plays a corresponding rôle for products and reciprocals. Several of the underlying ideas are rooted in algebraic combinatorics, especially the coalgebraic framework of [**16**], and have been applied in [**4**] to the study of quantum doubling constructions [**13**] in algebraic topology.

Our project has been stimulated by growing interest amongst combinatorialists and theoretical physicists [**12**] in the Landweber-Novikov algebra, and by positive feedback to the author's recent lectures on the subject. The audience at the University of Wales 1996 Summer School in Gregynog deserve special mention in this regard, particularly Francis Clarke and Peter Johnson; the topologists at Northwestern offered similar encouragement, crucially augmented by conversations with

1991 *Mathematics Subject Classification.* 05,55,57.

André Joyal. Much of the material here extends work with Victor Buchstaber [**4**] and William Schmitt [**16**], to whom the author offers his sincerest thanks for the pleasure of their collaboration, the depth of their insight, and the permission to include joint ideas.

We have attempted to make our presentation reasonably self-contained by summarizing relevant results from [**4**] in §§3 and 4. For readers who seek background information in algebra, combinatorics, and geometry, we recommend the authoritative texts of Kassel [**11**], Aigner [**2**], and Griffiths and Harris [**8**] respectively.

2. Formal power series

We begin by establishing notation and conventions for formal power series, basing our treatment on that of Henrici [**9**].

We assume given a commutative ring K with identity, known as the *scalars*, and consider the ring of formal power series $K[[T]]$ in the variable T over K. Any such power series $\sigma(T)$ consists of a sequence of scalars $(\sigma_0, \sigma_1, \dots)$, presented as the infinite sum

$$\sigma_0 + \sigma_1 T + \cdots + \sigma_k T^k + \cdots . \tag{2.1}$$

We emphasise that the coefficients σ_k may well be algebraically related in K, and that issues of convergence do not arise. Addition is taken termwise, and the product $\rho(T) \cdot \sigma(T)$ is given by the Cauchy product of the corresponding sequences. In the context of (2.1) this amounts to multiplying the appropriate sums and collecting together powers of T, and is clearly commutative. The ring of scalars is embedded in $K[[T]]$ as the subring of constant series, in which guise the identity 1 of K doubles as the identity of $K[[T]]$. In fact $K[[T]]$ is a K-algebra, and contains the polynomial algebra $K[T]$ as subalgebra.

We may represent each series $\sigma(T)$ by a matrix $C(\sigma)$, whose (j,k) entry consists of σ_{k-j} whenever $j \leq k$, and 0 otherwise, for all $j, k \geq 0$. Such matrices are called *semicirculant*, and form a commutative subring $\mathrm{SC}(K)$ of the ring $\mathrm{UT}(K)$ of upper triangular matrices. The matrix $C(\rho \cdot \sigma)$ of a product of power series is given by the product of matrices $C(\rho)C(\sigma)$, so the function $\sigma(T) \mapsto C(\sigma)$ defines an isomorphism $K[[T]] \to \mathrm{SC}(K)$ of K-algebras.

Every diagonal element of $C(\sigma)$ is σ_0, so that the matrix is invertible when and only when σ_0 is invertible in K. In consequence, $\sigma(T)$ admits a multiplicative inverse, or *reciprocal*, when and only when its constant term is invertible. The reciprocal $1/\sigma(T)$ may therefore be obtained by the same recursive procedure as is used for inverting matrices.

We may choose K to be a polynomial algebra $\mathbb{Z}[\alpha, \beta]$, generated by two sequences of indeterminates $(\alpha_0, \alpha_1, \dots)$ and $(\beta_0, \beta_1, \dots)$, where α_0 and β_0 are both invertible and usually taken to be 1. We write $\alpha(T) = \sum_{n \geq 0} \alpha_n T^n$ and $\beta(T) = \sum_{n \geq 0} \beta_n T^n$, noting that the coefficients α_n and β_n are algebraically independent by construction. Thus $\mathbb{Z}[\alpha, \beta]$ is the *universal example* of a ring equipped with two invertible formal power series (and therefore their product); for computations with $\alpha(T)$ and $\beta(T)$ over $\mathbb{Z}[\alpha, \beta]$ may be mapped to computations with $\rho(T)$ and $\sigma(T)$ over K under the ring homomorphism defined by $\alpha_n \mapsto \rho_n$ and $\beta_n \mapsto \sigma_n$, irrespective of any relations which hold amongst the ρ_n and σ_n. If we are interested only in reciprocation, it suffices to consider $\beta(T)$ over $\mathbb{Z}[\beta]$.

We now concentrate on those power series which lie in the principal ideal (T), and therefore have σ_0 zero; we denote such series generically by $s(T)$. For each

$p \geq 1$ we observe that the pth power again lies in the ideal, and write it as

$$s(T)^{\cdot p} = \sum_q s_{p,q} T^q, \tag{2.2}$$

observing that the $s_{p,q}$ are polynomial expressions in $s_1, \ldots, s_{q-p}$ (closely related to the partial Bell polynomials [**6**]); in particular, $s_{p,q}$ is zero for all $q < p$, and $s_{p,p} = s_1^p$. We may combine two such series by the operation of *substitution*, which we denote by $r(s(T))$, or $r \circ s(T)$. This is evaluated using the K-algebra operations in $K[[T]]$, and is not generally commutative. Clearly $r(s(T))$ is again divisible by T, which itself acts as a two sided compositional identity.

We represent each $s(T)$ by a matrix $M(s)$, whose (p,q) entry is $s_{p,q}$ for all $p, q \geq 1$. Of course $M(s)$ is upper triangular, and has increasing powers of s_1 down the diagonal. Straightforward computation reveals that $M(r \circ s)$ is the matrix product $M(r)M(s)$, so that the function $s(T) \mapsto M(s)$ gives a faithful representation of the semigroup (T) in $\mathrm{UT}(K)$; this representation seems to be due to Jabotinski [**10**]. It immediately implies that substitution is associative, which is an awkward calculation from first principles, and that $s(T)$ admits a substitutional inverse when and only when s_1 is invertible in K. We describe such a series as *reversible*, and refer to its inverse as the *reversion* of $s(T)$, written $s^{\langle -1 \rangle}(T)$. The set of reversible formal power series over K is therefore isomorphic to a non-abelian subgroup of $\mathrm{UT}(K)$, in which $M(s^{\langle -1 \rangle})$ is given by $M(s)^{-1}$.

We shall focus our attention on the subgroup $\mathrm{SR}(K)$ of *special* reversible series, for which s_1 is always 1. In any such series we reindex s_{n+1} as s_n for all $n \geq 0$, so that $s(T)$ assumes the form $T + s_1 T^2 + \cdots + s_n T^{n+1} + \cdots$. The *Lagrange inversion formula* originated in analytic function theory, and describes $s^{\langle -1 \rangle}(T)$ in terms of the reciprocal of $s(T)/T$ and its powers. It has spawned an enormous literature over the last 35 years, and been rediscovered many times.

Following our lead above, we may choose K to be a polynomial algebra $\mathbb{Z}[a,b]$ generated by two sequences of indeterminates $(a_1, a_2, \ldots)$ and $(b_1, b_2, \ldots)$, with identity $a_0 = b_0 = 1$; we write $a(T) = \sum_{n \geq 0} a_n T^{n+1}$ and $b(T) = \sum_{n \geq 0} b_n T^{n+1}$. The indeterminates a_n and b_n are algebraically independent by construction, and $\mathbb{Z}[a,b]$ is the universal example of a ring equipped with two reversible formal power series (either of which may be substituted in the other). If we are interested only in reversion, we restrict our attention to $b(T)$ over $\mathbb{Z}[b]$.

It is instructive to compute the upper left hand corners of the matrices $M(b)$ and $M(a \circ b)$, obtaining

$$\begin{pmatrix} 1 & b_1 & b_2 & b_3 & \ldots \\ 0 & 1 & 2b_1 & 2b_2 + b_1^2 & \ldots \\ 0 & 0 & 1 & 3b_1 & \ldots \\ 0 & 0 & 0 & 1 & \ldots \\ \vdots & \vdots & \vdots & \vdots & \vdots \end{pmatrix} \tag{2.3}$$

and

$$\begin{pmatrix} 1 & a_1 + b_1 & a_2 + 2a_1 b_1 + b_2 & a_3 + 3a_2 b_1 + a_1(2b_2 + b_1^2) + b_3 & \ldots \\ 0 & 1 & 2(a_1 + b_1) & 2a_2 + a_1^2 + 6a_1 b_1 + 2b_2 + b_1^2 & \ldots \\ 0 & 0 & 1 & 3(a_1 + b_1) & \ldots \end{pmatrix} \tag{2.4}$$

respectively; similarly, $M(b^{\langle -1\rangle})$ yields

$$\begin{pmatrix} 1 & -b_1 & 2b_1^2 - b_2 & -5b_1^3 + 5b_1b_2 - b_3 & \dots \\ 0 & 1 & -2b_1 & 5b_1^2 - 2b_2 & \dots \\ 0 & 0 & 1 & -3b_1 & \dots \\ 0 & 0 & 0 & 1 & \dots \\ \vdots & \vdots & \vdots & \vdots & \vdots \end{pmatrix}. \tag{2.5}$$

For general K, the matrices $M(s)$, $M(r \circ s)$, and $M(s^{\langle -1\rangle})$ arise from (2.3), (2.4), and (2.5) by substituting r for a and s for b.

3. Bounded flag manifolds

We now turn to geometry, introducing bounded flag manifolds and their associated Boolean algebras $\mathcal{X}(n)$ of nonsingular subvarieties.

We follow combinatorial convention by writing $[n]$ for the set of natural numbers $\{1, 2, \dots, n\}$, which becomes a partially ordered set (or *poset*, for short) when equipped with the standard linear ordering $<$. Every *interval* in $[n]$ has the form $[v, w]$ for some $1 \leq v \leq w \leq n$, and consists of all integers m satisfying $v \leq m \leq w$; our convention therefore dictates that $[1, w]$ and $[w]$ are synonymous. It is sometimes convenient to interpret $[0]$ as the empty set, and $[\infty]$ as the natural numbers.

We work with the *Boolean algebra* $2^{[n]}$ of finite subsets $Q \subseteq [n]$, which is a poset when ordered by inclusion; we write $|Q|$ for the cardinality of any such Q. For all $n \leq m$, the natural inclusions $[n] \subseteq [m]$ induce a direct system of embeddings $2^{[n]} \subseteq 2^{[m]}$, whose existence ensures that a finite set $Q \subseteq [\infty]$ may be assumed to lie in a specific interval $[m]$, for any m sufficiently large. For example, there are unique, mutually inverse, order-preserving bijections $\iota_Q \colon [\infty] \setminus Q \to [\infty]$ and $\iota^Q \colon [\infty] \to [\infty] \setminus Q$, which restrict to bijections between $[m] \setminus Q$ and $[m - |Q|]$ for any m sufficiently large. We decompose each Q into maximal subintervals as $Q = I(1) \cup \cdots \cup I(t)$, where $I(j) = [v(j), w(j)]$ for $1 \leq j \leq t$, and the decomposition is valid in any interval $[m] \supseteq Q$. It is occasionally convenient to let $I(j)^+$ denote the interval $[v(j), w(j) + 1]$.

We work in an ambient complex inner product space Z_{n+1}, which we assume to be invested with a preferred orthonormal basis $z_1, \dots, z_{n+1}$, and write Z_E for the subspace spanned by the vectors $\{z_e : e \in E\}$, given any $E \subseteq [n+1]$. We abbreviate $Z_{[w]}$ to Z_w for each $1 \leq w \leq n+1$, and write $CP(Z_E)$ for the projective space of lines in Z_E. We let $U - T$ denote the orthogonal complement of T in U for any subspaces $T < U$ of Z_{n+1}, and abuse notation by writing 0 for the subspace which consists only of the zero vector. A *bounded flag* U in Z_{n+1} is a sequence of proper subspaces

$$0 = U_0 < U_1 < \cdots < U_i < \cdots < U_n < U_{n+1} = Z_{n+1},$$

in which U_i contains the subspace Z_{i-1}, for every $1 \leq i \leq n+1$. The *standard* flag $Z_0 < \cdots < Z_i < \cdots < Z_{n+1}$ is therefore bounded.

We define the *bounded flag manifold* $B(Z_{n+1})$ to consist of all bounded flags in Z_{n+1}, for each $n \geq 0$; it is a nonsingular complex projective algebraic variety, and an n-dimensional subvariety of the usual flag manifold $F(Z_{n+1})$. We refer to **[4]** for further details, remarking that $B(Z_{n+1})$ was originally introduced by Bott and Samelson **[3]** as a homogeneous space, and subsequently described in **[15]** as an iterated sphere bundle. By way of example we note that $B(Z_1)$ consists solely of the standard flag, whilst $B(Z_2)$ is isomorphic to the projective line $CP(Z_2)$ with standard complex structure.

To each bounded flag U in Z_{n+1} we assign the *support* $S(U)$, given by $\{j \in [n] : U_j \neq Z_j\}$, and consider the subspace

$$e_Q = \{U \in B(Z_{n+1}) : S(U) = Q\} \tag{3.1}$$

for each subset $Q \subseteq [n]$; for example, $e_\varnothing$ consists solely of the standard flag. Then e_Q is an open topological $2|Q|$-cell, whose closure in $B(Z_{n+1})$ is

$$X_Q = \{U \in B(Z_{n+1}) : S(U) \subseteq Q\} = \bigcup_{R \subseteq Q} e_R.$$

In fact X_Q is the image of an algebraic embedding $i_Q\colon\ \times_j B(Z_{I(j)^+}) \to B(Z_{n+1})$, defined by choosing the kth component of $i_Q(U(1), \ldots, U(t))$ as

$$\begin{cases} Z_{v(j)-1} \oplus U(j)_{k-v(j)+1} & \text{if } k \in I(j) \\ Z_k & \text{otherwise,} \end{cases}$$

for $1 \leq k \leq n+1$. We deduce the following result of [**4**].

PROPOSITION 3.1. *For any $Q \subseteq [n]$, the subvariety X_Q is nonsingular and isomorphic to the cartesian product $\times_j B(Z_{I(j)^+})$.*

The cells e_Q provide a CW decomposition for $B(Z_{n+1})$ with 2^n cells, and define a basis for the integral homology groups $H_*(B(Z_{n+1}))$. We may interpret

$$\mathcal{X}(n) = \{X_Q : Q \subseteq [n]\}$$

as a Boolean algebra of nonsingular subvarieties of $B(Z_{n+1})$, ordered by inclusion, on which the support function S induces a bijection of Boolean algebras with $2^{[n]}$. For any $n \leq m$, the natural inclusions $[n] \subseteq [m]$ induce a direct system of embeddings $\mathcal{X}(n) \subseteq \mathcal{X}(m)$, and Proposition 3.1 ensures that the isomorphism class of each X_Q is independent of n. These embeddings are compatible with those of $2^{[n]}$ in $2^{[m]}$ described above.

To every bounded flag U in Z_{n+1} we may associate the lines $U_i - Z_{i-1}$ and $U_{i+1} - U_i$ in $Z_{[i,n+1]}$, for all $1 \leq i \leq n$. These define complex line bundles γ_i and ρ_i over $B(Z_{n+1})$, whose classifying maps combine to embed $B(Z_{n+1})$ algebraically into a product of complex projective spaces. Defining γ_0, γ_{n+1} and ρ_0 as 0, $\mathbb{C}$ and γ_1 respectively, we obtain

$$\gamma_i \oplus \rho_i \oplus \rho_{i+1} \oplus \cdots \oplus \rho_n \cong \mathbb{C}^{n-i+2} \tag{3.2}$$

for all $0 \leq i \leq n$.

As described in [**15**], the tangent bundle τ of $B(Z_{n+1})$ satisfies $\tau \oplus \mathbb{R} \cong (\oplus_{i=2}^{n+1} \gamma_i) \oplus \mathbb{R}$, so that (3.2) leads to isomorphisms

$$\nu \cong -\Big(\bigoplus_{i=2}^{n+1} \gamma_i\Big) \cong \bigoplus_{i=2}^{n} (i-1)\rho_i \tag{3.3}$$

for the stable normal bundle ν. We follow [**4**] in referring to the resulting U-structure on $B(Z_{n+1})$ as *basic*. We emphasise that these isomorphisms are of

$B(Z_{n+1})$ with the total space of the sphere bundle of $\gamma_1 \oplus \mathbb{R}$ over $B(Z_{[2,n+1]})$, and the basic U-structure then extends over the associated 3-disc bundle for all values of n; as a result, $B(Z_{n+1})$ represents zero in the complex cobordism ring Ω_*^U, and all Chern numbers of the basic U-structure vanish.

4. Algebraic constructions

In this section we perform various key algebraic constructions with the posets $\mathcal{X}(n)$; the crucial information is captured by an infinite matrix M_B, whose properties we explore in detail.

We begin by considering the free abelian group $\mathbb{Z}X_*$, generated by the isomorphism classes of the varieties X_Q as Q ranges over all finite subsets of $[\infty]$. To avoid elaborate notation we shall not distinguish between a variety and its isomorphism class, except in the case of $B(Z_{n+1})$ itself, where we write B_n for the class of $X_{[v,v+n-1]}$ given any positive integer v. Cartesian product invests $\mathbb{Z}X_*$ with the structure of commutative ring, and Proposition 3.1 implies that the class of X_Q is the monomial $\prod_{j=1}^t B_{[w(j)-v(j)+1]}$, in the notation of §3. In other words $\mathbb{Z}X_*$ is a polynomial algebra $\mathbb{Z}[B]$, generated by the sequence of isomorphism classes $(B_0, B_1, \dots)$, whose identity $B_0 = 1$ is represented by the singleton $X_\varnothing$.

Alternatively, we may consider the additive semigroup $[\infty]^*$, consisting of infinite sequences $\omega = (\omega_1, \omega_2, \dots)$ of nonnegative, eventually zero integers. As ω ranges over $[\infty]^*$, the monomials $B^\omega = \prod_j B_j^{\omega_j}$ range over a set of additive generators for $\mathbb{Z}X_*$. For every ω there are many possible finite sets $Q(\omega) \subseteq [\infty]$ with ω_j maximal intervals of cardinality j, for all $j \geq 0$; the isomorphism class of any $X_{Q(\omega)}$ represents the monomial B^ω in $\mathbb{Z}X_*$. By assigning the *dimension* $|\omega| = 2\sum j\omega_j$ to each monomial B^ω, we invest $\mathbb{Z}X_*$ with the structure of a graded ring, and in this parlance $|\omega|$ is the dimension of $X_{Q(\omega)}$ as a smooth manifold.

In order to record salient features of the Boolean algebras $\mathcal{X}(n)$, we introduce the infinite matrix M_B over the polynomial algebra $\mathbb{Z}X_*$. For all $p, q \geq 0$, the (p,q) entry consists of the formal sum of isomorphism classes $\sum X_Q$, taken over those subsets $Q \subseteq [p]$ of cardinality $p - q$. When expressed in terms of the B_n we obtain a homogeneous polynomial of dimension $2(p-q)$, confirming that M_B is lower triangular, and lies in the group $\mathrm{LT}(\mathbb{Z}X_*)$. For example, when $p = 3$ and $q = 1$, the subsets in question are $\{1,2\}$, $\{1,3\}$, and $\{2,3\}$, yielding isomorphism classes B_2, B_1^2, and B_2 respectively, and the $(3,1)$ entry of M_B is $2B_2 + B_1^2$. Continuing in this vein, we compute the upper left hand corner of M_B to be

$$\begin{pmatrix} 1 & 0 & 0 & 0 & \dots \\ B_1 & 1 & 0 & 0 & \dots \\ B_2 & 2B_1 & 1 & 0 & \dots \\ B_3 & 2B_2 + B_1^2 & 3B_1 & 1 & \dots \\ \vdots & \vdots & \vdots & \vdots & \end{pmatrix}. \tag{4.1}$$

Readers may enjoy comparing this matrix with $M(b)$ in (2.3).

The next step is to consider *pairs* of finite subsets $P \subseteq Q \subseteq [n]$, and to let $X(P,Q)$ denote the subvariety $X_{Q\setminus P}$ of $B(Z_{[n+1]\setminus P})$. With the assistance of the bijection ι_P we may identify $X_{Q\setminus P}$ with $X_{\iota_P(Q\setminus P)}$ in $B(Z_{n+1-|P|})$, and deduce that the isomorphism class of $X(P,Q)$ is independent of n, for n sufficiently large. In this notation $X(\varnothing, Q)$ is X_Q, whilst $X(Q,[n])$ is isomorphic to $B(Z_{n+1-|Q|})$.

In similar spirit we consider the free abelian group generated by *pairs* of isomorphism classes (X_P, X_Q), where P and Q range independently over finite subsets of $[\infty]$. This group is isomorphic to $\mathbb{Z}X_* \otimes \mathbb{Z}X_*$, and is a commutative ring with respect to cartesian product in each coordinate. As such it is a graded polynomial algebra $\mathbb{Z}[A, B]$, where A_m and B_n denote the classes of the pairs $(B(Z_{m+1}), \cdot)$ and $(\cdot, B(Z_{n+1}))$ respectively.

We define an infinite matrix $M_{A\circ B}$ in $\mathrm{LT}(\mathbb{Z}[A, B])$ by combining these constructions. For all $p, q \geq 0$, the (p, q) entry consists of the formal sum of isomorphism classes

$$\sum_{P \subseteq Q} (X(\varnothing, P), X(P, Q)), \tag{4.2}$$

taken over all $P \subseteq Q \subseteq [p]$ for which $|Q| = p - q$. For example, when $p = 3$ and $q = 1$, the appropriate subsets Q are $\{1,2\}$, $\{1,3\}$, and $\{2,3\}$, as before. For $\{1,2\}$, the intermediate subset P ranges over $\{1,2\}$, $\{1\}$, $\{2\}$, and $\varnothing$, which contribute monomials A_2, A_1B_1, A_1B_1 and B_2 respectively. Repeating for the other values of Q, we deduce that the $(3, 1)$ entry of $M_{A\circ B}$ is $2A_2 + A_1^2 + 6A_1B_1 + 2B_2 + B_1^2$; similar computations reveal the upper left hand corner of $M_{A\circ B}$ to be

$$\begin{pmatrix} 1 & 0 & 0 & 0 & \cdots \\ A_1 + B_1 & 1 & 0 & 0 & \cdots \\ A_2 + 2A_1B_1 + B_2 & 2(A_1 + B_1) & 1 & 0 & \cdots \\ \begin{matrix} A_3 + (A_1^2 + 2A_2)B_1 \\ +3A_1B_2 + B_3 \end{matrix} & \begin{matrix} 2A_2 + A_1^2 + 6A_1B_1 \\ +2B_2 + B_1^2 \end{matrix} & 3(A_1 + B_1) & 1 & \cdots \\ \vdots & \vdots & \vdots & \vdots & \end{pmatrix}.$$

Readers may compare *this* matrix with $M(a \circ b)$ in (2.4).

THEOREM 4.1. *The matrix $M_{A\circ B}$ is the product M_AM_B in* $\mathrm{LT}(\mathbb{Z}[A, B])$.

PROOF. For every $p \geq q \geq 0$, the (p, q) entry of M_AM_B is given by

$$\sum_{p \geq k \geq q} \Big(\sum_P (X_P, \cdot) \sum_R (\cdot, X_R)\Big), \tag{4.3}$$

where P ranges over subsets of $[p]$ of cardinality $p - k$ and R ranges over subsets of $[k]$ of cardinality $k - q$. But every variety $X(\varnothing, P)$ appearing in (4.2) may be paired bijectively with X_P in (4.3), and every variety $X(P, Q)$ may be paired with X_R, where $R = \iota_P(Q \setminus P)$; the latter pairing is also bijective, by setting $Q = P \cup \iota^P(R)$. The varieties so paired are isomorphic, making the sums (4.2) and (4.3) equal. □

We refer to a sequence of subvarieties $X_{Q_0} \subset X_{Q_1} \subset \cdots \subset X_{Q_\ell}$ in $\mathcal{X}(n)$ as a *chain* of *length* ℓ from X_{Q_0} to X_{Q_ℓ}; every such chain corresponds uniquely to the chain $Q_0 \subset Q_1 \subset \cdots \subset Q_\ell$ in $2^{[n]}$, written as $\gamma\colon Q_0 \rightsquigarrow Q_\ell$. By analogy with M_B, we construct an infinite matrix N_B which displays information concerning chains in the $\mathcal{X}(n)$. For each $p, q \geq 0$, the (p, q) entry consists of the formal sum of monomials

$$\sum (-1)^\ell \prod^{\ell} X(Q_{i-1}, Q_i), \tag{4.4}$$

$q = 1$ the relevant subsets $Q \subseteq [3]$ are $\{1,2\}$, $\{1,3\}$, and $\{2,3\}$. The first of these yields chains

$$\varnothing \subset \{1\} \subset \{1,2\}, \quad \varnothing \subset \{2\} \subset \{1,2\}, \quad \text{and} \quad \varnothing \subset \{1,2\},$$

in $2^{[3]}$, which contribute B_1^2, B_1^2 and $-B_2$ respectively to (4.4); the second yields chains which contribute B_1^2, B_1^2 and $-B_1^2$, and the third contributes B_1^2, B_1^2 and $-B_2$. The $(3,1)$ entry of N_B is therefore $5B_1^2 - 2B_2$. Continuing, we compute the upper left hand corner of N_B as

$$\begin{pmatrix} 1 & 0 & 0 & 0 & \dots \\ -B_1 & 1 & 0 & 0 & \dots \\ 2B_1^2 - B_2 & -2B_1 & 1 & 0 & \dots \\ -5B_1^3 + 5B_1B_2 - B_3 & 5B_1^2 - 2B_2 & -3B_1 & 1 & \dots \\ \vdots & \vdots & \vdots & \vdots & \end{pmatrix},$$

which compares with $M(b^{\langle -1 \rangle})$ in (2.5).

THEOREM 4.2. *The matrices M_B and N_B are mutually inverse in the group* $\mathrm{LT}(\mathbb{Z}[B])$.

PROOF. To demonstrate that $N_B M_B$ is the identity matrix, we need only check that its (p,q) entry is 0 whenever $p > q > 0$. The terms contributing to this entry consist of a chain from $X_\varnothing$ to X_Q, where $Q \subseteq [p]$ has cardinality $p - k$, together with a subvariety X_R, where $R \subseteq [k]$ has cardinality $k - q$, for some $p \geq k \geq q$. We partition these pairs into subsets Γ_0 and Γ_1, where all the former have $k = 0$ (and therefore $X_R = \cdot$), and all the latter have $k > 0$. Each pair in Γ_0 is therefore determined solely by a chain of the form $X_\varnothing \subset \cdots \subset X_{Q_{\ell-1}} \subset X_{Q_\ell}$ in $\mathcal{X}(p)$, to which we associate the pair in Γ_1 consisting of the chain $X_\varnothing \subset \cdots \subset X_{Q_{\ell-1}}$ in $\mathcal{X}(p)$, and the subvariety X_P in $\mathcal{X}(p - |Q_{\ell-1}|)$, where $P = \iota_{Q_{\ell-1}}(Q_\ell \setminus Q_{\ell-1})$. This association is bijective, since an inverse may be constructed by judicious application of $\iota^{Q_{\ell-1}}$.

Since $X(Q_{\ell-1}, Q_\ell)$ and X_P are isomorphic, we deduce that

$$(-1)^\ell \prod_{i=1}^{\ell} X(Q_{i-1}, Q_i) = (-X_P) \cdot (-1)^{\ell-1} \prod_{i=1}^{\ell-1} X(Q_{i-1}, Q_i)$$

in $\mathbb{Z}X_*$ and therefore that the combined contribution of the elements of Γ_0 and Γ_1 to the (p,q) entry is 0. □

For future reference, we note that the (p,q) entries of $M_{A \circ B}$ and N_B are homogeneous polynomials of dimension $2(p-q)$, in the variables A_m and B_n, and B_n respectively. We pay particular attention to the cases $q = 0$, and write the corresponding polynomials as $M_p(A, B)$ and $N_p(B)$, for all $p \geq 0$.

5. Reconciliation and reformulation

We now reconcile the constructions of §§2 and 4 by means of their matrix representations, and reformulate both in the context of Hopf algebras.

We consider the universal special reversible series $b(T)$ in $\mathrm{SR}(\mathbb{Z}[b])$, and its representing matrix $M(b)$ in $\mathrm{UT}(\mathbb{Z}[b])$. The ring isomorphism $\mathbb{Z}[b] \to \mathbb{Z}[B]$, induced by $b_n \mapsto B_n$ for each $n \geq 0$, defines a special reversible series $B(T) = \sum_{q \geq 0} B_q T^{q+1}$ over $\mathbb{Z}X_*$ (which serves equally well as the universal example). It is represented by the matrix $M(B)$ in $\mathrm{UT}(\mathbb{Z}[B])$, whose transpose turns out to be familiar! To analyze why, it is convenient to interpret transposition as an anti-isomorphism $\mathrm{UT}(\mathbb{Z}[B]) \to \mathrm{LT}(\mathbb{Z}[B])$.

PROPOSITION 5.1. *We have that* $M(B)^t = M_B$ *in* $\mathrm{LT}(\mathbb{Z}[B])$.

PROOF. For all $p, q \geq 1$, the (p, q) entry of $M(B)^t$ consists of the coefficient of T^p in $(T + B_1T^2 + \cdots + B_kT^{k+1} \dots)^q$. All monomials in this expansion have the form $B_{i_1}T^{i_1+1} \dots B_{i_q}T^{i_q+1}$, where one factor arises from each bracket, and we collect together those which satisfy $i_1 + \cdots + i_q = p - q$. We may express every such monomial as X_QT^p, where $Q \subseteq [p]$ is obtained by partitioning $[p]$ into successive subintervals of lengths $i_1 + 1$, $i_2 + 1$, ..., $i_q + 1$, and deleting the last (or only) element of each. Note that Q never contains the integer p, and that every subset of $[p-1]$ arises from one and only one monomial in the expansion; moreover, the cardinality of Q is $p - q$. The coefficient of T^p is therefore $\sum X_Q$, taken over all subsets $Q \subseteq [p-1]$ of cardinality $(p-1) - (q-1)$, as required. □

The reindexing in this proof arises because we chose $p, q \geq 1$ and $p, q \geq 0$ respectively, to index the entries of $M(b)$ and M_B.

COROLLARY 5.2. *In* $\mathrm{LT}(\mathbb{Z}[A, B])$, *we also have that*

$$M(B \circ A)^t = M_{A \circ B} \qquad \text{and} \qquad M(B^{\langle -1 \rangle})^t = N_B.$$

PROOF. We know that $M(B \circ A)^t = M(A)^t M(B)^t$, and it follows from Proposition 5.1 that we may rewrite the right hand side as $M_A M_B$; applying Theorem 4.1, we replace this by $M_{A \circ B}$ in turn, as required. The proof for $M(B^{\langle -1 \rangle})$ is similar, applying Theorem 4.2. □

These results exemplify our central theme, that computations with substitution and reversion of universal formal power series may be expressed in terms of the posets $\mathcal{X}(n)$, through the medium of the matrices the $M_{A \circ B}$ and N_B. The homogeneous polynomials $M_p(A, B)$ and $N_p(B)$ of §4 play a fundamental rôle, as ratified by the expressions

$$B(A(T)) = \sum_{p \geq 0} M_p(A, B) T^{p+1} \quad \text{and} \quad B^{\langle -1 \rangle}(T) = \sum_{p \geq 0} N_p(B) T^{p+1}. \tag{5.1}$$

For more general scalars, we make simple modifications to cope with power series $s(T)$ in $\mathrm{SR}(K)$. Specifically, we record the isomorphism class of each subvariety X_Q as a monomial $\prod_{j=1}^{t} s_{[w(j)-v(j)+1]}$ in K, and write M_s for the resulting matrix in $\mathrm{LT}(K)$, obtained from the $\mathcal{X}(n)$ by the procedure corresponding to that of (4.1). Of course M_s is obtained from the universal example M_B by the replacement of B with s, irrespective of any relations which may hold amongst the s_n. We deduce from Proposition 5.1 and Corollary 5.2 that

$$M(s)^t = M_s, \qquad M(s \circ r)^t = M_{r \circ s}, \qquad \text{and} \qquad M(s^{\langle -1 \rangle})^t = N_s \tag{5.2}$$

in $\mathrm{LT}(K)$. We may harmonize the order of composition (and eliminate transpositions) in these formulae by electing to substitute power series on the *right*, and to represent them by *lower* triangular matrices accordingly.

In the universal case, (5.1) and (5.2) may be translated into a more elegant by means of the respective equations

$$\delta(X_Q) = \sum_{\varnothing\subseteq P\subseteq Q} (X(\varnothing, P), X(P,Q)),$$

$$\chi(X_Q) = \sum_{\gamma\colon \varnothing\rightsquigarrow Q} (-1)^\ell \prod_{i=1}^{\ell} X(Q_{i-1}, Q_i), \qquad \text{and} \tag{5.3}$$

$$\epsilon(X_Q) = \begin{cases} X_Q & \text{if } Q = \varnothing \\ 0 & \text{otherwise} \end{cases}$$

on additive generators, and interpret $\mathbb{Z}[A, B]$ as $\mathbb{Z}[B]\otimes\mathbb{Z}[B]$ by rewriting A as $B\otimes 1$ and B as $1\otimes B$.

PROPOSITION 5.3. *The polynomial algebra $\mathbb{Z}[B]$ is a Hopf algebra with respect to coproduct δ, antipode χ, and counit ϵ.*

PROOF. We first we show that δ and χ are ring homomorphisms. As usual, we assume the decomposition of Q into maximal subintervals $I(j)$ for $1 \leq j \leq t$, and write $P(j)$ for $P \cap I(j)$, given any subset $P \subseteq Q$. Then we have

$$\sum_{\varnothing\subseteq P\subseteq Q} (X(\varnothing, P), X(P,Q)) = \prod_{j=1}^{t} \Big(\sum_{\varnothing\subseteq P(j)\subseteq I(j)} (X(\varnothing, P(j)), X(P(j), I(j))\Big),$$

which confirms that $\delta(X_Q) = \prod_{j=1}^{t} \delta(X_{I(j)})$, and

$$\sum_{\gamma\colon \varnothing\rightsquigarrow Q} (-1)^\ell \prod_{i=1}^{\ell} X(Q_{i-1}, Q_i) = \prod_{j=1}^{t} \Big(\sum_{\gamma(j)\colon \varnothing\rightsquigarrow I(j)} (-1)^\ell \prod_{i=1}^{\ell} X(Q_{i-1}(j), Q_i(j))\Big),$$

which confirms that $\chi(X_Q) = \prod_{j=1}^{t} \chi(X_{I(j)})$, as required; ϵ is multiplicative by definition.

To check that χ is indeed an antipode, we observe that

$$\mu\cdot\chi\otimes 1\cdot\delta(X_Q) = \sum_{\varnothing\subseteq P\subseteq Q} \Big(\sum_{\gamma\colon \varnothing\rightsquigarrow P} (-1)^\ell \prod_{i=1}^{\ell} X(P_{i-1}, P_i)\Big)\, X(P,Q)$$

in $\mathbb{Z}[B]$, where μ denotes the polynomial product. As in the proof of Theorem 4.2, and so long as $Q \neq \varnothing$, the contribution to this sum of each chain $\varnothing \subset \cdots \subset P_{\ell-1} \subset Q$ with $P = Q$ cancels that of $\varnothing \subset \cdots \subset P_{\ell-1}$ with $P = P_{\ell-1}$. We deduce that $\mu\cdot\chi\otimes 1\cdot\delta(X_Q)$ is 0 when $Q \neq \varnothing$, and is 1 when $Q = \varnothing$, as required. □

Now (5.1) and (5.2) encode the structure maps of $\mathbb{Z}[B]$, as

$$\delta(B(T)) = 1\otimes B(B(T)\otimes 1), \qquad \chi(B(T)) = B^{\langle -1\rangle}(T), \qquad \text{and} \qquad \epsilon(B(T)) = T \tag{5.4}$$

in terms of generating functions, and as

$$\delta(M_B) = M_{1\otimes B\circ B\otimes 1}, \quad \chi(M_B) = N_B, \quad \text{and} \quad \epsilon(M_B) = I \tag{5.5}$$

in terms of matrices, respectively.

When emphasizing these structures (as opposed to others which emerge below) we write $\mathbb{Z}^\circ X_*$ and $\mathbb{Z}^\circ[B]$, to acknowledge their origins in substitution. Cognoscenti will recognize $\mathbb{Z}^\circ[B]$ as the dual of the Landweber-Nóvikov algebra; for others, we provide an explanation in §6.

The expressions (5.3) are closely related to formulae of [**16**] and [**17**].

6. Complex cobordism

In this section we enrich our algebraic and geometric constructions, explaining how the correspondence between formal power series and the geometry of the varieties X_Q extends to the context of multiple complex cobordism theory.

Complex cobordism is, of course, based on the Thom spectrum MU. We refer to [**4**] for details of the double complex cobordism functor $\Omega^*_{2U}(\ )$, based on $2U = MU \wedge MU$, and (by extrapolation) of treble complex cobordism $\Omega^*_{3U}(\ )$, based on $3U = MU \wedge MU \wedge MU$. The corresponding coefficient rings are Ω^{2U}_* and Ω^{3U}_*, defined by cobordism classes of manifolds whose stable normal bundles split as $\nu \cong \nu_1 \oplus \nu_2$ and $\nu \cong \nu_1 \oplus \nu_2 \oplus \nu_3$ respectively; we refer to the ordered pair (ν_1, ν_2) as a $2U$-structure and the ordered triple (ν_1, ν_2, ν_3) as a $3U$-structure on the underlying manifold.

In §3 we invested each bounded flag manifold $B(Z_{n+1})$ with the basic, and bounding, U-structure $\nu \cong -(\oplus_{i=2}^{n+1} \gamma_i)$. So $B(Z_{n+1})$ also admits the basic $2U$-structure (γ, γ_1), where $\gamma = -(\oplus_{i=1}^{n+1} \gamma_i)$, and the basic $3U$-structures

$$(\gamma, \gamma_1, 0), \qquad (\gamma, 0, \gamma_1) \qquad \text{and} \qquad (0, \gamma, \gamma_1),$$

which we distinguish by the respective labels $(1, 2)$, $(1, 3)$, and $(2, 3)$. Applying Proposition 3.1, we extend these structures to each subvariety X_Q by cartesian product, writing γ_Q for the t-plane bundle $\gamma_1 \times \cdots \times \gamma_1$ over X_Q. We continue to refer to the resulting structures as basic; although the U-cobordism class of each X_Q is 0, the same is emphatically false in $2U$- and $3U$-cobordism, as we shall see.

It follows from the definitions that there are complex orientation classes x_1^{2U}, x_2^{2U} in $\Omega^2_{2U}(CP^\infty)$ and x_1^{3U}, x_2^{3U}, x_3^{3U} in $\Omega^2_{3U}(CP^\infty)$, and therefore expansions of the form

$$x_2^{2U} = \sum_{n \geq 0} b_n^U (x_1^{2U})^{n+1},$$

where b_n^U lies in Ω^{2U}_{2n}, and

$$x_2^{3U} = \sum_{n \geq 0} b_{n,2}^U (x_1^{3U})^{n+1}, \qquad x_3^{3U} = \sum_{n \geq 0} b_{n,3}^U (x_1^{3U})^{n+1},$$

$$\text{and} \qquad x_3^{3U} = \sum_{n \geq 0} b_{n,4}^U (x_2^{3U})^{n+1},$$

where $b_{n,2}^U$, $b_{n,3}^U$, and $b_{n,4}^U$ lie in Ω^{3U}_{2n}, for each $n \geq 0$. With these elements to hand, we may describe the $2U$- and $3U$-cobordism rings by embellishing results of [**4**].

THEOREM 6.1. *As Ω^U_*-algebras, we have that*

$$\Omega^{2U}_* \cong \Omega^U_*[b^U] \qquad \text{and} \qquad \Omega^{3U}_* \cong \Omega^U_*[b_{1,i}^U, b_{1,j}^U, b_{2,i}^U, b_{2,j}^U, \ldots],$$

where $i \neq j$ is any pair such that $2 \leq i, j \leq 4$; each b_n^U is represented by $B(Z_{n+1})$ with basic $2U$-structure, and each $b_{n,i}^U$ by $B(Z_{n+1})$ with the basic $3U$-structures $(1, 2)$, $(1, 3)$, and $(2, 3)$ respectively.

details, see [**14**], for example). The structural morphisms of coproduct, antiautomorphism, and counit are

$$(6.1)\qquad \delta\colon A_*^U \to A_*^U \otimes A_*^U, \quad \chi\colon A_*^U \to A_*^U \quad \text{and} \quad \epsilon\colon A_*^U \to \Omega_*^U$$

respectively, each of which restricts to the subalgebra $\mathbb{Z}[b^U]$ and turns it into a sub-Hopf algebra of A_*^U. This Hopf algebra is usually denoted by S_*, and is the integral dual of the Landweber-Novikov algebra; as such, the basis of monomials $(b^U)^\omega$ is dual to the basis of Landweber-Novikov cohomology operation s_ω in $A_U^{2|\omega|}$, as ω ranges over $[\infty]^*$.

By unravelling the definitions, we may interpret the morphisms of (6.1) as

$$\delta\colon \Omega_{2U}^* \to \Omega_{3U}^*, \quad \chi\colon \Omega_{2U}^* \to \Omega_{2U}^* \quad \text{and} \quad \epsilon\colon \Omega_{2U}^* \to \Omega_*^U,$$

and confirm that they are induced on the cobordism class of any $2U$-manifold $(M;\nu_1,\nu_2)$ by

$$(6.2)\qquad \begin{aligned} \delta(M;\nu_1,\nu_2) = (M;\nu_1,0,\nu_2), \qquad & \chi(M;\nu_1,\nu_2) = (M;\nu_2,\nu_1), \\ \text{and} \qquad & \epsilon(M;\nu_1,\nu_2) = (M;\nu_1 \oplus \nu_2) \end{aligned}$$

respectively. These restrict to the multiplicative generators b_n^U of S_* by the corresponding operations on the basic $2U$-structures on $B(Z_{n+1})$, and to additive generators by extension to the X_Q. We may now clarify the relationship between S_* and the Hopf algebra $\mathbb{Z}X_*$ of Theorem 5.3.

THEOREM 6.2. *The homomorphism $\mathbb{Z}X_* \to \Omega_*^{2U}$, induced by associating the basic $2U$-cobordism class to each isomorphism class X_Q, is an isomorphism of Hopf algebras onto S_*.*

PROOF. The homomorphism maps B_n to b_n^U for each $n \geq 0$, by Theorem 6.1, and so is an isomorphism of rings and preserves counits. Moreover, as explained in [**1**] for example, δ and χ are described on the generating function $b^U(T) = \sum_{n\geq 0} b_n^U T^{n+1}$ by

$$\delta(b^U(T)) = 1 \otimes b^U(b^U(T) \otimes 1) \qquad \text{and} \qquad \chi(b^U(T)) = (b^U)^{\langle -1\rangle}(T).$$

Comparing with (5.4), we conclude that the isomorphism also preserves δ and χ. □

Theorem 6.2 provides a deeper geometric structure to the calculus of formal power series, for the following reasons.

The generators of $\mathbb{Z}X_*$ are represented by connected algebraic varieties, whereas the linear combinations (5.3) arising in the descriptions of δ and χ (being purely formal) are not. In consequence, $\mathbb{Z}X_*$ provides little more than an alternative framework for the traditional procedures of substitution and reversion. In cobordism rings, however, we may construct interesting new elements by varying the normal structures on standard examples; we may also represent linear combinations by utilizing connected sums and reversed orientations. The first of these principles leads to the concise descriptions (6.2) of δ and χ, whilst the second allows us to represent the polynomials $M_p(A,B)$ and $N_p(B)$ of §4 by connected sums of appropriate X_Q with basic $3U$- and $2U$-structures. We are therefore presented with the delicate problem of reconciling the two viewpoints, which we pose as a corollary to Theorem 6.2.

COROLLARY 6.3. *There is a $3U$-cobordism between the bounded flag manifold $B(Z_{n+1})$ with $3U$-structure $(\gamma,0,\gamma_1)$ and the polynomial $M_n(A,B)$, where the As* *and Bs are invested with the 3U-structures* $(1,2)$ *and* $(2,3)$ *respectively. Similarly, there is a 2U-cobordism between* $B(Z_{n+1})$ *with 2U-structure* (γ_1, γ) *and the polynomial* $N_n(B)$ *with basic 2U-structure.*

PROOF. Combine Theorem 6.2 with the structure maps for $\mathbb{Z}X_*$ as described in (5.5), and for S_* as described in (6.2). □

We refer to the cobordisms of Corollary 6.3 as the *substitution cobordism* and the *reversion cobordism* respectively. We have succeeded in making them explicit only in the case $n = 1$ (an instructive exercise for readers), and it is a fascinating challenge to do better. A result of **[4]** asserts that the (p, q) entry of M_B may be represented in $\Omega^{2U}_{2(p-q)}$ by the cobordism class of $(X_{[q,p-1]}; \gamma - (q-1)\gamma_1, q\gamma_1)$ for any $p \geq q \geq 1$. Replacing each entry of M_A and M_B accordingly, we reduce $M_p(A, B)$ to a quadratic polynomial in the new classes; this may help to simplify the substitution cobordism.

7. Products and the Grassmannian

Although we began the study of formal power series with the straightforward commutative product, we have yet to discuss its geometric interpretation. We now make good this omission, drawing parallels with all aspects of our treatment of substitution and reversion, and employing the infinite complex Grassmannian BU.

For each poset $\mathcal{X}(n)$ we focus on a subset $\mathcal{C}(n)$, consisting of the elements of the chain $X_\varnothing \subset X_{[1]} \subset \cdots \subset X_{[n]}$ of length n; the support map defines an isomorphism of $\mathcal{C}(n)$ with the chain $\varnothing \subset [1] \subset \cdots \subset [n]$ in $2^{[n]}$. We develop the theory of these chains by analogy with that of the $\mathcal{X}(n)$ in earlier sections, allowing familiarity to breed substantial compression. We note that the direct system of inclusions $[n] \subseteq [m]$ induces compatible embeddings $\mathcal{C}(n) \subseteq \mathcal{C}(m)$, and let $\delta \colon [j_0] \rightsquigarrow [j_\ell]$ denote a generic chain of the form $[j_0] \subset [j_1] \subset \cdots \subset [j_\ell]$ in some suitably large $2^{[m]}$. The isomorphism class of each $X_{[j]}$ is B_j, and the $X_{[j]}$ therefore yield a set of polynomial generators for $\mathbb{Z}X_*$ as $[j]$ ranges over all $\mathcal{C}(n)$. If we wish to obtain *additive* generators for $\mathbb{Z}X_*$ in similar fashion, we have to enlarge our pool of posets to include products of the form $\mathcal{C}(n_1) \times \cdots \times \mathcal{C}(n_k) \subset \mathcal{X}(n_1 + \cdots + n_k)$.

We record the elements of the $\mathcal{C}(n)$ by means of a lower triangular matrix C_B in $\mathrm{LT}(\mathbb{Z}[B])$, whose (j, k) entry is $X_{[j-k]}$ for all $j \geq k \geq 0$. Similarly, we introduce $C_{A \cdot B}$ in $\mathrm{LT}(\mathbb{Z}[A, B])$ and D_B in $\mathrm{LT}(\mathbb{Z}[B])$ with (j, k) entries

$$\sum_{h=0}^{j-k} (X([0],[h]), X([h],[j-k])) \quad \text{and} \quad \sum_{\delta \colon [0] \rightsquigarrow [j-k]} (-1)^\ell \prod_{i=1}^{\ell} X([j_{i-1}],[j_i])$$

respectively. These expressions depend only on $j - k$, and consist of homogeneous polynomials $C_{j-k}(A, B)$ and $D_{j-k}(B)$ of dimension $2(j-k)$, in the generators A_m and B_n. For example, $C_{j-k}(A, B) = \sum_h A_h B_{j-k-h}$.

PROPOSITION 7.1. *The matrices* C_A *and* C_B *commute in* $\mathrm{LT}(\mathbb{Z}[A, B])$*, and* $C_A C_B = C_{A \cdot B}$*; the matrices* C_B *and* D_B *are mutually inverse in* $\mathrm{LT}(\mathbb{Z}[B])$.

To relate the matrices C_B to formal power series, we recall the universal invertible series $\beta(T)$ over $\mathbb{Z}[\beta]$, and its representing semicirculant $C(\beta)$ in $\mathrm{UT}(\mathbb{Z}[\beta])$. The multiplicative isomorphism $\mathbb{Z}[\beta] \to \mathbb{Z}[B]$, induced by $\beta_n \mapsto B_n$ for each $n \geq 0$, defines the invertible series $\widetilde{B}(T) = \sum_{k \geq 0} B_k T^k$ over $\mathbb{Z}X_*$ (which serves equally well as the universal example). We remark that

$$\widetilde{B}(T) = B(T)/T \tag{7.1}$$

in $\mathbb{Z}[B][[T]]$, where $B(T)$ is defined in §5, and that $\widetilde{B}(T)$ is represented by the semicirculant $C(\widetilde{B})$ in $\mathrm{UT}(\mathbb{Z}[B])$. In fact $C(\widetilde{B})^t = C_B$ by construction, which leads to the following corollary to Proposition 7.1.

COROLLARY 7.2. *In* $\mathrm{LT}(\mathbb{Z}[A, B])$, *we have that*

$$C(\widetilde{B})^t = C_B, \quad C(\widetilde{A} \cdot \widetilde{B})^t = C_{B \cdot A}, \quad \textit{and} \quad C(1/\widetilde{B})^t = D_B.$$

The commutativity of the product of power series therefore mirrors the relation $C_B C_A = C_A C_B$, and confirms that we may replace $C(\widetilde{B})$ by its transpose from the start; we then acquire an isomorphism of $\mathbb{Z}[B][[T]]$ with the semicirculant *lower* triangular matrices, and duly eliminate transpositions from Corollary 7.2.

These results allow us to compute products and reciprocals of the universal power series in terms of the posets $\mathcal{C}(n)$; in particular, the polynomials $C_k(A, B)$ and $D_k(B)$ appear in the expressions

$$\widetilde{A}(T)\widetilde{B}(T) = \sum_{k \geq 0} C_k(A, B) T^k \quad \text{and} \quad 1/\widetilde{B}(T) = \sum_{k \geq 0} D_k(B) T^k. \tag{7.2}$$

For power series $\sigma(T)$ over more general scalars K, we record the isomorphism class of each subvariety $X_{[j]}$ as σ_j, and write C_σ for the resulting matrix in $\mathrm{LT}(K)$. We may then generalize Corollary 7.2 to

$$C(\sigma)^t = C_\sigma, \quad C(\rho \cdot \sigma)^t = C_{\rho \cdot \sigma}, \quad \text{and} \quad C(1/\sigma)^t = D_\sigma.$$

The universal case is again linked to the theory of Hopf algebras. We define ring homomorphisms

$$\delta^\bullet \colon \mathbb{Z}[B] \to \mathbb{Z}[A, B], \quad \chi^\bullet \colon \mathbb{Z}[B] \to \mathbb{Z}[B], \quad \text{and} \quad \epsilon \colon \mathbb{Z}[B] \to \mathbb{Z}$$

by means of the respective equations

$$\delta^\bullet(X_{[n]}) = \sum_{i=0}^{n} (X([0], [i]), X([i], [n])),$$

$$\chi^\bullet(X_{[n]}) = \sum_{\delta \colon [0] \rightsquigarrow [n]} (-1)^\ell \prod_{i=1}^{\ell} X([j_{i-1}], [j_i]) \quad \text{and} \tag{7.3}$$

$$\epsilon(X_{[n]}) = \begin{cases} X_{[n]} & \text{if } n = 0 \\ 0 & \text{otherwise} \end{cases}$$

on polynomial generators, and replace $\mathbb{Z}[A, B]$ by $\mathbb{Z}[B] \otimes \mathbb{Z}[B]$, as before.

PROPOSITION 7.3. *The polynomial algebra* $\mathbb{Z}[B]$ *is a Hopf algebra with respect to the coproduct* $\delta^\bullet$, *antipode* $\chi^\bullet$, *and counit* ϵ.

PROOF. Since $\delta^\bullet$, $\chi^\bullet$, and ϵ are extended to monomials by multiplicativity, it remains only to establish that $\chi^\bullet$ is an antipode. This parallels the proof for χ in Proposition 5.3. $\square$

We denote this Hopf algebra by $\mathbb{Z}^\bullet X_*$ or $\mathbb{Z}^\bullet[B]$, whenever we wish to contrast it with $\mathbb{Z}^\circ X_*$ or $\mathbb{Z}^\circ[B]$ of Proposition 5.3; as algebras they are, of course, the same. Appealing to (7.2) and Corollary 7.2, the structure maps of Proposition 7.3 may be encoded in terms of generating functions by

$$\delta^\bullet(B(T)) = B(T) \otimes B(T), \quad \chi^\bullet(B(T)) = 1/B(T), \quad \text{and} \quad \epsilon(\widetilde{B}(T)) = 1, \tag{7.4}$$

and in terms of matrices by

$$\delta^\bullet(C_B) = C_{B\otimes 1\cdot 1\otimes B}, \qquad \chi^\bullet(C_B) = D_B, \qquad \text{and} \qquad \epsilon(C_B) = I.$$

It is possible to extend the entire analysis to additive generators by observing that $C_{B\cdot m} = C_B^m$ is obtained from m-fold cartesian products of posets $\mathcal{C}(n)$ by concentrating on chains which are themselves m-fold products.

We now relate the Hopf algebra $\mathbb{Z}^\bullet[B]$ to cobordism theory. The Whitney sum map $\oplus\colon BU \times BU \to BU$ turns $\Omega_*^U(BU_+)$ into an Ω_*^U-algebra, which is a polynomial algebra $\Omega_*^U[\beta^U]$ on generators β_n^U, where $\beta_0^U = 1$. The additive basis of monomials $(\beta^U)^\omega$ is dual to the basis of cobordism Chern classes c_ω in $\Omega_U^*(BU_+)$, as ω ranges over $[\infty]^*$. The Thom isomorphism $\Omega_*^U(MU) \to \Omega_*^U(BU_+)$ is a map of Ω_*^U-algebras, and equates $\Omega_*^U(BU_+)$ with both Ω_*^{2U} and A_*^U by virtue of §5.

LEMMA 7.4. *The Thom isomorphism $\Omega_*^{2U} \to \Omega_*^U(BU_+)$ identifies the $2U$-cobordism class of any $2U$-manifold $(M : \nu_1, \nu_2)$ with the U-cobordism class of $\nu_2\colon M \to BU$, where the latter M has the U-structure $\nu_1 \oplus \nu_2$.*

This lemma is proven in [**4**]; when combined with the result of [**15**] that β_n is represented by the classifying map $\gamma_1 : B(Z_{n+1}) \to BU$, it confirms that the Thom isomorphism acts by $b_n^U \mapsto \beta_n^U$.

The diagonal, orthogonal complement, and collapse maps

$$\Delta\colon BU \to BU \times BU, \qquad \perp\colon BU \to BU, \qquad \text{and} \qquad e\colon BU \to \cdot$$

induce Ω_*^U-algebra maps

$$\Delta_*\colon \Omega_*^U(BU_+) \to \Omega_*^U(BU_+) \otimes \Omega_*^U(BU_+), \qquad \perp_*\colon \Omega_*^U(BU_+) \to \Omega_*^U(BU_+),$$
$$\text{and} \quad e_*\colon \Omega_*^U(BU_+) \to \Omega_*^U.$$

Moreover, the composition $\oplus \cdot \perp \times 1 \cdot \Delta$ agrees with the projection e, so that $\Omega_*^U(BU_+)$ becomes an Ω_*^U-Hopf algebra with respect to the coproduct Δ_*, antipode $\perp_*$ and counit e_*. The subalgebra $\mathbb{Z}[\beta_U]$, which we abbreviate to R_* hereafter, is closed with respect to Δ_* and $\perp_*$, whilst e_* restricts as $R_* \to \mathbb{Z}$; in consequence, R_* is a sub-Hopf algebra over $\mathbb{Z}$ (isomorphic to the integral homology algebra $H_*(BU)$). The Thom isomorphism $A_*^U \to \Omega_*^U(BU_+)$ restricts to an isomorphism $t\colon S_* \to R_*$ of rings, but preserves neither coproducts nor antipodes.

THEOREM 7.5. *The homomorphism $\mathbb{Z}^\bullet X_* \to \Omega_*^U(BU_+)$, induced by associating the U-cobordism class $\gamma_Q\colon X_Q \to BU$ to each isomorphism class X_Q, is an isomorphism of Hopf algebras onto R_*.*

PROOF. The homomorphism maps B_n to β_n^U for each $n \geq 0$, and so is an isomorphism of rings and preserves counits.

Comparing with (7.4), we conclude that the isomorphism also maps $\delta^\bullet$ to Δ_* and $\chi^\bullet$ to $\perp_*$. □

The isomorphisms of Theorems 6.2 and 7.5 combine to produce a commutative diagram

(7.5)
$$\begin{array}{ccc} \mathbb{Z}^\circ X_* & \xrightarrow{\cong} & S_* < \Omega_*^U(MU) \\ {\scriptstyle 1}\downarrow & & \downarrow{\scriptstyle t} \\ \mathbb{Z}^\bullet X_* & \xrightarrow{\cong} & R_* < \Omega_*^U(BU_+). \end{array}$$

This compares the two Hopf algebra structures on $\mathbb{Z}X_*$, and the cobordism formulation of Lagrange inversion arises from the interplay between the two antipodes χ and $\chi^\bullet$.

PROPOSITION 7.6. *As elements of $\Omega_*^U(BU)$, there is a cobordism between $n+1$ times $\gamma\colon B(Z_{n+1}) \to BU$ and $-\gamma_1^{\oplus(n+1)}\colon B(Z_{n+1}) \to BU$.*

PROOF. By (5.1), Corollary 6.3, and Lemma 7.4, we may chase the coefficient of T^{n+1} in $B(T)$ around Diagram 7.5 to the U-cobordism class of $\gamma\colon B(Z_{n+1}) \to BU$; on the other hand, the Lagrange inversion formula [**9**] identifies this coefficient with $1/(n+1)$ times the coefficient of T^n in $(1/\widetilde{B})^{n+1}$. We may chase the latter around to $-\gamma_1^{\oplus(n+1)}\colon B(Z_{n+1}) \to BU$ by Theorem 7.5, and the result follows. □

We may reverse the argument of Proposition 7.6, and prove the Lagrange inversion formula by exhibiting a suitable *Lagrange cobordism*; this is simple when $n = 1$, but becomes more problematic as n increases. Nevertheless, manipulations with Chern classes confirm that such a cobordism does exist, and so provide some sort of hybrid proof. In the spirit of our calculus we would prefer to probe the underlying geometry by making the Lagrange cobordism explicit. We note in passing that the Chern class calculations have purely algebraic ramifications, allowing Lagrange inversion to be reformulated in the cohomology ring of $B(Z_{n+1})$; we discuss these matters elsewhere, together with more general forms of Lagrange inversion.

8. Faà di Bruno algebra

We conclude by summarizing modifications to the theory which deal with formal *divided* power series, otherwise known as *Hurwitz series* [**5**].

We assume (for notational convenience) that our ring of scalars K is free of additive torsion, and consider the K-algebra $K\{\{T\}\}$ of Hurwitz series over K, writing a typical element $\zeta(T)$ as $\sum_{k\geq 0} \zeta_k T^k/k!$. We emphasize that no scalar ζ_k need be divisible by $k!$ in K. Addition is taken termwise, and the product $\varepsilon(T)\cdot\zeta(T)$ is defined by the Cauchy product of the corresponding sequences, taking account of the extra factorials which arise from the denominators $k!$; clearly $K[[T]]$ is a subalgebra, and the reciprocal $1/\zeta(t)$ is again a Hurwitz series.

We concentrate on special reversible Hurwitz series which lie in the principal ideal $(T) < K\{\{T\}\}$, and therefore have ζ_0 zero; we denote such series generically by $f(T) = T + f_1T^2/2! + \cdots + f_nT^{n+1}/(n+1)! + \cdots$. For each $p \geq 1$ we observe that the pth divided power is again in (T), and write it as

(8.1)
$$f(T)^{\cdot p}/p! = \sum_p f_{p,q}T^q/q!,$$

where the $f_{p,q}$ are partial Bell polynomials in $f_1, \ldots, f_{q-p}$. Given two such series $e(T)$ and $f(T)$, we obtain a third by substitution, written as $e(f(T))$ or $e \circ f(T)$. The reversion $f^{\langle -1 \rangle}(T)$ also lies in (T) which therefore becomes a non-abelian group, with subgroup $\mathrm{SR}(K)$.

The universal ring of scalars is a polynomial algebra $\mathbb{Z}[g,h]$, generated by sequences of indeterminates $(g_0, g_1, \ldots)$ and $(h_0, h_1, \ldots)$, with $g_0 = h_0 = 1$; it admits Hurwitz series $g(T)$ and $h(T)$ and their reversions, plus the substitution of one in another. The ring homomorphism $\mathbb{Z}[g,h] \to \mathbb{Z}[a,b]$, which perceives the standard power series $a(T)$ and $b(T)$ as degenerate Hurwitz series, is therefore induced by $g_n \mapsto (n+1)!a_n$ and $h_n \mapsto (n+1)!b_n$.

To extend our geometric calculus to Hurwitz series, we construct a family of algebraic varieties $H(U)$, indexed by finite subsets U of $[\infty]$. Given $U = \{u(1), \ldots, u(c)\}$ of cardinality c, we consider a c-fold cartesian product $(CP^1)^{\times c}$, where the factors are indexed by the elements of U, and write $\eta_{u(j)}$ for the Hopf line bundle on the $u(j)$ factor, where $1 \leq j \leq c$. Then $H(U)$ is the Milnor hypersurface of degree $(1,1,\ldots,1)$, dual to the tensor product $\otimes_{j=1}^{n+1} \eta_{u(j)}$; to algebraic geometers, it is a hyperplane section and a nonsingular complex algebraic variety of dimension $c-1$. We continue to write the restriction of each Hopf bundle to $H(U)$ as $\eta_{u(j)}$, and abbreviate $H([n+1])$ to H_n in order to emphasize its dimension. Clearly H_0 is a singleton.

For each nonnegative integer n we consider the poset $\Pi(n)$ of partitions of the set $[n]$, ordered by refinement [**2**]; for each $n < m$ there is an inclusion $\Pi(n) \subset \Pi(m)$, obtained by adjoining singleton blocks $\{n+1\}, \ldots, \{m\}$ to all partitions of $[n]$. For every π in $\Pi(n+1)$ we define a further subvariety P_π of $(CP^1)^{\times(n+1)}$ as follows. We list the blocks of π as $\pi(j)$ for $1 \leq j \leq b(\pi)$, where each $\pi(j)$ is a subset of $[n+1]$ and $b(\pi)$ is the number of blocks, and construct P_π by successively restricting and dualizing the tensor products $\otimes_{\pi(j)} \eta_u$, for $1 \leq j \leq b(\pi)$. The result is independent of the order of dualization and has complex dimension $n+1-b(\pi)$; its structure is immediate.

PROPOSITION 8.1. *For any partition π of $[n+1]$, the subvariety P_π is nonsingular and isomorphic to the cartesian product $\times_j H(\pi(j))$.*

Singleton blocks contribute trivial cartesian factors to this decomposition, and the isomorphism type of P_π is therefore independent of large n.

We obtain posets $\mathcal{P}(n+1)$ of subvarieties of $(CP^1)^{\times\, n+1}$, ordered by $P_\nu < P_\pi$ whenever $\nu < \pi$ in $\Pi(n+1)$, and forming a direct system with respect to n. The *type* of π is the monomial $\prod_{j \geq 0} \phi_j^{w(j)}$, where $w(j)$ is the number of blocks of cardinality $j+1$ and $\phi_0 = 1$; by Proposition 8.1, it may be read off from the poset isomorphism $\Pi(n+1) \cong \mathcal{P}(n+1)$.

We now consider the free abelian group $\mathbb{Z}P_*$ generated by the isomorphism classes of all varieties P_π. Cartesian product invests $\mathbb{Z}P_*$ with the structure of commutative ring, and Proposition 8.1 implies that the class of P_π is the monomial $\prod_{j=1}^{b(\pi)} H_{|\pi(j)|-1}$. Thus $\mathbb{Z}P_*$ is a polynomial algebra $\mathbb{Z}[H]$, generated by the sequence of isomorphism classes $(H_0, H_1, \ldots)$, with identity $H_0 = 1$; it admits the Hurwitz series $H(T) = \sum_{q \geq 0} H_q T^{q+1}/(q+1)!$. We extend to the product $\mathbb{Z}P_* \otimes \mathbb{Z}P_*$ by

By analogy with (5.3) we turn $\mathbb{Z}P_*$ into a Hopf algebra, whose structure maps encode the combinatorics of the posets $\mathcal{P}(n)$. In turn, we may mimic (5.4) and (5.5) by describing δ, χ, and ϵ in terms of exponential generating functions or in terms of matrices; the former relates explicitly to computations with Hurwitz series, and yields

$$\delta(H(T)) = 1 \otimes H(H(T) \otimes 1), \qquad \chi(H(T)) = H^{\langle -1 \rangle}(T)$$
$$\text{and} \qquad \epsilon(H(T)) = T.$$

As a Hopf algebra, $\mathbb{Z}P_*$ is isomorphic to the Faà di Bruno algebra [**7**].

The ring homomorphism $\mathbb{Z}[H] \to \mathbb{Z}[B]$ of universal examples is induced by $H_n \mapsto (n+1)!B_n$, and embeds the Faà di Bruno algebra as a sub-Hopf algebra of the dual of the Landweber-Novikov algebra; sadly, we have no geometric interpretation for this map. Worse, there can be no $2U$-structure on H_n which renders it cobordant to $(n+1)!$ copies of B_n with basic $2U$-structure, since a simple calculation reveals that H_n can never be a U-boundary (except when $n = 1$ or 2). On the other hand, if we collapse Ω_*^{2U} to $H_*(MU)$ under the Thom map (which is an isomorphism on S_*), then H_n and $(n+1)!B_n$ may at least be made homologous, giving some degree of credence to the model. Nevertheless, our calculus for Hurwitz series remains at a rudimentary stage, and it is a stimulating challenge to find alternative realizations which lead to a full integration with the theory of §§5, 6 and 7. Usually, it is a problem to find natural representatives for a *divided* cobordism class; in this case, we are working with *multiples*.

References

[1] J. Frank Adams. *Stable Homotopy and Generalized Homology*. Chicago Lectures in Mathematics. University of Chicago Press, 1974.

[2] Martin Aigner. *Combinatorial Theory*. Springer-Verlag, 1979.

[3] Raoul Bott and Hans Samelson. Application of the theory of Morse to symmetric spaces. *American Journal of Mathematics*, 80:964–1029, 1958.

[4] Victor M. Buchstaber and Nigel Ray. Flag manifolds and the Landweber-Novikov algebra. Submitted, 1996.

[5] L. Carlitz. Some properties of Hurwitz series. *Duke Mathematical Journal*, 16:285–295, 1949.

[6] Louis Comtet. *Advanced Combinatorics*. D Reidel, 1974.

[7] Peter Doubilet, Gian-Carlo Rota, and Richard Stanley. On the Foundations of Combinatorial Theory VI: The Idea of Generating Function. In *Sixth Berkeley Symposium on Mathematical Statistics and Probability 2*, pages 267–318. University of California Press, 1972.

[8] Phillip Griffiths and Joseph Harris. *Principles of Algebraic Geometry*. Wiley Classics Library. John Wiley and Sons, 1994.

[9] Peter Henrici. *Applied and Computational Complex Analysis, Volume 1*. Wiley Classics Library. John Wiley and Sons, 1988.

[10] E. Jabotinski. Sur la représentation de la composition des fonctions par un produit de matrices. *Comptes Rendues*, 224:323–324, 1947.

[11] Christian Kassel. *Quantum Groups*, volume 155 of *Graduate Texts in Mathematics*. Springer Verlag, 1995.

[12] Toshiyuki Katsura, Yuji Shimizu, and Kenji Ueno. Complex cobordism ring and conformal field theory over **Z**. *Mathematische Annalen*, 291:551–571, 1991.

[13] Shahn Majid. *Foundations of Quantum Group Theory*. Cambridge University Press, 1995.

[14] Douglas C. Ravenel. *Complex Cobordism and Stable Homotopy Groups of Spheres*. Academic Press, 1986.

[15] Nigel Ray. On a construction in bordism theory. *Proceedings of the Edinburgh Mathematical Society*, 29:413–422, 1986.

[16] Nigel Ray and William Schmitt. Combinatorial models for coalgebraic structures. To appear; Advances in Mathematics, 1998.

[17] William Schmitt. Incidence Hopf algebras. *Journal of Pure and Applied Algebra*, 96:299–330, 1994.

DEPARTMENT OF MATHEMATICS, UNIVERSITY OF MANCHESTER, MANCHESTER M13 9PL, UK
E-mail address: nige@ma.man.ac.uk
URL: www.ma.man.ac.uk/ nige/

Contemporary Mathematics
Volume **220**, 1998

Notes on the Hopkins-Miller Theorem

Charles Rezk

ABSTRACT. We give an exposition of the proof of a theorem of Hopkins and Miller, that the spectra E_n admit an action of the Morava stabilizer group.

1. Introduction

These expository notes were originally written as an aid to the author while he was giving a course on the proof of the Hopkins-Miller theorem (Theorem 2.1) at Northwestern in Spring 1997. The aim was to describe the theorem, and give a sketch of the proof.

The aim was achieved, though only by glossing over certain points. In particular, several places rely on the existence of closed model category structures on various categories of spectra, and on subtle properties of this structure; I have not tried to deal with these issues. A few other points are left as references. At some points, the exposition is rather terse. Otherwise, the proof is as complete as I could make it.

The matter of the final part of these notes, on the construction of $\mathcal{A}_\infty$-ring structures, has to some extent been superseded by recent work of Hopkins and Goerss. They have developed more powerful obstruction machinery for constructing rigid algebraic structures on spectra, and are able to prove results about the existence of $\mathcal{E}_\infty$-structures as well as of $\mathcal{A}_\infty$-structures.

The material is based primarily on lectures and courses given by Mike Hopkins which I have attended at one time or another, and on discussions with Mike Hopkins and Haynes Miller. Also of great use were notes prepared by Lee Nave at a course given by Mike Hopkins at the University of Washington in the Spring of 1996.

All errors are mine, most of the true bits belong to others.

1.1. Organization. After giving a statement of the main result in Section 2, we outline Lubin-Tate's theory of universal deformations of formal groups over fields of positive characteristic, and its relation to homotopy theory, in Sections 3–7. In Sections 8–14 we summarize what we need about the theory of spectra and $\mathcal{A}_\infty$-ring spectra, and set up an obstruction theory and a spectral sequence for calculating maps between $\mathcal{A}_\infty$-ring spectra. In Sections 15–21 we give the calculation of the obstruction groups and the E_2-term of the spectral sequence,

1991 *Mathematics Subject Classification.* Primary 55N22; Secondary 55P42, 55S99.

assuming that we have an $\mathcal{A}_\infty$-ring spectrum structure on the spectra of interest; the obstruction groups vanish and the spectral sequence collapses for algebraic reasons. Finally, in Sections 22–23 we outline an obstruction theory for constructing an $\mathcal{A}_\infty$-ring spectrum structure, and compute the obstruction groups by reducing to the calculations already carried out.

2. The theorem

In this section we state the main theorem, modulo definitions of some of the key concepts. (Formal group laws are defined in Section 3, universal deformations of formal group laws are defined in Section 4, and $\mathcal{A}_\infty$-ring spectra are defined in Section 12.)

Let $\mathcal{FG}$ denote the category having as objects pairs (k,Γ), where k is a perfect field of finite characteristic p, and Γ is a formal group of height n over k, and with morphisms $\alpha\colon (k_1,\Gamma_1)\to(k_2,\Gamma_2)$ consisting of a pair (i,f), where i is a map $i\colon k_2\to k_1$ of rings and f is an isomorphism $f\colon \Gamma_1\to i^*\Gamma_2$ of formal group laws.

THEOREM 2.1 (Hopkins-Miller). *There exists a functor* $(k,\Gamma)\mapsto E_{k,\Gamma}$ *from* $\mathcal{FG}^{\mathrm{op}}$ *to the category of* $\mathcal{A}_\infty$*-ring spectra, such that*

1. $E_{k,\Gamma}$ *is a commutative ring spectrum,*
2. *there exists a unit in* $\pi_2 E_\Gamma$*,*
3. $\pi_{odd}E_{k,\Gamma}=0$*, from which it follows that* $E_{k,\Gamma}$ *is complex orientable,*
4. *and such that the corresponding formal group law over* $\pi_0 E_\Gamma$ *is the universal deformation of* (k,Γ).

In particular, when $(k,\Gamma)=(\mathbb{F}_{p^n},F_n)$ where F_n is the unique p-typical formal group law of height n with $[p]_\Gamma(x)=x^{p^n}$, then $E_{k,\Gamma}$ is called E_n, and

$$\pi_0 E_n = \mathbb{W}\mathbb{F}_{p^n}[\![u_1,\dots,u_{n-1}]\!].$$

The Hopkins-Miller theorem shows that this spectrum admits an action by the group of G automorphisms of $(\mathbb{F}_{p^n},F_n)$ "on the nose". This automorphism group is an extension of the Morava stabilizer group S_n by the Galois group of $\mathbb{F}_{p^n}$ over $\mathbb{F}_p$.

One application of the Hopkins-Miller theorem is the construction of "higher real K-theories", which are produced by taking the homotopy fixed points of E_n by the action of a finite subgroup of G. The fixed points of E_n by a maximal finite subgroup of G are usually denoted EO_n.

We will return to Theorem 2.1 in Section 7.

Part 1. Universal deformations

In the following sections, we define universal deformations and describe the Lubin-Tate moduli space. We use this to construct a functor from an appropriate category of formal group laws to the category of generalized homology theories (Section 6). It is this functor which the Hopkins-Miller theorem "lifts" to the category of $\mathcal{A}_\infty$-ring spectra.

3. Formal group laws

We define the category of formal group laws, and discuss some basic properties. More details on formal group laws may be found in [**7**], [**1**, part 2], and [**13**], among other places.

3.1. Definition of formal group laws. Let R be a commutative ring. A **formal group law** F over R is a formal power series $F(x,y) \in R[\![x,y]\!]$ such that

(a) $F(x,0) = x = F(0,x)$,
(b) $F(x,y) = F(y,x)$,
(c) $F(F(x,y),z) = F(x,F(y,z))$.

By (a) and (b) such a power series $F(x,y)$ necessarily has the form

$$F(x,y) = x + y + \sum_{i,j \geq 1} a_{ij} x^i y^j,$$

where $a_{ij} = a_{ji} \in R$. Sometimes I'll write $x \underset{F}{+} y$ for $F(x,y)$.

EXAMPLE 3.2. The additive formal group law is defined by $x \underset{G_a}{+} y = x + y$.

EXAMPLE 3.3. The multiplicative formal group law is defined by $1 + (x \underset{G_m}{+} y) = (1+x)(1+y)$.

EXAMPLE 3.4. Elliptic formal group laws. (See, e.g., Silverman [**17**].)

Let B is a complete local R-algebra with maximal ideal $\mathfrak{m}$; that is, $B \simeq \lim B/\mathfrak{m}^k$ and $\bigcap_k \mathfrak{m}^k = 0$. A formal group law F on R defines an abelian group structure on $\mathfrak{m}$. In particular, it defines an abelian group structure on the ideal of $R[\![x]\!]$ generated by x. We write $a \underset{F}{+} b$ for the sum of $a, b \in \mathfrak{m}$ in this group structure.

A **morphism** $f\colon F \to G$ of formal group laws over R is a formal power series $f(x) \in R[\![x]\!]$ such that $f(0) = 0$ and $f(x \underset{F}{+} y) = f(x) \underset{G}{+} f(y)$. The **isomorphisms** are precisely those $f(x) \in R[\![x]\!]$ for which $f(x) = ax + \cdots$ with a an invertible element of R. Such an isomorphism is called a **strict isomorphism** if $a = 1$.

Given a formal group law F over R and a power series $f(x) \in R[\![x]\!]$ with $f(0) = 0$ and with $f'(0)$ a unit in R, we can define a new formal group law G by

$$G(x,y) = f\left(F\left(f^{-1}(x), f^{-1}(y)\right)\right),$$

which comes with an isomorphism $f\colon F \to G$ of formal group laws.

I'll write $\mathrm{fgl}(R)$ for the groupoid consisting of all formal group laws on R and all isomorphisms between them. A map $\phi\colon R \to S$ of rings induces a functor $\phi^*\colon \mathrm{fgl}(R) \to \mathrm{fgl}(S)$ by sending

$$F(x,y) = \sum_{i,j} a_{ij} x^i y^j \qquad \text{to} \qquad \phi^* F(x,y) = \sum_{i,j} \phi(a_{ij}) x^i y^j.$$

For any formal group law, one can write the n-**series**

$$[n]_F(x) = x \underset{F}{+} \cdots \underset{F}{+} x,$$

by taking the n-fold sum of x in the group law F. The n-series is in fact an endomorphism of the formal group law F. Since $[n]_F(x) = nx + \cdots$, it is an automorphism if n is invertible in R.

3.5. Formal group laws over fields of characteristic p. We now consider formal group laws over a field k of characteristic p.

PROPOSITION 3.6. *If $f\colon F \to G$ is a homomorphism of formal group laws over* *for some $n \geq 0$ and some $g(x) \in k[\![x]\!]$ with $g(0) = 0$ and $g'(0) \neq 0$.*

For such a homomorphism $f\colon F \to G$ the largest such integer n is called the **height** of f (or if $f(x) = 0$, we say f has **infinite height**.) If $[p]_F(x)$ has height n then we say that the formal group law F has **height** n.

PROOF OF PROPOSITION 3.6. Since $f(F(x,y)) = G(f(x), f(y))$, by applying $\frac{\partial}{\partial y}$ and evaluating at $y = 0$ we see that

$$f'(x)F_2(x,0) = G_2(f(x),0)f'(0).$$

Since $F_2(x,0) = 1 +$ (higher terms), it has a multiplicative inverse, so we see that if $f'(0) = 0$ then $f'(x) = 0$, whence $f(x) = g(x^p)$ for some power series $g(x)$. The proof can be completed by induction if we can show that $g(x)$ is itself a morphism of formal group laws.

Let $\sigma\colon k \to k$ denote the homomorphism $x \mapsto x^p$. Then one can check that

$$F \xrightarrow{x^p} \sigma^* F \xrightarrow{g} G$$

is a diagram of formal group laws and homomorphisms. In fact,

$$g(\sigma^* F(x^p, y^p)) = g(F(x,y)^p) = f(F(x,y)) = G(f(x), f(y)) = G(g(x^p), g(y^p)),$$

whence

$$g(\sigma^* F(x,y)) = G(g(x), g(y)),$$

as desired. □

EXAMPLE 3.7. The additive formal group law has infinite height, since the p-series is $[p]_{G_a}(x) = px \equiv 0$. The multiplicative formal group law has height 1, since $[p]_{G_m}(x) = (1+x)^p - 1 \equiv 1 + x^p - 1 = x^p$. An elliptic formal group law has height 1 or height 2. (Silverman [**17**, Ch. 7].)

EXAMPLE 3.8. Over $k = \mathbb{F}_p$ there is a (p-typical) formal group law F_n (the **Honda** formal group law) with $[p]_{F_n}(x) = x^{p^n}$. (Ravenel [**13**, App. 2].)

4. The universal deformations

We define deformations and the deformation category, and introduce the universal deformations.

4.1. Deformations of a formal group law. Let (k, Γ) be a pair consisting of a field k of characteristic $p > 0$ and a formal group Γ over k. A **deformation** of (k, Γ) to a complete local ring B (with maximal ideal $\mathfrak{m} = \mathfrak{m}_B$ and projection $\pi\colon B \to B/\mathfrak{m}$) is a pair (G, i) consisting of a formal group law G over B and a homomorphism $i\colon k \to B/\mathfrak{m}_B$, such that $i^*\Gamma = \pi^* G$.

A morphism of deformations $(G_1, i_1) \to (G_2, i_2)$ is defined only when $i_1 = i_2$, in which case it consists of an isomorphism $f\colon G_1 \to G_2$ of formal group laws over B such that $\pi^* f$ is the identity map of $\pi^* G_1 = \pi^* G_2 = i_1^*\Gamma = i_2^*\Gamma$. That is,

$$f(x) \equiv x \mod \mathfrak{m}_B.$$

Such an isomorphism f is sometimes called a **$\star$-isomorphism**.

The category of deformations of Γ over B is denoted $\mathrm{Def}_\Gamma(B)$. It's nice to think of this category using the following picture.

$$\begin{array}{ccccc}
\mathrm{Def}_\Gamma(B) & & \longrightarrow & & \mathrm{fgl}(B) \\
\downarrow & & & & \downarrow \\
\coprod_{i\colon k \to B/\mathfrak{m}} \{\Gamma\} & \longrightarrow & \coprod_{i\colon k \to B/\mathfrak{m}} \mathrm{fgl}(k) & \longrightarrow & \mathrm{fgl}(B/\mathfrak{m})
\end{array}$$

This is a diagram of groupoids. The lower left-hand corner denotes the discrete groupoid, having as objects the set of maps $i\colon k \to B/\mathfrak{m}$, and as maps only identity maps. This groupoid sits inside a coproduct of copies of the groupoid $\mathrm{fgl}(k)$; the inclusion maps the "object" i to the pair (Γ, i), where G is a formal group law over k and $i\colon k \to B/\mathfrak{m}$ a homomorphism. The lower right-hand functor maps such objects to $\mathrm{fgl}(B/\mathfrak{m})$ by sending (G, i) to i^*G. The diagram commutes, and is a pull-back square of groupoids.

REMARK 4.2. Given a morphism $f\colon C \to D$ of groupoids, the simplicial nerve of f is a Kan fibration of simplicial sets if and only if f has the following property: for each object $c \in \mathrm{ob}\, C$ and morphism $\beta\colon fc \to d$ in D there exists an $\alpha\colon c \to d'$ in C such that $f\alpha = \beta$.

Thus, if I apply the classifying space functor to the above diagram, the square remains a pull-back square; furthermore, the right-hand vertical arrow becomes a fibration. Thus, the diagram induces a *homotopy* pull-back square of classifying spaces.

4.3. The Lubin-Tate theorem. Given $j\colon k \to B/\mathfrak{m}$, let $\mathrm{Def}_\Gamma(B)_j$ denote the full subgroupoid of $\mathrm{Def}_\Gamma(B)$ consisting of those deformations (G, i) for which $i = j$.

I will always feel free to replace a groupoid by its classifying space, and thus will speak of π_0 and π_1 of a groupoid, meaning the corresponding homotopy groups of its classifying space. Of course, if C is a groupoid then $\pi_0 C$ is just the set of isomorphism classes in C, and $\pi_1(C, X)$ (the fundamental group based at an object X) is the group of automorphisms of the object X in C.

THEOREM 4.4 (Lubin-Tate). **[10]** *If Γ is a formal group law of height $n < \infty$ over k, and if $D = (G, i)$ is some deformation of (k, Γ), then*

$$\pi_1\big(\mathrm{Def}_\Gamma(B)_i, D\big) = \{1\},$$

and there is a bijection

$$\pi_0\big(\mathrm{Def}_\Gamma(B)_i\big) \simeq \mathfrak{m}^{\times(n-1)}.$$

Furthermore, there exists a complete local ring $E(k, \Gamma)$, an isomorphism $i\colon k \xrightarrow{\sim} E(k, \Gamma)/\mathfrak{m}$, and a formal group law F on $E(k, \Gamma)$ such that the pair (F, id) is a ***universal deformation,*** *in the sense that the functor $B \mapsto \pi_0 \, \mathrm{Def}_\Gamma(B)$ is corepresented by the ring $E(k, \Gamma)$, so that a map $\phi\colon E(k, \Gamma) \to B$ corresponds to the isomorphism class of $\phi^* F$ in $\mathrm{Def}_\Gamma(B)$.*

An outline of the proof of Theorem 4.4 is given in Section 5.

Let me make clear what it means for (F, id) to be a universal deformation and formal group laws.

$$\begin{array}{ccc} E(k,\Gamma) & & B \\ \downarrow{\scriptstyle \pi} & & \downarrow{\scriptstyle \pi} \\ k & \xrightarrow{i} & B/\mathfrak{m} \end{array} \qquad \begin{array}{ccccc} F & & & & G \\ \wr\!\downarrow & & & & \wr\!\downarrow \\ \Gamma & \rightsquigarrow & i^*\Gamma & = & \pi^*G \end{array}$$

(The squiggly arrows are meant to represent the action of a functor on an object.) For (F, id) to be a universal deformation, there must exist a *unique* map $\phi\colon E(k,\Gamma) \to B$ such that the object (ϕ^*F, i) is isomorphic to (G, i) in the category $\mathrm{Def}_\Gamma(B)$. Then there exists a (necessarily *unique*) $\star$-isomorphism $g\colon G \to \phi^*F$. In other words, you get a unique pair of diagrams

$$\begin{array}{ccc} E(k,\Gamma) & \xrightarrow{\phi} & B \\ \downarrow{\scriptstyle \pi} & & \downarrow{\scriptstyle \pi} \\ k & \xrightarrow{i} & B/\mathfrak{m} \end{array} \qquad \begin{array}{ccccc} F & \rightsquigarrow & \phi^*F & \xleftarrow[g]{\sim} & G \\ \wr\!\downarrow & & \wr\!\downarrow & & \wr\!\downarrow \\ \Gamma & \rightsquigarrow & i^*\Gamma & = & \pi^*G \end{array}$$

where $g(x) \equiv x \mod \mathfrak{m}_B$.

The first part of Theorem 4.4 says that $\mathrm{fgl}(B) \to \mathrm{fgl}(B/\mathfrak{m})$ is a "covering space up to homotopy". Thus, an element in π_1 of the base "space" $\mathrm{fgl}(B/\mathfrak{m})$ induces an action on π_0 of the "fiber" $\mathrm{Def}_\Gamma(B)_i$. In particular, given $\gamma \in \pi_1(\mathrm{fgl}(k), \Gamma) \simeq \mathrm{Aut}(\Gamma)$, we obtain an induced map $\tilde{\gamma}\colon \pi_0 \mathrm{Def}_\Gamma(B)_i \to \pi_0 \mathrm{Def}_\Gamma(B)_i$ for each $i\colon k \to B/\mathfrak{m}$. Putting these together over all the i's we get a map $\tilde{\gamma}\colon \pi_0 \mathrm{Def}_\Gamma(B) \to \pi_0 \mathrm{Def}_\Gamma(B)$. It's easy to see that this action is natural in B; in other words, there is an induced action of $\mathrm{Aut}(\Gamma)$ on the functor $\pi_0 \mathrm{Def}_\Gamma(-)$, and hence on the ring $E(k,\Gamma)$ which represents the functor.

If $k = \mathbb{F}_{p^n}$, and $\Gamma = F_n$ is the Honda formal group law of height n, then $\mathrm{Aut}(F_n)$ contains the n-th Morava stabilizer group S_n as the group of strict automorphisms of F_n.

We can sometimes get a bigger group action on $E(k,\Gamma)$. Suppose k is a finite field, and fix a map $i_0\colon k \to B/\mathfrak{m}$. Then maps $i\colon k \to B/\mathfrak{m}$ can be factored $i = i_0\sigma^d$, where $\sigma\colon k \to k$ denotes $\sigma(x) = x^p$. If Γ is Galois invariant, i.e., if $\sigma^*\Gamma = \Gamma$ (for example, if Γ is the Honda formal group law), then there is an action of $\mathrm{Gal}(\mathbb{F}_{p^n}/\mathbb{F}_p) = C_n$ on $\mathrm{Def}_\Gamma(B)$, and hence on $E(k,\Gamma)$.

4.5. The ring $E(k,\Gamma)$. We can say something about the ring $E(k,\Gamma)$, at least if k is a perfect field. Since it is a complete local ring with residue field k, it must contain the Witt vectors $\mathbb{W}k$. For $k = \mathbb{F}_{p^n}$, $\mathbb{W}k = \mathbb{Z}_p[\zeta]$, where ζ is a primitive $(p^n - 1)$st root of unity. For a perfect field k the Witt ring $\mathbb{W}k$ has the following universal property; given a diagram

$$\begin{array}{ccc} \mathbb{W}k & \dashrightarrow & B \\ \downarrow & & \downarrow \\ k & \longrightarrow & B/\mathfrak{m} \end{array}$$

where B is a complete local ring and all maps are continuous homomorphisms, there is a unique dotted arrow making the diagram commute. (This is essentially [**16**, II.5].) This is easy to see when $k = \mathbb{F}_{p^n}$, since in this case the Teichmüller construction ensures that one can produce a canonical multiplicative section $k \to B$.

Thus $E(k,\Gamma)$ is a $\mathbb{W}k$-algebra, and in fact contains $\mathbb{W}k$ (because there exists at least one deformation of Γ to $\mathbb{W}k$, so that there exists a map from $E(k,\Gamma)$ back down to $\mathbb{W}k$.) Since a map $i\colon k \to B/\mathfrak{m}$ determines a unique map $\mathbb{W}k \to B$, making B also into a $\mathbb{W}k$-algebra, then the isomorphism of Theorem 4.4

$$\mathrm{Hom}_{\mathbb{W}k\text{-}\mathrm{alg}}(E(k,\Gamma),B) \simeq \pi_0 \mathrm{Def}_\Gamma(B)_i \simeq \mathfrak{m}^{\times(n-1)}$$

indicates that we should set

$$E(k,\Gamma) \simeq \mathbb{W}\mathbb{F}_{p^n}[\![u_1,\dots,u_{n-1}]\!].$$

4.6. Functoriality of the universal deformation. Consider a morphism $\alpha\colon (k_1,\Gamma_1) \to (k_2,\Gamma_2)$ in the category $\mathcal{FG}$; α is a pair (j,f) with $j\colon k_2 \to k_1$ and $f\colon \Gamma_1 \xrightarrow{\sim} j^*\Gamma_2$. Such a morphism induces a diagram

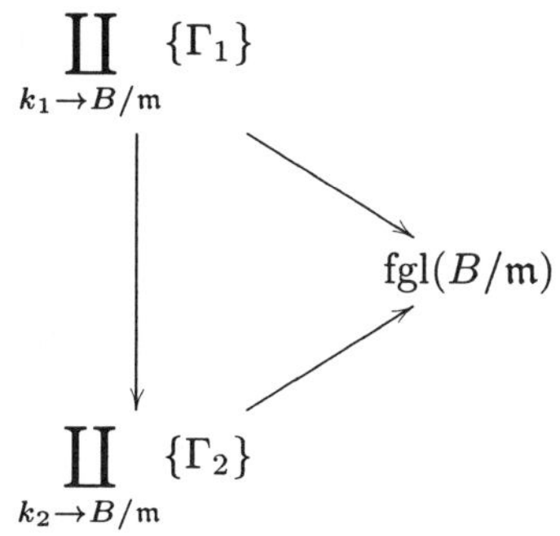

which commutes up to a homotopy (determined by f). This induces a map

$$\pi_0 \mathrm{Def}_{\Gamma_1}(B) \to \pi_0 \mathrm{Def}_{\Gamma_2}(B)$$

which is natural in B, and hence a map $E(k_2,\Gamma_2) \to E(k_1,\Gamma_1)$. Thus we obtain, according to Theorem 4.4, a functor

$$\mathcal{FG}^{\mathrm{op}} \to \{\text{complete local rings}\}$$

sending

$$(k,\Gamma) \mapsto E(k,\Gamma).$$

5. Construction of the universal deformation

I give a construction of the universal deformation, and at least indicate how the proof goes. I mainly follow Lubin-Tate's original proof [**10**], though I also borrow an observation of Drinfel'd [**4**].

5.1. Deformations to a trivial square 0 extension over k. I want to give at least an idea of how the proof goes. So what I'll do first is compute $\mathrm{Def}_\Gamma(k[\epsilon]/\epsilon^2)$.

Let $R = k[\epsilon]/\epsilon^2$. The inclusion $k \to R$ allows us to push a formal group law Γ on k to R; by abuse of notation, I will also refer to the pushed-forward-to-R formal group law as Γ. An arbitrary formal group law on R which lifts Γ must look something like

$$F(x,y) = x \underset{\Gamma}{+} y \underset{\Gamma}{+} \epsilon G(x,y)$$

where $G(x,y) \in k[\![x,y]\!]$. For F to really be a formal group law, we must also have

(a) $G(x,0) = 0 = G(0,x)$

(c) $G(x \underset{\Gamma}{+} y, z) + G(x,y) = G(x, y \underset{\Gamma}{+} z) + G(y,z)$.

Such a G is called a **Γ-symmetric 2-cocycle**. I'll write $Z^2_{\text{sym}}(\Gamma)$ for the set of such cocycles.

Now suppose I have another formal group law $F'(x,y) = x \underset{\Gamma}{+} y \underset{\Gamma}{+} \epsilon G'(x,y)$, and a map $f\colon F \to F'$ which reduces to the identity map mod ϵ. Such an f has the form,

$$f(x) = x \underset{\Gamma}{+} \epsilon g(x)$$

for some $g(x) \in k[\![x]\!]$ with $g(0) = 0$. We shall call such a g a **Γ-symmetric 1-cocycle**, and write $Z^1_{\text{sym}}(\Gamma)$ for the set of all such cocycles.

Then $f(x \underset{F}{+} y) = f(x) \underset{F'}{+} f(y)$ if and only if

$$G(x,y) + g(x \underset{\Gamma}{+} y) = G'(x,y) + g(x) + g(y).$$

This means that we can form a little chain complex:

$$0 \longrightarrow \left\{ \begin{array}{c} g(x) \in k[\![x]\!], \\ g(0) = 0 \end{array} \right\} \xrightarrow{\ \delta\ } \left\{ \begin{array}{c} G(x,y) \in k[\![x,y]\!], \\ G \text{ a } \Gamma\text{-symmetric 2-cocycle} \end{array} \right\} \longrightarrow 0$$

where $\delta g(x,y) = g(x) + g(y) - g(x \underset{\Gamma}{+} y)$. The homology of this chain complex is precisely the homotopy of $\mathrm{Def}_\Gamma(R)_{\mathrm{id}}$. That is, the cokernel of δ may be identified with the set of isomorphism classes in $\mathrm{Def}_\Gamma(R)_{\mathrm{id}}$, and the kernel can be identified as the group of automorphisms of any chosen object of $\mathrm{Def}_\Gamma(R)_{\mathrm{id}}$.

5.2. Lemmas about formal group laws. To compute this complex, we need a couple of lemmas. First, note that an **additive symmetric cocycle** is a symmetric cocycle for the additive group.

LEMMA 5.3. *An additive symmetric cocycle G(x,y), homogeneous of degree d, has the form $aC_d(x,y)$ for some $a \in k$, and where $C_d(x,y) \in \mathbb{Z}[\![x,y]\!]$ is defined by*

$$C_d(x,y) = \frac{1}{\lambda_d}(x^d + y^d - (x+y)^d)$$

with

$$\lambda_d = \begin{cases} 1 & \text{if } d \neq p^j \\ p & \text{if } d = p^j \end{cases}.$$

PROOF. See Fröhlich [**7**]. The proof is also given in [**1**, part 2]. □

LEMMA 5.4. *Let $\pi\colon R \to S$ be a surjective map of rings. If $F(x,y) \in R[\![x,y]\!]$ satisfies the axioms for a formal group law modulo terms of degree $(d+1)$, and if π^*F is a formal group law over S, then there exists a formal group law $F'(x,y) \in R[\![x,y]\!]$ such that $F' \equiv F \mod$ degree $(d+1)$ terms and such that $\pi^*F' = \pi^*F$.*

Lemma 5.4 really follows from the theorem of Lazard.

THEOREM 5.5 (Lazard). *There exists a ring $L \simeq \mathbb{Z}[s_1, s_2, \ldots]$ and a formal group law F over L such that the induced map*

$$\mathrm{Hom}_{\mathrm{Ring}}(L, R) \to \mathrm{ob}\,\mathrm{fgl}(R)$$

sending $f \mapsto f^*F$ is a bijection. Furthermore, if we write

$$x \underset{F}{+} y = x + y + \sum_{i,j} a_{ij} x^i y^j$$

then $a_{ij} \in L$ *is a polynomial in the variables* $s_1, \ldots, s_{i+j-1}$.

PROOF. See [**7**], [**1**, part 2] or [**13**, Appendix 2]. ☐

Lemma 5.3 shows that any Γ-symmetric 2-cocycle has the form

$$G_d(x,y) = aC_d(x,y) + \text{higher degree terms},$$

since the lowest degree term of a Γ-symmetric 2-cocycle must necessarily be an additive cocycle. Lemma 5.4 shows that such a Γ-symmetric 2-cocycle exists, since the expression

$$F(x,y) = x \underset{\Gamma}{+} y \underset{\Gamma}{+} \epsilon C_d(x,y)$$

is a formal group law modulo degree $(d+1)$, and modulo $\mathfrak{m}$ is the formal group law Γ.

There is an obvious filtration $\cdots \supset F_d Z^2_{\text{sym}}(\Gamma) \supset F_{d+1} Z^2_{\text{sym}}(\Gamma) \supset \cdots$ of the symmetric 2-cocycles in which $F_d Z^2_{\text{sym}}(\Gamma)$ consists of all elements whose lowest degree term is at least d. Thus we have shown that the associated quotients of this complete filtration $F_d Z^2_{\text{sym}} / F_{d+1} Z^2_{\text{sym}}$ are each isomorphic to k, generated by an element G_d.

Similarly, there is a filtration $\cdots \supset F_d Z^1_{\text{sym}}(\Gamma) \supset F_{d+1} Z^1_{\text{sym}}(\Gamma) \supset \cdots$ where $F_d Z^1_{\text{sym}}(\Gamma)$ consists of all elements whose lowest degree term is at least d. Thus each associated quotient $F_d Z^1_{\text{sym}} / F_{d+1} Z^1_{\text{sym}}$ is isomorphic to k, generated by an element $f_d(x) = x^d$. Thus $Z^\bullet_{\text{sym}}$ is a filtered chain complex, and all we need to do is calculate the associated spectral sequence!

Let's assume that

$$\Gamma(x,y) = x + y + bC_{p^n}(x,y) + \cdots.$$

We can do this without loss of generality, by the following lemma, and by the fact that isomorphic formal group laws have isomorphic complexes $Z^\bullet_{\text{sym}}$ of symmetric cocycles.

LEMMA 5.6. *A height* n *formal group law* Γ *over* k *is isomorphic to a formal group law* Γ′ *having the form*

$$\Gamma'(x,y) = x + y + bC_{p^n}(x,y) + \cdots,$$

with $b \neq 0$.

PROOF. This is because if Γ has the form

$$\Gamma(x,y) \equiv x + y + bC_k(x,y) + (x,y)^{k+1}$$

for $k \neq p^j$, we can let $f(x) = x + bx^k$ and $\Gamma'(x,y) = f(\Gamma(f^{-1}x, f^{-1}y))$, whence $f\colon \Gamma \to \Gamma'$ is an isomorphism of formal group laws, and

$$\Gamma'(x,y) \equiv x + y + (x,y)^{k+1}.$$

If Γ has the form

$$\Gamma(x,y) = x + y + bC_{p^j}(x,y) + (x,y)^{p^j+1}$$

one can show by Lemma 5.7 that $[p]_\Gamma(x) = bx^{p^j} + \cdots$, hence $b = 0$ or $j = n$. ☐

LEMMA 5.7. *If F is a formal group law with*

$$x \underset{F}{+} y = x + y + bC_k(x,y) + \cdots$$

then

$$[n]_F(x) = nx + b\{\frac{1}{\lambda_k}(n - n^k)\}x^k + \cdots.$$

5.8. Computation of the chain complex. Let $f_k(x) = x^k$. Then

$$\begin{aligned}(\delta f_k)(x,y) &= x^k + y^k - (x \underset{\Gamma}{+} y)^k \\ &= \begin{cases} C_k(x,y) + \cdots & \text{if } k \neq p^j, \text{ where } j \geq 0, \\ b^{p^j} C_{p^n}(x,y)^{p^j} + \cdots & \text{if } k = p^j, \text{ where } j \geq 0. \end{cases}\end{aligned}$$

Since $C_{p^n}(x,y) = C_p(x,y)^{p^{n-1}}$, we see that

$$\delta f_k = \begin{cases} G_k \bmod F_{k+1} Z^2_{\text{sym}} & \text{if } k \neq p^j, \\ G_{p^{n+j}} \bmod F_{p^{n+j}+1} Z^2_{\text{sym}} & \text{if } k = p^j. \end{cases}$$

Thus the cokernel of δ has dimension $(n-1)$ over k, generated by elements of the form

$$G'_{p^k}(x,y) = C_{p^k}(x,y) + \cdots$$

for $k = 1, \ldots, n-1$. The kernel of δ is trivial. Hence

$$\pi_0 \operatorname{Def}_\Gamma(R)_i = H^2_{\text{sym}}(\Gamma) \simeq k^{\times(n-1)},$$

and

$$\pi_1 \operatorname{Def}_\Gamma(R) \simeq 0.$$

5.9. The case of a general square-zero extension. The above argument proves the first part of the Lubin-Tate theorem for $B = k[\epsilon]/\epsilon^2$. The method of proof extends to prove a more general statement.

Let B be a complete local ring B with maximal ideal $\mathfrak{m}$ and residue field $k = B/\mathfrak{m}$. Let F_k be a formal group law over $B/\mathfrak{m}^k$ and let Γ denote its push-forward to k. Suppose F_{k+1} is a formal group law over $B/\mathfrak{m}^{k+1}$ which lifts F_k. Then any other formal group law F'_{k+1} over $B/\mathfrak{m}^{k+1}$ which lifts F_k has the form

$$F'_{k+1}(x,y) = F_k(x,y) \underset{F_{k+1}}{+} G(x,y)$$

where $G(x,y) \in Z^2_{\text{sym}}(\Gamma)\hat{\otimes}\mathfrak{m}^k/\mathfrak{m}^{k+1}$. Furthermore, an isomorphism $f\colon F_{k+1} \to F'_{k+1}$ which lifts the identity map of F_k has the form

$$f(x) = x \underset{F_{k+1}}{+} g(x)$$

where $g(x) \in Z^1_{\text{sym}}(\Gamma)\hat{\otimes}\mathfrak{m}^k/\mathfrak{m}^{k+1}$.

Thus the "fiber" of $\operatorname{fgl}(B/\mathfrak{m}^{k+1}) \to \operatorname{fgl}(B/\mathfrak{m}^k)$ over F_k has $H^2_{\text{sym}}(\Gamma) \otimes \mathfrak{m}^k/\mathfrak{m}^{k+1}$ as its set of components, and $H^1_{\text{sym}}(\Gamma) \otimes \mathfrak{m}^k/\mathfrak{m}^{k+1}$ as the set of isomorphisms of any object.

The above can be proved most easily by pulling back the map $\pi\colon B/\mathfrak{m}^{k+1} \to B/\mathfrak{m}^k$ along itself; the pulled-back map is a trivial square-0 extension, and the argument of the previous sections can be applied to it.

5.10. Construction of the universal deformation. To finish the proof, we need to construct the universal deformation. This will be a formal group law over the ring

$$E(k,\Gamma) = \mathbb{W}k[\![u_1,\dots,u_{n-1}]\!].$$

It turns out that it suffices to produce a formal group law over this ring which has the form

$$F(x,y) \equiv x \underset{\Gamma}{+} y \underset{\Gamma}{+} u_1 C_p(x,y) \underset{\Gamma}{+} \cdots \underset{\Gamma}{+} u_{n-1} C_{p^{n-1}}(x,y) \underset{\Gamma}{+} \cdots$$

modulo p and modulo $(u_1,\dots,u_{n-1})^2$. It is easy to construct such: we can define a formal group law over $k[\![u_1,\dots,u_{n-1}]\!]/(u_1,\dots,u_{n-1})^2$ by

$$x \underset{F}{+} y = x \underset{\Gamma}{+} y \underset{\Gamma}{+} u_1 G_p(x,y) \underset{\Gamma}{+} \cdots \underset{\Gamma}{+} u_{n-1} G_{p^{n-1}}(x,y),$$

where $G_{p^j}(x,y) \in Z^2_{\mathrm{sym}}(\Gamma)$ are the symmetric cocycles we constructed above. This group law then lifts up to $\mathbb{W}k[\![u_1,\dots,u_{n-1}]\!]$, by Lazard's Theorem 5.5.

The proof is completed by considering the map

$$\mathrm{Hom}_{\mathbb{W}k\text{-}\mathrm{alg}}(E(k,\Gamma),B) \to \pi_0 \operatorname{Def}_\Gamma(B)_i$$

which sends f to the isomorphism class of f^*F, where F is the above constructed formal group law. The left hand side is isomorphic as a set to $\mathfrak{m}^{\times(n-1)}$, which has a natural filtration by powers of $\mathfrak{m}$. The right hand side also has a filtration via the kernels of the maps $\pi_0 \operatorname{Def}_\Gamma(B)_i \to \pi_0 \operatorname{Def}_\Gamma(B/\mathfrak{m}^k)_i$, and one shows the map is an isomorphism by comparing filtration quotients; of course, if $B = k[\epsilon]/\epsilon^2$, then it is clearly an isomorphism by what we have shown.

EXAMPLE 5.11. Let $n=1$ and $k=\mathbb{F}_p$, and consider the multiplicative formal group law $G_m(x,y) = x+y+xy$ over $\mathbb{F}_p$. Then

$$E(\mathbb{F}_p, G_m) \simeq \mathbb{Z}_p$$

and we can take as the universal deformation $F(x,y) = x+y+xy$.

6. The homology theories $E_{k,\Gamma}$

In this section we use universal deformation theory to construct homology theory $E_{k,\Gamma}$ for each formal group law (k,Γ). In fact, we construct a functor E from the category $\mathcal{FG}$ of formal group laws to the category of homology theories.

6.1. Complex orientations and formal group laws. A reference for material in this section is from [**1**, part 2].

A ring spectrum E is **complex orientable** if there exists a cohomology class $x \in \tilde{E}^2\mathbb{CP}^\infty$ such that the restriction of this class to $\mathbb{CP}^1 \simeq S^2$ is the generator of $\tilde{E}^2 S^2 \simeq \pi_0 E$. Such a class determines a characteristic class $c_1(L) \in E^2(X)$ for a complex line bundle L over X.

For a complex oriented cohomology theory, $E^*\mathbb{CP}^\infty \simeq E^*[\![x]\!]$. Thus the orientation determines a formal group law F, induced by the map $\mathbb{CP}^\infty \times \mathbb{CP}^\infty \to \mathbb{CP}^\infty$ which corresponds to tensor product of line bundles. Thus

$$c_1(L \otimes L') = c_1(L) \underset{F}{+} c_1(L').$$

Note that this formal group law is of the form

for $a_{ij} \in \pi_{2(i+j-1)}E$. I'll call this a **degree -2 formal group law**, since the equation defining $F(x,y)$ is in degree -2. (That is, it's an element of $\pi_{-2}\mathcal{F}(\mathbb{CP}^\infty, E)$, where $\mathcal{F}(X,Y)$ denotes the function spectrum from X to Y.)

For example, the complex cobordism spectrum MU is complex orientable. The two inclusions of $\eta_L, \eta_R\colon MU \rightrightarrows MU \wedge MU$ induce two formal group laws over MU_*MU and a canonical strict isomorphism between them.

THEOREM 6.2 (Quillen). *The canonical formal group law F over MU_* is the universal formal group law (of degree -2) in the sense that, given a formal group law G of degree -2 over a graded ring R_*, there exists a unique map $\phi\colon MU_* \to R_*$ such that $\phi^*F = G$.*

Furthermore, the canonical isomorphism $f\colon \eta_R^*F \to \eta_L^*F$ over MU_*MU is the universal strict isomorphism in the sense that, given a strict isomorphism $g\colon G \to G'$ of formal group laws over R_*, there exists a unique map $\psi\colon MU_*MU \to R_*$ such that $\psi^*f = g$.

Sometimes we want to consider a cohomology theory E which is periodic; e.g., which has an element $u \in \pi_2 E$ which is invertible. Such an element will allow us to switch between degree -2 formal group laws and degree 0 ones (a **degree 0 formal group law** is one whose coefficients lie in degree 0).

Let R_* be a graded ring, such that R_2 contains an invertible element. Given a formal group law F defined over R_0 and a chosen invertible element $u_F \in R_2$, we produce a degree -2 formal group law as follows: define

$$\bar{F}(x,y) = u_F^{-1}F(u_F x, u_F y).$$

Thus if $F(x,y) = \sum a_{ij}x^iy^j$, then $\bar{F}(x,y) = \sum a_{ij}u_F^{i+j-1}x^iy^j$ is a degree -2 formal group law.

Given a morphism $g\colon F \to G$ of formal group laws over R_0, and chosen elements $u_F, u_G \in R_2$, we of course get corresponding formal group laws $\bar{F}$ and $\bar{G}$ of degree -2; let

$$\bar{g}(x) = u_G^{-1}g(u_F x).$$

Then $\bar{g}\colon \bar{F} \to \bar{G}$ is a morphism of degree -2 formal group laws. Note that $\bar{g}'(0) = u_G^{-1}u_F g'(0)$.

6.3. The Landweber exact functor theorem. Given $(k,\Gamma) \in \mathcal{FG}$ we define a functor $E_{k,\Gamma}$ on spaces by

$$(E_{k,\Gamma})_*(X) \simeq E(k,\Gamma)[u^{\pm}] \overset{\bar{F}}{\underset{MU_*}{\otimes}} MU_*(X).$$

The "$\bar{F}$" sitting on top of the tensor product symbol is meant to symbolize the fact that the tensor product is induced by the map $MU_* \to E(k,\Gamma)[u^{\pm}]$ which classifies $\bar{F}$, which is the degree -2 formal group law induced from F using $u_F = u$. That this is a homology theory comes from the Landweber exact functor theorem, as we will show below.

THEOREM 6.4 (Landweber). *Let R be a MU_* module. Suppose for each prime p and each n the map*

$$R/p,\dots,v_{n-1} \xrightarrow{\cdot v_n} R/p,\dots,v_{n-1}$$

is injective, where v_n is the coefficient of x^{p^n} in the p-series of the universal formal group law over MU_*. (That is, $p, v_1, v_2, \ldots$ is a regular sequence for R.) Then the functor

$$X \mapsto R_* \underset{MU_*}{\otimes} MU_*(X)$$

is a homology theory.

PROOF. Proved in [**8**]. □

PROPOSITION 6.5. *The map $\bar{F}\colon MU_* \to E(k,\Gamma)[u^{\pm}]$ which classifies a degree -2 universal deformation satisfies the hypotheses of the Landweber exact functor theorem.*

We will return to the proof of Proposition 6.5 in section 6.7.

REMARK 6.6.

1. If R_* is a ring with a degree -2 formal group law $\bar{F}$, we use the classifying map $MU_* \to R_*$ to produce the module structure on R_*.
2. It turns out that if we have two degree -2 formal group laws $\bar{F}$ and $\bar{G}$ over R_* which satisfy the Landweber condition, and are strictly isomorphic by a strict isomorphism $\bar{g}$, there is an induced equivalence of homology theories, via a map

$$R_* \overset{\bar{F}}{\underset{MU_*}{\otimes}} MU_*(X) \to R_* \overset{\bar{F}}{\underset{MU_*}{\otimes}} MU_*MU \underset{MU_*}{\otimes} MU_*(X) \to R_* \overset{\bar{G}}{\underset{MU_*}{\otimes}} MU_*(X)$$

where the left hand map is induced by the comultiplication

$$MU_* \to MU_*MU \underset{MU_*}{\otimes} MU_*X,$$

and the right hand map is induced by the map $MU_*MU \to R_*$ which classifies $\bar{g}$.

6.7. Construction of the functor. We define a functor

$$E\colon \mathcal{FG}^{\mathrm{op}} \to \{\text{homology theories}\},$$

as follows. For each object (k,Γ) in $\mathcal{FG}$, choose a universal deformation F of (k,Γ) to the ring $E(k,\Gamma)$, and let $E_{k,\Gamma}$ be the homology theory constructed above using the Landweber theorem. Let $E(k,\Gamma)_* = E(k,\Gamma)[u_F^{\pm}]$ denote the corresponding graded ring

Suppose we have a map $\alpha\colon (k_1,\Gamma_1) \to (k_2,\Gamma_2)$ in $\mathcal{FG}$; such an α consists of maps $i\colon k_2 \to k_1$ and $f\colon \Gamma_1 \to i^*\Gamma_2$. Given universal deformations F_1 of (k_1,Γ_1) and F_2 of (k_2,Γ_2), we get an induced map

$$\psi\colon E(k_2,\Gamma_2) \to E(k_1,\Gamma_1)$$

and an isomorphism

$$g\colon F_2 \to \psi^* F_1$$

such that $g \equiv f \mod \mathfrak{m}$. That is, (ψ, g) is the unique pair of maps fitting into the diagrams

$$E(k_2,\Gamma_2) \xrightarrow{\psi} E(k_1,\Gamma_1) \qquad F_2 \rightsquigarrow \psi^* F_2 \xleftarrow[g]{\sim} F_1$$

We extend ψ to a map $E(k_2,\Gamma_2)[u_{\bar{F}_2}^{\pm}] \to E(k_1,\Gamma_1)[u_{\bar{F}_1}^{\pm}]$ by sending

$$\psi(u_{F_2}) = g'(0)u_{F_1}.$$

The point here is that we have a diagram

$$\begin{array}{ccc} & & E(k_2,\Gamma_2)_* \\ & \overset{\bar{F}_2}{\nearrow} & \\ MU_* & & \downarrow \psi \\ & \underset{\bar{F}_1}{\searrow} & \\ & & E(k_1,\Gamma_1)_* \end{array}$$

which does *not* commute, since $\psi^*\bar{F}_2 \neq \bar{F}_1$. Instead, we may define $E_\alpha\colon E_{k_2,\Gamma_2} \to E_{k_1,\Gamma_1}$ to be the natural isomorphism of homology theories defined via the composite

$$E(k_2,\Gamma_2)_* \underset{MU_*}{\overset{\bar{F}_2}{\otimes}} MU_*(X) \to E(k_1,\Gamma_1)_* \underset{MU_*}{\overset{\psi^*\bar{F}_2}{\otimes}} MU_*(X) \xleftarrow{\sim} E(k_1,\Gamma_1)_* \underset{MU_*}{\overset{\bar{F}_1}{\otimes}} MU_*(X)$$

where the left hand arrow is induced by ψ, and the right hand arrow is induced by the strict isomorphism $\bar{g}\colon \bar{F}_1 \to \psi^*\bar{F}_2$ defined by

$$\bar{g}(x) = (\psi u_{F_2})^{-1} g(u_{F_1}x) = x + \cdots.$$

It is not too hard to show that E so defined is really a functor.

EXAMPLE 6.8. Let $(k,\Gamma) = (\mathbb{F}_p, G_m)$. Then $\operatorname{Aut}(k,\Gamma) \simeq \mathbb{Z}_p^\times$, where $n \in \mathbb{Z}_p$ corresponds to the map

$$[n]_\Gamma(x) = (1+x)^n - 1.$$

Since this has height one, $E(k,\Gamma) \simeq \mathbb{Z}_p$. We can use the multiplicative formal group law as its own universal lift:

$$x \underset{F}{+} y = x + y + xy.$$

This induces a degree -2 formal group law on $\mathbb{Z}_p[u^\pm]$ given by

$$x \underset{\bar{F}}{+} y = x + y + uxy.$$

The morphism $[n]_\Gamma(x)$ therefore lifts to an automorphism $[n]_F(x) = (1+x)^n - 1 = nx + \cdots$ of F, and hence induces an automorphism $[n]_{\bar{F}}(x) = u^{-1}\big((1+ux)^n - 1\big)$ of $\bar{F}$. We thus see that an automorphism $[n]_\Gamma$ of Γ lifts to a map $\psi_n\colon \mathbb{Z}_p[u^\pm] \to \mathbb{Z}_p[u^\pm]$ which sends $\psi_n(u) = nu$.

The theory $E_{\mathbb{F}_p,G_m}$ is precisely complex K-theory completed at p, and the ψ_n's are precisely the Adams operations.

6.9. Proof of Landweber exactness. It remains to show that the map $MU_* \to E(k,\Gamma)[u^\pm]$ is Landweber exact. It clearly suffices to show that that the sequence $p, w_1, w_2, \ldots$ is regular in $E(k,\Gamma)$, where

$$[p]_F(x) = px + \cdots + w_1x^p + \cdots + w_2x^{p^2} + \cdots.$$

(In fact, $w_i = v_iu^{1-p^i}$.)

The cokernel of multiplication by p is $k[\![u_1,\dots,u_{n-1}]\!]$. We need to compute the p-series of F over this ring. By Lemma 5.6 we can assume without loss of generality that $x \underset{\Gamma}{+} y = x + y + bC_{p^n}(x,y) + \cdots$, whence

$$x \underset{F}{+} y \equiv x + y + u_1 G_p(x,y) + \cdots + u_{n-1}G_{p^{n-1}}(x,y) + bC_{p^n}(x,y) + \cdots$$

modulo p and $(u_1,\dots,u_{n-1})^2$ for certain Γ-symmetric 2-cocycles G_i. Recall that

$$G_{p^i}(x,y) = C_{p^i}(x,y) + \text{higher terms}.$$

LEMMA 6.10. *We have that*

$$[p]_F(x) = u_i x^{p^i} + \cdots$$

modulo $p, u_1, \dots, u_{i-1}$ *and* $(u_i,\dots,u_{n-1})^2$, *and*

$$[p]_F(x) = bx^{p^n} + \cdots$$

modulo $p, u_1, \dots, u_{n-1}$.

PROOF. This is an immediate consequence of Lemma 5.7. □

Lemma 6.10 implies that the inclusion

$$k[\![w_1,\dots,w_{n-1}]\!] \to k[\![u_1,\dots,u_{n-1}]\!]$$

is in fact an isomorphism (since the Jacobian matrix $(\frac{\partial w_i}{\partial u_j})$ modulo $(u_1,\dots,u_{n-1})^2$ is upper triangular with ones on the diagonal by Lemma 6.10.) Clearly, the sequence $w_1, w_2, \dots$ is regular in this ring.

7. The construction

In the next few sections I will discuss the category of spectra (after Lewis-May-Steinberger), and categories of $\mathcal{A}_\infty$-ring spectra, denoted $\mathcal{A}_\infty$. The first thing to note is that $\mathcal{A}_\infty$ is a topological category; i.e., for any two objects, there is a topological space of maps. (See Section 12.)

Let $\mathcal{A}_\infty^{\mathrm{LT}} \subset \mathcal{A}_\infty$ denote the full topological subcategory consisting of $\mathcal{A}_\infty$-ring spectra E such that

1. E is cofibrant as an $\mathcal{A}_\infty$ ring, (see Section 12),
2. the underlying spectrum of E is a homotopy commutative ring spectrum, which
3. has $\pi_{\mathrm{odd}} E = 0$,
4. has an invertible element in $\pi_2 E$, and such that
5. $\pi_0 E$ is a complete local ring, with maximal ideal $\mathfrak{m}$, such that
6. $\pi_0 E/\mathfrak{m}$ is a perfect field of non-zero characteristic, for which
7. the degree 0 formal group law F over $\pi_0 E$, (which exists because E is complex orientable by properties 2 and 3, see [**1**, part 2]), is a universal deformation of its reduction modulo $\mathfrak{m}$.

Thus there is a functor

$$\mathcal{A}_\infty \supset \mathcal{A}_\infty^{\mathrm{LT}} \xrightarrow{\pi} \mathcal{FG}^{\mathrm{op}},$$

sending $E \mapsto (\pi_0 E/\mathfrak{m}, F)$.

THEOREM 7.1. *The functor π is a weak equivalence of topological categories. That is, π is surjective on isomorphism classes of $\mathcal{FG}$, and for each $E, F \in \mathcal{A}_\infty^{\mathrm{LT}}$ induces a weak equivalence*

$$\mathrm{map}_{\mathcal{A}_\infty}(E, F) \to \mathcal{FG}(\pi F, \pi E)$$

where the set on the right hand side is viewed as a space with the discrete topology.

PROOF. The theorem will follow from Corollary 14.6 and Corollary 22.7. □

Here's how you get Theorem 2.1 from Theorem 7.1. First, choose a small model for the category of formal group laws $\mathcal{FG}$; for example, we could fix an algebraically closed field K of characteristic p and require that $\mathcal{FG}$ only contain those objects (k, Γ) for which $k \subset K$. Likewise, choose a small model for $\mathcal{A}_\infty^{\mathrm{LT}}$. Now we can perform a "homotopy right Kan extension" of the topological functor $i\colon \mathcal{A}_\infty^{\mathrm{LT}} \to \mathcal{A}_\infty$ along the functor $\pi\colon \mathcal{A}_\infty^{\mathrm{LT}} \to \mathcal{FG}^{\mathrm{op}}$, obtaining a functor $\rho\colon \mathcal{FG}^{\mathrm{op}} \to \mathcal{A}_\infty$. (Homotopy right Kan extensions are a generalization of the notion of a homotopy inverse limit.) By some general nonsense about homotopy Kan extensions one discovers that ρ has the desired properties; in particular, if $(k, \Gamma) = \pi(E)$, then $\rho(k, \Gamma)$ is weakly equivalent to E.

Part 2. Spectra and $\mathcal{A}_\infty$-ring spectra

In the following sections I collect the results we need about spectra and $\mathcal{A}_\infty$-ring spectra. I only summarize the facts needed; one should consult the original sources for details and proofs of this material. The category of spectra I use is that of Lewis, May, and Steinberger [**9**]; in addition to their book, one should also consult the work of Elmendorf, Kriz, Mandell, and May [**6**], which refines and extends the constructions of [**9**]. (Note, however, that we make no use the theory of "S-modules" which is the main objective of [**6**].) In addition, I make use results of Hopkins, McClure, and Goerss on the closed model category structure for various categories of spectra; some of this work has not seen print.

I then (in Section 14) set up a spectral sequence and obstruction theory for computing maps between $\mathcal{A}_\infty$-ring spectra in the cases we need to consider.

8. Spectra

I give a sketch of the theory of spectra of Lewis, May, and Steinberger [**9**]; see also [**6**]. I will omit many details.

We work in the category of compactly generated weak Hausdorff spaces, denoted Top. The category of *pointed* compactly generated weak Hausdorff spaces is denoted $\mathcal{T}$. This means that

$$\mathrm{map}_*(X \wedge Y, Z) \simeq \mathrm{map}_*(X, \mathrm{map}_*(Y, Z)),$$

where map_* denotes the space of pointed maps with an appropriate topology.

A **universe** $\mathcal{U}$ is a countably dimensional real inner-product space. For example, $\mathbb{R}^\infty$ is a universe. An **indexing set** $\mathcal{A}$ in $\mathcal{U}$ is a collection of finite dimensional vector subspaces of $\mathcal{U}$ such that any finite dimensional subspace of $\mathcal{U}$ is contained in some element of $\mathcal{A}$. Thus, $\mathcal{A} = \{\mathbb{R}^0, \mathbb{R}^1, \mathbb{R}^2, \dots\}$ is a indexing set in $\mathbb{R}^\infty$. By abuse of notation, we let $\mathcal{U}$ denote the indexing set consisting of all finite dimensional subspaces of $\mathcal{U}$.

Let S^V denote the one-point compactification of the vector space V, and let $W - V$ denote the orthogonal complement of V in W. A **prespectrum on $\mathcal{A}$** is a collection of spaces X_V for each $V \in \mathcal{A}$ and maps

$$i_{V,W} \colon X_V \wedge S^{W-V} \to X_W$$

for each pair $V, W \in \mathcal{A}$ with $V \subset W$, such that for each $V_1 \subset V_2 \subset V_3$ the diagram

$$\begin{array}{ccc} S^{V_3-V_2} \wedge S^{V_2-V_1} \wedge X_{V_1} & \overset{\sim}{=\!=\!=} & S^{V_3-V_1} \wedge X_{V_1} \\ \downarrow & & \downarrow \\ S^{V_3} \wedge X_{V_2} & \longrightarrow & X_{V_3} \end{array}$$

commutes.

A map of prespectra $f \colon X \to Y$ is a collection of maps $f_V \colon X_V \to Y_V$ commuting with all the structure. We denote the category of prespectra by $\mathcal{PA}$.

Note that each structure map $i_{V,W}$ is adjoint to a map

$$\hat{\imath}_{V,W} \colon X_V \to \Omega^{W-V} X_W,$$

where $\Omega^V X = \mathrm{map}_*(S^V, X)$. A **spectrum on $\mathcal{A}$** is a prespectrum X on $\mathcal{A}$ for which each map $\hat{\imath}_{V,W}$ is a homeomorphism. The category of spectra $\mathcal{SA}$ is the full subcategory of $\mathcal{PA}$ consisting of the spectra.

Given two prespectra X and Y, we can topologize the pointed set of maps $\mathcal{PA}(X,Y)$ as a subspace of $\prod_{V \in \mathcal{A}} \mathrm{map}_*(X_V, Y_V)$. It is not hard to see that the composition maps $\mathcal{PA}(X,Y) \wedge \mathcal{PA}(Y,Z) \to \mathcal{PA}(X,Z)$ are continuous. Thus $\mathcal{PA}$ is a topological category. Likewise we set $\mathcal{SA}(X,Y) = \mathcal{PA}(X,Y)$, making $\mathcal{SA}$ into a topological category.

First we note that these indexing sets are not too important.

PROPOSITION 8.1. *If $\mathcal{A}$ is an indexing set of $\mathcal{U}$ then the functor $\mathcal{SU} \to \mathcal{SA}$ induced by restriction is an equivalence of categories.*

PROOF. Given $X \in \mathcal{SA}$, we can produce a spectrum $X' \in \mathcal{SU}$ by setting $X'_V = \Omega^{W-V} X_W$ where $W \in \mathcal{A}$ is some subspace of $\mathcal{U}$ containing V. It straightforward to check that this is well-defined, and produces the equivalence of categories. □

The following proposition says that any prespectrum can be turned into a spectrum. It is crucial and non-trivial.

PROPOSITION 8.2. [**9**, Appendix] *There is a pair of adjoint functors*

$$\ell \colon \mathcal{PA} \leftrightarrows \mathcal{SA} : r$$

where r is inclusion of subcategories. Furthermore, $\ell r \simeq 1$, and the functors ℓ and r are continuous (i.e., induce continuous maps on function spaces).

As a consequence, it follows that spectra have all small limits and colimits. Prespectra have limits and colimits which are computed space-wise. Limits of spectra can also be computed space-wise, since Ω commutes with limits of spaces. Colimits of spectra are computed by $\mathrm{colim}_I X \simeq \ell X'$ where X' is the prespectrum defined by $X'_V = \mathrm{colim}_I X_V$.

8.3. Constructions of spectra. Given a space X, define a prespectrum $\Sigma^\infty_{\text{pre}}X$ by

$$(\Sigma^\infty_{\text{pre}}X)_V = S^V \wedge X,$$

with the obvious structure maps, and define the **suspension spectrum** $\Sigma^\infty X$ by $\Sigma^\infty X = \ell(\Sigma^\infty_{\text{pre}}X)$. This has the property that

$$\mathcal{SU}(\Sigma^\infty X, Y) \simeq \text{map}_*(X, Y_0).$$

For each finite $V \subset \mathcal{U}$ we define a prespectrum S^{-V}_{pre} by

$$(S^{-V}_{\text{pre}})_W = \begin{cases} S^{W-V} & \text{if } V \subset W, \\ * & \text{otherwise,} \end{cases}$$

with the obvious structure maps. Define a **sphere spectrum** S^{-V} by $S^{-V} = \ell(S^{-V}_{\text{pre}})$. Such an object has the property that

$$\mathcal{SU}(S^{-V}, Y) \simeq Y_V.$$

For an *unpointed* space K and a spectrum X over $\mathcal{U}$ we define spectra $K \otimes X$ and X^K as follows. Let X^K be defined by

$$X^K = \text{map}(K, X_V).$$

It is not hard to check that X^K is a spectrum. Let $(K \otimes X)^{\text{pre}}$ be the prespectrum defined by

$$(K \otimes X)^{\text{pre}}_V = K_+ \wedge X_V,$$

and let $K \otimes X = \ell(K \otimes X)^{\text{pre}}$ denote the corresponding spectrum. We have the following fact.

PROPOSITION 8.4. [**9**, I, Prop. 3.3] *Given $K \in$ Top and $X, Y \in \mathcal{SU}$ there are natural isomorphisms*

$$\mathcal{SU}(K \otimes X, Y) \simeq \text{map}(K, \mathcal{SU}(X, Y)) \simeq \mathcal{SU}(X, Y^K).$$

This means we can define things like homotopies between maps. Given maps $f, g\colon X \to Y$ of spectra in $\mathcal{SU}$ a homotopy can be taken to be a map

$$H\colon I \to \mathcal{SU}(X, Y)$$

such that $H(0) = f$ and $H(0) = g$, or equivalently as a map

$$H'\colon I \otimes X \to Y$$

or a map

$$H''\colon X \to Y^I.$$

We can think of the object $I \otimes X$ as a cylinder on X, and Y^I as a spectrum of paths on Y. We can use these kinds of constructions to produce mapping-cylinders, path fibrations, and other nifty constructions.

There are constructions analogous to $K \otimes X$ and X^K which we can carry out when K is a *pointed* space. I only give one example: for a pointed space K and spectrum X let $(K \wedge X)^{\text{pre}}$ be the prespectrum defined by

$$(K \wedge X)^{\text{pre}}_V = K \wedge X_V,$$

and let $K \wedge X = \ell(K \wedge X)^{\text{pre}}$. Then we have

PROPOSITION 8.5. *Given $K \in \mathcal{T}$ and $X, Y \in \mathcal{SU}$ there is a natural isomorphism*

$$\mathcal{SU}(K \wedge X, Y) \simeq \text{map}_*(K, \mathcal{SU}(X, Y)).$$

8.6. Model category structure. We say a map $f\colon X \to Y$ in $\mathcal{SU}$ is a **weak equivalence** if for each $V \subset \mathcal{U}$ the induced map $f_V\colon X_V \to Y_V$ is a weak equivalence of pointed spaces. We can define **homotopy groups** of a spectrum X as follows: for each $n \in \mathbb{Z}$, choose a vector space $V \subset \mathcal{U}$ such that $n+\dim(V) = m \geq 0$, and write

$$\pi_n(X) = \pi_m(X_V).$$

One can show that this doesn't depend on the choice of V. Thus, a weak equivalence of spectra is just a map which induces isomorphisms on stable homotopy groups.

PROPOSITION 8.7. [**6**, Thm. 4.4] *The category of spectra $\mathcal{SU}$ admits the structure of a Quillen closed model category, such that a map $f\colon X \to Y$ is a*

1. *weak equivalence if each map $f_V\colon X_V \to Y_V$ is a weak equivalence of spaces,*
2. *fibration if each map $f_V\colon X_V \to Y_V$ is a Serre fibration,*
3. *cofibration if f is a retract of a map f' which is obtained by successively gluing on "cells" of the form*

$$\partial\Delta_n \otimes S^{-V} \to \Delta_n \otimes S^{-V}$$

where $n \geq 0$, $V \subset \mathcal{U}$, and Δ_n denotes the Euclidean n-simplex.

Furthermore, this structure gives $\mathcal{SU}$ the structure of a topological model category.

I won't say what all this means. Quillen closed model categories are defined in [**11**]. The existence of such a structure implies, among other things, that we can form a homotopy category $\operatorname{Ho}\mathcal{SU}$ by formally inverting the weak equivalences in $\mathcal{SU}$.

The statement that $\mathcal{SU}$ is a *topological* model category means that for a cofibration $i\colon X \to Y$ and a fibration $p\colon E \to F$ of spectra, the induced map of spaces

$$f\colon \mathcal{SU}(Y,E) \to \mathcal{SU}(X,E) \underset{\mathcal{SU}(X,F)}{\times} \mathcal{SU}(Y,F)$$

is a Serre fibration, and that if either i or p is also a weak equivalence, then so is f. (This is analogous to Quillen's notion of a **simplicial model category** [**11**].)

9. Smash products

In this section I develop the "external" smash product and related constructions.

9.1. Smash product. Given universes $\mathcal{U}$ and $\mathcal{U}'$, then the product space $\mathcal{U} \times \mathcal{U}'$ is another universe. Let

$$\mathcal{A} = \{V \times W \subset \mathcal{U} \times \mathcal{U}' \mid V \subset \mathcal{U}, V' \subset \mathcal{U}'\}.$$

This is an indexing set of $\mathcal{U} \times \mathcal{U}'$. Given $X \in \mathcal{SU}$ and $Y \in \mathcal{U}'$ define

$$(X \wedge Y)^{\mathrm{pre}}_{V\times W} = X_V \wedge Y_W;$$

this is a prespectrum over $\mathcal{A}$, where the structure maps are

$$S^{V'\times W' - V\times W} \wedge X_V \wedge Y_W \simeq S^{V'-V} \wedge S^{W'-W} \wedge X_V \wedge Y_W \to X_{V'} \wedge Y_{W'}.$$

Then let

$$X \wedge Y = \ell(X \wedge Y)^{\mathrm{pre}} \in \mathcal{SA} \simeq \mathcal{S}(\mathcal{U} \times \mathcal{U}').$$

Thus we obtain a smash product functor

PROPOSITION 9.2. *For $V \subset \mathcal{U}$ and $W \subset \mathcal{U}'$ there is a natural isomorphism*

$$S^{-V} \wedge S^{-W} \simeq S^{-V \times W}$$

of spectra over $\mathcal{U} \times \mathcal{U}'$.

PROOF. We compute on the level of prespectra, and then spectrify. If $V \subset V' \subset \mathcal{U}$ and $W \subset W' \subset U'$ we get

$$(S^{-V} \wedge S^{-W})^{\text{pre}}_{V' \times W'} \simeq S^{V'-V} \wedge S^{W'-W} \simeq S^{V' \times W' - V \times W} \simeq (S^{-V \times W})^{\text{pre}}_{V' \times W'}.$$

Hence the corresponding spectra are isomorphic. □

9.3. Function spectra. Let's fix a $Y \in \mathcal{SU}'$. Then the functor $- \wedge Y \colon \mathcal{SU} \to \mathcal{S}(\mathcal{U} \times \mathcal{U}')$ has a right adjoint, namely the function spectrum

$$F(Y, -) \colon \mathcal{S}(\mathcal{U} \times \mathcal{U}') \to \mathcal{SU}.$$

By this adjoint functor characterization, we see that for $V \subset \mathcal{U}$ we must have

$$F(Y, Z)_V \simeq \mathcal{SU}(S^{-V}, F(Y, Z)) \simeq \mathcal{S}(\mathcal{U} \times \mathcal{U}')(S^{-V} \wedge Y, Z),$$

and since we have already defined the smash product, we can use this observation to define $F(Y, Z)$ by the above formula. One has that for $V \subset W \subset \mathcal{U}$,

$$S^{W-V} \wedge S^{-W} \simeq S^{-V}$$

as one can see by checking this formula on the level of prespectra:

$$(S^{W-V} \wedge S^{-W})^{\text{pre}}_{W'} \simeq S^{W-V} \wedge S^{W'-W} \simeq S^{W'-V} \simeq (S^{-V}_{\text{pre}})_{W'}.$$

Thus the structure maps of $F(Y, Z)$ are defined to be the natural isomorphisms

$$\begin{aligned} F(Y, Z)_V &= \mathcal{S}(\mathcal{U} \times \mathcal{U}')(S^{-V} \wedge Y, Z) \\ &\simeq \mathcal{S}(\mathcal{U} \times \mathcal{U}')(S^{W-V} \wedge S^{-W} \wedge Y, Z) \\ &\simeq \Omega^{W-V} \mathcal{S}(\mathcal{U} \times \mathcal{U}')(S^{-W} \wedge Y, Z) \\ &= \Omega^{W-V} F(Y, Z)_W \end{aligned}$$

and we see that $F(Y, Z)$ is indeed a spectrum.

One can show that if K is an unpointed space and X a spectrum, then $X^K \simeq F(\Sigma^\infty K_+, X)$.

9.4. Associativity. The significant property of this category of spectra which we need is associativity of the smash product. If $X \in \mathcal{SU}_1$, $Y \in \mathcal{SU}_2$ and $Z \in \mathcal{SU}_3$, then there is a natural isomorphism

$$(X \wedge Y) \wedge Z \simeq X \wedge (Y \wedge Z)$$

of spectra over $\mathcal{U}_1 \times \mathcal{U}_2 \times \mathcal{U}_3$. In fact, both sides of the above equation are naturally isomorphic to the spectrification of the prespectrum defined by

$$V_1 \times V_2 \times V_3 \mapsto X_{V_1} \wedge Y_{V_2} \wedge Z_{V_3}.$$

10. Change of universe functors

I will talk about several different constructions which allow you to compare spectra which are defined over different universes. Some of the material of this section is in [**9**, Ch. II] and [**6**, App. A].

10.1. Push-forward and pull-back. Given a linear isometric embedding of universes $f\colon \mathcal{U} \to \mathcal{U}'$, that is, a vector space map which preserves the inner product, we produce functors

$$f_*\colon \mathcal{SU} \rightleftarrows \mathcal{SU}' : f^*$$

called **push-forward** and **pull-back**. Pull-back is easy; for $V \subset \mathcal{U}$ and $Y \in \mathcal{SU}'$ let

$$(f^*Y)_V = Y_{fV},$$

where $fV \subset \mathcal{U}'$ denotes the image of V under f. The natural homeomorphism $S^V \simeq S^{fV}$ induced by f produces isomorphisms

$$\Omega^{W-V}(f^*Y)_V \simeq \mathrm{map}_*(S^{W-V}, Y_{fV}) \simeq \mathrm{map}_*(S^{fW-fV}, Y_{fV}) \simeq Y_{fW} = (f^*Y)_W.$$

Thus f^*Y is a spectrum over $\mathcal{U}$.

Push-forward is a little more complicated. We define a pre-spectrum $(f_*X)^{\mathrm{pre}}$ as follows. For $W \subset \mathcal{U}'$ let

$$(f_*X)^{\mathrm{pre}}_W = S^{W-fV} \wedge X_V$$

where $V = f^*(fV \cap W)$. In other words, V is the largest subspace of $\mathcal{U}$ which maps into W. If $W \subset W'$ let $V' = f^*(fV' \cap W')$, whence we get structure maps

$$S^{W'-W} \wedge (f_*X)^{\mathrm{pre}}_W \simeq S^{W'-W} \wedge S^{W-fV} \wedge X_V \simeq S^{W'-fV'} \wedge S^{V'-V} \wedge X_V$$
$$\longrightarrow S^{W'-V'} \wedge X_V = (f_*X)^{\mathrm{pre}}_{W'}.$$

Of course we let $f_*X = \ell(f_*X)^{\mathrm{pre}} \in \mathcal{SU}'$.

We compute that $f_*S^{-V} \simeq S^{-fV}$, by a sequence of adjunctions

$$\mathcal{SU}'(f_*S^{-V}, Y) \simeq \mathcal{SU}(S^{-V}, f^*Y) \simeq (f^*Y)_V = Y_{fV} \simeq \mathcal{SU}' * S^{-fV}, Y).$$

Note that if the linear isometry $f\colon \mathcal{U} \to \mathcal{U}'$ is an *isomorphism* of universes, then the definition of f_*X simplifies to $(f_*X)_W = X_{f^*W}$, and so f_* and f^* form an adjoint equivalence of categories.

10.2. Many smash products. We can now use the push-forward functors to define many smash product functors from the category of spectra over $\mathcal{U}$ to itself. Let $\mathcal{U}^n = \mathcal{U} \times \cdots \times \mathcal{U}$. Then for $X_1, \ldots, X_n \in \mathcal{SU}$ and for some linear isometric map $f\colon \mathcal{U}^n \to \mathcal{U}$ we define a functor $\mathcal{SU}^n \to \mathcal{SU}$ by

$$X_1, \ldots, X_n \mapsto f_*(X_1 \wedge \cdots \wedge X_n).$$

Note that if $i\colon \mathcal{U}_1 \to \mathcal{U}_2$ and $j\colon \mathcal{U}_2 \to \mathcal{U}_3$ then $(j \circ i)_*X \simeq j_*i_*X$. Thus, given $f\colon \mathcal{U}^2 \to \mathcal{U}$ and $g\colon \mathcal{U}^2 \to \mathcal{U}$ we can compose smash product functors, so that

$$f_*(g_*(X \wedge Y) \wedge Z) \simeq (f \circ (g \times 1))_*(X \wedge Y \wedge Z)$$

where $f \circ (g \times 1))\colon \mathcal{U}^3 \to \mathcal{U}$ denotes the composite map on universes.

10.3. "Twisted" constructions involving the linear isometries space. Let $\mathcal{L}(\mathcal{U}, \mathcal{U}')$ denote the set of all linear isometries from $\mathcal{U}$ to $\mathcal{U}'$. Topologize this space by the compact-open topology. Then one has the following result.

PROPOSITION 10.4. [**9**, II, Lemma 1.5] *The space $\mathcal{L}(\mathcal{U}, \mathcal{U}')$ is contractible.*

For us, this means that we have a space $\mathcal{L}(\mathcal{U}^n,\mathcal{U})$ which parameterizes all possible ways of smashing n things together, and this parameter space is contractible. One would like to somehow glue "all" these smash-products together and get a super-universal smash product functor from $\mathcal{SU}$ to itself.

For the time being let's write $\mathcal{L}$ for $\mathcal{L}(\mathcal{U},\mathcal{U}')$. For maps of topological spaces $T \to \mathcal{L}$ and $T' \to \mathcal{L}$ let $\mathrm{map}_{\mathcal{L}}(T,T') \subset \mathrm{map}(T,T')$ denote the space of maps $T \to T'$ which commute with the projections to $\mathcal{L}$. Let Top denote the category of unpointed spaces, and let $(\mathrm{Top} \downarrow \mathcal{L})$ denote the category of spaces over $\mathcal{L}$.

PROPOSITION 10.5. *Given $t\colon T \to \mathcal{L} \in (\mathrm{Top} \downarrow \mathcal{L})$, $X \in \mathcal{SU}$ and $Y \in \mathcal{SU}'$ there exist functorial constructions*

- $T \rtimes X \in \mathcal{SU}'$,
- $Y^T \in \mathcal{SU}$,
- $\mathcal{S}(X,Y) \to \mathcal{L} \in (\mathrm{Top} \downarrow \mathcal{L})$,

such that there are natural isomorphisms

$$\mathcal{SU}'(T \rtimes X, Y) \simeq \mathrm{map}_{\mathcal{L}}\big(T, \mathcal{S}(X,Y)\big) \simeq \mathcal{SU}(X, Y^T),$$

and such that if $T = \{f\} \to \mathcal{L}$ is a single point, then

- $\{f\} \rtimes X \simeq f_* X$,
- $Y^{\{f\}} \simeq f^* Y$, *and*
- $\mathrm{map}_{\mathcal{L}}\big(\{f\}, \mathcal{S}(X,Y)\big) \simeq \mathcal{SU}'(f_*X, Y) \simeq \mathcal{SU}(X, f^*Y)$, *or in other words, the fiber of the map $\mathcal{S}(X,Y) \to \mathcal{L}(\mathcal{U},\mathcal{V})$ over the point f is homeomorphic to the above spaces.*

The construction $T \rtimes X$ is known as the **twisted half-smash product**.

First I'll outline a construction of $T \rtimes X$. First, suppose T is a compact space. For $V \subset \mathcal{U}$ let $T(V) \subset \mathcal{U}'$ denote the union of all $fV \subset \mathcal{U}'$ as f ranges over all elements in the image of $\xi\colon T \to \mathcal{L}$. Since T is compact there exists a finite dimensional vector subspace $W \subset \mathcal{U}'$ such that $T(V) \subset W$. Conversely, for each $W \subset \mathcal{U}'$ there is a largest $V \subset \mathcal{U}$ such that $T(V) \subset W$.

Given such V and W with $T(V) \subset W$, let $\xi(V) \to T$ denote the vector subbundle of the trivial bundle $W \times T \to T$ such that over $t \in T$ the fiber is $\xi(t)(V) \subset W$. Let $W - \xi(V)$ denote the orthogonal complement bundle of $\xi(V)$ in $W \times T$, and let $T^{W-\xi(V)}$ denote the Thom space. Define

$$(T \rtimes X)^{\mathrm{pre}}_W = T^{W-\xi(V)} \wedge X_V.$$

Suppose $W \subset W' \subset \mathcal{U}'$. Then there are corresponding spaces $V \subset V' \subset \mathcal{U}$ and sub-bundles $\xi(V) \subset \xi(V')$ of $W' \times T$. Then

$$(W' - W) \times (W - \xi(V)) \simeq W' - \xi(V) \simeq (W' - \xi(V')) \times (V' - V)$$

since $\xi(V') - \xi(V)$ can be identified with a trivial bundle with fiber $V' - V$. Thus structure maps are defined by

$$\begin{aligned} S^{W'-W} \wedge (T \rtimes X)^{\mathrm{pre}}_W &\simeq S^{W'-W} \wedge T^{W-\xi(V)} \wedge X_V \\ &\simeq T^{W'-\xi(V')} \wedge S^{V'-V} \wedge X_V \\ &\to T^{W'-\xi(V')} \wedge X_{V'} = (T \rtimes X)^{\mathrm{pre}}_{W'}. \end{aligned}$$

Of course, we now let $T \rtimes X = \ell(T \rtimes X)^{\mathrm{pre}}$. Note that if $T = \{f\} \in \mathcal{L}$ then we have just recovered the construction of $f_* X$.

If T is not compact, then we let

$$T \rtimes X = \operatorname{colim}_{K \subset T} K \rtimes X$$

as K ranges over all the compact subspaces of T.

For $\mathcal{S}(X, Y)$, we let this be equal *as a set* to

$$\mathcal{S}(X,Y) = \coprod_{f \in \mathcal{L}} \mathcal{SU}(X, f^*Y).$$

This has a natural map $\pi \colon \mathcal{S}(X,Y) \to \mathcal{L}$. We topologize it by giving it the largest topology such that a map $f \colon T \to \mathcal{S}(X,Y)$ is continuous map if and only if

1. the composite $\pi f \colon T \to \mathcal{L}$ is continuous, and
2. the corresponding map $T \rtimes X \to Y$ is a map of spectra.

This last is a construction of Elmendorf [**5**].

10.6. Applications of the twisted constructions. We need one more set of facts about all these constructions; namely, that they preserve features of the model category structure as best as is possible. Since I haven't been explicit about model category structures I won't give the general statement, but here are some consequences.

1. If $T \to \mathcal{L}$ with T a CW-complex, and $X \in \mathcal{SU}$ a cofibrant spectrum, then $T \rtimes X \in \mathcal{SU}'$ is cofibrant.
2. If $T \xrightarrow{f} T' \to \mathcal{L}$ is such that f is a weak equivalence between CW-complexes, and $X \in \mathcal{SU}$ is a cofibrant spectrum, then $T \rtimes X \to T' \rtimes X$ is a weak equivalence of spectra over $\mathcal{U}'$.
3. If $T \to \mathcal{L}$ with T a CW-complex, and $X \to Y$ a weak equivalence between cofibrant spectra over $\mathcal{U}$, then $T \rtimes X \to T \rtimes Y$ is a weak equivalence of spectra over $\mathcal{U}'$.
4. If $X \in \mathcal{SU}$ is a cofibrant spectrum, and $Y \to Y' \in \mathcal{SU}'$ is a weak equivalence of cofibrant spectra, then $X \wedge Y \to X \wedge Y'$ is a weak equivalence of spectra over $\mathcal{U} \times \mathcal{U}'$.
5. If $X \in \mathcal{SU}$ is a cofibrant spectrum, and $Y \in \mathcal{SU}'$, then $\mathcal{S}(X,Y) \to \mathcal{L}$ is a Serre fibration.

COROLLARY 10.7. *If $f, g \colon \mathcal{U} \to \mathcal{U}'$ are linear isometries, and $X \in \mathcal{SU}$ a cofibrant spectrum, then f_*X and g_*Y are weakly equivalent as spectra over $\mathcal{U}'$, and furthermore this weak equivalence may be chosen canonically up to homotopy.*

PROOF. Chose a path $\gamma \colon I \to \mathcal{L}$ between f and g (recall that $\mathcal{L}$ is contractible). Then we get a diagram

$$f_*X \simeq \{f\} \rtimes X \to I \rtimes X \leftarrow \{g\} \rtimes X$$

where each of the maps is a weak equivalence by (2) above, since the unit interval I is certainly a CW-complex. Similarly, given two homotopies γ and γ', an argument involving homotopies between homotopies shows that the weak equivalence is canonical up to homotopy. □

COROLLARY 10.8. *For any two universes $\mathcal{U}$ and $\mathcal{U}'$, the homotopy categories of spectra* $\operatorname{Ho}\mathcal{SU}$ *and* $\operatorname{Ho}\mathcal{SU}'$ *are* canonically *equivalent.*

11. Operads

I give a couple of definitions of operads.

11.1. Definition of operads. An **operad** in Top is a collection of spaces $\{C[n]\}_{n\geq 0}$ equipped with the following structure:

1. a right action of Σ_n on $C[n]$,
2. a distinguished point $1 \in C[1]$,
3. and for each sequence $I = (i_1, \ldots, i_k)$ of non-negative integers ($k \geq 0$), and $i_1 + \cdots + i_k = n$ a map
$$C[k] \times C[i_1] \times \cdots \times C[i_k] \to C[n].$$
Notationally, this map sends $f \in C[k]$, $g_j \in C[i_j]$ to $f\{g_1, \ldots, g_k\}$.

In addition, the following relations must hold.

1. $1\{g\} = g$ for $g \in C[n]$,
2. $f\{1, \ldots, 1\} = f$ for $f \in C[k]$,
3. for $f \in C[k]$, $g_j \in C[i_j]$, and $h_{j,\ell} \in C[m_{j,\ell}]$,
$$f\{g_1\{h_{1,1}, \ldots, h_{1,m_1}\}, \ldots, g_k\{h_{k,1}, \ldots, h_{k,m_k}\} = (f\{g_1, \ldots, g_k\})\{h_{1,1}, \ldots, h_{k,m_k}\}\},$$
4. $(f\sigma)\{g_1, \ldots, g_k\} = f\{g_{\sigma 1}, \ldots, g_{\sigma k}\}$ for $f \in C[k]$, $g_j \in C[i_j]$, and $\sigma \in \Sigma_k$,
5. $f\{g_1\sigma_1, \ldots, g_k\sigma_k\} = f\{g_1, \ldots, g_k\}\sigma_1 \times \cdots \times \sigma_k$ for $f \in C[k]$, $g_j \in C[i_j]$, and $\sigma_j \in \Sigma_{i_j}$.

A map of operads $C \to D$ is a collection of continuous maps $C[n] \to D[n]$ which commute with all the structure.

11.2. Examples of operads. Given a space $X \in$ Top, define $\mathcal{E}_X$ by
$$\mathcal{E}_X[n] = \mathrm{map}(X^n, X).$$
It is not hard to see that $\mathcal{E}_X$ is an operad, under composition of functions. It is called the **endomorphism operad** of X. Likewise, the spaces $\mathcal{L}[n] = \mathcal{L}(\mathcal{U}^n, \mathcal{U})$ also form an operad under composition, called the **linear isometries operad**.

Another example of an operad is the **associative operad**. Let $A[n]$ denote the set of monomials of length n in n distinct non-commuting formal variables $t_1, \ldots, t_n$. E.g., $A[2] = \{t_1t_2, t_2t_1\}$. Write $f(t_1, \ldots, t_n)$ for a general element of $A[n]$, and write $1 \in A[0]$ for the "empty" monomial, (not to be confused with the distinguished element of $A[1]$, which is denoted t_1). Then A is an operad, under "composition" of formal monomials; that is,
$$f\{g_1, \ldots, g_k\}(t_1, \ldots, t_n)$$
$$= f(g_1(t_1, \ldots, t_{i_1}), g_2(t_{i_1+1}, \ldots, t_{i_1+i_2}), \ldots, g_k(t_{i_1+\cdots+i_{k-1}+1}, \ldots, t_n)).$$
If X is a topological monoid, then you can produce a map $A \to \mathcal{E}_X$ of operads by evaluation, i.e.,
$$f(t_1, \ldots, t_n) \mapsto ((x_1, \ldots, x_n) \mapsto f(x_1, \ldots, x_n)).$$
In fact, giving the structure of a topological monoid on X is precisely equivalent to giving a map of operads from A to $\mathcal{E}_X$.

11.3. Σ-objects. I need a description of operads which shows off some of their other facets, and will be useful in Section 23. This definition works in any symmetric monoidal category. The primary example of interest here is the category of topological spaces, with cartesian product as monoidal product. To keep things simple, I state the definitions in terms of the category Top. But one should keep in mind that things work in a general symmetric monoidal category.

A **Σ-object** C is a collection of spaces $\{C[n]\}_{n\geq 0}$ such that $C[n]$ is equipped with a right action by Σ_n. Thus, any operad is a Σ-object, but not conversely.

Given a Σ-object C, define a functor $F_C\colon \mathrm{Top} \to \mathrm{Top}$ by

$$F_C(X) = \coprod_{n\geq 0} C[n] \underset{\Sigma_n}{\times} X^n.$$

Thus each Σ-object produces a functor; likewise, each map of Σ-objects produces a natural transformation of functors. It is sometimes helpful to think of F_C as a "power series" with coefficients $C[n]$.

Let I be the Σ-object defined by

$$I[n] = \begin{cases} \mathrm{pt} & \text{if } n = 1, \\ \varnothing & \text{otherwise.} \end{cases}$$

Then F_I is the identity functor.

Suppose D is another Σ-object. What is $F_C(X) \times F_D(X)$? We compute

$$\begin{aligned} F_C(X) \times F_D(X) &\simeq \left(\coprod_{m\geq 0} C[m] \underset{\Sigma_m}{\times} X^m\right) \times \left(\coprod_{n\geq 0} D[n] \underset{\Sigma_n}{\times} X^n\right) \\ &\simeq \coprod_{m,n\geq 0} (C[m] \times D[n]) \underset{\Sigma_m\times\Sigma_n}{\times} (X^m \times X^n) \\ &\simeq \coprod_{d\geq 0}\left(\coprod_{m+n=d} C[m] \times D[n] \underset{\Sigma_m\times\Sigma_n}{\times} \Sigma_d\right) \underset{\Sigma_d}{\times} X^d. \end{aligned}$$

Thus, if we define a new Σ-object $C \otimes D$ by

$$(C\otimes D)[d] = \coprod_{m+n=d} C[m] \times D[n] \underset{\Sigma_m\times\Sigma_n}{\times} \Sigma_d,$$

then we see that $F_C(X) \times F_D(X) \simeq F_{C\otimes D}(X)$. Note that this implies that $F_C(X)^{\times n} \simeq F_{C^{\otimes n}}(X)$.

We can produce other formulas of the same type. Thus, if K is a space, define a Σ-object $K \times C$ by $(K\times C)[n] = K \times C[n]$. Then $K \times F_C(X) \simeq F_{K\times C}(X)$. Also, if C_α is a diagram of Σ-objects, then $\operatorname{colim}(F_{C_\alpha}(X)) \simeq F_{\operatorname{colim} C_\alpha}(X)$.

The formula we are really after is the one for the composition of functors $F_C(F_D(X))$. We compute

$$\begin{aligned} F_C(F_D(X)) &\simeq \coprod_{m\geq 0} C[m] \underset{\Sigma_m}{\times} F_D(X)^m \\ &\simeq \coprod_{m\geq 0} C[m] \underset{\Sigma_m}{\times} F_{D^{\otimes m}}(X) \end{aligned}$$

Thus, if we set

$$C \circ D = \coprod_{m \geq 0} C[m] \underset{\Sigma_m}{\times} D^{\otimes m}$$

then we have $F_C \circ F_D \simeq F_{C \circ D}$. In full, this means

$$(C \circ D)[d] = \coprod_{m \geq 0} C[m] \underset{\Sigma_m}{\times} \left(\coprod_{i_1 + \cdots + i_m = d} D[i_1] \times \cdots \times D[i_m] \underset{\Sigma_{i_1} \times \cdots \times \Sigma_{i_m}}{\times} \Sigma_d \right).$$

This makes the category of Σ-objects into a monoidal category, where the monoidal structure corresponds to composition of functors.

11.4. Operads are monoids. Since the category of Σ-object is a monoidal category via the "$\circ$" product (but not symmetric monoidal, since $C \circ D \neq D \circ C$), we can think about monoids in this category. A **monoid** is just a Σ-object C equipped with maps $\eta \colon I \to C$ and $\mu \colon C \circ C \to C$ such that the following diagrams commute.

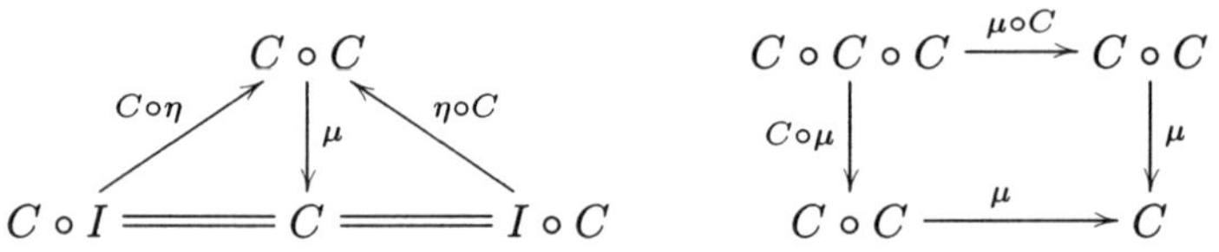

PROPOSITION 11.5. *The category of operads is equivalent to the category of Σ-object-monoids.*

PROOF. Suppose C is a Σ-object-monoid. Then the map $\eta \colon I \to C$ is just the choice of a distinguished point in $C[1]$, and the multiplication map $\mu \colon C \circ C \to C$ amounts to choosing a collection of maps

$$C[m] \times C[i_1] \times \cdots \times C[i_m] \to C[d]$$

which satisfy axioms (4) and (5) for an operad. The three remaining axioms (1), (2), and (3) follow from the fact that C is a monoid. The converse argument is similar. □

A Σ-object monoid C defines a functor $F_C \colon \mathrm{Top} \to \mathrm{Top}$ which is a triple. Thus, given an operad C, we can speak of algebras over the triple F_C.

PROPOSITION 11.6. *The category of C-algebras is equivalent to the category of F_C-algebras.*

This implies in particular that for a space X, $F_C(X) = \coprod C[m] \underset{\Sigma_m}{\times} X^n$ is a C-algebra. In fact, it is the free C-algebra on X, with the property that

$$C\text{-}\mathrm{alg}(F_C(X), Y) \simeq \mathrm{map}(X, Y).$$

For example, if A is the associative operad,

$$F_A(X) \simeq \coprod_{n \geq 0} X^n.$$

12. Algebra spectra

We discuss spectra which are also algebras over an operad.

12.1. Operads over the linear isometries operad. We want to have operads acting on spectra, rather than on spaces. First, note that there's a natural inclusion $\Sigma_n \subset \mathcal{L}(\mathcal{U}^n, \mathcal{U}^n)$. Let X be a spectrum over $\mathcal{U}$. For each $\sigma \in \Sigma_n$ there's a natural isomorphism $X^{(n)} \simeq \sigma_* X^{(n)}$ as follows. Recall that

$$(X^{(n)}_{\text{pre}})_{V_1 \times \cdots V_n} = X_{V_1} \wedge \cdots \wedge X_{V_n}.$$

This means that

$$(\sigma_* X^{(n)})^{\text{pre}}_{V_1 \times \cdots V_n} \simeq X_{V_{\sigma 1}} \wedge \cdots \wedge X_{V_{\sigma n}}.$$

There's a natural map $(X^{(n)})_{\text{pre}} \to (\sigma_* X^{(n)})_{\text{pre}}$ induced by permuting the factors, which is an isomorphism. This specifies the desired isomorphism of spectra.

The space $\mathcal{L}[n]$ gets a right Σ_n-action by permuting the source universe $\mathcal{U}^n$. Suppose $T \to \mathcal{L}[n]$ is a map of Σ_n-spaces. Then $T \ltimes X^{(n)}$ has a natural Σ_n-action, where $\sigma \in \Sigma_n$ acts via the composite of isomorphisms

$$T \ltimes X^{(n)} \simeq T \ltimes \sigma_* X^{(n)} \simeq T\sigma \ltimes X^{(n)} \simeq T \ltimes X^{(n)},$$

where $T\sigma$ denotes the space over $\mathcal{L}[n]$ defined by the composite map $T \to \mathcal{L}[n] \xrightarrow{\sigma} \mathcal{L}[n]$, and the right-hand map in the sequence is induced by the isomorphism $T\sigma \to T$ of spaces over $\mathcal{L}[n]$.

By a **Σ-object over the linear isometries operad**, I mean a map $C \to \mathcal{L}$ of Σ-objects. For $X \in \mathcal{SU}$ let $F_C(X)$ denote the spectrum

$$F_C(X) = \bigvee_{n \geq 0} C[n] \ltimes_{\Sigma_n} X^{(n)}.$$

Thus $F_C \colon \mathcal{SU} \to \mathcal{SU}$ is a functor. (Note: $C[0] \ltimes X^{(0)}$ is defined to be $\Sigma^\infty C[0]_+$.)

Given an **operad over the linear isometries operad**, i.e., an operad map $C \to \mathcal{L}$, the functor F_C is a triple on $\mathcal{SU}$, and we can speak of the **category of algebras over** C.

One example of an operad over $\mathcal{L}$ is the endomorphism operad of X. Define the endomorphism "operad" of a spectrum X to be an operad over $\mathcal{L}$, by

$$\mathcal{E}_X[n] = \mathcal{S}(X^{(n)}, X) \to \mathcal{L}[n].$$

Then one can show that a C-algebra structure on X is precisely the same as a map of operads $C \to \mathcal{E}_X$ making the diagram

$$\begin{array}{ccc} C & \longrightarrow & \mathcal{E}_X \\ & \searrow \quad \swarrow & \\ & \mathcal{L} & \end{array}$$

commute.

The main examples of operads we'll need are $\mathcal{A}_\infty$-operads. An $\mathcal{A}_\infty$ operad A is one for which the operad $\pi_0 A$ of path components, defined by $(\pi_0 A)[n] = \pi_0 A[n]$, is isomorphic to the associative operad, and each component of $A[n]$ is contractible. We call a spectrum which is an algebra over such an operad an $\mathcal{A}_\infty$-ring spectrum.

For example, there is an operad with $A[n] = \mathcal{L}[n] \times \Sigma_n$ which is an operad over the linear isometries operad; it induces a functor

$$F_A(X) \simeq \bigvee \mathcal{L}[n] \ltimes X^{(n)}.$$

In [**6**] the authors use this operad A to define their notion of $\mathcal{A}_\infty$-ring spectrum. It will be necessary for us to use a different $\mathcal{A}_\infty$-operad which has technical properties we will need. As will be noted below, this is not a major problem.

Note that if X is a spectrum, then $F_C(X)$ naturally is a C-algebra, and is in fact the **free C-algebra on X**, I'll denote the free C-algebra by $C(X)$. If Y is a C-algebra, I write UY for its underlying spectrum.

12.2. Properties of C-algebra spectra. The category of C-algebras has properties very much like those of the category of spectra $\mathcal{SU}$. I'll summarize them.

1. C-alg is a topological category. That is, for every two C-algebras X and Y the set of C-algebra maps $C\text{-}\mathrm{alg}(X,Y) \subset \mathcal{SU}(X,Y)$ can be topologized as a subspace of the space of spectrum maps.
2. The category of C-algebras has all small limits and colimits. Colimits of C-algebras can be quite complicated. But note that limits are easy. For example, if X and Y are C-algebras, and UX and UY denote the underlying spectra of X and Y, then $UX \times UY$ admits a natural C-algebra structure given by
$$F_C(UX \times UY) \to F_C UX \times F_C UY \to UX \times UY.$$
This is the product in the category of C-algebras. Thus, limits of C-algebras are computed on the underlying spectrum.
3. Given an unpointed space T and C-algebras X and Y, there are constructions $T \underset{C}{\otimes} X$ and Y^T, with the property that
$$C\text{-}\mathrm{alg}(T \underset{C}{\otimes} X, Y) \simeq \mathrm{map}\big(T, C\text{-}\mathrm{alg}(X,Y)\big) \simeq C\text{-}\mathrm{alg}(X, Y^T).$$
Furthermore, $U(Y^T) \simeq (UY)^T$. (But note that $U(T \underset{C}{\otimes} X) \not\simeq T \otimes UX$!) Let E be a C-algebra. Then $\pi_*(UE)^T = \pi_* U(E^T) \simeq E^{-*}T$. Thus, if E is an $\mathcal{A}_\infty$-ring spectrum, then E^T is an $\mathcal{A}_\infty$-ring spectrum whose homotopy is the cohomology of T.
4. The category of C-algebra spectra is a topological closed model category (this is non-trivial to prove). The weak equivalences and fibrations are taken to be those maps which are weak equivalences or fibrations on the underlying spectra.

Some of these results appear in EKMM [**6**]. The final result on the model category structure of C-algebras represents work of Hopkins. (EKMM prove the model category structure for certain special operads C, but not the general case.)

We note one more result, also due to Hopkins: the homotopy category of C-algebra spectra depends (up to equivalence) only on the weak homotopy type of the operad C. In particular, this means that all $\mathcal{A}_\infty$-operads give the same homotopy category of algebra spectra. Thus, we are always free to replace a given $\mathcal{A}_\infty$ operad A with an A' which is technically convenient for us.

13. Simplicial spectra

Let $\mathbf{\Delta}$ denote the category having as objects the finite ordered sets $[n] = \{0, 1, \dots, n\}$, and as morphisms those set maps $f\colon [m] \to [n]$ for which $i \le j$ implies $f(i) \le f(j)$. This category has a representation in Top which is easier to picture: there's a functor $\Delta\colon \mathbf{\Delta} \to \mathrm{Top}$ which sends $[n]$ to the Euclidean n-simplex

Δ_n which has vertices labelled $v_0, v_1, \ldots, v_n$, and which sends $f\colon [m] \to [n]$ to the affine map which sends v_i to $v_{f(i)}$.

A **simplicial object** in a category $\mathbf{C}$ is a functor $X_\bullet\colon \mathbf{\Delta}^{\mathrm{op}} \to \mathbf{C}$, and we call the category of such $s\mathbf{C}$. We define a realization functor $|-|\colon s\mathcal{SU} \to \mathcal{SU}$ as a quotient

$$\coprod_{n\geq 0} \Delta_n \otimes X_n / \sim .$$

More precisely, it is the coequalizer of

$$\coprod_{[m]\to[n]\in\mathbf{\Delta}} \Delta_m \otimes X_n \rightrightarrows \coprod_{[d]\in\mathbf{\Delta}} \Delta_d \otimes X_d,$$

in which the summand corresponding to $\delta\colon [m] \to [n] \in \mathbf{\Delta}$ is mapped by the top arrow to $\delta \otimes X_n$, and by the bottom arrow to $\Delta_m \otimes \delta$. Geometric realization is characterized by the fact that it is the left adjoint to a functor $\mathcal{SU} \to s\mathcal{SU}$ sending Y to $[n] \mapsto Y^{\Delta_n}$.

An **augmented simplicial object** is a simplicial object $X_\bullet$ together with a map $\epsilon\colon X_0 \to Y$ such that the two maps $X_1 \rightrightarrows X_0 \to Y$ are the same. Thus, if $X_\bullet \to Y$ is an augmented simplicial spectrum, then there's an induced map $|X_\bullet| \to Y$.

13.1. Facts about geometric realizations of simplicial spectra. We need the following facts [**6**].

1. If $X_\bullet \in s\mathcal{SU}$ and $Y_\bullet \in s\mathcal{SU}'$, then you can form a simplicial spectrum $X_\bullet \wedge Y_\bullet$ in $s\mathcal{S}(\mathcal{U} \times \mathcal{U}')$, defined by $[n] \mapsto X_n \wedge Y_n$. The crucial fact is that there's a natural isomorphism of spectra
$$|X_\bullet \wedge Y_\bullet| \simeq |X_\bullet| \wedge |Y_\bullet|.$$
2. If $T \to \mathcal{L}(\mathcal{U},\mathcal{U}')$ and $X_\bullet \in s\mathcal{SU}$, then there's a natural isomorphism
$$|T \rtimes X_\bullet| \simeq T \rtimes |X_\bullet|.$$
3. Geometric realization commutes with colimits.

So if $C \to \mathcal{L}$ is a Σ-object over $\mathcal{L}$, then there's a natural isomorphism

$$F_C|X_\bullet| \simeq |F_C(X_\bullet)|.$$

Now suppose $X_\bullet$ is a simplicial C-algebra for some operad C, i.e., a functor $X_\bullet\colon \mathbf{\Delta}^{\mathrm{op}} \to C\text{-}\mathrm{alg}$. Let $UX_\bullet$ denote the underlying simplicial spectrum. Then we can make $|UX_\bullet|$ into a C-algebra, by a map

$$F_C|UX_\bullet| \simeq |F_C UX_\bullet| \to |UX_\bullet|.$$

So the geometric realization of a simplicial C-algebra is again a simplicial C-algebra.

13.2. Resolutions of a C-algebra. Given a C-algebra X, let CX denote the free C-algebra on the underlying spectrum of X, so $UCX \simeq \bigvee_{n\geq 0} C[n] \underset{\Sigma_n}{\rtimes} X^{(n)}$. Then we get an augmented simplicial object

$$X \leftarrow CX \leftleftarrows CCX \cdots$$

which I'll denote by $C^{\bullet+1}X$. (By abuse of notation I omit the forget functor U.)

We need a model category fact: for good (i.e., cofibrant) choices of X and C, itself a cofibrant C-algebra, and that $C^{\bullet+1}X$ is "Reedy cofibrant" in the category of simplicial C-algebras. This means in particular that

$$C\text{-}\operatorname{alg}(X, E) \to C\text{-}\operatorname{alg}(|C^{\bullet+1}X|, E)$$

is a weak equivalence of spaces. We will compute $\mathcal{A}_\infty$ maps between $\mathcal{A}_\infty$-ring spectra using a spectral sequence based on this resolution, using an $\mathcal{A}_\infty$-operad C.

REMARK 13.3. It is important that we be able to produce a Reedy cofibrant resolution, as we will see below. In order to produce such it is convenient to have an operad C for which a cofibrant C-algebra X has an underlying spectrum UX which is cofibrant as a spectrum. It turns out that this will be the case when C is a *cofibrant operad*; these will be discussed in Section 22.2. Thus, we will assume henceforth that C is a cofibrant $\mathcal{A}_\infty$-operad.

The proof that a cofibrant operad suffices is quite simple, so we sketch it here (using notions from Section 22.2). Let Y be a cofibrant C-algebra over a cofibrant operad C. It will suffice to can produce a C-algebra X for which UX is cofibrant as a spectrum together with a trivial fibration $p\colon X \to Y$ of C-algebras, for then the theory of model categories tells us that p admits a section, and so UY is a retract of the cofibrant spectrum UX and thus cofibrant.

We produce such an X by means of a dirty trick. Start with a trivial fibration of *spectra* $p\colon X \to UY$ in which X is a cofibrant spectrum. We need to put a C-algebra structure on X so that p becomes a map of C-algebras. Consider the diagram

$$\begin{array}{ccccc}
 & & \{\mathcal{S}(X^{(n)}, X) \times_{\mathcal{S}(X^{(n)},Y)} \mathcal{S}(Y^{(n)}, Y)\} & \longrightarrow & \{\mathcal{S}(X^{(n)}, X)\} \\
 & \nearrow & {\scriptstyle f}\downarrow & & \downarrow{\scriptstyle g} \\
C & \longrightarrow & \{\mathcal{S}(Y^{(n)}, Y)\} & \longrightarrow & \{\mathcal{S}(X^{(n)}, Y)\}
\end{array}$$

The map g is bunch of trivial fibrations of spaces (one for each $n \geq 0$), because $X^{(n)}$ is a cofibrant spectrum and p is a trivial fibration of spectra. One can show that the pull-back f is actually a map between *operads*; since f is a trivial fibration space-wise, it is a trivial fibration of operads, and the cofibrancy property of C implies that the dotted arrow exists as a map of operads. You can check that such a dotted arrow gives precisely a C-algebra structure on X making p into a map of C-algebras, as desired!

14. Homotopy spectral sequence

We describe the Bousfield-Kan homotopy spectral sequence of a pointed cosimplicial space, and the Bousfield obstruction theory for an unpointed cosimplicial space.

14.1. Spectral sequence. A functor $\mathbf{\Delta} \to \mathbf{C}$ is called a **cosimplicial object** in $\mathbf{C}$. Thus

$$[n] \mapsto Y^n = C\text{-}\operatorname{alg}(C^{n+1}X, E)$$

is a cosimplicial space $Y^\bullet$.

If $X_\bullet$ is a simplicial object in some category, there is a "skeletal" filtration

$$X_0 \simeq |X_\bullet|_0 \to |X_\bullet|_1 \to |X_\bullet|_2 \to \cdots \to |X_\bullet|$$

where $|X_\bullet|_k$ is a quotient of $\coprod_{0\leq n\leq k} \Delta_n \otimes X_n$. This gives a tower

$$C\text{-}\operatorname{alg}(|X_\bullet|_0, E) \leftarrow C\text{-}\operatorname{alg}(|X_\bullet|_1, E) \leftarrow \cdots \leftarrow C\text{-}\operatorname{alg}(|X_\bullet|, E)$$

of spaces. If $X_\bullet$ is "Reedy cofibrant", then the induced tower is a tower of fibrations.

The spaces in the tower, called the Tot-tower, can be derived directly from the cosimplicial space $Y^\bullet$. The total space $\operatorname{Tot}(Y^\bullet)$ is a subspace of $\prod_{n\geq 0} \operatorname{map}(\Delta_n, Y^n)$. The kth approximation to the total space $\operatorname{Tot}^k(Y^\bullet)$ is the corresponding subspace of $\prod_{0\leq n\leq k} \operatorname{map}(\Delta_n, Y^n)$. One has

$$C\text{-}\operatorname{alg}(|C^{\bullet+1}|_k, E) \simeq \operatorname{Tot}^k Y^\bullet.$$

(The Reedy cofibrancy condition on $X_\bullet$ is the exact condition needed for the cosimplicial space $Y^\bullet$ to be fibrant in the sense of Bousfield and Kan [**3**].)

Suppose we have chosen a C-algebra map $f\colon X \to E$. This gives us base-points $C^{n+1}X \to X \to E$ in Y^n, and a base-point in $\operatorname{Tot} Y^\bullet$. The Bousfield-Kan spectral sequence of a pointed cosimplicial space [**3**] is a second quadrant spectral sequence "converging" to $\pi_*(\operatorname{Tot} Y^\bullet, f)$, where

$$E_2^{s,t} = \pi^s\pi_t(Y^\bullet, f) \Longrightarrow \pi_{t-s}(\operatorname{Tot} Y^\bullet, f).$$

There are differentials

$$d_r\colon E_r^{s,t} \to E_r^{s+r,t+r-1}.$$

Some comment on the E_2-term is appropriate. $E_2^{*,t}$ is the "cohomotopy" of the cosimplicial object

$$\pi_t Y^0 \rightrightarrows \pi_t Y^1 \mathrel{\substack{\longrightarrow\\ \longleftarrow\\ \longrightarrow\\ \longleftarrow\\ \longrightarrow}} \pi_t Y^2 \mathrel{\substack{\longrightarrow\\ \longleftarrow\\ \longrightarrow\\ \longleftarrow\\ \longrightarrow\\ \longleftarrow\\ \longrightarrow}} \cdots$$

where homotopy groups are taken at the base-point determined by f. If $t \geq 2$, then the cohomotopy of the cosimplicial object is just the cohomology of the corresponding co-chain complex. The same also holds for $t = 1$ if all Y^s are look spaces. The "corner" $E_2^{0,0}$ is only a pointed set, defined as the equalizer

$$E_2^{0,0} = \operatorname{Ker}(\pi_0 Y^0 \rightrightarrows \pi_0 Y^1).$$

14.2. Obstruction theory. The above spectral sequence can be defined once we have chosen a point in $\operatorname{Tot} Y^\bullet$. But what if we don't know whether such a point exists (this is one of the things we want to prove). There is an obstruction theory, due to Bousfield [**2**], which lets us solve this problem.

First, suppose we choose a point $y_0 \in \operatorname{Tot}^0 Y^\bullet \simeq Y^0$. This defines a class $[y_0] \in \pi_0 Y^0$, which we can think of as $E_1^{0,0}$. We can lift y_0 to $\operatorname{Tot}^1 Y^\bullet$ if and only if $[y_0]$ is in the equalizer of $\pi_0 Y^0 \to \pi_0 Y^0$. That is, y_0 lifts only if the class $[y_0]$ "survives" to $E_2^{0,0}$.

The next step is to lift to $\operatorname{Tot}^2 Y^\bullet$. The problem here involves extending a map from $\partial\Delta_2 \to Y^2$ to all of Δ_2; it turns out that if a point $y_0 \in \operatorname{Tot}^0 Y^\bullet$ can be lifted to $\operatorname{Tot}^1 Y^\bullet$, then the obstruction to lifting y_0 to $\operatorname{Tot}^2 Y^\bullet$ lies in $E_2^{2,1}$ (at least if all the $\pi_1 Y^s$ act trivially on $\pi_t Y^s$, so that this group is well defined; there is a description of this obstruction even when this is not true, but then the obstruction is just a relation, rather than a map).

This process continues; to lift to $\operatorname{Tot}^3 Y^\bullet$ we have to show that an obstruction

THEOREM 14.3 (Bousfield [**2**, 6.1]). *If $Y^\bullet$ is a cosimplicial space, and given a vertex $y_0 \in Y^0$ such that $[y_0] \in \operatorname{Ker}(\pi_0 Y^0 \rightrightarrows \pi_0 Y^1)$, and if*

$$E_2^{s,s-1} = 0$$

for $s \geq 2$, then y_0 lifts to $\operatorname{Tot} Y^\bullet$.

14.4. Computing spaces of maps of $\mathcal{A}_\infty$ algebras. Let E and F be $\mathcal{A}_\infty$-ring spectra (over our chosen $\mathcal{A}_\infty$-operad C), such that the underlying spectra of E and F are equivalent to E_{k_1,Γ_1} and E_{k_2,Γ_2} respectively. Suppose furthermore that F is cofibrant as a C-algebra spectrum. Then we can construct a simplicial resolution of F, and thereby get an obstruction theory and potentially a spectral sequence computing the homotopy of $C\text{-}\operatorname{alg}(F, E)$.

THEOREM 14.5. *In the above situation*

(1a)

$$E_2^{0,0} \simeq \operatorname{Hom}_{E_*\text{-}\mathrm{alg}}(E_*F, E_*),$$

(1b)

$$E_2^{s,t} \simeq \operatorname{Der}^s_{E_*\text{-}\mathrm{alg}}(E_*F, E_{*+t}), \qquad \textit{for } t - s \geq -1, t > 0.$$

(2a)

$$\operatorname{Hom}_{E_*\text{-}\mathrm{alg}}(E_*F, E_*) \simeq \mathcal{FG}\big((k_1, \Gamma_1), (k_2, \Gamma_2)\big),$$

(2b)

$$\operatorname{Der}^s_{E_*\text{-}\mathrm{alg}}(E_*F, E_{*+t}) \simeq 0, \qquad \textit{for all } s, t \geq 0.$$

PROOF. Part (1a) is proved in Section 16. Part (1b) is proved in Section 18. Part (2a) is proved in Section 17. Part (2b) is proved in Sections 19, 20, and 21. □

COROLLARY 14.6. *The topological functor*

$$\mathcal{A}_\infty^{\mathrm{LT}} \to \mathcal{FG}^{\mathrm{op}}$$

is fully faithful (i.e., induces weak equivalence on mapping spaces for all objects of $\mathcal{A}_\infty^{\mathrm{LT}}$.)

PROOF. Given objects $E, F \in \mathcal{A}_\infty^{\mathrm{LT}}$ which map to formal group laws (k_1, Γ_1) and (k_2, Γ_2) respectively, we see by Theorem 14.3 and Theorem 14.5 that $\mathcal{A}_\infty(F, E)$ is a homotopy discrete space with $\pi_0 \mathcal{A}_\infty(F, E) \simeq \mathcal{FG}\big((k_1, \Gamma_1), (k_2, \Gamma_2)\big)$. □

Part 3. Calculations

In the following sections we compute the obstruction groups and E_2-term of the spectral sequence described in Section 14. These calculations will also be used in Section 23 when we construct $\mathcal{A}_\infty$-structures.

15. Preliminary calculations

Let E and F be Landweber exact spectra. Then

$$E_*F \simeq E_* \underset{MU_*}{\otimes} MU_*MU \underset{MU_*}{\otimes} F_*.$$

PROPOSITION 15.1. *For Landweber exact theories E_* and F_*, the module E_*F is flat over E_*.*

PROOF. It suffices to show that E_*MU is flat as an MU_* module. That is, we must show that the functor

$$\{MU_*\text{-modules}\} \longrightarrow \{E_*\text{-modules}\}$$

defined by $N \mapsto E_*MU \underset{MU_*}{\otimes} N$ is exact. But this functor factors

$$\{MU_*\text{-modules}\} \longrightarrow \{MU_*MU\text{-comodules}\} \longrightarrow \{E_*\text{-modules}\},$$

sending

$$N \mapsto MU_*MU \underset{MU_*}{\otimes} N \mapsto E_* \underset{MU_*}{\otimes} MU_*MU \underset{MU_*}{\otimes} N \simeq E_*MU \underset{MU_*}{\otimes} N,$$

where the left-hand functor is exact since MU_*MU is a flat module over MU_*, and the right-hand functor is exact by Landweber's theorem. □

15.2. Universal coefficient and Künneth theorems. Let E be a ring-spectrum, and X and Y arbitrary spectra. One would like to have a **Künneth spectral sequence**

$$\mathrm{Tor}_s^{E_*}(E_*X, E_{*+t}Y) \Longrightarrow E_{s+t}(X \wedge Y)$$

and a **coefficient spectral sequence**

$$\mathrm{Ext}_{E_*}^s(E_*X, E_{*+t}) \Longrightarrow E^{t-s}X.$$

This doesn't happen in general, but it does happen in many cases for good spectra E. Adams [**1**, part 3] gave a general criterion for the existence of such spectral sequences. We need Adams' criterion in one case; it is a nice observation of Hopkins.

PROPOSITION 15.3. *If E is a Landweber exact theory, then there are Künneth and universal coefficient spectral sequences as described above.*

PROOF. Adams constructs the spectral sequences for E using the fact that for any arbitrary spectrum X there exists a map $P \to X$ of spectra such that

(a) $E_*P \to E_*X$ is surjective,
(b) E_*P is projective over E_*,
(c) $E_*(P \wedge Y) \simeq E_*P \underset{E_*}{\otimes} E_*Y$ (for the Künneth theorem), and
(d) $E^*P \simeq \mathrm{Hom}_{E_*}(E_*P, E_*)$ (for the coefficient theorem).

In the case when $E = MU$, Adams shows that it suffices to take P to be an appropriate wedge of Spanier-Whitehead duals of Thom spectra of finite complex Grassmanians.

If $P \to X$ is the map Adams constructed for MU, and E is a Landweber exact theory, then properties (a), (b), (c), and (d) hold for E: (a) and (b) follow because tensoring preserves surjections and projectives, (c) is trivial, and (d) follows because P is a wedge of finite spectra T, and because $E^*T \simeq E_*DT$ is a free module over E_* by computation, since DT is a Thom spectrum of a finite complex Grassmanian. □

COROLLARY 15.4. *If E and F_i $(i = 1, \dots, n)$ are Landweber exact spectra, and $F \simeq F_1 \wedge \cdots \wedge F_n$, then E_*F is flat over E_*, and*

$$E_*F \simeq E_*F_1 \underset{E_*}{\otimes} \cdots \underset{E_*}{\otimes} E_*F_n.$$

PROPOSITION 15.5. *If $E = E_{k_1,\Gamma_1}$ and $F = E_{k_2,\Gamma_2}$, then the evaluation map*

PROOF. By the universal coefficient theorem, we only need to prove that the Ext^s groups vanish for $s > 0$. This follows from Lemma 15.6. □

LEMMA 15.6. *Let M be a flat E_*-module. Then*

$$\mathrm{Ext}^s_{E_*}(M, E_*) \simeq 0$$

for $s > 0$.

PROOF. Let $\mathfrak{m} = (p, u_1, \ldots, u_{n-1})$ denote the maximal ideal of E_*. We first prove that $\mathrm{Ext}^s_{E_*}(M, E_*/\mathfrak{m}^d) \simeq 0$ for each d.

For $d = 1$ we have

$$\mathrm{Ext}^s_{E_*}(M, E_*/\mathfrak{m}) \simeq \mathrm{Ext}^s_{E_*/\mathfrak{m}}(M \underset{E_*}{\otimes} E_*/\mathfrak{m}, E_*/\mathfrak{m})$$

by flat base change. Since $E_*/\mathfrak{m} \simeq k[u^{\pm}]$ is injective as a graded module over itself, these Ext groups are 0 for $s > 0$.

Note that $\mathfrak{m}^d/\mathfrak{m}^{d+1}$ is a finite $k[u^{\pm}]$-vector space, so $\mathrm{Ext}^s_{E_*}(M, \mathfrak{m}^d/\mathfrak{m}^{d+1}) \simeq 0$ for $s > 0$. Thus for $d > 1$, the exact sequence

$$0 \to \mathfrak{m}^d/\mathfrak{m}^{d+1} \to E_*/\mathfrak{m}^{d+1} \to E_*/\mathfrak{m}^d \to 0$$

shows that $\mathrm{Ext}^s_{E_*}(M, E_*/\mathfrak{m}^{d+1}) = 0$ for $s > 0$, as well as showing that the map $\mathrm{Hom}_{E_*}(M, E_*/\mathfrak{m}^{d+1}) \to \mathrm{Hom}_{E_*}(M, E_*/\mathfrak{m}^d)$ is surjective.

Finally, use the exact sequence

$$0 \to E_* \to \prod E_*/\mathfrak{m}^d \xrightarrow{1-S} \prod E_*/\mathfrak{m}^d \to 0$$

in $\mathrm{Ext}^s_{E_*}(M, -)$ to prove the lemma. (Here S denotes the "shift" map.) □

REMARK 15.7. Note that Lemma 15.6 and hence Proposition 15.5 (and, as we shall see, Lemma 18.4 and Corollary 19.3) can fail for an arbitrary Landweber exact theory E_*. For example, if $E = E(n)$, the Johnson-Wilson theory with $E(n)_* \simeq \mathbb{Z}_{(p)}[v_1, \ldots, v_{n-1}, v_n^{\pm}]$, then it is not the case that $\mathrm{Ext}^1_{E_*}(E_*E, E_*)$ vanishes, even though E_*E is flat over E_*.

16. Identification of $E_2^{0,0}$

In this section we show that

$$E_2^{0,0} \simeq \mathrm{Hom}_{E_*\text{-}\mathrm{alg}}(E_*F, E_*),$$

proving part (1a) of Theorem 14.5

We need to compute the equalizer of the pair of maps

$$\pi_0 C\text{-}\mathrm{alg}(CF, E) \rightrightarrows \pi_0 C\text{-}\mathrm{alg}(C^2F, E).$$

First note that for any C-algebra Y there is a natural map

$$\pi_0 C\text{-}\mathrm{alg}(Y, E) \longrightarrow \mathrm{Hom}_{E_*\text{-}\mathrm{alg}}(E_*Y, E_*).$$

This map sends a C-algebra map $f\colon Y \to E$ to the map

$$E_*Y \xrightarrow{E_*f} E_*E \to E_*.$$

Note that E_*Y is in fact an associative E_* algebra, since the $\mathcal{A}_\infty$-structure on Y induces maps $E_*Y \underset{E_*}{\otimes} E_*Y \to E_*(Y \wedge Y) \to E_*Y$.

Now suppose $Y = CX$ for some spectrum X. Given an E_* module M, let TM denote the E_* tensor algebra on M; i.e.,

$$TM = \bigoplus_{n\geq 0} M^{\otimes n} \simeq E_* \oplus M \oplus (M \underset{E_*}{\otimes} M) \oplus \cdots .$$

At the level of the homotopy category of spectra,

$$Y = CX \simeq \bigvee_{n\geq 0} X^{(n)} \simeq S^0 \vee X \vee X^{(2)} \vee \cdots .$$

The inclusion $X \to CX$ induces a map $E_*X \to E_*CX$ of E_*-modules, and hence a map $\alpha_X : T(E_*X) \to E_*CX$ of algebras.

LEMMA 16.1. *If E_*X is flat over E_*, then α_X is an isomorphism.*

PROOF. This is clear from Corollary 15.4. □

In particular Lemma 16.1 holds for $X = C^nF$.

Now consider the commutative diagram

$$\begin{array}{ccc} \pi_0 C\text{-}\operatorname{alg}(CX, E) & \longrightarrow & \operatorname{Hom}_{E_*\text{-}\mathrm{alg}}(E_*CX, E_*) \\ \big\downarrow & & \big\downarrow \\ [X, E] & \longrightarrow & \operatorname{Hom}_{E_*}(E_*X, E_*) \end{array}$$

The left-hand vertical arrow is an isomorphism, since CX is the free C-algebra on X. If E_*X is flat over E_*, the bottom and right-hand vertical arrows are isomorphisms by Proposition 15.5 and Lemma 16.1; hence the top arrow is also an isomorphism. Thus, we have reduced our problem to that of computing the equalizer of the pair of maps

$$\operatorname{Hom}_{E_*\text{-}\mathrm{alg}}\left(T(E_*F), E_*\right) \rightrightarrows \operatorname{Hom}_{E_*\text{-}\mathrm{alg}}\left(T^2(E_*F), E_*\right) .$$

But this is precisely just the equalizer of

$$\operatorname{Hom}_{E_*}(E_*F, E_*) \rightrightarrows \operatorname{Hom}_{E_*}\left(T(E_*F), E_*\right)$$

where the top arrow sends $f\colon E_*F \to E_*$ to the composite

$$T(E_*F) \to E_*F \xrightarrow{f} E_*$$

and the bottom arrow sends it to the composite

$$T(E_*F) \xrightarrow{T(f)} T(E_*) \to E_*.$$

Thus

$$E_2^{0,0} = \operatorname{Hom}_{E_*\text{-}\mathrm{alg}}(E_*F, E_*).$$

17. Computation of $\operatorname{Hom}_{E_*\text{-}\mathrm{alg}}(E_*F, E_*)$

In this section we show that

$$\operatorname{Hom}_{E_*\text{-}\mathrm{alg}}(E_*F, E_*) \simeq \mathcal{FG}\big((k_1, \Gamma_1), (k_2, \Gamma_2)\big),$$

which proves part (2a) of Theorem 14.5.

Let $E = E_{k_1,\Gamma_1}$, $F = E_{k_2,\Gamma_2}$. There are universal deformations (G_e, i_e) of the formal groups (k_e, Γ_e), where G_e is a degree 0 formal group law over the ring

Recall from Section 6 that chosen invertible elements $u_{G_e} \in \pi_2 E_{k_e,\Gamma_e}$ allow us to produce degree -2-formal group laws $\bar{G}_e$.

We need to understand the set

$$\mathrm{Hom}_{E_*\text{-alg}}(E_*F, B_*).$$

Recall from Section 15 that

$$E_*F \simeq E_* \underset{MU_*}{\otimes} MU_*MU \underset{MU_*}{\otimes} F_*.$$

Thus by Theorem 6.2 a map $E_*F \to B_*$ of commutative graded rings corresponds precisely to the following data: a map $\phi\colon E_* \to B_*$, a map $\psi\colon F_* \to B_*$, and a *strict* isomorphism $\bar{g}\colon \psi^*\bar{G}_2 \to \phi^*\bar{G}_1$ of formal group laws. Represent such data by the triple $(\phi, \bar{g}, \psi)$.

Suppose we fix once and for all an E_*-algebra structure on B_*, in other words, a map ϕ. Then elements of the set $\mathrm{Hom}_{E_*\text{-alg}}(E_*F, B_*)$ correspond to data $(\psi, \bar{g})$. That is,

$$\mathrm{Hom}_{E_*\text{-alg}}(E_*F, B_*) \simeq \{(\psi\colon F_* \to B_*,\ \bar{g}\colon \psi^*\bar{G}_2 \xrightarrow{\sim} \phi^*\bar{G}_1)\}.$$

Write $\psi_0\colon F_0 \to B_0$ for the restriction of ψ to degree 0. Since $F_* \simeq F_0[u_{G_2}^{\pm}]$, the set of homomorphisms $\psi\colon F_* \to B_*$ corresponds bijectively to the set of pairs $(\psi_0, \psi(u_{G_2}))$ where ψ_0 is a map $F_0 \to B_0$ and $\psi(u_{G_2})$ is an invertible element in B_2.

Recall from Section 6 that the strict isomorphism $\bar{g}\colon \psi^*\bar{G}_2 \to \phi^*\bar{G}_1$ of degree -2 formal group laws and the elements $\phi(u_{G_1})$ and $\psi(u_{G_2})$ induce an isomorphism $g\colon \psi_0^*G_2 \to \phi_0^*G_1$ of degree 0 formal group laws, not necessarily strict, by $g(x) = \phi(u_{G_1})\bar{g}(\psi(u_{G_2})^{-1}x)$. Note that if $\bar{g}(x) = x + a_2x^2 + a_3x^3 + \cdots$, then

$$g(x) = (\phi(u_{G_1})\psi(u_{G_2})^{-1})x + a_2(\phi(u_{G_1})\psi(u_{G_2})^{-2})x^2 + \cdots.$$

Thus, given $g(x)$ and $\phi(u_{G_1})$, we can recover $\psi(u_{G_2})$ (since $\phi(u_{G_1}) = g'(0)\psi(u_{G_2})$), and hence recover $\bar{g}(x)$.

Putting this together, we see that the data $(\psi, \bar{g})$ corresponds bijectively to data (ψ_0, g). Thus, we have proved

PROPOSITION 17.1. *There is an isomorphism of sets*

$$\mathrm{Hom}_{E_*\text{-alg}}(E_*F, B_*) \simeq \{(\psi_0\colon F_0 \to B_0, g\colon \psi^*G_2 \xrightarrow{\sim} \phi^*G_1)\}.$$

REMARK 17.2. There is another way to describe this set. It is easy to see that $E_*F \simeq E_0F[u_{G_1}]$, so that

$$\mathrm{Hom}_{E_*\text{-alg}}(E_*F, B_*) \simeq \mathrm{Hom}_{E_0\text{-alg}}(E_0F, B_0).$$

Thus, a map from E_0F to a ring B corresponds precisely to: a map $\phi\colon E_0 \to B$, a map $\psi\colon F_0 \to B$, and an isomorphism $g\colon \psi^*G_2 \xrightarrow{\sim} \phi^*G_1$ (not necessarily strict).

Equivalently, there is a square diagram of groupoids

$$\begin{array}{ccc} \mathrm{Hom}_{\mathrm{Ring}}(E_0F, B) & \longrightarrow & \mathrm{Hom}_{\mathrm{Ring}}(E_0, B) \\ \downarrow & & \downarrow \\ \mathrm{Hom}_{\mathrm{Ring}}(F_0, B) & \longrightarrow & \mathrm{fgl}(B) \end{array}$$

commuting up to a natural isomorphism, which is a homotopy pull-back square. (A set S is regarded as a groupoid having as objects the set S, and as maps only identity maps.) This way of viewing things will come in handy in Section 21.

Suppose now that $B = B_0$ is a *complete* local ring, and that B is a *continuous* E_0-algebra, i.e., the map $\phi_0\colon E_0 \to B$ is a continuous map between complete local rings. The continuity hypothesis is equivalent to requiring that $\phi_0(\mathfrak{m}_E) \subset \mathfrak{m}_B$, where $\mathfrak{m}_E = (p, u_1, \ldots, u_{n-1}) \subset E_0$ is the maximal ideal. Given some E_0-algebra map $E_0F \to B$, I would like to know that the induced map $F_0 \to E_0F \to B$ between complete rings is continuous. This is ensured by the following well-known lemma.

LEMMA 17.3. *Let $\mathfrak{m}_E \subset E_0$ and $\mathfrak{m}_F \subset F_0$ be the maximal ideals of these rings. Then $\mathfrak{m}_E E_0F = \mathfrak{m}_F E_0F$ as ideals of E_0F.*

PROOF. The ideal $\mathfrak{m}_E$ is generated by the coefficients of the terms of the p-series $[p]_{G_1}(x)$ which are of degree less than p^n. Likewise, the ideal $\mathfrak{m}_F$ is generated by the coefficients of the terms of the p-series $[p]_{G_2}(x)$ which are of degree less than p^n. There is an isomorphism of formal group laws $g\colon G_1 \to G_2$ defined over the ring E_0F, and hence an equation $g \circ [p]_{G_1} = [p]_{G_2} \circ g$. The lemma follows easily. □

We have fixed a map $\phi_0\colon E_0 \to B$ and hence a formal group law ϕ^*G_1 over B which is a deformation of (k_1, Γ_1) to B. Maps $\psi\colon F_0 \to B$ classify $\star$-isomorphism classes of deformations of (k_2, Γ_2) to B. Consider the diagram of groupoids

$$\begin{array}{ccc} P & \longrightarrow & \{\phi^*G_1\} \\ \downarrow & & \downarrow \\ \mathrm{Def}_{\Gamma_2}(B) & \longrightarrow & \mathrm{fgl}(B) \\ \downarrow & & \downarrow \\ \coprod_{j\colon k_2 \to B/\mathfrak{m}_B} \{i^*\Gamma_2\} & \longrightarrow & \mathrm{fgl}(B/\mathfrak{m}_B) \end{array}$$

in which both squares are homotopy-pullback squares of groupoids. We get two descriptions of P from this diagram. The first comes from the top square: since $\mathrm{Def}_{\Gamma_2}(B)$ is homotopy discrete and $\pi_0 \mathrm{Def}_{\Gamma_2}(B) \simeq \mathrm{Hom}^{\mathrm{cts}}_{\mathrm{Ring}}(F_0, B)$, we see that P is equivalent to a disjoint union of a bunch of loop-spaces of $\mathrm{fgl}(B)$, and in fact P is homotopy discrete with

$$\pi_0 P \simeq \coprod_{\psi\colon F_0 \xrightarrow{\mathrm{cts}} B} \mathrm{Hom}_{\mathrm{fgl}(B)}(\phi^*G_1, \psi^*G_2).$$

Thus, $P \simeq \mathrm{Hom}_{E_*\text{-}\mathrm{alg}}(E_*F, B_*)$ by Proposition 17.1 and Lemma 17.3.

The second description comes from the big rectangle. This also describes P as a union of loop-spaces, and shows that P is homotopy discrete with

$$\pi_0 P \simeq \coprod_{j\colon k_2 \to B/\mathfrak{m}_B} \mathrm{Hom}_{\mathrm{fgl}(B/\mathfrak{m}_B)}(i^*\Gamma_1, j^*\Gamma_2),$$

where $i\colon k_1 \to B/\mathfrak{m}_B$ is the map induced by $\phi_0\colon E_0 \to B$ modulo maximal ideals. Thus we conclude

$$\mathrm{Hom}_{E_*\text{-}\mathrm{alg}}(E_*F, B_*) \simeq P \simeq \mathcal{FG}\big((B/\mathfrak{m}_B, i^*\Gamma_1), (k_2, \Gamma_2)\big).$$

as desired.

REMARK 17.4. Note that if we let $B_* = E_*/\mathfrak{m}^d$ for any $d \geq 1$ we still get an isomorphism

$$\mathrm{Hom}_{E_*\text{-}\mathrm{alg}}(E_*F, E_*/\mathfrak{m}^d) \simeq \mathcal{FG}\big((k_1, \Gamma_1), (k_2, \Gamma_2)\big).$$

In particular, this holds for $d = 1$. If we restrict to degree 0 and then tensor down via the map $E_0 \to E_0/\mathfrak{m} = k_1$, we see that

$$\mathrm{Hom}_{k_1\text{-}\mathrm{alg}}(E_0F/\mathfrak{m}, k_1) \simeq \mathcal{FG}\big((k_1, \Gamma_1), (k_2, \Gamma_2)\big)$$

where $\mathfrak{m}$ denotes the ideal generated by $p, u_1, \ldots, u_{n-1} \in E_0 \subset E_0F$. This means one can give a description of $E_0F/\mathfrak{m}$ in many cases using formal group law theory.

For example, if $\Gamma_1 = \Gamma_2 = \Gamma$ is the Honda formal group law, so that $E = F = E_n$, then $\mathcal{FG}\big((\mathbb{F}_{p^n}, \Gamma), (\mathbb{F}_{p^n}, \Gamma)\big)$ is isomorphic to the set of pairs (σ, f) where $\sigma\colon \mathbb{F}_{p^n} \to \mathbb{F}_{p^n}$ is a Galois automorphism, and $f(x) = e_0x \underset{\Gamma}{+} e_1x^p \underset{\Gamma}{+} e_2x^{p^2} \underset{\Gamma}{+} \cdots$ is a power series with coefficients $e_i \in \mathbb{F}_{p^n}$, with $e_0 \neq 0$ (see [**13**, App. 2]). This shows that

$$E_0F/\mathfrak{m} \simeq (\mathbb{F}_{p^n} \underset{\mathbb{F}_p}{\otimes} \mathbb{F}_{p^n})[e_0, e_1, e_2, \ldots]/(e_0^{p^n-1} - 1,\ e_1^{p^n} - e_1,\ e_2^{p^n} - e_2,\ \ldots).$$

(Note that a map of $\mathbb{F}_{p^n}$-algebras $\mathbb{F}_{p^n} \underset{\mathbb{F}_p}{\otimes} \mathbb{F}_{p^n} \to \mathbb{F}_{p^n}$ corresponds precisely to a map of rings $\sigma\colon \mathbb{F}_{p^n} \to \mathbb{F}_{p^n}$.)

18. Derivations of associative algebras

In this section we define the derived functors of associative algebra derivations and show part (1b) of Theorem 14.5.

18.1. A complex for the E_2-term. Suppose Y is a C-algebra, and choose a map $f\colon Y \to E$. The set $\pi_t(C\text{-}\mathrm{alg}(Y, E), f)$ is the same as homotopy classes of maps $S^t \to C\text{-}\mathrm{alg}(Y, E)$ which send the base-point of the sphere to f, and where homotopies are base-point preserving.

$$\begin{array}{ccc} \mathrm{pt} & \xrightarrow{f} & C\text{-}\mathrm{alg}(Y, E) \\ \big\downarrow & \overset{g}{\nearrow} & \\ S^t & & \end{array}$$

By adjunction, these are the same as homotopy classes of lifts in the diagram

$$\begin{array}{ccc} & & E^{S^t} \\ & \overset{\hat{g}}{\nearrow} & \big\downarrow \\ Y & \xrightarrow{f} & E \end{array}$$

of C-algebras, where homotopies are ones which are fiberwise over f. Recall from Section 12.2 that E^{S^t} is a C-algebra whose underlying spectrum is isomorphic to $F(\Sigma^\infty S^t_+, E) \simeq E \vee \Sigma^{-t}E$. Thus $\pi_*E^{S^t} \simeq E^{-*}S^t \simeq E_*[\epsilon]/\epsilon^2$, with $|\epsilon| = -t$. The spectrum E^{S^t} is naturally a ring spectrum over E. Applying $E_*(-)$ everywhere and using the ring structure of E_* gives us a diagram of algebras

$$\begin{array}{ccccc} & & E_*E^{S^t} & \longrightarrow & E_*[\epsilon]/\epsilon^2 \\ & {\scriptstyle E_*\hat{g}} \nearrow & \downarrow & & \downarrow \\ E_*Y & \xrightarrow{E_*f} & E_*E & \longrightarrow & E_* \end{array}$$

and thus we get a map

$$\pi_t\left(C\text{-}\operatorname{alg}(Y,E),f\right) \to \operatorname{Hom}_{(E_*\text{-}\operatorname{alg}\downarrow E_*)}(E_*Y, E_*[\epsilon_t]/\epsilon_t^2),$$

where the right hand side denotes maps in the category of E_*-algebras augmented over E_*. Note that E_*Y is augmented over E_*, via the composite $E_*Y \to E_*E \to E_*$; this augmentation itself depends on the choice of $f\colon Y \to E$.

Now suppose $Y = CX$ where E_*X is a flat E_*-module. Then using the commutative diagram

$$\begin{array}{ccc} \pi_0 C\text{-}\operatorname{alg}(CX, E^{S^t}) & \longrightarrow & \operatorname{Hom}_{E_*\text{-}\operatorname{alg}}\left(T(E_*X), E_*[\epsilon_t]/\epsilon_t^2\right) \\ \downarrow & & \downarrow \\ [X, E \vee \Sigma^{-t}E] & \longrightarrow & \operatorname{Hom}_{E_*}(E_*X, E_* \oplus E_{*+t}) \end{array}$$

and an argument similar to that of Section 16 one shows that the top horizontal map is a bijection. Thus, we have

$$E_2^{s,t} \simeq H^s\left[\operatorname{Hom}_{(E_*\text{-}\operatorname{alg}\downarrow E_*)}\left(T^{\bullet+1}(E_*F), E_*[\epsilon_t]/\epsilon_t^2\right)\right],$$

for $t > 0$ and $t - s \geq -1$.

18.2. Associative algebra derivations. Let A be a commutative ring, and suppose R is an associative A-algebra. Let M be an R-bimodule relative to A; that is, M is a module over the ring $R^e = R \underset{A}{\otimes} R^{\mathrm{op}}$. Then form a ring $R \oplus M$ with multiplication given by

$$(r,m)\cdot(r',m') = (rr', rm' + mr')$$

for $r, r' \in R$, $m, m' \in M$. The inclusion map $\eta\colon R \to R \oplus M$ and the projection map $\pi\colon R \oplus M \to R$ are algebra maps, and $M \subset R \oplus M$ is an ideal such that $M^2 = 0$.

Consider sections $s\colon R \to R \oplus M$ of the projection π. Such an $s = (\mathrm{id}, D)$ where $D\colon R \to M$ is an A-module map such that

$$D(rr') = r(Dr') + (Dr)r'$$

for $r, r' \in R$. Such a map $D\colon R \to M$ is called an **A-derivation of R into M**. The set of A-derivations is written $\operatorname{Der}_A(R,M)$, whence

$$\operatorname{Hom}_{(A\text{-}\operatorname{alg}\downarrow R)}(R, R \oplus M) \simeq \operatorname{Der}_A(R,M).$$

More generally, let $\phi\colon R \to S$ be a map of associative A-algebras, and let M be an S-bimodule. The map ϕ makes M into an R-bimodule by restriction. Hence

$$\operatorname{Hom}_{(A\text{-}\operatorname{alg}\downarrow S)}(R, S \oplus M) \simeq \operatorname{Der}_A(R,M).$$

Note that $\mathrm{Der}_A(R,M)$ is naturally an A-module. If $A = S = E_*$, $R = E_*Y$, and $M = E_{*+t}$, then this module is

$$\mathrm{Der}_{E_*}(E_*Y, E_{*+t}) \simeq \mathrm{Hom}_{(E_*\text{-}\mathrm{alg}\downarrow E_*)}(E_*Y, E_*[\epsilon_t]/\epsilon_t^2).$$

Thus,

$$E_2^{s,t} \simeq H^s\left[\mathrm{Der}_{E_*}(T^{\bullet+1}(E_*F), E_{*+t})\right].$$

18.3. Derived functors of associative algebra derivations. Let sA-alg denote the category of simplicial associative A-algebras. This has a model category structure as follows. The **weak equivalences** $f\colon R_\bullet \to S_\bullet$ are maps which induce isomorphisms on homology of the associated chain-complexes. A **cofibration** f is a retract of a free map. A **free map** $f\colon R_\bullet \to S_\bullet$ is a map which in each simplicial degree is an inclusion

$$R_n \to S_n \simeq R_n\langle X_n\rangle,$$

where $R\langle X\rangle$ denotes the algebra obtained from R by adjoining the elements of the set X as free generators, and such that the the sets $X_\bullet$ are closed under the degeneracy operators of $S_\bullet$. (See Quillen [**11**].)

A **cofibrant resolution** of a simplicial algebra $R_\bullet$ is a diagram $A \xrightarrow{i} P_\bullet \xrightarrow{f} R_\bullet$, where i is a cofibration, and f is a weak equivalence. There is a functorial cofibrant resolution $P_\bullet$ of R, given by $P_n = F^{n+1}(R)$, where FB denotes the free associative A-algebra on the underlying set of R. Another resolution is the complex $Q_n = T^{n+1}(R)$. It is a cofibrant resolution if R is a projective A-module; this is easy to see if R is a free A-module, since if X is a set of module generators of R, then the set of monomials in X is a set of module generators of TR.

Given an associative A-algebra R over S, and an S-bimodule M, define

$$\mathrm{Der}_A^s(R,M) = H^s\left[\mathrm{Der}_A(P_\bullet, M)\right],$$

the **s-th derived functor of derivations** from B to M, where $P_\bullet \to R$ is a cofibrant resolution of R. The derived functors do not depend on the choice of cofibrant resolution.

Our complex $T^{n+1}(E_*F)$ of E_*-algebras is *not* cofibrant, since E_*F is *not* projective over E_*; nonetheless, we will show below that this complex computes derived functors of derivations.

Suppose $R = TN$, where N is an A-module, and let M be an S-bimodule. Then if $P_\bullet \to N$ is a free resolution of N by a simplicial A-module, then $TP_\bullet$ is a cofibrant resolution of $R = TN$. Then, since

$$\mathrm{Der}_A(TN, M) \simeq \mathrm{Hom}_{(A\text{-}\mathrm{alg}\downarrow S)}(TN, S \oplus M) \simeq \mathrm{Hom}_A(N, M)$$

it is not hard to see that

$$\mathrm{Der}_A^s(R,M) \simeq \mathrm{Ext}_A^s(N,M).$$

Thus, Lemma 15.6 gives

LEMMA 18.4. *If we take $A = S = E_*$, $M = E_{*+t}$, and $R = TN$ where N is a flat E_*-module, then*

$$\mathrm{Der}_{E_*}^s(TN, E_{*+t}) = 0, \quad s > 0.$$

18.5. The E_2-term is a derived functor. Suppose $R_\bullet$ is a simplicial A-algebra, augmented over S, and let M be an S-module Then there is a spectral sequence of A-modules

$$E_1^{p,q} = \mathrm{Der}^p_A(R_q, M) \Longrightarrow \mathrm{Der}^{p+q}_A(R_\bullet, M).$$

This spectral sequence is constructed by producing a double complex $P_{p,q}$ of A-algebras, in which each row $P_{\bullet,q} \to R_q$ is a cofibrant resolution of R_q, and so that the diagonal is a cofibrant resolution of $R_\bullet$. Such a resolution can be constructed by the functorial cofibrant resolution $P_{p,q} = F^{p+1}R_q$.

Now letting $A = S = E_*$, $M = E_{*+t}$ and $R_q = T^{q+1}(E_*F)$ we see from Lemma 18.4 applied to $N = E_*F$ that the spectral sequence at the E_1-term is

$$E_1^{p,q} = \mathrm{Der}^p_{E_*}(T^{q+1}(E_*F), E_{*+t}) = \begin{cases} 0 & \text{if } p > 0, \\ \mathrm{Der}_{E_*}(T^{q+1}(E_*F), E_{*+t}) & \text{if } p = 0. \end{cases}$$

Thus the spectral sequence is concentrated in a single column of the E_1-term, and collapses at E_2. In fact, $E_1^{0,*}$ is precisely the cochain complex whose cohomology we want to compute, and by the convergence of the spectral sequence, what it computes is $\mathrm{Der}^s_{E_*}(E_*F, E_{*+t})$.

19. Change of ground ring

In this section we show that in order to prove that $\mathrm{Der}^s_{E_*}(E_*F, E_{*+t})$ vanishes, it suffices to show that $\mathrm{Der}^s_{E_0/\mathfrak{m}}(E_0F/\mathfrak{m}, E_t/\mathfrak{m})$ vanishes.

19.1. Associative algebra differentials. Given an A-algebra R, define $I_{R/A}$ to be an R-bimodule equipped with a A-derivation $d\colon R \to I_{R/A}$ which induces an isomorphism

$$\mathrm{Der}_A(R, M) \simeq \mathrm{Hom}_{R^e}(I_{R/A}, M).$$

Such an object always exists: $I_{R/A}$ can be constructed as the cokernel of δ in

$$R \underset{A}{\otimes} R \underset{A}{\otimes} R \underset{A}{\otimes} R \xrightarrow{\delta} R \underset{A}{\otimes} R \underset{A}{\otimes} R \xrightarrow{p} I_{R/A} \to 0$$

where

$$\delta(a \otimes b \otimes c \otimes d) = ab \otimes c \otimes d - a \otimes bc \otimes d + a \otimes b \otimes cd.$$

The universal derivation $d\colon R \to I_{R/A}$ is given by $d(r) = p(1 \otimes r \otimes 1)$. By the bar complex, we see that $I_{R/A}$ is also the kernel of μ in

$$0 \to I_{R/A} \to R \underset{A}{\otimes} R \xrightarrow{\mu} R.$$

We call $I_{R/A}$ the R-bimodule of **associative algebra differentials** of R.

Let $R = TN$ be the tensor algebra on an A-module N. Then derivations $D\colon TN \to M$ to some R-bimodule M correspond bijectively to maps

$$\mathrm{Hom}_{(A\text{-}\mathrm{alg}\downarrow R)}(TN, R \oplus M) \simeq \mathrm{Hom}_A(N, M),$$

and hence

$$\mathrm{Der}_A(TN, M) \simeq \mathrm{Hom}_A(N, M).$$

Since $I_{TN/A}$ should be an R-bimodule, we see that

$$I_{TN/A} \simeq TN \underset{A}{\otimes} N \underset{A}{\otimes} TN.$$

Suppose R is an associative A-algebra. Define

$$D_{R/A} = I_{P_\bullet/A} \underset{P_\bullet^e}{\otimes} R^e$$

for some choice $P_\bullet \to R$ of cofibrant resolution of R. Thus $D_{R/A}$ is a complex of projective R-bimodules, and

$$\operatorname{Ext}^s_{R^e}(D_{R/A}, M) \simeq H^s\big[\operatorname{Hom}_{R^e}(I_{P_\bullet/A} \underset{P_\bullet^e}{\otimes} R^e, M)\big]$$
$$\simeq H^s\big[\operatorname{Der}_A(P_\bullet, M)\big] \simeq \operatorname{Der}^s_A(R, M).$$

Here "Ext" really means some kind of "hyper-Ext".

We therefore get a coefficient long exact sequence for derived functors of derivations.

PROPOSITION 19.2. *Let $0 \to M' \to M \to M'' \to 0$ be a short exact sequence of R-bimodules. Then there is a long exact sequence*

$$0 \to \operatorname{Der}_A(R, M') \to \operatorname{Der}_A(R, M) \to \operatorname{Der}_A(R, M'') \to \operatorname{Der}^1_A(R, M) \to \cdots.$$

COROLLARY 19.3. *If $\operatorname{Der}^s_{E_*}(E_*F, E_{*+t}/\mathfrak{m}) = 0$ for all s, then*

$$\operatorname{Der}^s_{E_*}(E_*F, E_{*+t}) = 0$$

for all s.

PROOF. The proof is similar to that of Lemma 15.6. □

19.4. Change of ground ring. Suppose $A \to A'$ is a map of commutative rings. Then given an A-algebra R we get an induced A'-algebra $R' = R \underset{A}{\otimes} A'$.

PROPOSITION 19.5. *If R is flat as an A-module, and M is an R'-bimodule, then*

$$\operatorname{Der}^s_A(R, M) \simeq \operatorname{Der}^s_{A'}(R', M).$$

PROOF. Let $P_\bullet \to R$ be a cofibrant A-algebra resolution of R. Then $P'_\bullet = P_\bullet \underset{A}{\otimes} A'$ is a cofibrant A'-algebra, since $A\langle X\rangle \underset{A}{\otimes} A' \simeq A'\langle X\rangle$ for a set X. Since $P_\bullet$ is also a complex of projective A-modules, it forms a projective A-module resolution of R. Thus, if R is flat, $P'_\bullet$ is a cofibrant resolution of R'. □

Using $E_* \to E_*/\mathfrak{m}$ and the flatness of E_*F over E_*, we get

COROLLARY 19.6. *There is an isomorphism*

$$\operatorname{Der}^s_{E_*}(E_*F, E_{*+t}/\mathfrak{m}) \simeq \operatorname{Der}^s_{E_*}(E_*F/\mathfrak{m}, E_{*+t}/\mathfrak{m}).$$

Using $E_0/\mathfrak{m} \to E_*/\mathfrak{m}$ and the fact that $E_0/\mathfrak{m}$ is a field, we get

COROLLARY 19.7. *There is an isomorphism*

$$\operatorname{Der}^s_{E_*}(E_*F/\mathfrak{m}, E_{*+t}/\mathfrak{m}) \simeq \operatorname{Der}^s_{E_0/\mathfrak{m}}(E_0F/\mathfrak{m}, E_t/\mathfrak{m}).$$

20. Reduction to commutative algebras

20.1. Commutative algebra derivations. We now let B be a commutative A-algebra, and let M be a B-module. Then $B \oplus M$ can be given the structure of an A-algebra augmented over B. The sections of this augmentation correspond to (commutative) A-algebra derivations of B into M. A commutative algebra derivation is the same thing as an associative algebra derivation for which M is viewed as a B-bimodule with the same B-module structure on both sides.

There are natural isomorphisms

$$\mathrm{Der}_A(B, M) \simeq \mathrm{Hom}_{B^e}(I_{B/A}, M) \simeq \mathrm{Hom}_B(I_{B/A} \underset{B^e}{\otimes} B, M)$$

using the map $B^e = B \underset{A}{\otimes} B \to B$. We call $\Omega_{B/A} = I_{B/A} \underset{B^e}{\otimes} B$ the B-module of **commutative algebra differentials** on B; there is a universal derivation $d\colon B \to \Omega_{B/A}$. Because of the universal property of $\Omega_{B/A}$, we see that $\Omega_{B/A}$ may also be described as the free B-module on elements dx for $x \in B$, modulo the relations $da = 0$ for $a \in A$ and $d(xy) = xdy + ydx$. In particular, if $B = A[X]$ is a polynomial algebra generated by a set X, then $\Omega_{B/A}$ is a free B-module generated by the set of symbols $\{dx \mid x \in X\}$.

The category of simplicial commutative algebras has a model category structure, in which a **weak equivalence** $B_\bullet \to B'_\bullet$ is a map which induces an isomorphism on the homology of the associated chain complex, and a **cofibration** is a retract of a free map. A **free map** $f\colon B_\bullet \to B'_\bullet$ is a map which in each simplicial degree is an inclusion

$$B_n \to B'_n \simeq B_n[X_n],$$

where $B[X]$ denotes the polynomial algebra over B generated by elements of X, and such that the sets $X_\bullet$ are closed under degeneracy operators of $B'_\bullet$. (See Quillen [**11**].)

Thus, we can define derived functors of commutative algebra derivations, called **André-Quillen cohomology**, by

$$H^s_A(B, M) = H^s\bigl[\mathrm{Der}_A(Q_\bullet, M)\bigr]$$

where $Q_\bullet \to B$ is a cofibrant resolution of B as a commutative algebra. (This is Quillen's construction of André-Quillen cohomology; see Quillen [**12**].)

If $Q_\bullet \to B$ is a cofibrant commutative algebra resolution of B, write $L_{B/A} = \Omega_{Q_\bullet/A} \underset{Q_\bullet}{\otimes} B$. Then

$$H^s_A(B, M) \simeq H^s\bigl[\mathrm{Hom}_B(L_{B/A}, M)\bigr].$$

20.2. Comparing associative and commutative derivations. Let's fix a commutative A-algebra B and a B-module M. Then let

$$\Lambda_{B/A} \simeq D_{B/A} \underset{B^e}{\otimes} B.$$

Thus, $\Lambda_{-/A}$ is a construction which produces a simplicial B-module from the commutative A-algebra B. It has the property that

$$\mathrm{Der}^s_A(B, M) \simeq H^s\bigl[\mathrm{Hom}_B(\Lambda_{B/A}, M)\bigr].$$

On the other hand, we have the gadget $L_{B/A}$ which is another simplicial B-module.

PROPOSITION 20.3. *If B is a commutative ring which is flat over A, then there is a first quadrant spectral sequence*

$$H_s\left(\Lambda_B^{t+1}(L_{B/A})\right) \Longrightarrow H_{s+t}\Lambda_{B/A}.$$

(Here Λ_B^d denotes the dth exterior power functor; $\Lambda_B^{t+1}L_{B/A}$ is the simplicial B-module obtained by applying the $(t+1)$st exterior power functor degree-wise to $L_{B/A}$.)

COROLLARY 20.4. *Let $A = E_0/\mathfrak{m} = k_1$ and $B = E_0F/\mathfrak{m}$. If $L_{B/A}$ is acyclic, then $\Lambda_{B/A}$ is acyclic, whence*

$$\mathrm{Der}^s_{E_0/\mathfrak{m}}(E_0F/\mathfrak{m}, E_t) = 0.$$

PROOF. Since $L_{B/A}$ is an acyclic simplicial module which is degree-wise projective, it has a simplicial contracting homotopy. Hence so does $\Lambda_B^{t+1}L_{B/A}$, whence the result follows by Proposition 20.3. □

20.5. Construction of the spectral sequence. First, let $Q_\bullet \to B$ be a vertical cofibrant resolution of B as a *commutative* algebra; without loss of generality we can assume that each Q_p is a polynomial algebra. Now, let $P_{p,\bullet} \to Q_p$ be a horizontal functorial cofibrant resolution of Q_p as an *associative* algebra. Hence, we get a double complex $P_{\bullet,\bullet}$ of associative algebras; the diagonal of this double complex is a cofibrant resolution of B as an associative algebra.

We apply the functor $P \mapsto I_{P/A} \underset{P^e}{\otimes} B$ to each spot in the double complex. The total homology of the resulting complex of B modules is precisely $H_*(\Lambda_{B/A})$. Taking horizontal homology in the complex we get

$$E_1^{p,q} = H_q(I_{P_{p,\bullet}} \underset{P^e_{p,\bullet}}{\otimes} B) \simeq H_q(\Lambda_{Q_p/A} \underset{Q_p}{\otimes} B).$$

LEMMA 20.6. *Given a flat commutative A-algebra Q there is a natural isomorphism*

$$\mathrm{Tor}^{Q^e}_{q+1}(Q,Q) \xrightarrow{\sim} H_q(\Lambda_{Q/A}),$$

and a natural map

$$i\colon \Omega_{Q/A} \xrightarrow{\sim} H_0\Lambda_{Q/A}.$$

Lemma 20.6 is proved below.

Note that $\mathrm{Tor}^{Q^e}_*(Q,Q)$ has the structure of a graded commutative algebra. This algebra structure arises from the fact that there are algebra maps $Q \to Q^e = Q \underset{A}{\otimes} Q \to Q$. Furthermore, if Q is a polynomial algebra, then $\mathrm{Tor}^{Q^e}_*(Q,Q)$ is an exterior algebra generated by Tor_1. Thus, in the case that Q is a polynomial algebra over A, (whence Q^e is a polynomial algebra over Q,) the natural map i extends in a natural way to an isomorphism of algebras

$$\Lambda_B^{*+1}(\Omega_{Q/A}) \xrightarrow{\sim} \mathrm{Tor}^{Q^e}_{*+1}(Q,Q) \xrightarrow{\sim} H_*(\Lambda_{Q/A}).$$

In particular,

$$H_q(\Lambda_{Q_p/A}) \simeq \Lambda_{Q_p}^{q+1}(\Omega_{Q_p/A})$$

whence

$$E_1^{p,q} \simeq \Lambda_{Q_p}^{q+1}(\Omega_{Q_p/A}) \underset{Q_p}{\otimes} B \simeq \Lambda_B^{q+1}(\Omega_{Q_p/A} \underset{Q_p}{\otimes} B).$$

(This last follows because $\Lambda^*_{Q_p}(\Omega_{Q_p/A})$ is a projective Q_p-module.) The right-hand side of this equation is the pth degree of $\Lambda^{q+1}_B L_{B/A}$, and thus Proposition 20.3 holds.

PROOF OF LEMMA 20.6. Recall that $\Lambda_{Q/A} = I_{P_\bullet/A} \underset{P^e_\bullet}{\otimes} Q$ where $P_\bullet \to Q$ is a cofibrant resolution of Q by associative algebras, and that there are exact sequences

$$\begin{array}{ccccccccc}
0 & \longrightarrow & I_{P_\bullet/A} & \longrightarrow & P_\bullet \underset{A}{\otimes} P_\bullet & \longrightarrow & P_\bullet & \longrightarrow & 0 \\
 & & \Big\downarrow{\scriptstyle f} & & \Big\downarrow & & \Big\downarrow & & \\
0 & \longrightarrow & I_{Q/A} & \longrightarrow & Q \underset{A}{\otimes} Q & \longrightarrow & Q & \longrightarrow & 0
\end{array}$$

The right-hand vertical arrow induces an isomorphism on homology by construction. Since Q is flat over A the middle vertical arrow also induces an equivalence on homology. Hence by the 5-lemma the map f is an equivalence.

This map f factors into a sequence of maps of simplicial modules

$$I_{P_\bullet/A} \simeq I_{P_\bullet/A} \underset{P^e_\bullet}{\otimes} P^e_\bullet \to I_{P_\bullet/A} \underset{P^e_\bullet}{\otimes} Q^e \to I_{Q/A}.$$

The left-hand arrow in this sequence is an equivalence by a straightforward spectral sequence argument (since $I_{P_\bullet/A}$ is a projective $P^e_\bullet$-module). Since the composite arrow is an equivalence, it follows that $I_{P_\bullet/A} \underset{P^e_\bullet}{\otimes} Q^e$ is a projective Q-bimodule resolution of $I_{Q/A}$.

Hence the complex $I_{P_\bullet/A} \underset{P^e_\bullet}{\otimes} Q^e \underset{Q^e}{\otimes} Q \simeq I_{P_\bullet/A} \underset{P^e_\bullet}{\otimes} Q$ (which is precisely $\Lambda_{Q/A}$) computes $\mathrm{Tor}^{Q^e}_*(I_{Q/A}, Q)$. Using the long exact sequence of Tor applied to the sequence $0 \to I_{Q/A} \to Q^e \to Q \to 0$, and the fact that $\mathrm{Tor}^{Q^e}_*(Q^e, Q)$ vanishes for $* > 0$, we see that

$$\mathrm{Tor}^{Q^e}_{q+1}(Q, Q) \xrightarrow{\sim} H_q(\Lambda_{Q/A})$$

as desired. □

21. End of the calculation

In this section we show that the André-Quillen cohomology groups, denoted $H^s_{E_0/\mathfrak{m}}(E_0F/\mathfrak{m}, E_t)$, vanish, which by Corollary 20.4, Corollary 19.4, Corollary 19.6, and Corollary 19.3 proves part (2b) of Theorem 14.5.

21.1. Differentials and the Frobenius map. Suppose that A is an $\mathbb{F}_p$-algebra, and B an A-algebra. Then the map $\sigma_B \colon B \to B$ defined by $\sigma_B(x) = x^p$ is a homomorphism. For an algebra map $A \to B$ we get a commutative diagram

$$\begin{array}{ccc}
B & \xrightarrow{\sigma_B} & B \\
\Big\uparrow & & \Big\uparrow \\
A & \xrightarrow{\sigma_A} & A
\end{array}$$

Of course, σ_B is not in general a map of A-algebras. Let σ^*B denote the pushout of commutative rings in

$$\begin{array}{ccc} B & \longrightarrow & \sigma^*B \\ \uparrow & & \uparrow \\ A & \xrightarrow{\sigma_A} & A \end{array}$$

Then there is an A-algebra map $\phi\colon \sigma^*B \to B$ (the **Frobenius map**). If $B = A[X]$ is a polynomial algebra over A generated by the set X, then $\sigma^*B \simeq A[X]$, and ϕ sends $x \in X$ to x^p.

PROPOSITION 21.2. *For any A-algebra B as above, the map*

$$L_{\sigma^*B/A} \to L_{B/A}$$

induced by $\phi\colon \sigma^*B \to B$ is the 0 *map.*

PROOF. Let $Q_\bullet \to B$ be a cofibrant commutative A-algebra resolution of B. We may suppose that $Q_n = A[X_n]$ for some set X_n. Then $\sigma^*Q_\bullet \to \sigma^*B$ is a commutative A-algebra resolution of σ^*B. If we apply the differentials functor $\Omega_{-/A}$ we get induced maps which send

$$dx \mapsto d\phi(x) = d(x^p) = px^{p-1}dx = 0$$

for each $x \in X_n \subset \sigma^*Q_n$. Thus, on the level of complexes, we get the zero map $L_{\sigma^*B/A} \to L_{B/A}$. □

COROLLARY 21.3. *If B is an A-algebra and if the maps σ_A and σ_B are isomorphisms, then $H_*(L_{B/A}) \equiv 0$.*

PROOF. If σ_A and σ_B are isomorphisms, then $\phi\colon \sigma^*B \to B$ is an isomorphism of A-algebras, whence $L\phi$ is an equivalence. But $L\phi \sim 0$ by Proposition 21.2. □

21.4. The Frobenius map on $E_0F/\mathfrak{m}$. Recall from Remark 17.4 that if $E = F = E_n$, then one can compute that

$$B = E_0F/\mathfrak{m} \simeq (\mathbb{F}_{p^n} \underset{\mathbb{F}_p}{\otimes} \mathbb{F}_{p^n})[e_0, e_1, e_2, \dots]/(e_0^{p^n-1} - 1,\ e_1^{p^n} - e_1,\ e_2^{p^n} - e_2,\ \dots).$$

Clearly the Frobenius maps on $A = \mathbb{F}_{p^n}$ and on B are isomorphisms, so in this case we are done.

More generally, let $E = E_{k_1,\Gamma_1}$ and $F = E_{k_2,\Gamma_2}$, and let $A = E_0/\mathfrak{m} = k_1$, $B = E_0F/\mathfrak{m}$ and let Q be a commutative ring of characteristic p. Then given a map of rings $f\colon B \to Q$ there is a commutative diagram

$$\begin{array}{ccc} B & \xrightarrow{\sigma_B} & B \\ \downarrow{\scriptstyle f} & & \downarrow{\scriptstyle f} \\ Q & \xrightarrow{\sigma_Q} & Q \end{array}$$

In other words, there are two maps

$$(\sigma_B)_*, (\sigma_Q)^*\colon \operatorname{Hom}_{\mathrm{Ring}}(B,Q) \rightrightarrows \operatorname{Hom}_{\mathrm{Ring}}(B,Q);$$

the top map takes f to $\sigma_Q f$, the bottom map to $f\sigma_B$, and $(\sigma_B)_* = (\sigma_Q)^*$. Thus to show that σ_B is an isomorphism, it suffices to show that $(\sigma_Q)^*$ is an isomorphism for all Q.

Recall (from Remark 17.2) our homotopy pull-back square

$$\begin{array}{ccc} \mathrm{Hom}_{\mathrm{Ring}}(E_0F/\mathfrak{m}, Q) & \longrightarrow & \mathrm{Hom}_{\mathrm{Ring}}(k_2, Q) \\ \downarrow & & \downarrow \\ \mathrm{Hom}_{\mathrm{Ring}}(k_1, Q) & \longrightarrow & \mathrm{fgl}(Q) \end{array}$$

which is functorial in Q. The automorphism σ_Q induces a map between such squares. Since k_1 and k_2 are perfect fields, the induced pth-power maps $(\sigma_Q)^*$ on $\mathrm{Hom}_{\mathrm{Ring}}(k_1, Q)$ and $\mathrm{Hom}_{\mathrm{Ring}}(k_2, Q)$ are isomorphisms.

Let $\mathrm{fgl}_n(Q) \subset \mathrm{fgl}(Q)$ denote the full subgroupoid of formal group laws of height n. (Although Q is only a ring of characteristic p, and not in general a field, we define height as before: a formal group law F has height n if $[p]_F(x) = ax^{p^n} + \cdots$ where $a \in Q$ is an *invertible* element. Under this definition it is possible for a formal group law over Q to not have a well-defined height. Note however that if $f\colon F \to G$ is an isomorphism of formal group laws over Q, and if F is of height n, then G is also of height n, since $[p]_G = f \circ [p]_F \circ f^{-1}$.)

PROPOSITION 21.5. *There is a natural isomorphism of the functor*

$$(\sigma_Q^n)^*\colon \mathrm{fgl}_n(Q) \to \mathrm{fgl}_n(Q)$$

with the identity functor, whence σ_Q^* is an equivalence of categories.

COROLLARY 21.6. *The Frobenius map on $E_0F/\mathfrak{m}$ is an isomorphism.*

Recall that there is a homomorphism $\alpha_F\colon F \to \sigma^* F$ of formal group laws given by $\alpha_F(x) = x^p$.

LEMMA 21.7. *Given a map $f\colon F \to G$ of formal group laws over a ring of characteristic p, such that $f'(0) = 0$, there exists a unique map $g\colon \sigma^*F \to G$ of formal group laws such that $g \circ \alpha_F = f$.*

PROOF. Because $f(F(x,y)) = G(fx, fy)$, applying $\frac{\partial}{\partial y}$ and setting $y = 0$ gives $f'(x)F_2(x,0) = G_2(fx,0)f'(0)$, whence $f'(x) = 0$. Thus $f(x) = g(x^p)$ for a (unique) power series g. Since

$$g(\sigma^*F(x^p, y^p)) = g(F(x,y)^p) = f(F(x,y)) = G(fx, fy) = G(g(x^p), G(y^p))$$

we see that $g(x \underset{F}{+} y) = g(x) \underset{G}{+} g(y)$. □

PROOF OF PROPOSITION 21.5. If F is a formal group law of height n, then the p-series $[p]_F(x)\colon F \to F$ has the form $[p]_F(x) = ax^{p^n} + \cdots$, with a invertible. Thus by Lemma 21.7 there exists a unique isomorphism $\beta_F\colon (\sigma^n)^*F \to F$ such that $\beta_F \circ \alpha_F^n = [p]_F$.

If $h\colon F \to G$ is a map between height n formal group laws, we see that

$$h(\beta_F(x^{p^n})) = h([p]_F(x)) = [p]_G(h(x)) = \beta_G(h(x)^{p^n}) = \beta_G(h(x^{p^n})),$$

whence $h \circ \beta_F = \beta_G \circ h$, whence $\beta\colon (\sigma^n)^* \to \mathrm{id}$ is an natural isomorphism of functors. Thus $(\sigma^n)^*$ is an equivalence of categories. □

Part 4. Construction of the $\mathcal{A}_\infty$-ring structures

In the following sections we show that the homology theories introduced in Section 6 admit $\mathcal{A}_\infty$-ring structures, by showing that certain obstructions to the existence of such structures vanish. This technique originated in work of Robinson [**15**] on the existence of $\mathcal{A}_\infty$-structures for Morava K-theories (see also Remark 23.8).

22. Producing an algebra structure

We still need to show that for each formal group law (k, Γ) the ring spectrum $E = E_{k,\Gamma}$ admits an $\mathcal{A}_\infty$-structure. That is, for an appropriate $\mathcal{A}_\infty$-operad C, we want a map of operads making the diagram

$$\begin{array}{ccc} C & \dashrightarrow & \mathcal{E}_E \\ & \searrow \quad \swarrow & \\ & \mathcal{L} & \end{array}$$

commute. Recall from Section 12.1 that $\mathcal{E}_E$ is an operad with $\mathcal{E}_E[n] = \mathcal{S}(E^{(n)}, E)$.

We will choose our operad C so that it is a cofibrant $\mathcal{A}_\infty$ operad.

22.1. Free operads. If A is a Σ-object in some symmetric monoidal category, the **free operad** $\mathcal{F}A$ on A is an operad equipped with a map of Σ-objects $A \to \mathcal{F}A$ inducing an isomorphism

$$\mathrm{Hom}_{\mathrm{oper}}(\mathcal{F}A, C) \simeq \mathrm{Hom}_\Sigma(A, C),$$

for each operad C. There is a free operad functor which is left adjoint to the forgetful functor from operads to Σ-objects.

Since operads are monoids in the category of Σ-objects, the free operad should really be a kind of "free monoid". This is true, but the formula for such a free monoid is somewhat strange. Morally speaking,

$$\mathcal{F}A = I \amalg A \circ (I \amalg A \circ (I \amalg A \circ (\cdots))).$$

More precisely, $\mathcal{F}A$ can be written as a countable colimit of expressions of the above form which terminate at some finite depth.

The reason for this strange formula is that the monoidal structure $\circ$ on Σ-objects is not "bi-distributive"; rather, it distributes over coproducts on the right, but not on the left. I.e., in general $(A \amalg B) \circ C \simeq (A \circ C) \amalg (B \circ C)$, but $A \circ (B \amalg C) \not\simeq (A \circ B) \amalg (A \circ C)$.

22.2. Model category for operads. We give a model category structure for the category $s\,\mathrm{oper}$ of simplicial operads.

Say a Σ-object A is **Σ-free** if each $A[n]$ is free as a Σ_n-simplicial set. We define a closed model category for simplicial operads by letting $f\colon A \to B$ be

1. a **weak equivalence** if each map $f[n]\colon A[n] \to B[n]$ is a weak equivalence on the underlying simplicial set, and
2. a **cofibration** if it is a retract of a free map. A **free map** is a map f of the form
$$f_n\colon A_n \to B_n \simeq A_n \underset{\mathrm{oper}}{\amalg} \mathcal{F}X_n$$
where X_n is a discrete, Σ-free Σ-object, and such that the subobjects $X_n \subset B_n$ are closed under degeneracy operators.

That this indeed gives a model category structure follows from general results of Quillen [**11**].

To produce our desired operad C, first let $A \to \mathcal{L}$ be an arbitrary $\mathcal{A}_\infty$-operad over the linear isometries operad. Let $\operatorname{Sing} A$ be the simplicial operad obtained by taking singular simplices, and let $\tilde{C} \to \operatorname{Sing} A$ be a cofibrant resolution of $\operatorname{Sing} A$; we can choose $\tilde{C}$ so that $\tilde{C}_n = \mathcal{F}K_n$ for some Σ-free Σ-object K_n. Then by adjunction we get a map $C = |\tilde{C}| \to A$ which is a weak equivalence of topological operads.

22.3. The spectral sequence. We want to show that $\operatorname{map}_{(\mathrm{oper}\downarrow\mathcal{L})}(C, \mathcal{E}_E)$ is non-empty. The simplicial object $\tilde{C}_n = \mathcal{F}K_n$ leads to a cosimplicial space $Y^\bullet$, with

$$Y^n = \operatorname{map}_{(\mathrm{oper}\downarrow\mathcal{L})}(\mathcal{F}K_n, \mathcal{E}_E),$$

with $\operatorname{Tot} Y^\bullet = \operatorname{map}_{(\mathrm{oper}\downarrow\mathcal{L})}(|\tilde{C}|, \mathcal{E}_E)$. Thus, if we choose an element $\sigma \in \pi_0 Y^0 \rightrightarrows \pi_0 Y^1$, there is an obstruction theory / spectral sequence of the cosimplicial space (see Section 14) with

$$E_2^{s,t} = \pi^s \pi_t\big(\operatorname{map}_{(\mathrm{oper}\downarrow\mathcal{L})}(\mathcal{F}K_n, \mathcal{E}_E), \sigma)\big).$$

22.4. Identification of $E_2^{0,0}$. We need to compute the equalizer of the pair of arrows

$$\pi_0 \operatorname{map}_{(\mathrm{oper}\downarrow\mathcal{L})}(\mathcal{F}K_0, \mathcal{E}_E) \rightrightarrows \pi_0 \operatorname{map}_{(\mathrm{oper}\downarrow\mathcal{L})}(\mathcal{F}K_1, \mathcal{E}_E).$$

Note that for any operad B over $\mathcal{L}$ there is a natural map

$$\pi_0 \operatorname{map}_{(\mathrm{oper}\downarrow\mathcal{L})}(B, \mathcal{E}_E) \to \operatorname{Hom}_{\mathrm{oper}}(\pi_0 B, \pi_0 \mathcal{E}_E)$$

which sends an operad map $f\colon B \to \mathcal{E}_E$ over $\mathcal{L}$ to the map $\pi_0 f\colon \pi_0 B \to \mathcal{E}_E$, where $\pi_0 B$ denotes the discrete operad with $(\pi_0 B)[n] = \pi_0(B[n])$.

Since $\mathcal{E}_E[n] \simeq \mathcal{S}(E^{(n)}, E)$, we see that $\pi_0\mathcal{E}_E[n] \simeq [E^{(n)}, E]$. In fact, by Proposition 15.5,

$$\pi_0\mathcal{E}_E[n] \simeq \operatorname{Hom}_{E_*}((E_*E)^{\otimes n}, E_*).$$

Now suppose $B = \mathcal{F}K$ for some discrete Σ-object K equipped with a map $K \to \mathcal{L}$. Such a K is necessarily Σ-free, since $\mathcal{L}$ is. There is a commutative diagram

$$\begin{array}{ccc}
\pi_0 \operatorname{map}_{(\mathrm{oper}\downarrow\mathcal{L})}(\mathcal{F}K, \mathcal{E}_E) & \longrightarrow & \operatorname{Hom}_{\mathrm{oper}}(\mathcal{F}K, \pi_0\mathcal{E}_E) \\
\downarrow & & \downarrow \\
\pi_0 \operatorname{map}_{(\Sigma\downarrow\mathcal{L})}(K, \mathcal{E}_E) & \longrightarrow & \operatorname{Hom}_{\Sigma}(K, \pi_0\mathcal{E}_E)
\end{array}$$

The vertical arrows are isomorphisms by the characteristic property of a free operad. The lower horizontal arrow is also an isomorphism, since K is discrete and Σ-free. Thus the upper horizontal arrow is an isomorphism. Thus we see that $E_2^{0,0}$ is isomorphic to the equalizer of the pair of arrows

$$E_2^{0,0} \to \operatorname{Hom}_{\mathrm{oper}}(\mathcal{F}K_0, \pi_0\mathcal{E}_E) \rightrightarrows \operatorname{Hom}_{\mathrm{oper}}(\mathcal{F}K_1, \pi_0\mathcal{E}_E).$$

Since $\pi_0 C$, the coequalizer of $\mathcal{F}K_1 \rightrightarrows \mathcal{F}K_0$, is the associative monoid operad, it follows that $E_2^{0,0} \simeq \operatorname{Hom}_{\mathrm{oper}}(\pi_0 C, \pi_0\mathcal{E}_E)$. This is precisely the set of ring spectrum structures on E; that is, an associative ring spectrum structure on E (in the "classical" sense, rather than in the $\mathcal{A}_\infty$ sense) is precisely equivalent to a map from the

22.5. Identification of E_2-term. Let us now choose the element $\sigma \in E_2^{0,0}$ corresponding to the "usual" ring spectrum structure on E. We obtain as before an obstruction theory / spectral sequence with

$$E_1^{s,t} \simeq \pi_t\big(\mathrm{map}_{(\mathrm{oper}\downarrow\mathcal{L})}(\mathcal{F}K_s, \mathcal{E}_E), \sigma\big)$$

and "converging" to $\pi_{t-s}\big(\mathrm{map}_{(\mathrm{oper}\downarrow\mathcal{L})}(C, \mathcal{E}_E), \sigma\big)$.

THEOREM 22.6. *The E_2-term of this spectral sequence is given by*

$$E_2^{s,t} = \begin{cases} 0 & \text{if } s > 0, \\ [\Sigma^t E, E] & \text{if } s = 0, t > 0. \end{cases}$$

PROOF. This follows from the discussion of Section 23.1, Theorem 23.5, Remark 23.6, and Proposition 23.7. □

In particular, the obstruction groups $E_2^{s,s-1}$ vanish identically, and so

COROLLARY 22.7. *The spectrum E admits an A_∞-ring structure whose underlying ring structure is equivalent to the usual one.*

23. Computation of algebra-structure obstruction groups

If B is an operad and K is an unpointed space, let $\mathrm{map}(K, B)$ denote the operad with $\mathrm{map}(K, B)[n] = \mathrm{map}(K, B[n])$, and with operad structure given by "point-wise composition". Now if given a map $B \to \mathcal{L}$ of operads, define an operad B^K over the linear isometries operad via the pullback square

$$\begin{array}{ccc} B^K & \longrightarrow & \mathrm{map}(K, B) \\ \downarrow & & \downarrow \\ \mathcal{L} = \mathrm{map}(\mathrm{pt}, \mathcal{L}) & \longrightarrow & \mathrm{map}(K, \mathcal{L}) \end{array}$$

To prove Theorem 22.6, we must first compute the E_1-term. Suppose that $B \to \mathcal{L}$ is a map of operads, and $\sigma\colon B \to \mathcal{E}_E$ is some fixed map of operads over $\mathcal{L}$. Then the set

$$\pi_t\big(\mathrm{map}_{(\mathrm{oper}\downarrow\mathcal{L})}(B, \mathcal{E}_E), \sigma\big)$$

is the same as the set of homotopy classes of maps $S^t \to \mathrm{map}_{(\mathrm{oper}\downarrow\mathcal{L})}(B, \mathcal{E}_E)$ which send the base-point of the sphere to σ, and where homotopies are base-point preserving.

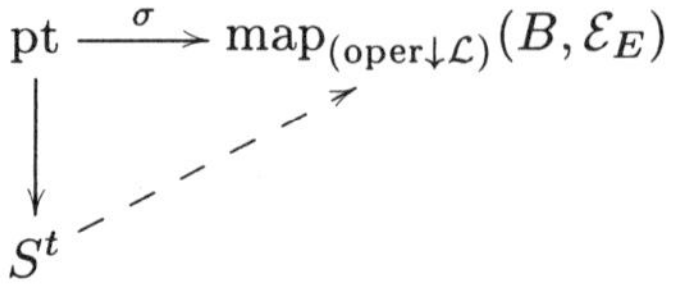

By adjunction, these are the same as homotopy classes of lifts in the diagram

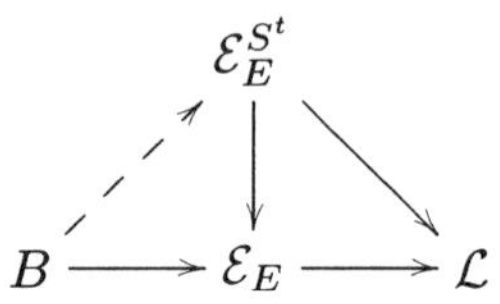

That is, there is a natural isomorphism

$$\pi_t\big(\mathrm{map}_{(\mathrm{oper}\downarrow\mathcal{L})}(B, \mathcal{E}_E), \sigma\big) \to \pi_0\big(\mathrm{map}_{(\mathrm{oper}\downarrow\mathcal{E}_E)}(B, \mathcal{E}_E^{S^t})\big).$$

There is a natural map

$$\pi_0 \operatorname{map}_{(\mathrm{oper}\downarrow\mathcal{L})}(B, \mathcal{E}_E^{S^t}) \to \operatorname{Hom}_{\mathrm{oper}}(\pi_0 B, \pi_0\mathcal{E}_E).$$

If $B = \mathcal{F}K$ and K is discrete and Σ-free there is a square

$$\begin{array}{ccc} \pi_0 \operatorname{map}_{(\mathrm{oper}\downarrow\mathcal{L})}(\mathcal{F}K, \mathcal{E}_E^{S^t}) & \longrightarrow & \operatorname{Hom}_{\mathrm{oper}}(\mathcal{F}K, \pi_0\mathcal{E}_E^{S^t}) \\ \downarrow & & \downarrow \\ \pi_0 \operatorname{map}_{(\Sigma\downarrow\mathcal{L})}(K, \mathcal{E}_E^{S^t}) & \longrightarrow & \operatorname{Hom}_{\mathrm{oper}}(K, \pi_0\mathcal{E}_E^{S^t}) \end{array}$$

The vertical arrows are isomorphisms by the universal property of free operads. The lower horizontal arrow is an isomorphism since K is discrete and Σ-free. Thus the top horizontal arrow is an isomorphism. This isomorphism, and the fact that the map $\mathcal{E}_E^{S^t} \to \mathcal{E}_E$ is a split fibration (using the base-point of S^t), implies that the map

$$\pi_0\big(\operatorname{map}_{(\mathrm{oper}\downarrow\mathcal{E}_E)}(\mathcal{F}K, \mathcal{E}_E^{S^t})\big) \to \operatorname{map}_{(\mathrm{oper}\downarrow\pi_0\mathcal{E}_E)}(\mathcal{F}K, \pi_0\mathcal{E}_E^{S^t})$$

is an isomorphism.

Since $\mathcal{E}_E^{S^t}[n]$ is weakly equivalent to $\operatorname{map}(S^t, \mathcal{E}_E[n])$, it follows that

$$\pi_0\mathcal{E}_E^{S^t}[n] = \operatorname{Hom}_{E_*}\big((E_*E)^{\otimes n}, E_*[\epsilon]/\epsilon^2\big),$$

where $|\epsilon| = -t$; furthermore, the projection map $\pi_0\mathcal{E}_E^{S^t} \to \pi_0\mathcal{E}_E$ sends ϵ to 0.

Since the pinch map $S^t \to S^t \vee S^t$ makes the sphere into a cogroup up to homotopy, the induced map of operads

$$\pi_0\mathcal{E}_E^{S^t} \underset{\pi_0\mathcal{E}_E}{\times} \pi_0\mathcal{E}_E^{S^t} \simeq \pi_0\mathcal{E}_E^{S^t\vee S^t} \to \pi_0\mathcal{E}_E^{S^t}$$

makes $\pi_0\mathcal{E}_E^{S^t}$ into an abelian group object in the category of operads over $\pi_0\mathcal{E}_E$.

23.1. Quillen cohomology of an operad. Suppose $A \to B$ is a map of discrete operads. Let $C_\bullet$ be a cofibrant resolution of a discrete operad A. Let $N \to B$ be an abelian group operad over an operad B. Then we define the **Quillen cohomology of an operad** by

$$H^s(A, N) = H^s\big[\operatorname{Hom}_{(\mathrm{oper}\downarrow B)}(C_\bullet, N)\big].$$

In our case we take A to be the associative operad, $B = \pi_0\mathcal{E}_E$, and we take $N = \pi_0\mathcal{E}_E^{S^t}$ for $t > 0$. Thus we have that

$$E_2^{s,t} = H^s(A, \pi_0\mathcal{E}_E^{S^t}).$$

23.2. Bimodules. Let A be an operad. An **A-bimodule** is a Σ-object M equipped with maps $\ell\colon A \circ M \to M$ and $r\colon M \circ A \to M$ such that the diagrams

$$\begin{array}{ccc} A\circ A\circ M & \xrightarrow{\mu\circ M} & A\circ M \\ \big\downarrow{\scriptstyle A\circ\ell} & & \big\downarrow{\scriptstyle \ell} \\ A\circ M & \xrightarrow{\ \ell\ } & M \end{array} \qquad \begin{array}{ccc} M\circ A\circ A & \xrightarrow{M\circ\mu} & M\circ A \\ \big\downarrow{\scriptstyle r\circ A} & & \big\downarrow{\scriptstyle r} \\ M\circ A & \xrightarrow{\ r\ } & M \end{array}$$

and

$$\begin{array}{ccccc} A \circ M & \xleftarrow{\eta \circ M} & M & \xrightarrow{M \circ \eta} & M \circ A \\ & \searrow^{\ell} & \| & \swarrow_{r} & \\ & & M & & \end{array}$$

commute.

There are several examples of bimodules.

1. A is itself an A-bimodule.
2. If K is a Σ-object, then $A \circ K \circ A$ is an A-bimodule, called the free A-bimodule on K
3. If X and Y are objects, define a Σ-object $\mathfrak{hom}(X, Y)$ by
$$\mathfrak{hom}(X, Y)[n] = \operatorname{Hom}(X^{\otimes n}, Y).$$
If $X = Y$, then $\mathfrak{hom}(X, X)$ is the endomorphism operad of X. If X and Y are A-algebras, then $\mathfrak{hom}(X, Y)$ is an A-bimodule.

An A-bimodule M induces a functor from A-algebras to A-algebras as follows. Define $M \underset{A}{\circ} (X)$ by the coequalizer

$$M \underset{A}{\circ} (X) = \operatorname{Cok}\left((M \circ A)(X) \rightrightarrows M(X)\right).$$

Note that if $M = A$, then $A \underset{A}{\circ} (X) \simeq X$; i.e., the bimodule A induces the identity functor on A-algebras. Likewise, if $M = A^{\circ d}$, then $A^{\circ d} \underset{A}{\circ} (X) \simeq A^{d-1}(X)$. There is an adjunction

$$\operatorname{Hom}_{A\text{-bimod}}(M, \mathfrak{hom}(X, Y)) \simeq \operatorname{Hom}_{A\text{-alg}}\left(M \underset{A}{\circ} (X), Y\right). \tag{23.3}$$

23.4. Bar complex. If M and N are A-bimodules, let $\mathcal{B}(M, A, N)$ denote the simplicial object

$$M \circ N \leftleftarrows M \circ A \circ N \;\substack{\longleftarrow\\ \longleftarrow\\ \longleftarrow}\; M \circ A \circ A \circ N \;\substack{\longleftarrow\\ \longleftarrow\\ \longleftarrow\\ \longleftarrow}\; \cdots$$

In particular, $\mathcal{B}(A, A, A)$ is a resolution of A by free A-bimodules.

If $N \to B$ is a coefficient system (i.e., an abelian group object over B), then we define the **Hochschild cohomology** of A by

$$HH^s(A, N) = H^s\left[\operatorname{Hom}_{(A\text{-bimod}\downarrow B)}\left(\mathcal{B}(A, A, A), N\right)\right].$$

The fact that relates the Quillen cohomology and Hochschild cohomology of an operad is the following.

THEOREM 23.5. **[14]** *If A is a discrete, Σ-free operad, and N is a coefficient system for A, then there is an exact sequence*

$$0 \to HH^0(A, N) \to HH^0(I, N) \to H^0(A, N) \to HH^1(A, N) \to 0$$

and isomorphisms

$$H^s(A, N) \simeq HH^{s+1}(A, N)$$

for $s > 0$.

REMARK 23.6. Recall that I denotes the "trivial operad". Since $I \circ I \simeq I$, the bar complex $\mathcal{B}(I, I, I)$ is constant, so the Hochschild cohomology of I is concentrated in degree 0. In the case of interest to us, in which the coefficient system N is $\pi_0 \mathcal{E}_E^{S^t} \to \pi_0 \mathcal{E}_E$ pulled back along $I \to \pi_0 \mathcal{E}_E$, it is trivial to show that

$$\begin{aligned} HH^0(I, N) &\simeq \operatorname{Hom}_{(I\text{-bimod}\downarrow \pi_0 \mathcal{E}_E)}(I, \pi_0 \mathcal{E}_E^{S^t}) \\ &\simeq \operatorname{Hom}_{E_*}(E_* E, E_{*+t}) \\ &\simeq [\Sigma^t E, E], \end{aligned}$$

and that

$$HH^s(I, N) = 0 \quad \text{for } s > 0.$$

PROPOSITION 23.7. *If A is the associative operad, then $HH^s(A, \pi_0 \mathcal{E}_E^{S^t}) \simeq 0$ for all $s \geq 0$ and $t > 0$.*

PROOF. These cohomology groups are computed using the co-chain complex

$$C^n = \operatorname{Hom}_{(A\text{-bimod}\downarrow \pi_0 \mathcal{E}_E)}(A^{\circ n+2}, \pi_0 \mathcal{E}_E^{S^t}).$$

Note that

$$\begin{aligned} \operatorname{Hom}_{A\text{-bimod}}(A^{\circ n+2}, \pi_0 \mathcal{E}_E^{S^t}) &\simeq \operatorname{Hom}_{A\text{-bimod}}\big(A^{\circ n+2}, \mathfrak{hom}_{E_*}(E_* E, E_*[\epsilon]/\epsilon^2)\big) \\ &\simeq \operatorname{Hom}_{A\text{-alg}}(A^{\circ n+2} \underset{A}{\circ} (E_* E), E_*[\epsilon]/\epsilon^2) \\ &\simeq \operatorname{Hom}_{E_*\text{-alg}}\big(T^{n+1}(E_* E), E_*[\epsilon]/\epsilon^2)\big), \end{aligned}$$

using the adjunction of (23.3). Likewise,

$$\operatorname{Hom}_{A\text{-bimod}}(A^{\circ n+2}, \pi_0 \mathcal{E}_E) \simeq \operatorname{Hom}_{E_*\text{-alg}}\big(T^{n+1}(E_* E), E_*\big).$$

Thus,

$$C^n \simeq \operatorname{Hom}_{(E_*\text{-alg}\downarrow E_*)}(T^{n+1}(E_* E), E_*[\epsilon]/\epsilon^2),$$

and as we know from Sections 20 and 21, this complex computes $\operatorname{Der}^s_{E_*}(E_* E, E_{*+t})$, which vanishes. ☐

REMARK 23.8. Compare this to Robinson's calculation [**15**]. If we had taken $E \simeq K(n)$, the nth Morava K-theory, and if p is odd (so that $K(n)$ is a commutative ring spectrum), then the proof of Proposition 23.7 shows that $HH^s(A, \pi_0 \mathcal{E}_E^{S^t})$ is the Quillen cohomology of the $K(n)$-algebra $K(n)_* K(n)$ with coefficients in the module $K(n)_{*+t}$. By a result of Quillen [**12**], these groups are the same up to a shift in dimension as the Hochschild cohomology of the *algebra* $K(n)_* K(n)$; this recovers Robinson's calculations of the obstruction groups for this spectrum.

The Hochschild cohomolgy of $K(n)_* K(n)$ is in general non 0. However, the algebra $K(n)_* K(n) \simeq \Sigma(n) \otimes \Lambda(n)$, where $\Lambda(n)$ is an exterior algebra on n-generators, and the Hochschild cohomology of $K(n)_* K(n)$ originates almost entirely from this exterior factor. By replacing $K(n)$ with $E_{k,\Gamma}$ we get rid of this contribution.

PROOF OF THEOREM 23.5. Let $C_\bullet \to A$ be a cofibrant resolution of the operad A by free operads; thus, $C_n = \mathcal{F} K_n$. We obtain a double complex $D_{m,n} = \mathcal{B}(C_n, C_n, C_n)_m$ of bimodules using the bar construction. The diagonal of $D_{\bullet,\bullet}$ is a $C_\bullet$-bimodule resolving $C_\bullet$. If N is a coefficient system for A, then by applying Hochschild cohomology we obtain a spectral sequence

LEMMA 23.9. *Let C be a discrete operad. Then there is a sequence*

$$0 \to HH^0(C,N) \to HH^0(I,N) \to H^0(C,N) \to HH^1(C,N) \to 0$$

natural in C; furthermore, if $C = \mathcal{F}K$ for some Σ-object K, the sequence is exact, and $HH^s(C,N) \simeq 0$ for $s > 1$.

PROOF. See [**14**]. □

References

1. J.F. Adams, *Stable homotopy and generalised homology theories*, University of Chicago Press, 1973.
2. A. K. Bousfield, *Homotopy spectral sequences and obstructions*, Israel J. Math. **66** (1989), 1–3.
3. A. K. Bousfield and D. M. Kan, *Homotopy limits, completions and localizations*, Lecture Notes in Mathematics 304, Springer-Verlag, 1972.
4. V. G. Drinfel'd, *Elliptic modules*, Math. USSR-Sb. **23** (1974), 561–592.
5. A. D. Elmendorf, *The Grassmanian geometry of spectra*, Journal of Pure and Applied Algebra **54** (1988), 37–94.
6. A. D. Elmendorf, I. Kriz, M. A. Mandell, and J. P. May, *Rings, modules, and algebras in stable homotopy theory*, Mathematical Surveys and Monographs 47, AMS, 1997.
7. A. Fröhlich, *Formal groups*, Lecture Notes in Mathematics 74, Springer-Verlag, 1968.
8. P. S. Landweber, *Homological properties of comodules over $MU_*(MU)$ and $BP_*(BP)$*, American Journal of Mathematics **98** (1976), 591–610.
9. L. G. Lewis, J. P. May, and M. Steinberger, *Equivariant stable homotopy theory*, Lecture Notes in Mathematics 1213, Springer-Verlag, 1986.
10. J. Lubin and J. Tate, *Formal moduli for one-parameter formal lie groups*, Bull. Soc. math. France **94** (1966), 49–60.
11. D. G. Quillen, *Homotopical algebra*, Lecture Notes in Mathematics 43, Springer-Verlag, 1967.
12. ———, *On the (co-)homology of commutative rings*, Applications of Categorical Algebra, Proceedings in the Symposia in Pure Mathematics, AMS, 1970.
13. D. C. Ravenel, *Complex cobordism and stable homotopy groups of spheres*, Academic Press, 1986.
14. C. Rezk, *Operads and spaces of algebra structures*, PhD dissertation, MIT, 1995.
15. A. Robinson, *Obstruction theory and the strict associativity of Morava K-theories*, Advances in Homotopy Theory, Cambridge University Press, 1989, pp. 143–152.
16. J.-P. Serre, *Local fields*, Graduate Texts in Mathematics, Springer-Verlag, 1979.
17. J. H. Silverman, *The arithmetic of elliptic curves*, Graduate Texts in Mathematics, Springer-Verlag, 1986.

DEPARTMENT OF MATHEMATICS, NORTHWESTERN UNIVERSITY, EVANSTON, IL 60208
E-mail address: `rezk@math.nwu.edu`

Contemporary Mathematics
Volume **220**, 1998

Commutative Morava homology Hopf algebras

Hal Sadofsky and W. Stephen Wilson

ABSTRACT. We give the Dieudonné module theory for $\mathbf{Z}/2(p^n - 1)$-graded bicommutative Hopf algebras over $\mathbf{F}_p$. These objects arise as the Morava K-theory of homotopy associative, homotopy commutative H-spaces.

1. Introduction

In [**DG70**, V], Demazure and Gabriel classify commutative unipotent algebraic groups over a perfect field of characteristic p in terms of Dieudonné modules. Making appropriate translations of terminology, this classifies a category of Hopf algebras in terms of Dieudonné-modules.

Bousfield, in an appendix to [**Bou96b**], proves many of the results of interest in an accessible way, adapting and clarifying [**DG70**, V] for his specific purposes. In particular, he gives much of the structure for $\mathbf{Z}/(2)$-graded bicommutative Hopf algebras over $\mathbf{F}_p$. He needs these results for his work with mod p K-theory. We are interested in similar results for Morava K-theory. The Morava K-theory of homotopy commutative H-spaces gives rise to bicommutative $\mathbf{Z}/2(p^n - 1)$-graded Hopf algebras satisfying certain conditions. Having a Dieudonné module theory for these Hopf algebras gives us a great deal of control over their structure. When the grading is forgotten then Bousfield's results can be applied, so our contribution is really to unravel the gradings in our case and make appropriate definitions so that Bousfield's proofs go through with little or no modification.

Specifically, we want to study the category $\mathcal{C}(n)$ of *commutative Morava Hopf algebras.* These are bicommutative, biassociative, Hopf algebras over $\mathbf{F}_p$ which are graded over $\mathbf{Z}/2(p^n - 1)$ $(n > 0)$ and have an exhaustive primitive filtration. This last condition, *an exhaustive primitive filtration*, means that each element maps trivially under some iterate of the reduced coproduct (into the tensor product of the cokernel of coaugmentation ideal). (Bousfield calls this property "irreducible," a term we wish to reserve for a Hopf algebra with no proper non-trivial sub-Hopf algebras). Bousfield's work analyzes these Hopf algebras quite thoroughly by reducing the $\mathbf{Z}/(2)$-graded case to the ungraded case by splitting off the sub-algebra generated by odd-dimensional primitives. The study of these Hopf algebras in [**HRW98**] concentrates on important consequences of the grading. Our goal here is to put the grading into Bousfield's results.

1991 *Mathematics Subject Classification.* Primary 16W30, 55N20.

[**HRW98**] shows $\mathcal{C}(n)$ splits (for odd primes) as the product of two categories; $\mathcal{OC}(n)$, whose objects are Hopf algebras generated by primitives in odd degrees, and hence primitively generated exterior algebras, and $\mathcal{EC}(n)$, whose objects are all concentrated in even degrees. (Note that [**Bou96b**] proves the same result for the $\mathbf{Z}/(2)$-graded case, and his proof also holds without alteration for $\mathcal{C}(n)$.) So, our real category of study will be $\mathcal{EC}(n)$.

In [**Bou96a**], Bousfield shows the $\mathbf{Z}/(2)$-graded case is an abelian category, but again, his proof works just as well for $\mathcal{C}(n)$. Our contribution is to define a functor to Dieudonné modules, to which we again appeal to [**Bou96b**] with small alterations.

We define a *Morava-Dieudonné module* to be a $\mathbf{Z}/(p^n-1)$-graded abelian group M_* with endomorphisms $F : M_t \to M_{pt}$, and $V : M_t \to M_{t/p}$ (note that in $\mathbf{Z}/(p^n-1)$ every t/p exists) such that:

(i) $FV = p = VF$;
(ii) for each $x \in M$, there exists a $q \geq 1$ with $V^q(x) = 0$.

We denote the category of Morava-Dieudonné modules by $\mathcal{MD}(n)_*$. Note the definition implies the modules are all p-torsion, i.e. $p^q(x) = (FV)^q(x) = F^qV^q(x) = 0$ where we have used (i) and (ii).

Our main theorem is:

THEOREM 1.1. *There is a functor $m_* : \mathcal{EC}(n) \to \mathcal{MD}(n)_*$ which is an equivalence of abelian categories.*

Let $\mathbf{Z}/(n)$ act on $\mathbf{Z}/(p^n-1)$ by $k * j = jp^k$. The orbits are given by $\gamma = \{j, jp, jp^2, jp^3, \ldots, jp^n = j\}$. We can see with no work that $\mathcal{MD}(n)_*$ splits as a product of categories $\mathcal{MD}_\gamma(n)_*$ where all elements are in the degrees contained in γ. This translates into a similar theorem, a major result of [**HRW98**], about $\mathcal{EC}(n)$.

COROLLARY 1.2 ([**HRW98**], Theorem 1.13). *As categories:*

$$\mathcal{EC}(n) \simeq \prod_\gamma \mathcal{EC}(n)_\gamma$$

where $\mathcal{EC}(n)_\gamma$ consists of Hopf algebras in $\mathcal{EC}(n)$ with all primitives in degrees 2γ.

This result was proven in [**HRW98**] so it could be combined with results of [**RW80**] to show that there is only the trivial map from $K(n)_*(K(\mathbf{Z}/(p^i), s))$ to $K(n)_*(K(\mathbf{Z}/(p^j), t))$ when $s \neq t$.

The motivation for us is Morava K-theory, $K(n)_*(-)$, which is a generalized homology theory with a Künneth isomorphism. The coefficient ring is $K(n)_* \simeq \mathbf{F}_p[v_n, v_n^{-1}]$ where the degree of v_n is $2(p^n-1)$. By setting $v_n = 1$ we get a $\mathbf{Z}/2(p^n-1)$-graded theory, $\overline{K(n)}_*(-)$. If a space X is connected and the loop space of an H-space then $\overline{K(n)}_*(X)$ is in $\mathcal{C}(n)$ for odd primes (there are some complications with commutativity for $p=2$).

In [**RW80**] the Morava K-theory of all (abelian) Eilenberg-Mac Lane spaces was computed. They all, with the exception of $K(\mathbf{Z}, 1)$, the circle, lie in $\mathcal{EC}(n)$. Let $I = (i_1, \ldots, i_q)$, $0 \leq i_1 < i_2 < \cdots < i_q < n$. When $i_1 > 0$ define $s(I) = (i_1-1, i_2-1, \ldots, i_q-1)$ and when $i_q < n-1$ define $s^{-1}(I) = (i_1+1, i_2+1, \ldots, i_q+1)$. We can just read off the Morava-Dieudonné module stucture from [**RW80**].

THEOREM 1.3. *The Morava-Dieudonné module,* $m_*(\overline{K(n)}_*(K(\mathbf{Z}/(p^j),q)))$, *is a free* $\mathbf{Z}/(p^j)$ *module on generators* $a_I = a_{(i_1)}a_{(i_2)}\cdots a_{(i_q)}$, $I = (i_1,\dots,i_q)$, $0 \le i_1 < i_2 < \cdots < i_q < n$ *in degree* $p^{i_1}+p^{i_2}+\cdots+p^{i_q}$ *with*

$$V(a_I) = \begin{cases} a_{s(I)} & \text{if } i_1 > 0 \\ (-1)^{q-1} p a_{(i_2-1)}a_{(i_3-1)}\cdots a_{(n-1)} & \text{if } i_1 = 0 \end{cases}$$

and

$$F(a_I) = \begin{cases} p a_{s^{-1}(I)} & \text{if } i_q < n-1 \\ (-1)^{q-1} a_{(0)}a_{(i_1+1)}\cdots a_{(i_{q-1}+1)} & \text{if } i_q = n-1. \end{cases}$$

We have, from [**RW80**, Corollary 13.1], a short exact sequence of Hopf algebras:

$$\mathbf{F}_p \to \overline{K(n)}_*(K(\mathbf{Z}/(p^i),q)) \to \overline{K(n)}_*(K(\mathbf{Z}/(p^{i+j}),q)) \to \overline{K(n)}_*(K(\mathbf{Z}/(p^j),q)) \to \mathbf{F}_p.$$

This translates into the most obvious exact sequence in the Morava-Dieudonné modules. For the surjection, [**RW80**, Proposition 11.4] is enough and for the injection we need that and [**RW80**, Theorem 11.1(c)].

Since

$$\lim_{\to} \overline{K(n)}_*(K(\mathbf{Z}/(p^i),q)) \simeq \overline{K(n)}_*(K(\mathbf{Q}/\mathbf{Z}_{(p)},q)) \simeq \overline{K(n)}_*(K(\mathbf{Z},q+1))$$

the above map shows the Morava-Dieudonné module for this is a bunch of copies of $\mathbf{Q}/\mathbf{Z}_{(p)}$ and gives both F and V.

The description of the Morava K-theory of Eilenberg-Mac Lane spaces in terms of Morava-Dieudonné modules is much simpler than the Hopf algebra description of [**RW80**]. This is because it is no longer necessary to know the order of truncation of the algebra generators to give a precise description (as it is for Hopf algebras). The order of F on generators can be calculated from the description of the Morava-Dieudonné module structure, but it is not necessary to know it to give the description. This is a definite improvement because the orders are probably wrong in [**RW80**].

Note that because Morava K-theory has a Künneth isomorphism the Morava-Dieudonné module of the Morava K-theory of a finite product of Eilenberg-Mac Lane spaces is just the direct product of the respective Morava-Dieudonné modules.

We define an *irreducible* Morava Hopf algebra as one with no proper non-trivial sub-Hopf algebras. Since we are working with an abelian category, it is clear that any finite Morava Hopf algebra can be filtered so that the quotients are irreducible. The remaining interesting problem is to identify these finite irreducibles. This is easiest to do in the category of Morava-Dieudonné modules. We need only work in the category $\mathcal{EC}(n)_\gamma$ because of Corollary 1.2. Let $|\gamma|$ be the order of the orbit γ.

THEOREM 1.4. *Every finite irreducible Morava-Dieudonné module can be filtered so the quotients are irreducible Morava-Dieudonné modules. The finite irreducible Morava-Dieudonné modules in* $\mathcal{EC}_\gamma(n)$ *are:*

(i) *An* $\mathbf{F}_p$ *(in any degree in* γ*) with* F *and* V *trivial.*

(ii) *A* q *dimensional vector space over* $\mathbf{F}_p$ *in each degree in* γ *with* V *and* $f(F^{|\gamma|})$ *acting trivially (*F *is an isomorphism) where* f *is a primitive polynomial over* $\mathbf{F}_p$ *(of degree* q*).*

is the more complicated $P(x)/(f(F^{|\gamma|})(x))$ where again x is primitive. This last relation just tells us how to write $x^{p^{q|\gamma|}}$ in terms of $x^{p^{i|\gamma|}}$ for $i < q$. The primitive nature of the polynomial prevents us from finding sub-Hopf algebras of this.

This theory can be generalized to a work over a perfect field of characteristic p, see [**Rav75**]. If we do this over the agebraic closure of $\mathbf{F}_p$, then the irreducibles are much simplified. In fact, none of the more exotic irreducible Hopf algebras have been spotted in nature, raising the question of whether they can arise as part of the Morava K-theory of a space or not.

There is an additional theorem worth pointing out. Under very mild conditions, see [**Bou96b**, Appendix B, Theorem B.4], Sweedler proves what we would call a dual Borel theorem (about the coalgebra structure) in [**Swe67a**] and [**Swe67b**].

We remark that our classification of Hopf algebras constructs Dieudonné modules by, as usual, considering maps of some variant of the Witt ring (see section 2 for the variant suited to the problem in this paper) into the Hopf algebra under consideration. There is another context familiar to algebraic topologists where objects called Dieudonné modules arise. This is in the classification of formal groups over a field k of characteristic p. We would like to point out the similarities of these situations as the authors have not seen this made explicit elsewhere.

Recall that bicommutative Hopf k-algebras are abelian group objects in the category of commutative k-algebras. The Witt algebra (or as in this paper, some variant of it) is a particular bicommutative Hopf k-algebra that serves as a generator for this category. Our classification comes about by taking maps of Hopf algebras from the Witt algebra to some other Hopf algebra and considering this group as a "Diedonné module" - a module over the ring of self maps of the Witt Hopf algebra.

Commutative formal groups over k are, on the other hand, abelian group objects in the category of formal schemes over k. There is a special commutative formal group (infinite dimensional) called the Witt vectors. (The Witt algebra is essentially the representing ring of the Witt vectors in the category of k-algebras.) To classify formal groups [**Laz75**], we use the "curves" functor which is maps of formal groups from the Witt vectors (as a formal group) to the formal group under consideration. This gives us a "Diedonné module" which is a module over the self maps of the Witt vectors. Because the Witt vectors are represented by the Witt Hopf algebra, the Diedonné module is also a module over the self maps of the Witt Hopf algebra.

Background. A Dieudonné module theory has long been developed for graded Hopf algebras, [**Sch70**]. The authors had been talking about this project for some time when the second author was decimated by the chairmanship of his department. By the time he had recovered, the two papers, [**Bou96b**] and [**Bou96a**], were available and the project was easy to complete by co-opting the proofs from Bousfield, [**Bou96b**]. Pete Bousfield politely but firmly declined the authors' invitation to be a coauthor. The authors regret his decision and make no claim to originality. We feel the theorems should be made available however. Following a suggestion of Igor Kriz's we were developing the general Dieudonné module theory for Hopf rings only to find that Paul Goerss had already done it. We thank the referee for his helpful suggestions.

2. Differences from Bousfield

At the suggestion of the referee we assume the basic facts about the elementary Hopf algebras which we use without a detailed review. Some basic and specific references are [**Wit36**], [**Ser67**], [**Kra72**], [**HRW98**], and [**Bou96b**].

There is a bicommutative Hopf algebra $W = \mathbf{Z}[x_0, x_1, \dots]$ with $x_0^{p^n} + px_1^{p^{n-1}} + \dots + p^{n-1}x_{n-1}^p + p^n x_n$ primitive. We can grade W by setting $|x_0| = 2t$ and $|x_i| = p^i 2t$. The primitives are homogeneous, so we get a graded Hopf algebra which we call W_t. From it we get Hopf algebras graded over $\mathbf{Z}/2(p^n - 1)$ by defining

$$W(s,t) \equiv \mathbf{F}_p[x_0, x_1, \dots, x_s] \subset W_t \otimes \mathbf{F}_p.$$

Bicommutative Hopf algebras over perfect fields of characteristic p come with two self maps, the *Frobenius* F and the *Verschiebung*, V. The Frobenius is the pth power map, i.e. $F(x) = x^p$, which is a map of Hopf algebras since we are in characteristic p, but not of graded Hopf algebras. The Verschiebung can be defined as the dual of the Frobenius map on the dual Hopf algebra, as in [**HRW98**], or in terms of the coproduct, as in [**Bou96b**]. It gives a map of Hopf algebras, but not a map of graded Hopf algebras since $|V(x)| = |x|/p$. We have $V(x_i) = x_{i-1}$ in $W(s,t)$. When x is primitive, $V(x) = 0$, so V is 0 on primitively generated bicommutative Hopf algebras. Clearly $FV = VF$ since V is a map of algebras. $\mathrm{Hom}_{\mathcal{EC}(n)}(A, B)$ is an abelian group with addition of f and g defined using the coproduct on A and the product on B:

$$A \xrightarrow{\psi} A \otimes A \xrightarrow{f \otimes g} B \otimes B \longrightarrow B.$$

It is standard that $FV = p$ in the group $\mathrm{Hom}_{\mathcal{EC}(n)}(A, B)$.

The proof of Theorem 1.1 is the same as Bousfield's, [**Bou96b**], and at the suggestion of the referee has not been included here. We must include the definition of the functor though because here we do differ from Bousfield.

There is a surjective map $W(s, t/p) \to W(s-1, t)$ which takes the primitive generator in degree $2t/p$ to zero.

We use the sequence:

$$W(0,t) \leftarrow W(1, t/p) \leftarrow \cdots \leftarrow W(s, t/p^s) \leftarrow W(s+1, t/p^{s+1}) \quad \cdots$$

to define

$$m_t : \mathcal{EC}(n) \longrightarrow \mathcal{MD}(n)_t$$

by

$$m_t(A) = \varinjlim \mathrm{Hom}(W(s, t/p^s), A).$$

REMARK 2.1. To get the equivalence of $\mathcal{EC}(n)_\gamma$ and $\mathcal{MD}(n)_\gamma$ we need only restrict our attention to the t in γ. In particular, when $\gamma = \{0\}$ we are precisely in the ungraded case of [**Bou96b**] except with a slightly modified definition of the functor between them.

To complete our definition of $m_* : \mathcal{EC}(n) \to \mathcal{MD}(n)_*$ we need to define the action of F and V on $m_*(A)$. We require a minor departure from the standard definition in order to preserve grading. In our case we use the fact that the image of F on $W(s,t)$ is canonically isomorphic to $W(s, pt)$, i.e. the composite

$$W(s,t) \xrightarrow{\cong} W(s,pt) \hookrightarrow W(s,t)$$

We get a map of sequences

$$\begin{array}{ccccccccc} W(0,t) & \leftarrow & W(1,t/p) & \leftarrow & \cdots & \leftarrow & W(s,t/p^s) & \leftarrow & W(s+1,t/p^{s+1}) & \cdots \\ \uparrow & & \uparrow & & & & \uparrow & & \uparrow & \\ W(0,pt) & \leftarrow & W(1,t) & \leftarrow & \cdots & \leftarrow & W(s,t/p^{s-1}) & \leftarrow & W(s+1,t/p^s) & \cdots \end{array}$$

which induces the map F

$$m_t(A) = \varinjlim \mathrm{Hom}(W(s,t/p^s),A) \longrightarrow m_{pt}(A) = \varinjlim \mathrm{Hom}(W(s,pt/p^s),A).$$

We now need the map V. The image of V acting on $W(s,t)$ is the sub-Hopf algebra $W(s-1,t)$. There is a canonical map of $W(s,t/p)$ which surjects to this image. In particular, this gives us maps

$$W(s,t/p) \to W(s-1,t) \subset W(s,t).$$

The corresponding map of sequences

$$\begin{array}{ccccccccc} W(0,t) & \leftarrow & W(1,t/p) & \leftarrow & \cdots & \leftarrow & W(s,t/p^s) & \leftarrow & W(s+1,t/p^{s+1}) & \cdots \\ \uparrow & & \uparrow & & & & \uparrow & & \uparrow & \\ W(0,t/p) & \leftarrow & W(1,t/p^2) & \leftarrow & \cdots & \leftarrow & W(s,t/p^{s+1}) & \leftarrow & W(s+1,t/p^{s+2}) & \cdots \end{array}$$

induces V

$$m_t(A) = \varinjlim \mathrm{Hom}(W(s,t/p^s),A) \longrightarrow m_{t/p}(A) = \varinjlim \mathrm{Hom}(W(s,(t/p)/p^s),A).$$

3. Irreducibles

PROOF OF THEOREM 1.4. We construct the filtration of an arbitrary finite Morava-Dieudonné module and see what we get. Assuming that M_* is finite, we look at the kernel of V on it. This is always a nontrivial sub-Morava-Dieudonné module because V iterated enough times must be zero. Of course this kernel has trivial V action so we have reduced our problem to filtering a Morava-Dieudonné module with $V = 0$. This is a $\mathbf{F}_p$ vector space because $p = FV = 0$.

If F has a kernel then the kernel is a sub-Morava-Dieudonné module with trivial F and V action. Any element of this generates a sub-module of type (i) of the Theorem. We have now reduced our problem to the case where V acts trivially and F has no kernel, i.e. where F is an isomorphism of finite dimensional vector spaces. Thus the dimensions of M_t for each $t \in \gamma$ must be the same.

Pick any $x \in M_*$. Because M_* is finite dimensional there must be some minimal u such that $F^u(x)$ can be written in terms of $F^i(x)$ for $i < u$. We first claim that u is a multiple of $|\gamma|$. If not, then $F^u(x)$ is not in the same degree as x and the above relation does not include $x = F^0(x)$. Since F is an isomorphism, we can divide the relation by some power of F until x is in the relation. This contradicts the minimality of u.

Our relation can now be written as $f(F^{|\gamma|})(x) = 0$ because elements in dimensions other than that of x cannot be part of the relation. Factor f into primitive polynomials $f_1 f_2 \dots f_k$. The element $y = f_2 \dots f_k(F^{|\gamma|})(x)$ is in $M_{|x|}$. It has the property that $f_1(F^{|\gamma|})(y) = 0$ and the vector space $\{y, F^{|\gamma|}(y), F^{2|\gamma|}(y), F^{3|\gamma|}(y), \dots, F^{(q-1)|\gamma|}(y)\}$ (where q is the degree of f_1) is an irreducible sub-vector space of $M_{|x|}$ under the action of $F^{|\gamma|}$. We now take the sub-Morava-Dieudonné module generated by y. This completes our proof and gives our modules of type (ii).

Note that we have just reproven the fundamental structure theorem for finite modules over the P.I.D. $\mathbf{F}_p[z]$ (where $z = F^{|\gamma|}$) and we could have just referred to it instead. □

References

[Bou96a] A. K. Bousfield. On λ-rings and the K-theory of infinite loop spaces. *K-Theory*, 10:1–30, 1996.

[Bou96b] A. K. Bousfield. On p-adic λ-rings and the K-theory of H-spaces. *Mathematische Zeitschrift*, 223:483–519, 1996.

[DG70] M. Demazure and P. Gabriel. *Groupes algébriques*. North-Holland, Amsterdam, 1970.

[HRW98] M. J. Hopkins, D. C. Ravenel, and W. S. Wilson. Morava Hopf algebras and spaces $K(n)$ equivalent to finite Postnikov systems. In S. O. Kochman and P. Selick, editors, *Stable and Unstable Homotopy*, volume 19 of *The Fields Institute for Research in Mathematical Sciences Communications Series*, pages 137–163, Providence, R.I., 1998. American Mathematical Society.

[Kra72] D. Kraines. Approximations to self dual Hopf algebras. *American Journal of Mathematics*, 94(4):963–973, 1972.

[Laz75] M. Lazard. *Commutative Formal Groups*, volume 443 of *Lecture Notes in Mathematics*. Springer-Verlag, New York, 1975.

[Rav75] D. C. Ravenel. Dieudonné modules for Abelian Hopf algebras. In D. Davis, editor, *Reunion Sobre Teoria de homotopia, Universidad de Northwestern, Agosto 1974*, number 1 in Serie notas de matemática y simposia, pages 177–194, Mexico, D.F., 1975. Sociedad Matematica Mexicana.

[RW80] D. C. Ravenel and W. S. Wilson. The Morava K-theories of Eilenberg-Mac Lane spaces and the Conner-Floyd conjecture. *American Journal of Mathematics*, 102:691–748, 1980.

[Sch70] C. Schoeller. Etude de la categorie des algebres de Hopf commutatives connexes sur un corps. *Manuscripta Math.*, 3:133–155, 1970.

[Ser67] J.-P. Serre. *Représentations Linéaires des Groupes Finis*. Hermann, Paris, 1967.

[Swe67a] M. E. Sweedler. Hopf algebras with one group-like element. *Transactions of the American Mathematical Society*, 127:515–526, 1967.

[Swe67b] M. E. Sweedler. Weakening a theorem on divided powers. *Transactions of the American Mathematical Society*, 154:427–428, 1967.

[Wit36] E. Witt. Zyklische körper und algebren der charakteristik p vom grad p^n. *J. Reine ang. Math.*, 176:126–140, 1936.

University of Oregon, Eugene, OR, 97403
E-mail address: sadofsky@math.uoregon.edu

Johns Hopkins University, Baltimore, Maryland 21218
E-mail address: wsw@math.jhu.edu

Contemporary Mathematics
Volume **220**, 1998

Northwestern 1997 Algebraic Topology Conference - Problems

1. (F. R. Cohen) How does Jie Wu's description of the homotopy groups of the 3-sphere as the centers of certain combinatorially given groups fit with periodicity? Is there a reasonable group theoretic description of the v_1-periodic summand of the homotopy? Can one use the intrinsic group theory to obtain the exponents for the classical homotopy groups?

2. (F. R. Cohen) What are the indecomposable factors of the double loop space of a 2-connected mod-p Moore space?

3. (F. R. Cohen) Fix a prime p. Let $X_p(R^n, k)$ denote the space of ordered k-tuples of points in Euclidean space of dimension n, with the property that k is greater than p and the centroids of distinct subsets of cardinality p are themselves distinct.

What is the homology of

$$X_p(R^n, k)/\Sigma_k$$

with $\mathbb{F}_p$ coefficients with either the trivial representation or the sign representation? If the "answers" are the natural ones, then this calculation solves problem (11).

4. (Octav Cornea) Let $(W; V_0, V_1)$ be a smooth, compact cobordism and assume that there exists a smooth function $f : W \longrightarrow \mathbf{R}$ which is maximal, regular and constant on V_1 and minimal, regular and constant on V_0 and which has exactly one critical point.

It is easy to see that *the quotient W/V_0 has the homotopy type of a co-H-space.* One possible proof is to use the proposition 2.1 in [**2**] which shows that for a certain collar $U \subset W$, $U = V_0 \times [0,1]$ of V_0 and some disk $D \subset W$ around the critical point we have that W is a homotopy retract mod V_0 of $U \bigcup D$. We may assume that the intersection of D with U is an NDR. Hence, W/V_0 is the homotopy retract of $(U \bigcup D)/V_0$ which is a suspension. For other variants see [**4, 5**].

Problem. *Are there examples (W, f) as above such that the quotient W/V_0 does not have the homotopy type of a suspension ?*

Remarks. 1. If the function f is analytic with respect to some chart around the critical point, then the quotient W/V_0 has the homotopy type of a suspension (see [**3**]).

2. Assume M is also Riemannian. The homotopy type of W/V_0 is the Conley index [1] of the critical point of f viewed as an invariant set for the flow induced

References

[1] Ch.Conley, *Isolated invariant sets and the Morse index*, CBMS Regional Conf.Ser. in Math 38, Amer. Math. Soc., Providence, RI, 1976.
[2] O.Cornea, *Some properties of the relative Lusternik-Schnirelmann category*, To appear in the Fields Institute Proceedings.
[3] O.Cornea, *Spanier-Whitehead duality and critical points*, Preprint, Spring 1997.
[4] E.N.Dancer, *Degenerate critical points, homotopy indeces and Morse inequalities* J. Reine Ang. Math. 350 (1984), 1-22
[5] J.Pears, *Degenerate Critical Points and the Conley Index*, Thesis, University of Edinburgh, 1995

5. (Brayton Gray) It is well known that $\Sigma\Omega\Sigma X \simeq \bigvee \Sigma X^{(i)}$ and $\Sigma^\infty\Omega^n\Sigma^n X \simeq \bigvee \Sigma^\infty D_{n,i}(X)$. The question was raised in the 1974 topology conference at Northwestern as to whether one could prove that

$$\Sigma^n\Omega^n\Sigma^n X \simeq \bigvee D_{n,i}(X).$$

This turned out to be false in case $\tilde{H}_k(X;\mathbb{Z}/2\mathbb{Z}) \neq 0$ for some k.

Strong Conjecture Localize away from 2. Then

$$\Sigma^2\Omega^2\Sigma^2 X \simeq \bigvee \Sigma^2 D_{2,i}(X)$$

at least in the case where $X = S^{2n-1}$.

Weak Conjecture There is a map

$$\Sigma^2\Omega^2 S^{2n+1} \longrightarrow \Sigma^2 D_{2,p}(S^{2n-1}) = S^{2np} \cup_p e^{2np+1} = P^{2np+1}(p)$$

for $p > 2$.

6. (Brayton Gray) Let

$$W_n \longrightarrow S^{2n-1} \longrightarrow \Omega^2 S^{2n+1} \longrightarrow BW_n$$

be the double suspension fibration sequence. The weak conjecture implies that BW_n is a retract of $\Omega^2 P^{2np+1}(p)$ and that $BW_n = \Omega T_{2np-1}$, where T_{2np-1} are certain spaces constructed by Anick [] and further developed by Anick and Gray and discussed elsewhere in this volume. Localized at $p > 2$, BW_n is an H-space and it is homotopy commutative and homotopy associative if $p > 3$.

Question For which n is BW_n and H-space when localized at 2? The space $BW_1 \simeq \Omega^2 S^3\langle 3\rangle$ clearly is an H-space, and it is conceivable that BW_2 is homotopy equivalent to the loop space of the fiber of the map

$$\alpha : S^5 \longrightarrow SU(3)\langle 3\rangle,$$

where α induces an isomorphism on homology in dimension 5.

7. (Hans-Werner Henn) Let G_n denote the group $GL_n(\mathbb{Z}[\frac{1}{2}])$ and D_n its subgroup of diagonal matrices so that $D_n \cong (\mathbb{Z}[\frac{1}{2}]^\times)^n \cong \mathbb{Z}^n \times (\mathbb{Z}/2)^n$. The image of the restriction map

$$\rho_n : H^*(G_n;\mathbb{F}_2) \longrightarrow H^*(D_n;\mathbb{F}_2)$$

is isomorphic to $\mathbb{F}_2[w_1,\dots,w_n] \otimes \Lambda(e_1,\dots,e_{2n-1})$ [**4**]. Furthermore, ρ_n is known to be injective for $n \leq 3$ by [**4**] and [**2**]. By recent work of Dwyer [**1**] ρ_n fails to be injective if $n \geq 32$ (and work in progress with J. Lannes [**3**] shows that it even fails if $n \geq 14$.). However, the reasoning in [**1**] as well as in [**3**] is very indirect and doesn't produce any explicit elements in the kernel of of ρ_n.

Problem: Construct explicit elements in the kernel of ρ_n.

Remark: Injectivity of ρ_n amounts to an unstable version of the Lichtenbaum-Quillen conjecture for $\mathbb{Z}[\frac{1}{2}]$ at the prime 2 (cf. [**1**]), so the problem addresses the failure of such a conjecture.

References

[1] W.G. Dwyer, *Exotic cohomology for* $GL_n(\mathbb{Z}[1/2])$, Preprint 1996

[2] H.-W. Henn, *The cohomology of* $SL(3, \mathbb{Z}[1/2])$, Heft Nr. **136**, Forschergruppe Topologie und nichtkommutative Geometrie Univ. Heidelberg, Mai 1996; to appear in K-Theory

[3] H.-W. Henn and J. Lannes, work in progress.

[4] S. Mitchell, *On the plus construction for* $BGL\mathbb{Z}[\frac{1}{2}]$, Math.Zeit. **209** (1992), 205-222

8. (Mark Hovey) What are the possible supports of $K(n)$-local spectra? Recall that the support of X is the set of j such that $K(j)_*X$ is nonzero. Neil Strickland and I conjecture that the only possible supports of $K(n)$-local spectra are the sets $\{m, m+1, ..., n\}$ for some m. Note that of course this is the support of $L_{K(n)}F(m)$ where $F(m)$ is a finite spectrum of type m. The Bousfield class of a $K(n)$-local spectrum is completely determined by its support, so this is equivalent to classifying $K(n)$-local spectra up to Bousfield class.

9. (Mark Hovey) What are the invariant radical ideals in $E(n)_*$ and E_*? Here E is the Morava E-theory spectrum whose coefficients are power series over the Witt vectors of F_{p^n} on $n-1$ indeterminants $u_1, ..., u_{n-1}$ of degree 0 and which has a unit u in degree 2. Naturally we expect the invariant ideals to be the I_j for $0 <= j <= n$. However, the Landweber filtration theorem does not imply this directly. No general version of the Landweber filtration theorem is known for $E(n)_*E(n)$-comodules, though there are versions that hold with some restriction. In particular, is every invariant radical ideal prime?

10. (Ran Levi) Let G be a finite group and let p be a prime number. Then the loop space on its p-completed classifying space $\Omega BG^\wedge_p$ is a disjoint union of some power of p copies of $\Omega BO^pG^\wedge_p$, where O^pG is the maximal normal p-perfect subgroup of G. This last loop space has the the property of being a retract of the loop space on the p-completion of a finite complex. Define the p-essential dimension of G, $ed_p(G)$ to be the minimal dimension of such a complex. It is clearly an invariant of the isomorphism type of G or more generally, of the homotopy type of $BG^\wedge_p$.

Question: What is the meaning of the p-essential dimension of a group?

Notice that $ed_p(G) = 0$ if and only if the subgroup O^pG contains no p-torsion, in which case $BG^\wedge_p$ is the classifying space of a finite p-group. Also if the Sylow p-subgroup of O^pG is cyclic then $ed_p(G)/2$ divides $p-1$.

11. (Ran Levi) Let G be a finite group. Does the p-completed classifying space $BG^\wedge_p$ have a homotopy exponent? The answer is known to be yes in many examples, but is not known in general. It is also known that $\Omega BG^\wedge_p$ does not have a multiplicative

Remark: A negative answer will contradict the Moore finite exponent conjecture (for any 1-connected space X of the homotopy type of a finite CW complex the homotopy groups of X have an exponent if and only if the rational homotopy of X is a finite dimensional vector space).

12. (Norihiko Minami) Recently we obtained the following result:

THEOREM 1. *If the odd primary Mahowald element η_j, which is detected by $h_0 h_j$ in the classical Adams spectral sequence, exists, then it always has (a nonzero multiple of) an extension of β_1 as its "Hopf invariant."*

Here we mean the "Hopf invariant" by the p-th James-Hopf-Snaith invariant associated with a meta-stable double suspension sequence:

$$j_p : \pi_{i+(2k+3)}(S^{2k+3}) \to \pi^s_{(i+1)-q(1+k)}(M(p)),$$

where $M(p)$ is the mod p Moore spectrum with $\pi^s_0(M(p)) \cong \mathbb{Z}/p$.

COROLLARY 2. *Suppose an odd primary η_j element exists. Then there exists a stablc map*

$$\bar{\eta}_j : S^{q(p^j+1)-2} \to P^{(q(p^j+1)-2)-(qp-2)} = P^{q(p^j-p+1)},$$

whose composite with the canonical collapsing map

$$P^{q(p^j-p+1)} \to S^{q(p^j-p+1)}$$

is a non-zero multiple of β_1.

Furthermore, any such $\bar{\eta}_j$ is detected by the EO_{p-1}-Hurewicz map

$$\pi^s_*(P^{q(p^j-p+1)}) \to (EO_{p-1})_*(P^{q(p^j-p+1)}),$$

where EO_{p-1} is the spectrum constructed by Hopkins-Miller.

PROOF. For the second claim, we use the fact that the EO_{p-1} Hurewicz map of S^0 detects β_1. □

REMARK 3. *For $p = 2$, Mahowald proved the 2-primary η_j element has ν as its Hopf invariant. Our argument can be generalized to obtain a simpler proof of this result.*

However, Corollary 2 does not hold for $p = 2$. This is essentially because the KO-Hurewicz map

$$\pi^s_*(S^0) \to KO_*(S^0)$$

does not detect ν.

Recalling the current situation of the Adams 2-line elements:

p	2	3	5
θ_j?	???	??	No (Ravenel)
η_j?	Yes (Mahowald)	??	?

and a well-known result

$$\theta_j \text{ exits} \Longrightarrow \eta_{j-1} \text{ exists},$$

we wonder if odd primary η_j element really exists? Motivated by these, we are tempted to ask

QUESTION 4. *Is it possible that $(EO_{p-1})_*(EO_{p-1})$-operations impose any restrictions on such a $\bar{\eta}_j \in \pi_*^s(P^{q(p^j-p+1)})$ as in Corollary 2? More generally, can we impose any useful restriction on the EO_{p-1}-Hurewicz image*

$$\pi_*^s(P^{q(p^j-p+1)}) \to (EO_{p-1})_*(P^{q(p^j-p+1)}),$$

using $(EO_{p-1})_*(EO_{p-1})$-operations?

REMARK 5. *Such a study is essentially the odd primary analogue, in the context of EO_{p-1}, of Mahowald's Annal's paper.*

13. (Stewart Priddy) Is $K(Z/2, \mathrm{n})$ stably indecomposable for $n > 1$?

14. (Stewart Priddy) Is group cohomology isomorphic to Lie algebra cohomology

$$H^*(N; Z) \approx H^*(LN; Z)?$$

Here N is a finitely generated torsion free discrete group and LN is the Lie algebra associated to the sub quotients of the upper central series by Malcev. LN is a finitely generated torsion free abelian group. Over the rationals the rings are known to be isomorphic since the completed universal enveloping algebra and group rings are isomorphic via the exponential map.

In case $N = U_n(Z)$ the upper triangular matrices over Z with ones down the diagonal then $LN = < e_{ij} : 1 \le i \le j \le n >$ with the bracket

$$[e_{ij}, e_{kl}] = \begin{cases} e_{il} \text{ if } j = k \\ -e_{kj} \text{ if } i = l \\ 0 \text{ otherwise} \end{cases}$$

15. (Laurence R. Taylor)

We begin with some notation. For any space X, let Sq_2 denote the composition $H_4(X; \mathbf{Z}) \to H_4(X; \mathbf{Z}/2\mathbf{Z}) \to H_2(X; \mathbf{Z}/2\mathbf{Z})$ where the last map is the dual to the squaring map. For any group G, let K_1 denote the kernel of the epimorphism, $G \to H_1(G; \mathbf{Z}/2\mathbf{Z})$. Let K_2 denote the kernel of the epimorphism $K_1 \to H_1(K_1; \mathbf{Z}/2\mathbf{Z})$. Note K_2 is normal in G and define $\hat{G} = G/K_2$. Let $\pi\colon \hat{X} \to X$ denote the regular cover of X with deck transformations $\hat{G}$, $G = \pi_1(X)$.

Problem: Does there exist a space X and a class $x \in H_4(X; \mathbf{Z})$ such that $\mathrm{Sq}_2(x) \notin \mathrm{Im}\,\pi$?

Remark: Whenever $x \in \mathrm{Im}\,\pi$, there exists a surface F and a map $f\colon F \to X$ such that $f_*[F] = \mathrm{Sq}_2(x)$ and such that the map

$$f_*\colon H_1(F; \mathbf{Z}/2\mathbf{Z}) \to H_1(X; \mathbf{Z}/2\mathbf{Z})$$

is the 0 map. This is of interest in the theory of 4–manifolds.

Selected Titles in This Series

(*Continued from the front of this publication*)

(See the AMS catalog for earlier titles)

Selected Titles in This Series

220 **Mark Mahowald and Stewart Priddy, Editors,** Homotopy theory via algebraic geometry and group representations, 1998

219 **Marc Henneaux, Joseph Krasil′shchik, and Alexandre Vinogradov, Editors,** Secondary calculus and cohomological physics, 1998

218 **Jan Mandel, Charbel Farhat, and Xiao-Chuan Cai, Editors,** Domain decomposition methods 10, 1998

217 **Eric Carlen, Evans M. Harrell, and Michael Loss, Editors,** Advances in differential equations and mathematical physics, 1998

216 **Akram Aldroubi and EnBing Lin, Editors,** Wavelets, multiwavelets, and their applications, 1998

215 **M. G. Nerurkar, D. P. Dokken, and D. B. Ellis, Editors,** Topological dynamics and applications, 1998

214 **Lewis A. Coburn and Marc A. Rieffel, Editors,** Perspectives on quantization, 1998

213 **Farhad Jafari, Barbara D. MacCluer, Carl C. Cowen, and A. Duane Porter, Editors,** Studies on composition operators, 1998

212 **E. Ramírez de Arellano, N. Salinas, M. V. Shapiro, and N. L. Vasilevski, Editors,** Operator theory for complex and hypercomplex analysis, 1998

211 **Józef Dodziuk and Linda Keen, Editors,** Lipa's legacy: Proceedings from the Bers Colloquium, 1997

210 **V. Kumar Murty and Michel Waldschmidt, Editors,** Number theory, 1998

209 **Steven Cox and Irena Lasiecka, Editors,** Optimization methods in partial differential equations, 1997

208 **Michel L. Lapidus, Lawrence H. Harper, and Adolfo J. Rumbos, Editors,** Harmonic analysis and nonlinear differential equations: A volume in honor of Victor L. Shapiro, 1997

207 **Yujiro Kawamata and Vyacheslav V. Shokurov, Editors,** Birational algebraic geometry: A conference on algebraic geometry in memory of Wei-Liang Chow (1911–1995), 1997

206 **Adam Korányi, Editor,** Harmonic functions on trees and buildings, 1997

205 **Paulo D. Cordaro and Howard Jacobowitz, Editors,** Multidimensional complex analysis and partial differential equations: A collection of papers in honor of François Treves, 1997

204 **Yair Censor and Simeon Reich, Editors,** Recent developments in optimization theory and nonlinear analysis, 1997

203 **Hanna Nencka and Jean-Pierre Bourguignon, Editors,** Geometry and nature: In memory of W. K. Clifford, 1997

202 **Jean-Louis Loday, James D. Stasheff, and Alexander A. Voronov, Editors,** Operads: Proceedings of Renaissance Conferences, 1997

201 **J. R. Quine and Peter Sarnak, Editors,** Extremal Riemann surfaces, 1997

200 **F. Dias, J.-M. Ghidaglia, and J.-C. Saut, Editors,** Mathematical problems in the theory of water waves, 1996

199 **G. Banaszak, W. Gajda, and P. Krasoń, Editors,** Algebraic K-theory, 1996

198 **Donald G. Saari and Zhihong Xia, Editors,** Hamiltonian dynamics and celestial mechanics, 1996

197 **J. E. Bonin, J. G. Oxley, and B. Servatius, Editors,** Matroid theory, 1996

196 **David Bao, Shiing-shen Chern, and Zhongmin Shen, Editors,** Finsler geometry, 1996

195 **Warren Dicks and Enric Ventura,** The group fixed by a family of injective